The Discovery of Isotopes

Michael Thoennessen

The Discovery of Isotopes

A Complete Compilation

Springer

Michael Thoennessen
Department of Physics and Astronomy
Michigan State University
East Lansing, MI
USA

ISBN 978-3-319-31761-8 ISBN 978-3-319-31763-2 (eBook)
DOI 10.1007/978-3-319-31763-2

Library of Congress Control Number: 2016935977

Printed on acid-free paper

This Springer imprint is published by Springer Nature
The registered company is Springer International Publishing AG Switzerland

To my wonderful wife Ute

Preface

The idea to document the discovery of all isotopes was not an ad hoc decision, but it developed over several years. In 2003, I was asked to write a review article about the drip-lines which was published a year later (Rep. Prog. Phys. **67** (2004) 1187). The article contained tables of the most neutron- and proton-rich known nuclides at the drip-lines covering elements from magnesium through berkelium and from beryllium through calcium for the proton- and neutron-rich nuclides, respectively.

In the fall semester of 2007, I gave Josh Ginepro—an undergraduate student in the Professorial Assistantship Program of the Honors College (PAPHC) at Michigan State University—the task to search the Nuclear Science References (NSR) data base of the National Nuclear Data Center (NNDC) for the earliest experimental papers for all isotopes. In the following summer four undergraduate students from the Research Experience for Undergraduates (REU) program and one student from the MSU High School Honors Science Program each selected one element and wrote short paragraphs describing the discovery of its isotopes. Only then did we realize the momentous task of covering all of the almost 3000 isotopes of the 118 elements.

For each element these paragraphs were subsequently published in a series of papers in Atomic Data and Nuclear Data Tables. In the following years several other MSU as well as REU students contributed to the project and the last manuscript was submitted in November 2011 and published in 2013.

A summary of the project has been published in another review article in 2013 covering all discoveries until the end of 2012 (Rep. Prog. Phys. **76** (2013) 056301). An overview of the project and the latest discoveries are available online: http://www.nscl.msu.edu/~thoennes/isotopes/.

East Lansing, MI, USA
November 2015

Michael Thoennessen

Acknowledgments

The following undergraduate students were involved in the project: S. Amos (REU, 2009), A. Bury (HSHSP, 2008), J. Claes (MSU, 2010), A. Fritsch (REU, 2008), C. Fry (REU, 2011), K. Garofali (MSU, 2010), J.Q. Ginepro (PAPHC, 2007–2008), J.L. Gross (MSU, 2009–2011), M. Heim (REU, 2008), J. Kathawa (MSU, 2010–2011), E. May (PAPHC, 2010–2011), D. Meierfrankenfeld (MSU, 2009), A. Nystrom (REU, 2010), A. Parker (REU, 2010), R. Robinson (MSU, 2010–2011), A. Schuh (REU, 2008), A. Shore (REU, 2008), T. Szymanski (MSU, 2009).

The project has been funded by the NSF under grants: PHY06-06007 (NSCL), PHY11-02511 (NSCL), PHY07-54541 (REU), PHY10-62410 (REU), the High School Honors Science Program (HSHSP), and the Professorial Assistantship Program of the Honors College (PAPHC) at MSU.

I would like to thank Dieter Ackermann, Ani Aprahamian, John Arblaster, Peter Armbruster, Bertram Blank, Joseph Cerny, Zdenek Cimpl, David Darling, Christoph Düllmann, Juan Flegenheimer, Hans Geissel, Brandon Hamer, John Hardy, Charlie Holbrow, John Kelley, Karlheinz Langanke, Filip Kondev, Toshiyuko Kubo, Jan Kurcewicz, Shaohua Liu, Walter Loveland, Joseph Magill, Murray Martin, Geoff McBride, Carola Pomplun, Christopher Parkyn, Boris Pritychenko, Helge Ravn, Alby Reid, Ross Rosenwald, Krzystof Rykaczewski, Hideyuki Sakai, Darek Seweryniak, Balraj Singh, Alejandro Sonzogni, Zsolt Soti, Mark Stoyer, Louis Taff, Oleg Tarasov, Brett Thornton, Antonio Villari, Bill Walters, and many others for fruitful discussions, positive feedback and suggestions. In addition, I would like to thank the many anonymous referees of the papers. Their criticism and comments significantly improved the papers and helped to develop a consistent set of criteria. I am also thankful to David Schultz, the editor of Atomic Data and Nuclear Data Tables, who did a fantastic job in handling and carefully editing our manuscripts. Furthermore many thanks to the Michigan State University library staff for their tremendous help in finding even the most obscure articles. Last but not least, I would like to thank Ute Thoennessen for proofreading the manuscript and Sigurd Hofmann for very thoughtful and detailed suggestions during the final review.

Contents

Chapter 1
Introduction

Isotopes are at the heart of nuclear physics. They represent the several thousand different ways that protons and neutrons can combine to form different types of nuclei. The term isotope is in this context actually not quite correct. The general term for a specific combination of neutrons and protons is called a nuclide. Strictly speaking, isotopes are only nuclides with the same number of protons but different numbers of neutrons. However, over the years it has become customary to use the term isotope also in the more general sense of a nuclide. Thus, rather than adhering to the correct terminology all the time, in the following the term isotope will also be used synonymously with the term nuclide.

The force responsible for binding neutrons and protons together into isotopes is the strong force which is one of the four fundamental forces in nature and which is still not fully understood. Similarly to Mendeleev's sorting of the elements in a periodic table was helpful to understand the atoms, the sorting of the nuclides in a chart as a function of neutrons and protons reveals features and properties of the strong force.

In one of the first charts presented, the nuclides were ordered with the number of neutrons on the vertical axis and the number of protons on the horizontal axis (Fig. 1.1) [1]. In this representation, all nuclides on a vertical line correspond to the same element. Several immediate questions arise: What are the limits of existence? What is the largest number of protons (neutrons) that nature will hold together? How many (or few) neutrons are necessary to hold a nuclide together for a given number of protons?

The answers to these questions are essential to the understanding of the strong force. In order to map out the chart of nuclides it is necessary to either find new nuclides in nature or produce them in the laboratory. Although started more than over 100 years ago this task is still not completed. Presently there are about 3000 different nuclides known. They range from 1 to 118 protons and 1 to 177 neutrons.

Only 254 of these nuclides are stable or at least no decay has ever been observed. Another 34 nuclides are considered primordial nuclides which means the nuclei that are observed today could have already existed prior to the formation of the Earth.

M. Thoennessen, *The Discovery of Isotopes*,
DOI 10.1007/978-3-319-31763-2_1

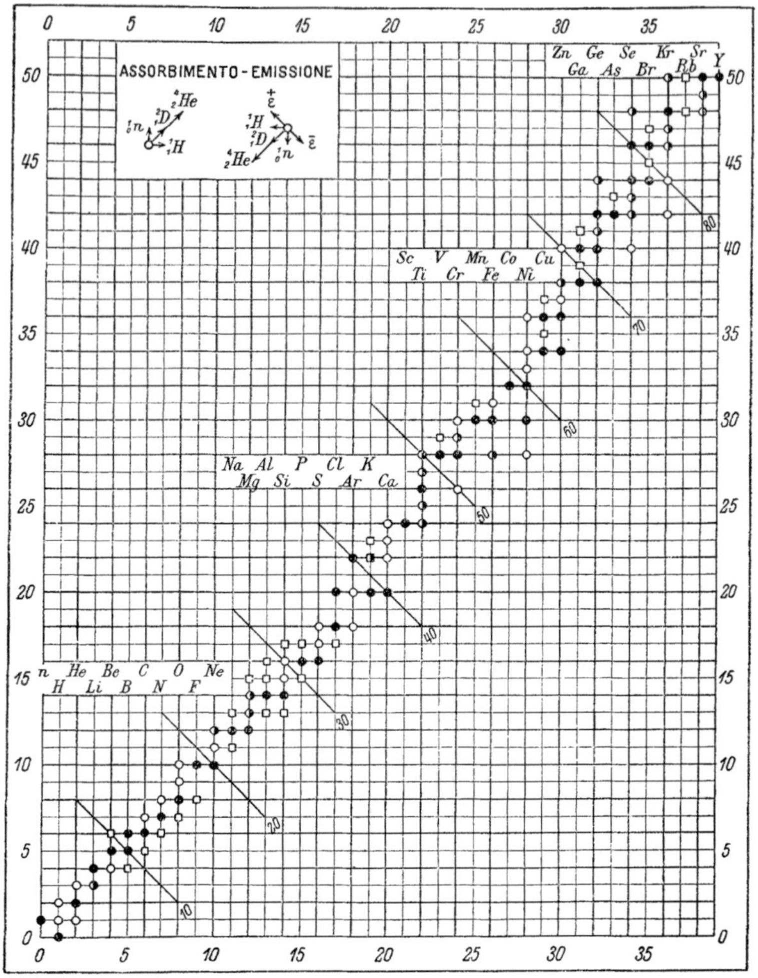

Fig. 1.1 First page of one of the first charts of nuclides published in 1935 by Fea [1]. Different elements (number of protons) are listed along the horizontal axis and different isotopes for each element are displayed along the vertical axis. This ordering of the axes is reversed with respect to the present day chart of nuclides (With kind permission from Springer Science and Business Media: G. Fea, *Tabelle riassuntive e bibliografia delle trasmutazioni artificiali*, Nuovo Cimento 12 (1935) 368)

They are radioactive and decay with half-lives of more than 80 million years into other nuclides. Another approximately 50 radioactive nuclides can be found naturally on Earth and are decay products of the primordial nuclides or were produced by cosmic rays. All other nuclides were produced in the laboratory or nuclear reactors and explosions.

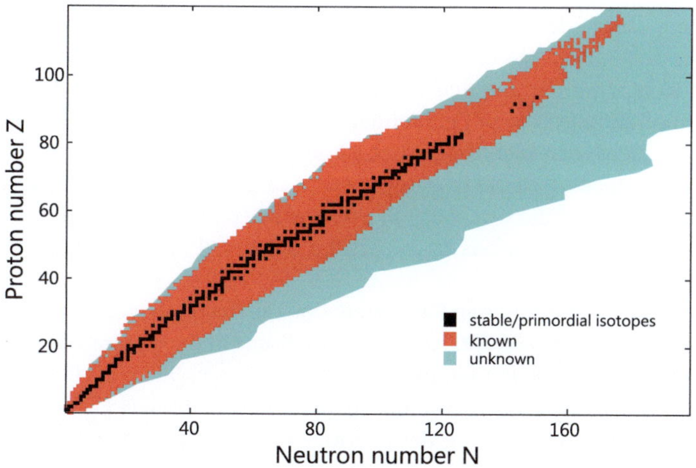

Fig. 1.2 Present day chart of nuclides. Stable and primordial isotopes are shown by *black squares*. Currently known isotopes are shown in *red* and the *light blue* area indicates isotopes that might exist but still have to be discovered

Figure 1.2 shows the present day chart of nuclides where the axes have been reversed so that each horizontal line now corresponds to a certain element. The black squares correspond to the stable and primordial isotopes while the red area represents all presently known radioactive isotopes. Recently it was estimated that about 7000 bound nuclides exist [2] which are shown by the light blue area. As can be seen in the figure the mapping of the nuclear landscape is by far from complete especially in the region of neutron-rich isotopes.

While new isotopes continue to be discovered it is worthwhile to review and summarize the current status. The most up-to-date knowledge of nuclear properties like existence, mass, half-life, etc., are continuously updated in databases (see for example [3]), however, a comprehensive overview of the discovery of the isotopes does not exist.

The present book fills this void by documenting the discovery of all isotopes presently known. It represents a historical review from the beginning of nuclear physics when the existence of isotopes (and nuclei) was not even known to the most recent large scale experiments where several tens of new isotopes are produced within a single fragmentation reaction.

Such a documentation is not only important for the historical record but also for an overview of the current status of which isotopes are known and which discoveries are still controversial.

The recognition for the discovery of a new element is well established with strict criteria set by the International Union of Pure and Applied Chemistry (IUPAC) and the International Union of Pure and Applied Physics (IUPAP) [4, 5]. Such formal rules and official certification do not exist for the discovery of isotopes and the criteria are certainly debatable [6].

It is impossible to define a common set of criteria for the discovery of all isotopes. They differ for stable, radioactive and short-lived unbound nuclides and change with time as new production and detection methods are being developed. For the present compilation the overall guidelines for assigning credit of an isotope discovery were (1) clear identification, either through decay-curves and relationships to other known isotopes, particle or γ-ray spectra, or unique mass and Z-identification, and (2) publication of the discovery in a scientific (refereed) journal. Further more specific criteria for the various cases will be discussed in the introductions of the individual chapters.

The compilation presented in the current book is based on work performed over a period of four years between 2007 and 2011 [6]. Some of the more specific criteria had changed during this time so that not all isotopes were treated the same. In addition, many more nuclides were observed during and after the project [7, 8].

All assignments made during the previous project were reviewed and common criteria were applied for each of the different production and detection methods. New discoveries up to the end of 2015 were included.

The discoveries are presented approximately in historical order starting with the isotopes which were first observed by their radioactive decay at which time the origin of the radiation was not known. It is followed by the discovery of stable isotopes and the first production of new isotopes with nuclear reactions. Starting with the discovery of fission in the beginning of 1939, the isotope discoveries are sorted by their production and detection methods.

References

1. G. Fea, Nuovo Cimento **12**, 368 (1935)
2. J. Erler et al., Nature **486**, 509 (2012)
3. ENSDF, Evaluated Nuclear Structure Data File, maintained by the National Nuclear Data Center at Brookhaven National Laboratory, published in Nuclear Data Sheets (Academic Press, Elsevier Science), http://www.nndc.bnl.gov/ensdf/
4. IUPAC Transfermium Working Group, Pure Appl. Chem. **63**, 879 (1991)
5. R.C. Barber et al., Prog. Part. Nucl. Phys. **29**, 453 (1992)
6. M. Thoennessen, Rep. Prog. Phys. **76**, 056301 (2013)
7. M. Thoennessen, Int. J. Mod. Phys. E **23**, 1430002 (2014)
8. M. Thoennessen, Int. J. Mod. Phys. E **24**, 1530002 (2015)

Chapter 2
Radioactive Decay Chains

2.1 The Discovery of the Existence of Isotopes

Technically the discovery of isotopes should not begin before the concept of isotopes had been developed. However, since the discovery of radioactivity in 1896 many isotopes of the three naturally occurring radioactive decay chains were uniquely identified as separate substances without knowing the underlying physics. These so-called radio-elements were characterized by the type and energy of the emitted radiation and the half-life of their decay.

The detailed study and classification of the various substances and their relationships allowed Soddy [1] and Fajans independently in early 1913 [2] to finally understand and accept the fact that substances with different characteristics existed which were chemically identical. Soddy placed the substances in a two-dimensional grid of mass versus "relative number of negative electrons" [1, 3]. Each column represented an element so that the different isotopes for a given element lined up vertically and the rows corresponded to the mass. It can be considered as the first precursor of the chart of nuclides. It is interesting to note that a month earlier Russell had already placed the radio-elements correctly into element columns, however, he did not arrange them by mass [4]. Figure 2.1 shows the three radioactive decay chains as presented by Soddy in September 1913 at the 83rd meeting of the British Association for the Advancement of Science in Birmingham [5].

Soddy introduced the term isotope in December 1913 [6]. The name was derived from the Greek words iso (same) and tope (place) because the different radio-elements occupied the same place in the periodic table of elements.

Thus in principle Soddy should be credited with the discovery of all the isotopes he correctly identified and placed in his chart. However, it seems more appropriate to credit the discovery to the researchers who measured and characterized the unique properties of these isotopes for the first time although they did not yet understand the implications of their measurements.

M. Thoennessen, *The Discovery of Isotopes*,
DOI 10.1007/978-3-319-31763-2_2

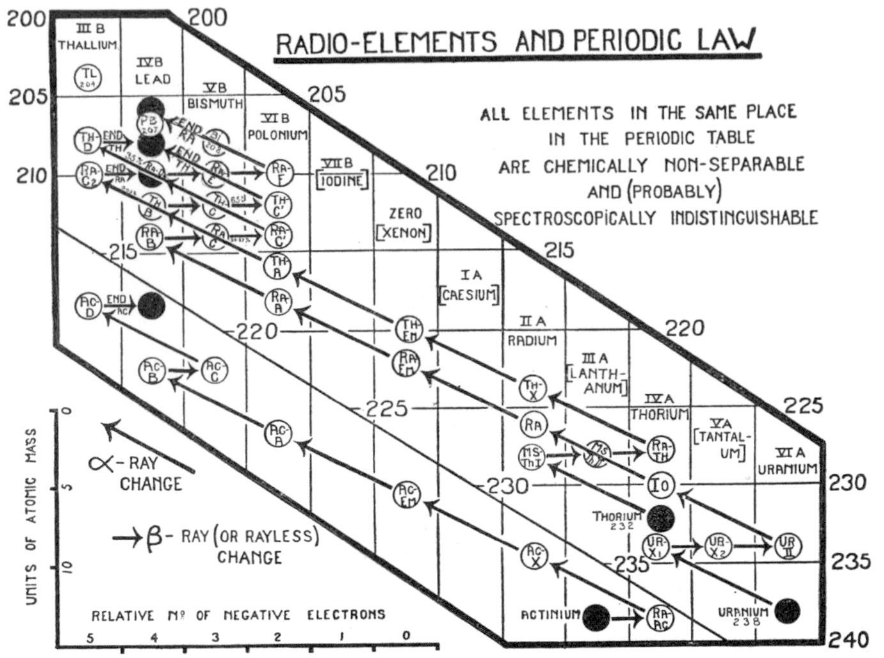

Fig. 2.1 Classification of the radio-elements presented by Soddy at the 83rd meeting of British Association for the Advancement of Science in Birmingham on September 16, 1913 [5]

In the following the discovery of the 46 radioactive isotopes of the three naturally occurring radioactive decay chains will be described divided into three time periods: (1) 1896–1904 when essentially nothing was known about the radiation and the exponential decay law was discovered [7], (2) 1904–1912 after Rutherford proposed the decay chain connections [8] and before Soddy established the concept of isotopes and (3) 1913–1961 after isotopes were established and the few missing connections and last branches of the decays were measured.

For the research of the isotopes of the radioactive decay chains several books and articles were consulted, for example, the 1908 edition of "Gmelin–Kraut's Handbuch der anorganischen Chemie" [9], Soddy's 1911 book "The Chemistry of the Radio-elements" [10], the 1913 edition of Rutherford's book "Radioactive Substances and their Radiations" [11], and the 1933 article by Mary Elvira Weeks "The Discovery of the Elements. XIX. The Radioactive Elements" published in the Journal of Chemical Education [12]. In addition, the Wikipedia page on the radioactive decay chains was a good starting point [13].

2.2 Discovery of Radioactivity

The isotopes characterized by their decay within the first eight years of the discovery of radioactivity between 1896 and 1904 are listed in Table 2.1.

The credit for the discovery of the first two radioactive isotopes is granted purely on the observation of radiation. In 1896, Becquerel discovered radiation itself by placing uranium salt on a photographic plate covered with thick paper. When he developed the plate, he could see the silhouette of the salt. He concluded that the uranium salt emits radiation which passes through paper [14]. This radiation was mainly due to ^{238}U.[1] Only in 1931, Aston demonstrated that ^{238}U was the principal isotope of naturally occurring uranium [26].

Becquerel's observation of this new type of radiation which came only a few months after Roentgen discovered X-rays [27] was called uranium rays. Subsequently, many other substances were studied to search for a similar effect and it took two years before Schmidt in Erlangen, Germany, succeeded with thorium compounds. In addition to photographic plates he studied the effect of these thorium rays on an electroscope [15]. The radiation can be attributed to ^{232}Th. Schmidt even estimated the atomic weight (atomic or mass number of the isotope) of the sources

Table 2.1 Isotopes observed during the initial studies of the radio-elements between 1896 and 1904. The submission date, first author and reference of the publication, as well as the method of identification or the measured half-life, and the presently accepted half-life are listed

Date	First author	Ref.	Identification/$T_{1/2}$	Present $T_{1/2}$	Isotope
02/24/1896[a]	Becquerel	[14]	Radiation	$4.468 \pm 0.003 \times 10^9$ y	^{238}U
03/24/1898	Schmidt	[15]	Radiation	$1.405 \pm 0.006 \times 10^{10}$ y	^{232}Th
07/18/1898[a]	Curie	[16]	Chem. sep.	138.376 ± 0.002 d	^{210}Po
12/26/1898[a]	Curie	[17]	Chem. sep.	1600 ± 7 y	^{226}Ra
09/13/1899	Rutherford	[7]	~1 min	55.6 ± 0.1 s	^{220}Rn
11/06/1899[a]	Curie	[18]	Few days	3.8235 ± 0.0003 d	^{222}Rn
11/22/1899	Rutherford	[19]	11 h	10.64 ± 0.01 h	^{212}Pb
05/03/1900	Crookes	[20]	Chem. sep.	24.10 ± 0.03 d	^{234}Th
11/01/1900	Hofmann	[21]	Chem. sep.	22.20 ± 0.22 y	^{210}Pb
09/15/1902	Rutherford	[22]	~4 d	3.66 ± 0.04 d	^{224}Ra
02/16/1903[a]	Debierne	[23]	Few seconds	3.96 ± 0.01 s	^{219}Rn
12/22/1903	Rutherford	[24]	3 min	3.098 ± 0.012 min	^{218}Po
			34 min	26.8 ± 0.9 min	^{214}Pb
			28 min	19.9 ± 0.4 min	^{214}Bi
2/1/1904	Debierne	[25]	40 min	36.1 ± 0.2 min	^{211}Pb

[a]Date of presentation

[1]In modern nomenclature the element is uniquely identified by its symbol corresponding to a specific number of protons (Z). The superscript indicates the mass number (A) of the isotope. The number of neutrons can then be calculated as N = A−Z.

of the uranium and thorium rays: "It seems that these are related to the high atomic weight, 240 for uranium and 232 for thorium" [15]. Less than three months later [28] M. Curie independently reported the radioactivity of thorium [29]. Aston demonstrated only in 1932 that ^{232}Th constituted the only thorium isotope of primordial origin [30].

While Becquerel and Schmidt primarily studied the radiation itself, Pierre and Marie Curie together with Bemond concentrated on the source of the radiation. In 1898, they analyzed the properties of a metal that they extracted from the original substance (pitchblende) which contained the radiation and concluded that it must correspond to a new element: "If the existence of this new metal is confirmed, we suggest to call it polonium, named after the country of origin of one of us" [16]. This discovery of the element polonium was due to the radiation of ^{210}Po. Earlier M. Curie had already suggested the presence of a new element by observing strong activities in pitchblende and chalcolite [29]. In 1904/1905 Rutherford showed that the polonium activity, radio-tellurium (discovered in 1902 by Marckwald [31, 32]) and his radium F (originally named radium E) were the same (^{210}Po) [33–35].

The Curies continued their chemical analyses of the radioactive substances and discovered the next new element only a few months later: "The new radio-active substance which we have found has all the chemical appearances of almost pure barium: it is neither precipitated by hydrogen sulfide nor by the ammonium sulfide, nor with ammonia; the sulfate is insoluble in water and in acids; the carbonate is insoluble in water, the chloride, very soluble in water, is insoluble in concentrated hydrochloric acid and alcohol. Finally, the spectrum of this substance is easily recognizable as barium... The various reasons we have enumerated lead us to believe that the new radioactive substance contains a new element, which we propose to name radium" [17]. The radiation of this new element was due to ^{226}Ra.

The next important step was the realization that the radiation decreased with time. In 1899 Rutherford observed that thorium compounds emitted a radioactive substance which he called emanation. He noticed that "the intensity of the radiation has fallen to one-half its value after an interval of about one minute" [7]. His detailed measurements allowed him to derive the exponential decay law. The emanation corresponded to ^{220}Rn and the measured half-life was quite accurate. The current accepted value is 55.6 ± 0.1 s.

About at the same time Pierre and Marie Curie also reported a time dependence of radiation in the discovery of another radon isotope (^{222}Rn) originating from radium: "Subtracting the contribution of the activated plate due to the radioactive substance, it remains radioactive for several days. However, the induced radioactivity is decreasing, first very rapidly, then slower and slower and tends to disappear asymptotically" [18]. They did, however, not quantify the time dependence any further. While Rutherford submitted his paper on September 13, the submission date of the Curie paper which appeared on November 6, was not given. Thus they probably should share the credit for the discovery of the element radon. It should be mentioned that sometimes the credit [36] is given to Ernst Dorn for his work on the emanation [37].

Only a couple of months later Rutherford detected another activity in the decay of thorium. From the decay curve shown in Fig. 2.2 he extracted a half-life of 11 h. This

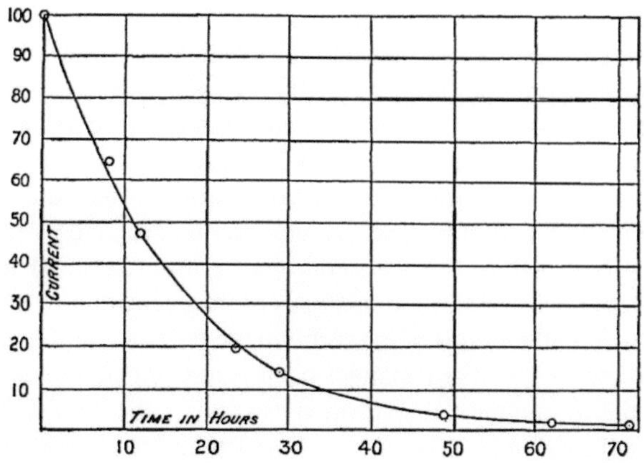

Fig. 2.2 Decay curve of radioactivity observed by Rutherford in 1899 within the decay chain of thorium [19]. The activity corresponds to ^{212}Pb (Reprinted by permission of Taylor and Francis Ltd, www.tandfonline.com, E. Rutherford, *Radioactivity produced in substances by the action of thorium compounds*, Philosophical Magazine 49 (1900) 161.)

activity was the direct cause of the previously discovered emanation [19]. Rutherford called this activity "excited radioactivity" and later on attributed it to ThA (^{212}Pb) [8]. In 1902 Rutherford and Soddy continued their study of thorium and observed an activity with a half-life of about 4 days which they named ThX (^{224}Ra) [22].

Two more radioactive substances were discovered in 1900 and identified by performing chemical separations and proving that they were different from previously observed radioactive materials. Crookes ensured that his new substance separated from uranium nitrate was different from polonium and radium and called it UrX [20]. UrX was later renamed UrX_1 and corresponds to ^{234}Th. Hofmann and Strauss separated a radioactive substance from various uranium samples and determined that the radioactivity was still present when the substance consisted of lead only [21]. Later on it was realized that this "radiolead" was the same as the independently discovered RaD [33–35] corresponding to ^{210}Pb. Aston's 1927 claim of stable ^{210}Pb [38] was later shown to be incorrect [39].

Once the concept of the radioactive decay law had become established it is reasonable to require a fairly accurate half-life measurement as a criterion to claim the discovery of a new isotope. This does not necessarily apply to the discovery of an element and affects the discovery of actinium. In 1902 Giesel reported the observation of a new active substance which could not be related to either uranium, radium or thorium [40]. Two years later he named this new element emanium [41]. It turned out that emanium was the same substance that Debierne had earlier called actinium [42] and Debierne is generally credited with the discovery of actinium. However, Kirby argued that Debierne could not have observed actinium in his first papers [42, 43] and thus credit should be given to Giesel [44]. Giesel did not measure the half-life of actinium so he is not credited for the discovery of the isotope (^{227}Ac) responsible for the activity. The half-life of ^{227}Ac was only measured in 1911 by M. Curie [45] who was credited with the discovery of this isotope (see next section).

A few months later Giesel then also reported a third type of emanation which he determined to be different from radium emanation and thorium emanation, however, he did not measure the half-life [46]. A month later Debierne independently observed the actinium emanation (^{219}Rn) and reported that it disappeared to half its value within a few seconds [23]. Debierne followed up his studies and a year later observed an additional activity: "I also found the radioactive decay law caused by the actinium emanation by measuring the induced activity in the same set-up, from the moment the emanation has ceased to act: The decay is regular, it reaches its half value in 40 min" [25]. Rutherford reproduced the observation by Debierne and named it actinium A [8]. It was later reclassified as AcB and corresponds to ^{211}Pb.

Rutherford's experiments with radium resulted in the discovery of three new activities in 1903: "An analysis of the decay curves of excited activity, produced for different intervals of exposure in the presence of the emanation, shows that there are three well-marked changes occurring in emanation X of radium. In the first change, half the matter is transformed in 3 min; in the second, half in 34 min; and in the third, half in 28 min" [24]. At the time Rutherford and Barnes did not name the activities which correspond to ^{218}Po, ^{214}Pb, and ^{214}Bi, respectively. Rutherford and Soddy had reported these activities labeled radium-excited activity I, II, and III already a year earlier without measuring the half-lives [47].

2.3 Discovery of Radioactive Decay Chains

At the beginning of 1904 the decay of 16 different substances (including actinium) had been measured. In his Bakerian lecture on May 19, Rutherford reported the discovery of two additional isotopes from thorium emanation (^{212}Bi) and actinium emanation (^{211}Bi) and presented for the first time a comprehensive scheme of four separate decay chains as shown in Fig. 2.3. In addition to placing the known activities into the chains, Rutherford had some evidence for the existence of AcX (^{223}Ra) but referred to a later paper for the results and labeled it with a question mark in the figure [8]. The half-life of actinium was not yet known at the time.

Although the scheme was simple and offered valuable guidance for future researchers, the sequential naming of A, B, C, etc. was unfortunate. At the time the scheme was incomplete and several intermediate decay products were missing. As more and more additional isotopes were discovered in the following years, it became necessary to rename some of the isotopes which led to confusion until in 1913 Soddy introduced the concept of isotopes. Until then another 17 new isotopes were discovered (see Table 2.2).

The predicted actinium X activity (^{223}Ra) was discovered within the year by Godlewski: "The activity increased in the first day after removal to about 15 % of its original value, and then decayed with the time according to an exponential law, falling to half value in about ten days" [48]. A year earlier Giesel had reported a new substance separated from emanium (actinium) [41] without any further more detailed measurements [63].

Three additional isotopes had to be inserted into the actinium chain. In 1906, Hahn discovered radio-actinium (^{227}Th) in between Ac and AcX [50] and two years later Hahn and Meitner determined that AcC (^{207}Tl) was radioactive and not the end of the

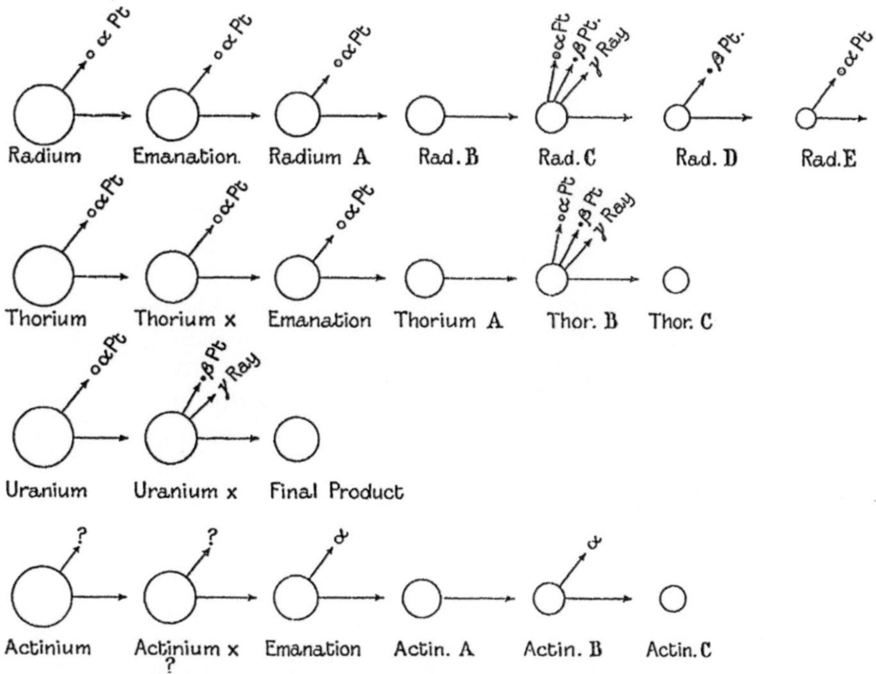

Fig. 2.3 Rutherford's radioactive decay chains based on the 18 radio-elements (including actinium but not actinium X) known by 1904 [8]

decay chain [55]. Geiger then added a new isotope (^{215}Po) between the emanation and AcA in 1911: "It has been shown that the actinium emanation is complex, consisting of two products each of which emits α rays. The first one—the emanation—with a period of 4 seconds emits α rays of a range of 5.7 cm. The second product emits α rays of range 6.5 cm, and has a period of the order of 1/500 of a second" [60]. It was the first time that such a short half-life was measured. Earlier Geiger and Marsden had established an upper half-life limit of 0.1 s for this activity [58]. The addition of a new AcA required the renaming of the subsequent isotopes including Hahn and Meitner's AcC which was renamed AcD and later AcC".

Another member of this series was discovered by Antonoff in 1911, but he did not connect it to the actinium chain [59]. Since he observed the new activity from a uranium nitrate sample he speculated that it was another decay branch of uranium naming it UY (^{231}Th). In the same year, M. Curie measured the half-life of ^{227}Ac [45]. This actinium activity had been known since 1902 (see previous section). It is interesting to note that Marie Curie is listed as Mme. P. Curie as the author of the paper. For all the isotopes of the actinium series see Fig. 2.6.

Five of the six additional isotopes of the thorium decay chain were discovered by Hahn. In 1905, he observed another substance that decayed to thorium emanation which he called radiothorium (^{228}Th). Hahn was not able to extract the half-life; he only stated that the radioactivity did not show any reduction after two months [49].

Table 2.2 Isotopes of the radio-elements discovered between 1904 and 1912. The submission date, first author and reference of the publication, as well as decay chain, the measured half-life, and the presently accepted half-life are listed

Date	First author	Ref.	Chain	Measured $T_{1/2}$	Present $T_{1/2}$	Isotope
08/20/1904	Rutherford	[8]	Ac	1.5 min	2.14 ± 0.02 min	^{211}Bi
			Th	55 min	60.55 ± 0.06 min	^{212}Bi
01/02/1905	Godlewski	[48]	Ac	10 d	11.43 ± 0.05 d	^{223}Ra
01/24/1905	Rutherford	[34]	Ra	6 d	5.012 ± 0.005 d	^{210}Bi
03/07/1905	Hahn	[49]	Th	"long"	1.9116 ± 0.0016 y	^{228}Th
03/27/1906	Hahn	[50]	Ac	20 d	18.68 ± 0.09 d	^{227}Th
04/07/1906	Hahn	[51]	Th	~sec.	0.299 ± 0.002 μs	^{212}Po
03/23/1907	Hahn	[52]	Th	7 y	5.75 ± 0.03 y	^{228}Ra
09/09/1907	Boltwood	[53]	U/Ra	1900 y	75380 ± 300 y	^{230}Th
03/07/1908	Hahn	[54]	Th	6.5 h	6.15 ± 0.02 h	^{228}Ac
08/28/1908	Hahn	[55]	Ac	5.1 min	4.77 ± 0.03 min	^{207}Tl
01/22/1909	Hahn	[56]	Th	3.1 min	3.053 ± 0.004 min	^{208}Tl
09/20/1909	Hahn	[57]	Ra	1−2.5 min	1.3 ± 0.3 min	^{210}Tl
11/14/1909	Geiger	[58]	Th	0.14 s	0.145 ± 0.002 s	^{216}Po
June 1911	Antonoff	[59]	Ac	1.5 d	25.52 ± 0.01 h	^{231}Th
July 1911[a]	Geiger	[60]	Ac	2 ms	1.781 ± 0.004 ms	^{215}Po
08/27/1911	M. Curie	[45]	Ac	21 y	21.772 ± 0.003 y	^{227}Ac
March 1912[a]	Geiger	[61]	U/Ra	200,000 y	$245{,}500 \pm 600$ y	^{234}U
06/07/1912	Fajans	[62]	Ra	10^{-6} s	164.3 ± 2.0 μs	^{214}Po

[a]Publication date

Subsequently, Hahn determined that radiothorium is a decay product of thorium and that two other isotopes—mesothorium I (^{228}Ra [52, 64]) and mesothorium II (^{228}Ac [54])—exist between these two isotopes.

Hahn also extended the thorium chain when he discovered in 1906 that ThA emitted α-particles with two distinctly different energies [51]. Following Rutherford's convention he named the new activity thorium C (^{212}Po). A few years later Hahn observed together with Meitner yet another activity towards the end of the thorium decay chain. They determined the half-life of the activity accurately but assumed incorrectly that it sequentially followed the decay of ThB and ThC [56]. The logical conclusion was to name the new activity ThD. Later it was reclassified as ThC" (^{208}Tl).

The last isotope within the thorium chain was identified by Geiger and Marsden in 1909 with a new method to measure very short times between two successive α-particle emissions. They let the radioactive substance diffuse between two closely spaced scintillation screens which were viewed from opposite ends by two microscopes: "The scintillations, which were viewed by two observers on the corresponding screens over a period of several minutes, were registered via an electromagnetically operated pen onto the same paperstrip of a chronograph. The velocity of the paperstrip was in most cases 0.5 cm per second." With this set-up they were able to determine a lifetime of about 0.2 s from the thorium emanation [58]. A year later Rutherford and Geiger placed this activity (^{216}Po) between the thorium emanation and ThA [65]. At the same time they suggested that Hahn's ThB and ThC activities were not sequential. As a consequence they renamed the old ThA to ThB, and indicated that ThC consists of ThC$_1$, and ThC$_2$. To confuse the situation even further, later on ThC$_2$ and ThD were renamed one more time to ThC' and ThC".

The thorium decay chain was completely known at the time when Soddy developed the isotope concept. Figure 2.4 shows all radio-elements of the chain. Isotopes known to Rutherford in 1904 and his suggested decay paths are shown in black. Isotopes known to Soddy in 1913 and the modifications to the decay paths are shown in red.

The radium decay chain was even more complicated than the other two. It already contained two more isotopes at the time Rutherford presented his overview in 1904. A year later Rutherford reported that he had overlooked an activity in RaD which he then named D$_1$ (^{210}Bi) [34]. In a paper submitted four months later Rutherford renamed radium D$_1$ to radium E and the old radium E to radium F [35]. In October 1904, Hofmann et al. [66] had identified a β-emitter chemically separated from radiolead (radium D, ^{210}Pb) which Rutherford acknowledged [35], however, the extracted half-life of about six weeks was incorrect.

Shortly after Hahn and Meitner encountered in 1909 the difficulty explaining the two different α-decays in ThC (see above [56]) they determined that RaC also consists of two substances which they called RaC$_1$ and RaC$_2$ [57]. RaC$_2$ was later called RaC" (^{210}Tl) while RaC$_1$ remained RaC. Hahn and Meitner also indicated that there might be yet another unknown activity present. This was confirmed by Fajans three years later who deduced that RaC$_1$ (or RaC) branches off into RaC$_2$ and RaD—via the new RaC' (^{214}Po)—with a ratio of 3/10,000 which agrees with the currently accepted ratio of 2.1/10,000 [62]. His estimate of an extremely short half-life (10^{-6} s) was too short and is actually about 100 times larger.

The other remaining open question of the radium decay chain was its relation to the uranium chain. It had been realized early on that due to the relatively short half-life radium could not be the origin of a radioactive decay chain and Soddy proposed a uranium as the ultimate origin of the radium series in 1905 [67]. The first step bridging the gap between radium and uranium was presented by Boltwood in 1907: "...there is good cause for believing that uranium minerals contain an element emitting α rays, which is different from the other elements that have been identified, which produces no emanation, and which resembles thorium in its chemical properties" [53]. Later in the year Boltwood suggested the name "ionium" (^{230}Th) for the new substance [68, 69]. His estimate of the half-life (1900 y) was somewhat short.

In 1912 Geiger and Nuttall discovered a second uranium isotope (^{234}U): "Uranium I therefore, which has a period of 5×10^9 years, emits α-particles of range 2.5 cm in air at atmospheric pressure and at 15 °C, and is followed by another α-ray product, uranium II, which has a period of about 2×10^6 years and emits α-particles of range 2.9 cm" [61]. Earlier, the existence of a second α emitting uranium isotope was suspected from the number of emitted α-particles per uranium atom [70, 71]. Geiger and Nuttall did not make the connection to ionium. For all the isotopes of the combined radium/uranium series see Fig. 2.7.

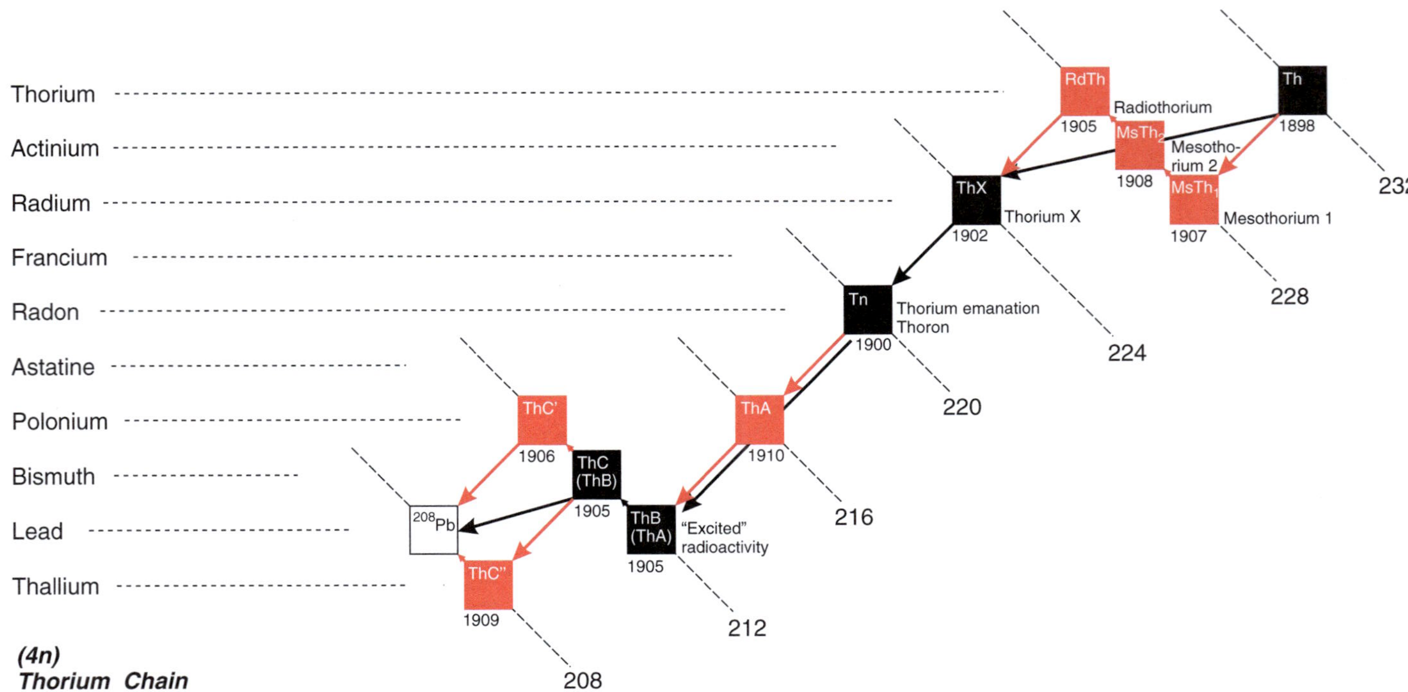

Fig. 2.4 Isotopes and decay paths of the thorium decay chain. The chain includes isotopes which have a mass number which is a multiple of four (A = 4n, n = 52–58). Isotopes known to Rutherford in 1904 and his suggested decay paths are shown in *black*. Isotopes known to Soddy in 1913 and the modifications to the decay paths are shown in *red*

2.4 Completing the Radioactive Decay Chains

In the February 1913 issue of the Chemical News, Soddy sorted the radio-elements for the first time in a chart as a function of mass and charge [1]. He placed the 25 known isotopes of the uranium (14) and thorium (11) decay series correctly and predicted the existence of eka-tantalum (^{234}Pa) connecting the radium with the uranium series. The existence of ^{234}Pa had also independently been predicted as UX_2 by Fajans [2] and Russell [4]. However, he incorrectly placed the 8 at the time known isotopes of the actinium series at the same location as the uranium series starting at ^{230}Th [1, 3].

Eka-tantalum or UX_2 (^{234}Pa) was discovered by Fajans and Göhring already one month later [72]. In a subsequent paper with Beer, Fajans realized that it corresponded to the first observation of a new element and named it brevium: "The reason to name the short-lived element which O. Göhring and I discovered is that it is the only element which fits in the location of the fifth group of the last horizontal row of the periodic table. However, for the radioactive characterization the name UrX2 is preferred, because it immediately clarifies its genetic relation" [73]. Fajans and Göhring's discovery also corresponds to the first observation of an isomeric excited state. The ground-state of ^{234}Pa was discovered by Hahn in 1921 who named it at first Z [74] and then UZ [75].

Following the discovery of eka-tantalum, Soddy realized that the radium and actinium chains were not identical starting from ionium/radioactinium and subsequently placed the actinium series separately from the other two without any mass assignment (see Fig. 2.1 [5]). In the 1913 Chemical Society Annual Report he proposed two solutions shown in Fig. 2.5. Both options predicted the existence of another isotope UZ connecting UY with actinium. In the first and second solution actinium would have an atomic weight of 230 and 226, respectively [76]. The connection of UY to the actinium series had previously already been suggested by Antonoff [77].

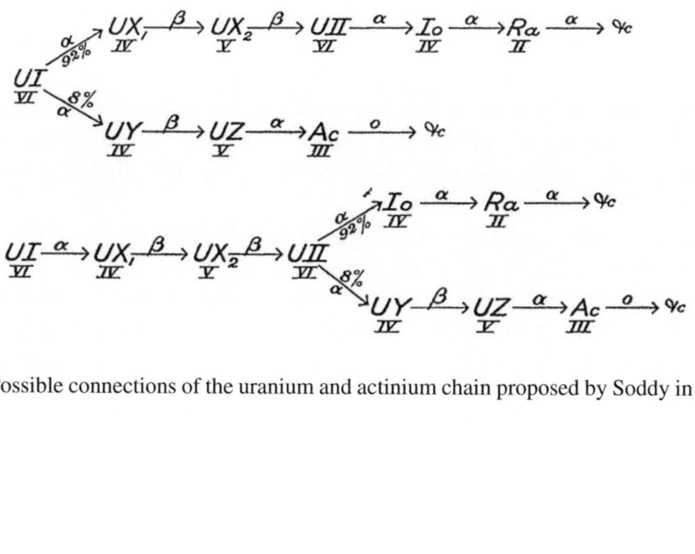

Fig. 2.5 Possible connections of the uranium and actinium chain proposed by Soddy in 1913 [76]

Table 2.3 Isotopes of the radio-elements discovered between 1913 and 1961. The submission date, first author and reference of the publication, as well as decay chain, the identification method or measured half-life, and the presently accepted half-life are listed

Date	First author	Ref.	Chain	Identification/$T_{1/2}$	Present $T_{1/2}$	Isotope
03/15/1913	Fajans	[72]	U/Ra	1.1 min[a]	1.17 ± 0.03 min	^{234}Pa
09/03/1913	Marsden	[78]	Ac	α-particle range	0.516 ± 0.003 s	^{211}Po
03/16/1918	Hahn	[79]	Ac	1,200–180,000 y	$32{,}760 \pm 110$ y	^{231}Pa
07/12/1935	Dempster	[80]	Ac	Mass-spectroscopy	7.04×10^8 y	^{235}U[b]
10/21/1935[c]	Preiswerk	[81]	U/Ra	4 min	4.202 ± 0.011 min	^{206}Tl
07/03/1939	Hulubei	[82]	U/Ra	X-rays	1.5 ± 0.3 s	^{218}At
01/09/1939	Perey	[83]	Ac	21 ± 1 min	22.00 ± 0.07 min	^{223}Fr
12/30/1943	Karlik	[84]	Ac	α-particle range	100 ± 20 μs	^{215}At
05/20/1948	Studier	[85]	U/Ra	Sequential α-decay	35 ms	^{218}Rn[d]
12/17/1952	Hyde	[86]	Ac	8 min	7.6 ± 0.2 min	^{215}Bi
			Ac	0.9 min	56 ± 3 s	^{219}At
04/29/1961	Nurmia	[87]	U/Ra	7.5 ± 1.0 min	8.32 ± 0.07 min	^{206}Hg

[a]Isomeric state
[b]See also Sect. 3.2
[c]Date of presentation
[d]See also Sect. 9.3

With the exception of the placement and the ultimate origin of the actinium series all main decay branches of the three decay chains were established by early 1913. It took another 48 years before the last isotope of one of these three naturally occurring decay chains was identified. Table 2.3 lists the 12 remaining isotopes discovered between 1913 and 1961.

The proposed isotope UZ (^{231}Pa) was discovered in 1918 by Hahn and Meitner [79]. A few months earlier Soddy and Cranston had demonstrated the existence of UZ but did not measure any properties of UZ or its decay. In contrast, Hahn and Meitner had determined the range of the α-particles and determined a half-life between 1,200 and 180,000 years. They are generally given equal credit for the discovery [12]. Hahn and Meitner named the new substance protactinium [79, 88], however, Soddy quoted them with the name protoactinium [89, 90]. Both names were subsequently in use until the name protactinium was officially selected at the 15th IUPAC conference in Amsterdam in 1949 [91, 92].

Once it was demonstrated that brevium (^{234}Pa) and protactinium (^{231}Pa) belonged to the same element, Fajans withdraw the right to name the element: "Protactinium occurs in the periodic system in the same location as uranium X_2, and since it is the by far longer-lived element which determines the name, the term brevium is no longer needed" [93]. In 1973, Fajans and Morris reflected on the discovery and naming of protactinium [94].

The discovery of UZ (^{231}Pa) still did not resolve the question of the placement of the actinium series. In his Nobel lecture in 1922 Soddy mentioned the possibility that it could be a completely independent series which initiated from another uranium isotope [95]. In 1929 Rutherford predicted the existence of actino-uranium (^{235}U) [96] based on Aston's extrapolation of 231 for the mass of protactinium as the precursor of ^{207}Pb [97]. It was experimentally only resolved with the mass determination of ^{231}Ac in 1934 [98] and the discovery of actino-uranium (^{235}U) in 1935 by Dempster in his mass spectrograph [80].

In addition to solving this major mystery of the origin of the actinium series, several small side branches of this decay chain were observed after Soddy published his chart in 1913.

At the end of 1913 Marsden and Wilson detected another α-decay from AcC which had been expected in analogy to the thorium and radium series: "The results showed that in addition to the α-particles of actinium C with a range of 5.4 cm, a small number, about 1 in 600, can penetrate as far as about 6.45 cm. Special experiments showed that the long-range α-particles could not be due to radium or thorium impurity, and they must therefore be attributed to the expected new branch product" [78]. This activity—later named AcC'—corresponds to ^{211}Po.

The next branch point was reported 26 years later and corresponded to the discovery of the new element francium. Perey discovered ^{223}Fr as an α-decay branch of ^{227}Ac in 1939 [83]. She determined the branching to be 1 % [99]. In 1952, Hyde and Ghiorso explored the continuation of this parallel branch, explored the α-decay of ^{223}Fr, and discovered ^{219}At and ^{215}Bi: "The observed branching rate is ca. 4×10^{-5}, and the At^{219} daughter decays predominantly by the emission of 6.27-Mev alpha-particles with a half-life of 0.9 min to the new isotope Bi^{215}, which in turn emits β^- particles with a half-life of 8 min" [86]. In the same paper, Hyde and Ghiorso questioned Karlik and Bernert's claim for the observation of ^{215}At in the actinium series. Karlik and Bernert had deduced the discovery of ^{215}Ac from the measurement of the α-particle energy. The value agreed with the extrapolated value for mass number 215 in a plot of the decay energy as a function of the mass number for atomic number 85 and they quoted a β-decay branch of 5×10^{-4} % from AcA (^{215}Po) [84]. In a note added in proof Hyde and Ghiorso mentioned that the work by Karlik and Bernert on ^{215}At had been confirmed [100].

The discovery of ^{219}At and ^{215}Bi by Hyde and Ghiorso completed the actinium series in 1953. The complete chain is shown in Fig. 2.6. Isotopes known to Rutherford in 1904 and his suggested decay paths are shown in black. Isotopes known to Soddy in 1913 and the modifications to the decay paths are shown in red. All subsequently discovered isotopes and decay paths are shown in light blue.

Once the connection between the uranium and radium series had been established only four additional small side branches were discovered. The complete chain is shown in Fig. 2.7 and the explanation for the different colors are the same as in Fig. 2.6. For ^{234}Pa the discovery of the isomer (UX_2) and the ground-state (UX) are indicated. Two of the isotopes (^{206}Tl and ^{218}Rn) were discovered before they were identified as a decay branch in the uranium chain. For these isotopes the date of discovery as well as the date when they were linked to the chain are listed.

The discovery of ^{218}At has been extensively debated. In an overview paper in 2010, Thornton and Burdette [101] pointed out that Hulubei and Cauchois reported evidence for the observation of ^{218}At already in 1939. Hulubei and Cauchois had

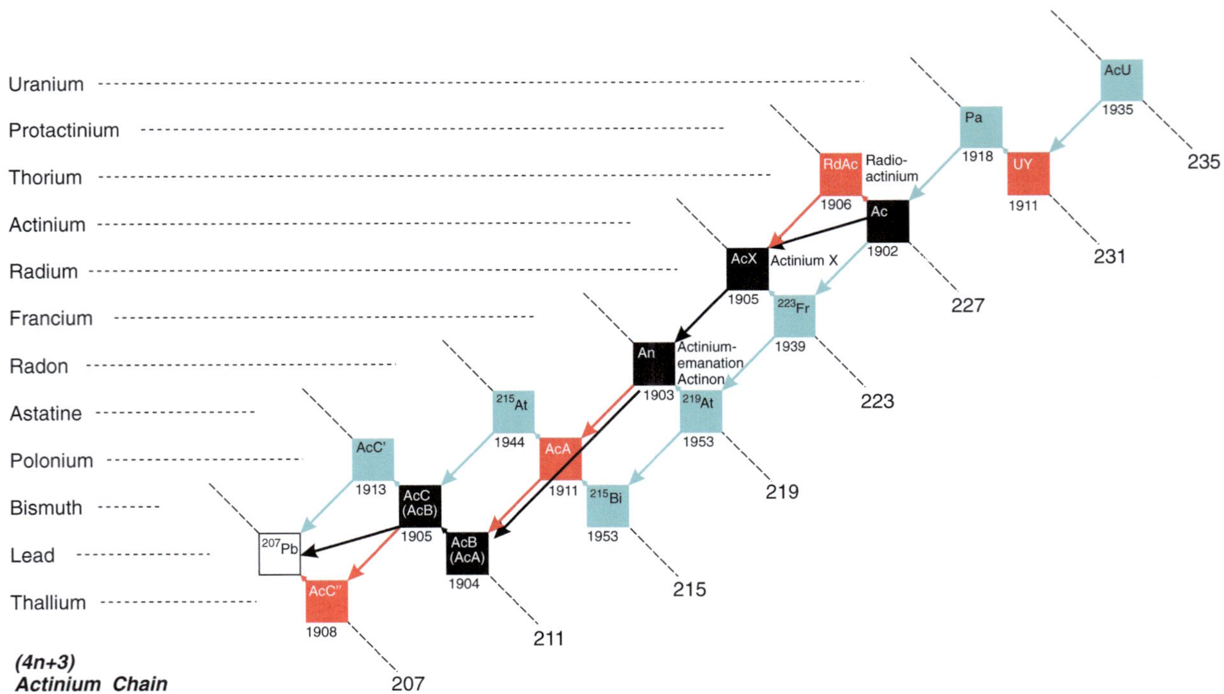

Fig. 2.6 Isotopes and decay paths of the actinium decay chain (A = 4n + 3, n = 51–58). Isotopes known to Rutherford in 1904 and his suggested decay paths are shown in *black*. Isotopes known to Soddy in 1913 and the modifications to the decay paths are shown in *red*. All subsequently discovered isotopes and decay paths are shown in *light blue*

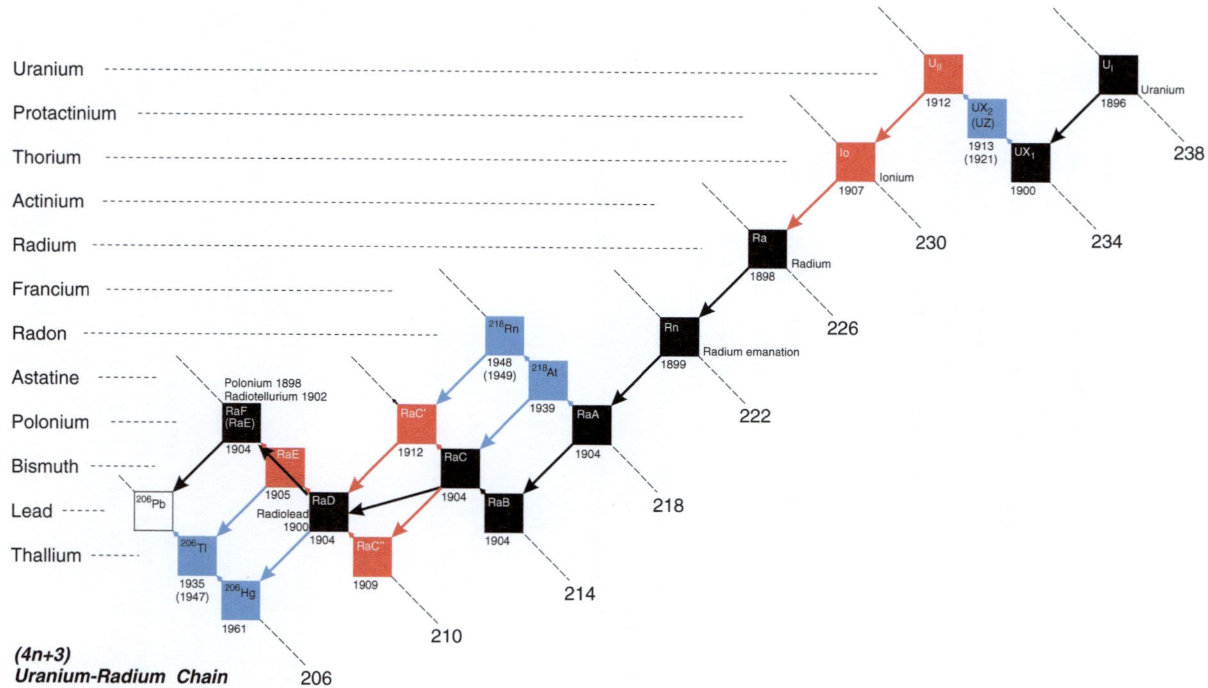

Fig. 2.7 Isotopes and decay paths of the uranium/radium decay chain (A = 4n + 2, n = 51 − 59). Isotopes known to Rutherford in 1904 and his suggested decay paths are shown in *black*. Isotopes known to Soddy in 1913 and the modifications to the decay paths are shown in *red*. All subsequently discovered isotopes and decay paths are shown in *light blue*

reported the observation of three X-rays which were close to the predicted values for eka-iodine (astatine) [82]. These X-rays were observed from a ^{222}Rn sample and the astatine lines could only originate from ^{218}At populated either by α and β-decay through ^{218}Po or by β and α-decay through ^{222}Fr. The acceptance of the Hulubei and Cauchois discovery of ^{218}At places the discovery of this isotope before the discovery of the element astatine. The discovery of astatine has been credited [102] to Corson, MacKenzie, and Segrè for their observation of the α-decay of ^{211}At in 1940 [103, 104]. It is justified to give Hulubei and Cauchois credit for the discovery of the isotope but not the element because they lacked chemical proof [101].

The observation of a weak β-decay branch of ^{218}At in 1949 linked ^{218}Rn to the uranium decay chain. Walen reported a 1 % branch in a paper submitted on November 24, 1948 claiming the discovery of the new isotope ^{218}Rn [105]. He probably was not aware of the paper by Studier and Hyde submitted six months earlier. They had identified ^{218}Rn in the artificially produced protactinium decay chain beginning at ^{230}Pa (see Sect. 9.3) [85].

In 1947, Broda and Feather reported a 5×10^{-5} % α-decay branch from RaE (^{210}Bi) linking ^{206}Tl to the uranium series [106]. At time the existence of ^{206}Tl was well established. It had been discovered in 1935 by Preiswerk and von Halban when they irradiated metallic thallium and thallium nitrate with neutrons from an 800 mCi radon-beryllium source. ^{206}Tl was formed in the neutron capture reaction on ^{205}Tl and decayed to ^{206}Pb with a half-life of 4 min [81].

The last isotope of the radium series was discovered in 1961 by Nurmia et al.. ^{206}Hg was observed in the α-decay of ^{210}Pb with a branching ratio of $1.8 \pm 0.5 \times 10^{-6}$ % and decayed with a half-life of 7.5 ± 1.0 min to ^{206}Tl [87].

The fourth possible radioactive decay chain with A = 4n + 1 does not exist in nature. It is called the neptunium series after the isotope with the longest half-life (^{237}Np, $T_{1/2} = 2.1 \times 10^6$ y) discovered in 1948 by Wahl and Seaborg [107] (see Sect. 5.2). The majority of the chain was discovered by Hagemann et al. in 1947 [108] (see Sect. 7.2).

References

1. F. Soddy, Chem. News **107**, 97 (1913)
2. K. Fajans, Phys. Z. **14**, 131 (1913)
3. F. Soddy, Jahrb. Radioaktiv. **10**, 188 (1913)
4. A.S. Russell, Chem. News **107**, 49 (1913)
5. F. Soddy, Rep. Br. Assoc. Adv. Sci. **83**, 445 (1913)
6. F. Soddy, Nature **92**, 399 (1913)
7. E. Rutherford, Phil. Mag. **49**, 1 (1900)
8. E. Rutherford, Phil. Trans. Roy. Soc. Lond. A **204**, 169 (1905)
9. *Gmelin-Kraut's Handbuch der anorganischen Chemie*, Siebente Auflage, Band III, Abteilung 2, herausgegeben von C. Friedheim (Heidelberg 1908)
10. F. Soddy, *The Chemistry of the Radio-elements* (Longmans, Green and Co., London, 1911)
11. E. Rutherford, *Radioactive Substances and their Radiations* (Cambridge University Press, New York, 1913)

12. M.E. Weeks, J. Chem. Educ. **10**, 79 (1933)
13. http://en.wikipedia.org/wiki/Decay_chain
14. H. Becquerel, Compt. Rend. Acad. Sci. **122**, 420 (1896)
15. G.C. Schmidt, Ann. Physik **65**, 141 (1898)
16. P. Curie, M. Curie, Compt. Rend. Acad. Sci. **127**, 175 (1898)
17. P. Curie, M. Curie, G. Bemont, Compt. Rend. Acad. Sci. **127**, 1215 (1898)
18. P. Curie, M. Curie, Compt. Rend. Acad. Sci. **129**, 714 (1899)
19. E. Rutherford, Phil. Mag. **49**, 161 (1900)
20. W. Crookes, Proc. Roy. Soc. A **66**, 409 (1900)
21. K.A. Hofmann, E. Strauss, Ber. Deuts. Chem. Ges. **33**, 3126 (1900)
22. E. Rutherford, F. Soddy, Phil. Mag. **4**, 370 (1902)
23. A. Debierne, Compt. Rend. Acad. Sci. **136**, 446 (1903)
24. E. Rutherford, H.T. Barnes, Phil. Mag. **7**, 202 (1904)
25. A. Debierne, Compt. Rend. Acad. Sci. **138**, 411 (1904)
26. F.W. Aston, Nature **128**, 725 (1931)
27. W.C. Roentgen, Ueber eine neue Art von Strahlen. Sitzungsberichte der Physikalisch-medizinischen Gesellschaft zu Würzburg **29**, 132 (1895)
28. L. Badash, J. Chem. Educ. **43**, 219 (1966)
29. M. Curie, Compt. Rend. Acad. Sci. **126**, 1101 (1898)
30. F.W. Aston, Nature **129**, 649 (1932)
31. W. Marckwald, Ber. Deuts. Chem. Ges. **35**, 4239 (1902)
32. W. Marckwald, Phys. Z. **4**, 51 (1902)
33. E. Rutherford, Phil. Mag. **8**, 636 (1904)
34. E. Rutherford, Nature **71**, 341 (1905)
35. E. Rutherford, Phil. Mag. **10**, 290 (1905)
36. J.R. Partington, Nature **179**, 912 (1957)
37. F.E. Dorn, Abhandlungen der Naturforschenden Gesellschaft zu Halle **23**, 1 (1901)
38. F.W. Aston, Nature **120**, 224 (1927)
39. K.T. Bainbridge, E.B. Jordan, Phys. Rev. **50**, 282 (1936)
40. F. Giesel, Ber. Deuts. Chem. Ges. **35**, 3608 (1902)
41. F. Giesel, Ber. Deuts. Chem. Ges. **37**, 3963 (1904)
42. A. Debierne, Compt. Rend. Acad. Sci. **129**, 593 (1899)
43. A. Debierne, Compt. Rend. Acad. Sci. **130**, 906 (1900)
44. H.W. Kirby, ISIS **62**, 290 (1971)
45. M. Curie, Radium (Paris) **8**, 353 (1911)
46. F. Giesel, Ber. Deuts. Chem. Ges. **36**, 342 (1903)
47. E. Rutherford, F. Soddy, Phil. Mag. **5**, 576 (1903)
48. T. Godlewski, Nature **71**, 294 (1905)
49. O. Hahn, Proc. Roy. Soc. A **76**, 115 (1905)
50. O. Hahn, Nature **73**, 559 (1906)
51. O. Hahn, Phys. Z. **7**, 412 (1906)
52. O. Hahn, Ber. Deuts. Chem. Ges. **40**, 1462 (1907)
53. B.B. Boltwood, Nature **76**, 544 (1907)
54. O. Hahn, Phys. Z. **9**, 246 (1908)
55. O. Hahn, L. Meitner, Phys. Z. **9**, 649 (1908)
56. O. Hahn, L. Meitner, Verh. Deutsch. Phys. Gesell. **11**, 55 (1909)
57. O. Hahn, L. Meitner, Phys. Z. **10**, 697 (1909)
58. H. Geiger, E. Marsden, Phys. Z. **11**, 7 (1910)
59. G.N. Antonoff, Phil. Mag. **22**, 419 (1911)
60. H. Geiger, Phil. Mag. **22**, 201 (1911)
61. H. Geiger, J.M. Nuttall, Phil. Mag. **23**, 439 (1912)
62. K. Fajans, Phys. Z. **13**, 699 (1912)
63. T. Godlewski, Phil. Mag. **10**, 35 (1905)
64. O. Hahn, Phys. Z. **8**, 277 (1907)

65. E. Rutherford, H. Geiger, Phil. Mag. **22**, 621 (1911)
66. K.A. Hofmann, L. Gonder, V. Wölfl, Ann. Physik **320**, 615 (1904)
67. F. Soddy, Phil. Mag. **9**, 768 (1905)
68. B.B. Boltwood, Am. J. Sci **24**, 370 (1907)
69. B.B. Boltwood, Phys. Z. **8**, 884 (1907)
70. B.B. Boltwood, Am. J. Sci. **25**, 269 (1908)
71. H. Geiger, E. Rutherford, Phil. Mag. **20**, 691 (1910)
72. K. Fajans, O. Göhring, Naturwiss. **1**, 339 (1913)
73. K. Fajans, P. Beer, Ber. Deuts. Chem. Ges. **46**, 3486 (1913)
74. O. Hahn, Naturwiss. **9**, 84 (1921)
75. O. Hahn, Ber. Deuts. Chem. Ges. **54**, 1131 (1921)
76. F. Soddy, Chem. Soc. Ann. Rep. **10**, 262 (1913)
77. G.N. Antonoff, Phil. Mag. **26**, 1058 (1913)
78. E. Marsden, R.H. Wilson, Nature **92**, 29 (1913)
79. O. Hahn, L. Meitner, Phys. Z. **19**, 208 (1918)
80. A.J. Dempster, Nature **136**, 180 (1935)
81. P. Preiswerk, H. von Halban, Comput. Rend. Acad. Sci. **201**, 722 (1935)
82. H. Hulubei, Y. Cauchois, Compt. Rend. Acad. Sci. **209**, 39 (1939)
83. M. Perey, Compt. Rend. Acad. Sci. **208**, 97 (1939)
84. B. Karlik, T. Bernert, Naturwiss. **32**, 44 (1944)
85. M.H. Studier, E.K. Hyde, Phys. Rev. **74**, 591 (1948)
86. E.K. Hyde, A. Ghiorso, Phys. Rev. **90**, 267 (1953)
87. M.J. Nurmia et al., Nature **190**, 427 (1961)
88. L. Meitner, Naturwiss. **6**, 324 (1918)
89. F. Soddy, J. Chem. Soc. Abstr. **114**, ii 345 (1918)
90. F. Soddy, Annu. Rep. Prog. Chem. **15**, 195 (1918)
91. IUPAC: Commission de Nomenclature de Chimie Inorganique, in *Comptes Rendus de la Quinzième Conférence* (Amsterdam, 1949)
92. W.H. Koppenol, Helv. Phys. Acta **88**, 95 (2005)
93. K. Fajans, *Radioaktivität und die neueste Entwicklung der Lehre von den chemischen Elementen*, 2nd edn. (Friedr. Vieweg and Sohn, Braunschweig, 1920)
94. K. Fajans, D.F.C. Morris, Nature **244**, 137 (1973)
95. F. Soddy, *Nobel Lectures, Chemistry 1901–1921* (Elsevier Publishing Company, Amsterdam, 1966)
96. E. Rutherford, Nature **123**, 313 (1929)
97. F.W. Aston, Nature **123**, 313 (1929)
98. A.V. Grosse, J. Phys. Chem. **38**, 487 (1934)
99. M. Perey, J. Phys. Radium **10**, 435 (1939)
100. P. Avignon, J. Phys. Radium **11**, 521 (1950)
101. B.F. Thornton, S.C. Burdette, Bull. Hist. Chem. **35**(2), 86 (2010)
102. F.A. Paneth, Nature **159**, 8 (1947)
103. D.R. Corson, K.R. MacKenzie, E. Segrè, Phys. Rev. **57**, 459 (1940)
104. D.R. Corson, K.R. MacKenzie, E. Segrè, Phys. Rev. **58**, 672 (1940)
105. R.J. Walen, J. Phys. Radium **10**, 95 (1949)
106. E. Broda, N. Feather, Proc. Roy. Soc. A **190**, 20 (1947)
107. A.C. Wahl, G.T. Seaborg, Phys. Rev. **73**, 940 (1948)
108. F. Hagemann et al., Phys. Rev. **72**, 252 (1947)

Chapter 3
Isotopes of Stable Elements

3.1 Discovery of Isotopes in Stable Elements

Similar to the discovery of isotopes of the first radio-elements, the first identification of a stable isotope precedes the discovery of isotopes itself. Rutherford's 1908 publication of "The Charge and Nature of the α-Particle" can be considered as the first identification of the ^{4}He isotope [1]. Rutherford placed an α-particle source of "radium C" (^{214}Bi) inside an electromagnetic field to measure its charge. From the known E/M ratio the mass was deduced: "We have already seen that the evidence is strongly in favor of the view that $E = 2e$. Consequently $M = 3.84\,m$, i.e., the atomic weight of an α-particle is 3.84. The atomic weight of the helium atom is 3.96. Taking into account probable experimental errors in the estimates of the value of E/M for the α-particle, we may conclude that an α-particle is a helium atom, or, to be more precise, the α-particle, after it has lost its positive charge, is a helium atom."

The first indications that there might be two separate isotopes for a given element was presented by J.J. Thomson in 1913 [2, 3]. He used an electric discharge tube to ionize particles (rays of positive electricity or anode rays) from a neon gas and ions leaving through an aperture then passed through an electric and magnetic field and were deflected depending on their mass and velocity. Ions with the same mass and charge formed parabolas on a photographic plate at the end of the device: "The photograph shows that, in addition to helium and neon, there is another gas with an atomic weight about 22. This gas has been found in every specimen of neon which has been examined, including a very carefully purified sample prepared by Mr. E.W. Watson and a specimen very kindly supplied by M. Claud, of Paris... The substance giving the line 22 also occurs with a double charge, giving a line for which $m/e = 11$. There can, therefore, I think, be little doubt that what has been called neon is not a simple gas but a mixture of two gases, one of which has an atomic weight about 20 and the other about 22" [3]. The photographic plate exhibiting the line corresponding to an atomic weight of 22 is shown in Fig. 3.1 [4].

Thomson at the time did not make the connection to the isotope concept which was being developed by Soddy around the same time. It has been argued that the

M. Thoennessen, *The Discovery of Isotopes*,
DOI 10.1007/978-3-319-31763-2_3

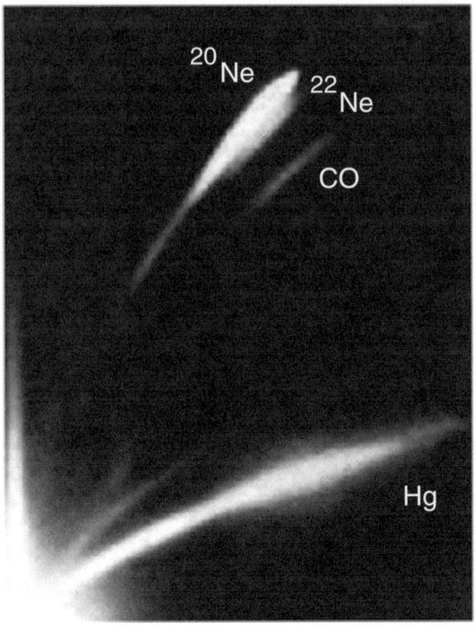

Fig. 3.1 Exposure of a photographic plate which was struck by ions from Thomson's positive ray experiment. The faint line corresponding to mass 20 is barely resolved from the intense ^{22}Ne line [4]

credit for the discovery of the two isotopes in neon (^{20}Ne and ^{22}Ne) should be given to Francis Aston who was Thomson's assistant [5–7]. At the 1913 meeting of British Association for the Advancement of Science (the same meeting where Soddy presented the isotope concept) Aston presented his paper "A new elementary constituent of the atmosphere" where he stated: "...evidence has now been obtained that atmospheric neon is not homogeneous, but consists of a mixture of two elements of approximate atomic weights, 19.9 and 22.1 respectively... The two elements appear to be identical in all their properties except atomic weight" [8].

Thomson did not readily accept the isotopes concept and as late as 1921 argued that the heavier neon could (although not probable) be due to a hydride (H_2Ne) [9]. Thus, it is justified to credit Aston with the discovery of these two neon isotopes.

The method of positive ray analysis was further improved and refined primarily during the following years by two competing groups: Francis Aston at the Cavendish Laboratory at the University of Cambridge and Arthur Dempster at the Ryerson Physical Laboratory of the University of Chicago. Both groups developed mass spectrometers with Dempster succeeding first in 1918 [10] followed by Aston a year later [11]. These two groups dominated the field of mass spectroscopy and identification of isotopes for the next twenty years. Between them they discovered over 80 % of all stable isotopes. Good references for the discovery of the stable isotopes were the second edition of Aston's book "Mass Spectra and Isotopes" [12], the report on "Relative Abundancies of the Elements" by Bainbridge and Nier [13] and the book "The History of Early Nuclear Physics (1896–1931)" by Mladenovic [14].

3.2 Dempster's Mass Spectrographs

Dempster discovered 40 isotopes of 21 different elements with his mass spectrographs between 1917 and 1939. They are listed in Table 3.1. The table includes the first identification of ^{235}U establishing it as the origin of the actinium radioactive decay chain (see Sect. 2.4) [15]. Figure 3.2 [10] shows a schematic picture of Dempster's first mass spectrograph. Charged ions (positive rays) were accelerated by an electric field and then deflected by 180° through a magnetic field. Quantitative spectra could be recorded by changing the potential of the electric field. In contrast to Thomson's gas discharge tube, Dempster used surface ionization to generate the positive rays. Thus he had the advantage of being able to analyze solid materials which were difficult to obtain in gaseous form. The mass resolution of the spectrograph was 1 in 100.

In his first paper Dempster presented results for hydrogen, oxygen, sodium and potassium and it can be argued that these measurements represent the first identification of the isotopes ^{1}H, ^{16}O, ^{23}Na, and ^{39}K. Dempster realized that his device was capable of separating isotopes and attempted to measure magnesium and chlorine which had been suggested by Soddy to consist of a mixture of isotopes [16]. However, due to mercury contamination he did not succeed in extracting spectra of these elements with sufficient resolution to separate the isotopes.

In 1920 Dempster succeeded to separate different isotopes from the same element for the first time. He identified the three magnesium isotopes ^{24}Mg, ^{25}Mg, and ^{26}Mg [17]. Two years later he separated two calcium isotopes in (^{40}Ca and ^{44}Ca) and four zinc isotopes (^{64}Zn, ^{66}Zn, ^{68}Zn, and ^{70}Zn) [18]. The later ones he actually had incorrectly reported as ^{63}Zn, ^{65}Zn, ^{67}Zn, and ^{69}Zn a year earlier [27].

Table 3.1 Isotopes discovered between 1917 and 1939 by Dempster. The isotopes are sorted by the date of submission

Date	Ref.	Isotopes
10/20/1917	[10]	^{1}H, ^{16}O, ^{23}Na, ^{39}K
12/10/1920[a]	[17]	^{24}Mg, ^{25}Mg, ^{26}Mg
07/21/1922	[18]	^{40}Ca, ^{44}Ca, ^{64}Zn, ^{66}Zn, ^{68}Zn, ^{70}Zn
05/23/1935	[19]	^{192}Pt, ^{198}Pt
06/12/1935	[20]	^{102}Pd, ^{104}Pd, ^{105}Pd, ^{106}Pd, ^{108}Pd, ^{110}Pd, ^{197}Au
07/12/1935	[15]	^{235}U[b]
05/25/1936	[21]	^{130}Ba, ^{132}Ba, ^{136}Ce, ^{138}Ce
06/23/1936	[22]	^{120}Te
02/01/1937	[23]	^{148}Nd, ^{150}Nd
11/02/1937	[24]	^{180}W
02/24/1938	[25]	^{152}Gd, ^{154}Gd, ^{158}Dy, ^{160}Dy, ^{162}Er, ^{164}Er, ^{168}Yb, ^{170}Yb
03/31/1939	[26]	^{174}Hf

[a] Publication date
[b] Also see Sect. 2.4

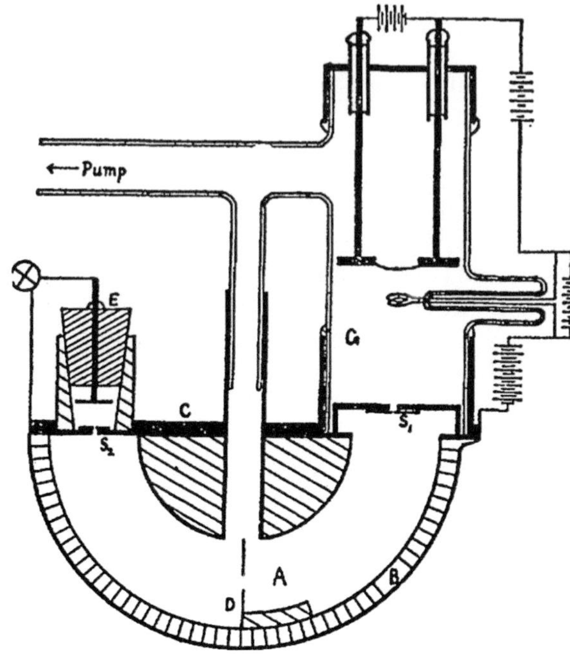

Fig. 3.2 Schematic of the first mass spectrograph developed by Dempster [10]. The ions are accelerated in the glass tube G and then enter the magnetic field (A) through slit S_1. They are deflected by 180°, exit the magnet through another set of slits (S_2) and the intensity is measured by an electrometer (Reprinted figure with permission from A.J. Dempster, *A new method of positive ray analysis*, Phys. Rev. 11 (1918) 316, Copyright 1918 by the American Physical Society.)

It then took another 13 years before Dempster discovered his next isotopes after he built a new improved mass spectrograph. The new spectrograph was a double focusing device which combined Aston's method with his own method in order to achieve better resolution. In addition he added a high frequency oscillating spark source which could ionize elements which were not accessible before.

On May 23rd 1935 he submitted the discovery of five platinum isotopes (^{192}Pt, ^{194}Pt, ^{195}Pt, ^{196}Pt, and ^{198}Pt) [19], however, unbeknownst to him the discovery of the three middle isotopes had been submitted less than two weeks earlier (5/11/1935) by Fuchs and Kopfermann [28] (see Sect. 3.5). In the same year he also discovered all six stable palladium isotopes (^{102}Pd, ^{104}Pd, ^{105}Pd, ^{106}Pd, ^{108}Pd, and ^{110}Pd) and determined that gold had only one (^{197}Au) stable isotope [20]. Due to the sensitivity of his mass spectrograph Dempster was then able to discover another 17 isotopes with very small abundances mostly in the rare earth region between 1936 and 1939.

3.3 Aston's Mass Spectrographs

Aston used a different method to separate isotopes in his mass spectrometer. He deflected the ions with consecutive electric and magnetic fields in opposite directions in such a way that ions with the same mass-over-charge ratio would focus to

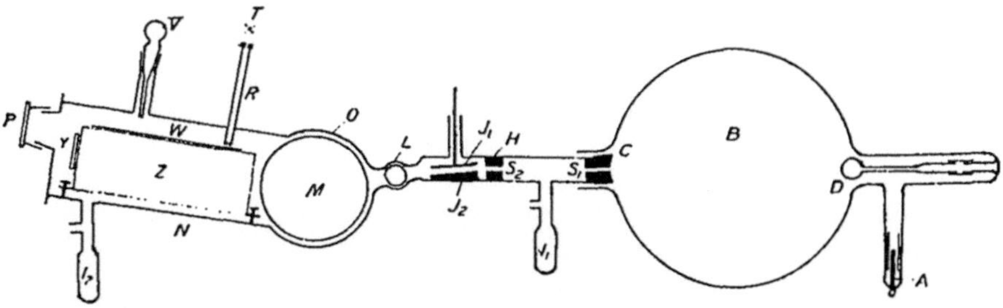

Fig. 3.3 Schematic diagram of Aston's first mass spectrometer [29]. The discharge tube is labeled as B, the electric field is applied between J_1 and J_2, and M represents the location of the electromagnet. The deflected ions then illuminate the photographic plate W (Reprinted by permission of Taylor & Francis Ltd, www.tandfonline.com, F.W. Aston, *The mass-spectra of chemical elements*, Philosophical Magazine 39 (1920) 611.)

the same point independent of velocity. Figure 3.3 shows the schematic diagram of the mass spectrometer [11, 29] which had a resolving power of about 1 in 130 [30]. After confirming the existence of the two different neon isotopes [31, 32] in 1919, Aston went on to discover 100 new isotopes of 47 different elements with this spectrograph during the next five years. These isotopes are listed in chronological order in Table 3.2.

Aston continued to use a gas discharge tube as a source so he predominantly analyzed elements which were easily available as gases, like the noble gases, halogens and alkali metals. The identification of the two stable lithium isotopes ^{6}Li and ^{7}Li was achieved with the old parabola methods with J.J. Thomson's son George Paget Thomson as a collaborator [37]. When Aston investigated krypton he observed for the first time more than two isotopes of an element. Figure 3.4 shows the line spectrum of the six krypton isotopes discovered by Aston [51].

Aston was extremely careful with his measurements and very cautious with claiming the discovery of new isotopes. Most of the isotopes which he initially only mentioned as possible were later confirmed. For example during his investigations of neon he stated in 1920: "On the clearest spectra obtained with neon there are distinct indications of a line corresponding to an isotope of mass 21. This line is extremely faint, so that if this constituent exists its proportion would be very small, probably well under 1 %" [32]. The existence of ^{21}Ne was established in 1928 by Hogness and Kvalnes [52].

In 1921 Aston also noticed a possible line for ^{30}Si which was confirmed by Mulliken three years later [53]. Initial hints of ^{128}Xe and ^{130}Xe [36], ^{54}Fe [41], and

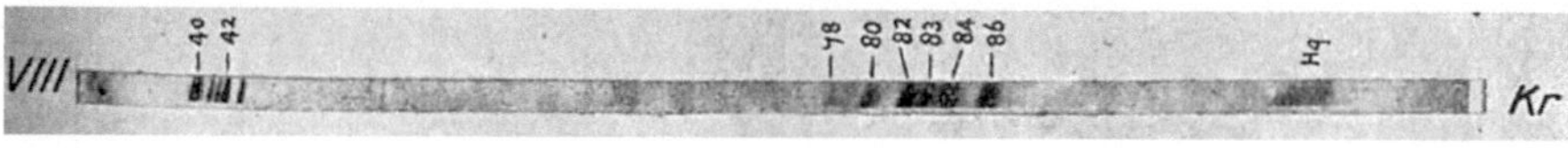

Fig. 3.4 Mass spectrum of krypton observed by Aston in 1920 [51] (Reprinted by permission from Macmillan Publishers Ltd: F.W. Aston, *Isotopes and atomic weights*, Nature 105 (1920) 617, copyright 1920.)

Table 3.2 Isotopes discovered between 1919 and 1924 by Aston using his first mass spectrometer. The isotopes are sorted by the date of submission

Date	Ref.	Isotopes
09/16/1913[a]	[8]	^{20}Ne, ^{22}Ne[a]
12/06/1919	[33]	^{12}C, ^{35}Cl, ^{37}Cl
02/25/1920	[34]	^{14}N, ^{36}Ar, ^{40}Ar, ^{78}Kr, ^{80}Kr, ^{82}Kr, ^{83}Kr, ^{84}Kr, ^{86}Kr, ^{202}Hg, ^{204}Hg
06/20/1920	[35]	^{10}B, ^{11}B, ^{19}F, ^{28}Si, ^{29}Si, ^{31}P, ^{32}S, ^{75}As, ^{79}Br, ^{81}Br
11/30/1920	[36]	^{127}I, ^{129}Xe, ^{131}Xe, ^{132}Xe, ^{134}Xe, ^{136}Xe
02/01/1921	[37][b]	^{6}Li, ^{7}Li
03/12/1921	[38]	^{41}K, ^{85}Rb, ^{87}Rb, ^{133}Cs
06/10/1921	[39]	^{58}Ni, ^{60}Ni
06/07/1922	[40]	^{116}Sn, ^{117}Sn, ^{118}Sn, ^{119}Sn, ^{120}Sn, ^{122}Sn, ^{124}Sn, ^{128}Xe, ^{130}Xe
08/26/1922	[41]	^{56}Fe
11/06/1922	[42]	^{27}Al, ^{74}Se, ^{76}Se, ^{77}Se, ^{78}Se, ^{80}Se, ^{82}Se, ^{124}Xe, ^{126}Xe
11/16/1922	[43]	^{121}Sb, ^{123}Sb
05/23/1923	[44]	^{70}Ge, ^{72}Ge, ^{74}Ge
07/25/1923	[45]	^{63}Cu, ^{65}Cu
09/18/1923	[46]	^{45}Sc, ^{48}Ti, ^{51}V, ^{52}Cr, ^{55}Mn, ^{59}Co, ^{69}Ga, ^{71}Ga, ^{89}Y, ^{88}Sr, ^{107}Ag, ^{109}Ag
01/23/1924	[47]	^{115}In
06/02/1924	[48]	^{54}Fe, ^{86}Sr, ^{138}Ba, ^{139}La, ^{141}Pr
08/08/1924	[49]	^{90}Zr, ^{92}Zr, ^{94}Zr, ^{140}Ce, ^{142}Ce, ^{142}Nd, ^{144}Nd, ^{146}Nd
11/04/1924	[50]	^{110}Cd, ^{111}Cd, ^{112}Cd, ^{113}Cd, ^{114}Cd, ^{116}Cd, ^{126}Te, ^{128}Te, ^{130}Te, ^{209}Bi

[a]Discovered with the parabola method
[b]With G.P. Thomson

^{145}Nd [49] were confirmed later by Aston himself in Refs. [40], [48], and [54], respectively. Aston demonstrated in 1922 that a possible ^{39}Cl line [51] reported earlier was not present [40]. In addition, Aston shifted his provisional assignment of five xenon isotopes as 128, 130, 131, 133, and 135 in early 1920 [34] by one mass unit to 129, 131, 132, 134, and 136 later in the year [36].

When Aston reported the discovery of ^{63}Cu and ^{65}Cu in 1923 [45] he corrected earlier results by Dempster who had claimed the observation of three copper isotopes (^{62}Cu, ^{64}Cu, and ^{66}Cu) [55] only a month earlier.

It appears that Aston made only one mistake during the measurements with his first mass spectrograph. His observation of a stable ^{121}Sn isotope first reported in 1922 [40] and reaffirmed in 1927 [30] turned out to be incorrect. In 1936 Bainbridge demonstrated that ^{121}Sn was not stable [56].

As Aston measured more and more new isotopes he realized that his experiments were limited by the resolution and sensitivity of the mass spectrometer. So already in 1921 he started the initial work on a new device. The overall design was essentially the same but all individual components were newly constructed. The resolving power was about a factor of five (1 in 650) better than the first spectrograph [30] which was then sufficient to separate mass lines for even the heaviest nuclides.

Table 3.3 Isotopes discovered by Aston with his second mass spectrograph between 1925 and 1935. The isotopes are sorted by the date of submission

Date	Ref.	Isotopes
08/01/1925	[57]	^{198}Hg, ^{199}Hg, ^{200}Hg, ^{201}Hg
06/12/1926	[58]	^{33}S, ^{34}S
07/30/1927	[59]	^{196}Hg, ^{206}Pb, ^{207}Pb, ^{208}Pb
12/31/1927[a]	[60]	^{112}Sn, ^{114}Sn, ^{115}Sn
08/22/1928	[61]	^{67}Zn
07/26/1930	[62]	^{50}Cr, ^{53}Cr, ^{54}Cr
08/21/1930	[63]	^{92}Mo, ^{94}Mo, ^{95}Mo, ^{96}Mo, ^{97}Mo, ^{98}Mo, ^{100}Mo
11/28/1930	[64]	^{182}W, ^{183}W, ^{184}W, ^{186}W
01/31/1931	[65]	^{96}Ru, ^{99}Ru, ^{100}Ru, ^{101}Ru, ^{102}Ru, ^{104}Ru, ^{186}Os, ^{187}Os, ^{188}Os, ^{189}Os, ^{190}Os, ^{192}Os
03/31/1931	[66]	^{185}Re, ^{187}Re
05/29/1931	[67]	^{73}Ge, ^{76}Ge, ^{125}Te
07/21/1931	[68]	^{87}Sr, ^{135}Ba, ^{136}Ba, ^{137}Ba
07/11/1932	[69]	^{93}Nb, ^{181}Ta
12/02/1933	[54]	^{143}Nd, ^{145}Nd, ^{147}Sm, ^{148}Sm, ^{149}Sm, ^{152}Sm, ^{154}Sm, ^{151}Eu, ^{153}Eu, ^{155}Gd, ^{156}Gd, ^{157}Gd, ^{158}Gd, ^{160}Gd, ^{159}Tb
02/17/1934	[70]	^{161}Dy, ^{162}Dy, ^{163}Dy, ^{164}Dy, ^{165}Ho, ^{166}Er, ^{167}Er, ^{168}Er, ^{170}Er, ^{169}Tm, ^{171}Yb, ^{172}Yb, ^{173}Yb, ^{174}Yb, ^{176}Yb, ^{175}Lu
04/21/1934	[71]	^{42}Ca, ^{43}Ca, ^{46}Ti, ^{47}Ti, ^{49}Ti, ^{50}Ti, ^{91}Zr, ^{96}Zr, ^{103}Rh, ^{144}Sm, ^{150}Sm, ^{176}Hf, ^{177}Hf, ^{178}Hf, ^{179}Hf, ^{180}Hf
07/21/1934	[72]	^{61}Ni, ^{62}Ni, ^{106}Cd, ^{108}Cd
01/24/1935	[73]	^{57}Fe

[a]Publication date

During the next ten years Aston investigated 22 new elements discovering 72 isotopes. He also revisited 14 elements and with the improved resolution of the new spectrograph he was able to identify another 33 new isotopes for a total of 103. The isotopes are listed in chronological order in Table 3.3.

Aston was not as accurate reporting the results from his second spectrograph as he was with the first device. This may be due to the fact that most of the isotopes with the largest abundances had been discovered and it became more and more difficult to identify very rare isotopes. The claims for the existence of twelve isotopes turned out to be incorrect.

When he studied lead and identified ^{206}Pb, ^{207}Pb, and ^{208}Pb, he also saw "an exceedingly faint line at 209" and evidence for the isotopes 203, 204 and 205 [59]. Five years later he confirmed the apparent existence of these isotopes and claimed the observation of ^{210}Pb [74]. In the meantime (two months earlier), Schüler and Jones had already reported the existence of ^{204}Pb [75]. However, subsequently the observations of ^{205}Pb, ^{209}Pb, and ^{210}Pb were shown to be incorrect by Bainbridge and Jordan [56]. ^{203}Pb could also not be confirmed and Nier set a limit of 5 % of Aston's value [76].

The apparent discovery of two isotopes of the element mercury (^{197}Hg and ^{203}Hg) [77], was also proven to be incorrect by Bainbridge and Jordan [56] and Nier [78], respectively. In addition, in the same paper Bainbridge and Jordan could not find any evidence for Aston's claim of ^{115}Cd [72].

The observation of two additional zinc isotopes (^{65}Zn and ^{69}Zn) [61] in 1928 was also incorrect [79]. In 1931 Aston reported five new germanium isotopes [80] out of which only two were correct (^{73}Ge and ^{76}Ge). However, this paper should not be credited with the discovery of these isotopes because the relative abundances were also incorrect (^{75}Ge was stronger than ^{76}Ge). In a subsequent paper [67] Aston measured the correct abundances for ^{73}Ge and ^{76}Ge, but still claimed the existence of ^{71}Ge, ^{75}Ge, and ^{77}Ge. Their non-existence was shown two years later by Bainbridge [81] suggesting that Aston's observations were due to hydrides of ^{70}Ge, ^{74}Ge, and ^{76}Ge [82].

The investigation of ruthenium yielded in addition to six clearly identified isotopes the "possibility of an extremely faint seventh" with mass number 98 [65]. It took another 13 years before this observation was verified by Ewald [83].

When Aston published the discovery of two additional nickel isotopes (^{61}Ni and ^{62}Ni), he also suggested the existence of two more isotopes: "Lines at 56 and 64 present to less than 1 % are probably due to isotopes, but this is not yet certain" [72]. A year later deGier and Zeeman could confirm ^{64}Ni but failed to detect ^{56}Ni [84].

Finally it should be mentioned that Aston claimed the discovery of ^{108}Cd disregarding an earlier measurement by Svensson [85]: "Svensson has already claimed to have demonstrated the presence of isotopes 118 and 108 by observations on the band spectra of cadmium hydride. The new mass-spectra show that the former cannot be present even to the extend of 0.3 %. This discrepancy, and his failure to detect the more abundant isotope 106, suggest that the arguments on which his claim is based are not strictly valid" [72].

Overall Aston discovered a total of 205 isotopes, and with the exception of ^{6}Li and ^{7}Li, he performed all experiments by himself without any coauthors listed on the publications.

3.4 Other Mass Spectrographs

Although Dempster and Aston discovered the majority of the stable isotopes, many other groups developed mass spectrographs and made significant contributions to the exploration of stable isotopes. The discovery is only the first step which provides a method to produce and study these isotopes further. The subsequent measurements of properties like masses and abundances reveal important information about the underlying structure and force that hold nuclides together.

Table 3.4 Isotopes discovered between 1921 and 1949 with other mass spectrographs. The submission date, first author and reference of the publication, as well as the type of spectrometer are listed

Date	First author	Ref.	Spectrometer	Isotopes
07/17/1908	Rutherford	[1]	E/M measurement	$^{4}He^{a}$
05/23/1921	Thomson	[86]	Parabola method	^{9}Be
08/13/1928	Hogness	[52]	180° magnetic analyser	^{21}Ne
03/04/1932	Bainbridge	[87]	180° magnetic analyser	^{122}Te, ^{123}Te, ^{124}Te
02/24/1934	Zeeman	[88]	Parabola method	^{38}Ar
07/15/1935	Nier	[89]	180° magnetic analyser	^{40}K
09/28/1935	deGier	[84]	Parabola method	^{64}Ni
10/26/1935	deGier	[90]	Parabola method	^{58}Fe
04/28/1936	Blewitt	[91]	180° magnetic analyser	^{84}Sr, ^{134}Ba
09/28/1937	Nier	[92]	180° magnetic analyser	^{184}Os
12/23/1937	Nier	[93]	180° magnetic analyser	^{36}S, ^{46}Ca , ^{48}Ca
09/09/1943	Ewald	[83]	Double focussing	^{98}Ru
08/08/1947	Inghram	[94]	60° single focussing	^{138}La
10/14/1948	Hess	[95]	60° single focussing	^{156}Dy
03/22/1949	Duckworth	[96]	Double focussing	^{190}Pt
10/24/1949	Hess	[97]	60° single focussing	^{50}V
	Leland	[98]	60° single focussing	

[a] Charge measured in an electromagnetic field and the mass was determined from the known charge-to-mass ratio

In addition to the work at the University of Cambridge by Aston and the University of Chicago by Dempster, mass spectrometers at nine different institutions in Europe and the U.S. contributed to the discovery of new isotopes. Table 3.4 lists the isotopes in chronological order together with the first author and the type of spectrometer.

Two groups continued to identify new isotopes with J.J. Thomson's original parabola method. At Cambridge, G.P. Thomson determined that ^{9}Be was the only stable beryllium isotope [86]. Zeeman and deGier discovered ^{38}Ar [88], ^{64}Ni [84], and ^{58}Fe [90] at the University of Amsterdam using the method still in the early 1930s. Aston had suggested the presence of ^{64}Ni [72] but had dismissed indications for ^{58}Fe: "Line 58 was present but weakened as the work proceeded and was most probably due to traces of nickel still left in the tube" [73].

In 1928, at the chemical laboratories of the University of California, Hogness and Kvalnes established the existence of ^{21}Ne [52] for which Aston had observed an "extremely faint" line in 1920 [32]. When Bainbridge from the Bartol Research Foundation of the Franklin Institute in Swarthmore, Pennsylvania, extended the number of stable tellurium isotopes with the observation of ^{122}Te, ^{123}Te, and ^{124}Te he also found "indications of an extremely faint isotope at 127" [87]. However, this isotope was found not to be stable a few years later by Dempster [22].

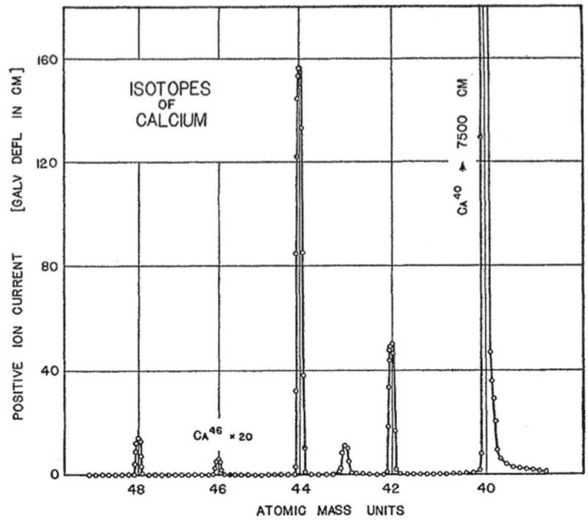

Fig. 3.5 Mass spectrum of calcium isotopes measured by Nier in 1937 including the discovery of ^{46}Ca and ^{48}Ca [93] (Reprinted figure with permission from A.O. Nier, *The isotopic constitution of calcium, titanium, sulphur and argon*, Phys. Rev. 53 (1938) 282, Copyright 1938 by the American Physical Society.)

As an example of the impressive improvements of the various types of new mass spectrographs developed in the 1930s Fig. 3.5 shows the mass spectrum of calcium published by Nier in 1938 [93]. With his spectrograph at Harvard University he was able to cleanly identify the new isotopes ^{46}Ca and ^{48}Ca where ^{46}Ca has only an abundance of 0.0033 %.

The last five stable isotopes were discovered in the 1940s. The existence of a stable ^{98}Ru isotope demonstrated by Ewald in 1943 at the Kaiser Wilhelm-Institut für Chemie in Berlin-Dahlem, Germany [83], had previously been indicated by Aston in 1931 [65]. The discovery of the essentially "stable" last isotope was simultaneously reported by Hess and Inghram at Argonne National Laboratory [97] and by Leland at the University of Minnesota [98]. The papers were both submitted on October 24, 1949 and published in the same issue of Physical Review. ^{50}V is actually unstable with a half-life of 1.4×10^{17} y.

3.5 Molecular Spectroscopy and Nuclear Reactions

Soon after the discovery of the isotope concept Bohr predicted within his atomic model that the presence of isotopes should be observable in molecular spectra. The frequencies of the vibrating atoms should depend on the mass of the atoms [7]. The effect was experimentally confirmed for the first time in 1920 essentially simultaneously by Loomis at New York University [99] and by Kratzer at the University of Göttingen [7]. From 1924 through 1935 14 new isotopes were discovered with this method at nine different institutions (see Table 3.5).

Table 3.5 Stable isotopes discovered by molecular spectroscopy and in nuclear reactions. The submission date, first author and reference of the publication, as well as the method of discovery are listed

Date	First author	Ref.	Method	Isotopes
03/22/1924[a]	Mulliken	[53]	Molecular spec.	^{30}Si
12/17/1924	Blackett	[100]	Nuclear reaction	^{17}O
01/14/1929	Giauque	[101]	Molecular spec.	^{18}O
06/24/1929	King	[102, 103]	Molecular spec.	^{13}C
11/26/1929	Naude	[104]	Molecular spec.	^{15}N
03/10/1931	Schüler	[105]	Molecular spec.	^{203}Tl, ^{205}Tl
12/05/1931	Urey	[106]	Molecular spec.	^{2}H
02/01/1932	Schüler	[75]	Molecular spec.	^{204}Pb
03/09/1934	Oliphant	[107]	Nuclear reaction	^{3}He
03/22/1934	Wehrli	[108]	Molecular spec.	^{113}In
05/11/1935	Fuchs	[28]	Molecular spec.	^{194}Pt, ^{195}Pt, ^{196}Pt
07/20/1935[a]	Marsh	[109]	Nuclear reaction	^{176}Lu
08/07/1935	Venkatesachar	[110]	Molecular spec.	^{191}Ir, ^{193}Ir
10/22/1937	Oldenberg	[111]	Nuclear reaction	^{180}Ta

[a] Publication date

The first discovery of a new isotope with the method of molecular spectroscopy was the identification of ^{30}Si by Mulliken 1924. He identified the isotope in the spectrum of silicon nitrate [53]. Previously, Aston had only indicated the possibility of a stable ^{30}Si isotope [35].

The heavy oxygen isotope ^{18}O was first observed in atmospheric absorption spectra [101]. The discovery of ^{13}C was described in a paper that was submitted simultaneously to Physical Review [102] and Nature [103]. Physical Review published the article 12 days before Nature.

As mentioned in the previous Sect. 3.3, Schüler reported the discovery of ^{204}Pb [75] before Aston could verify [74] his earlier observation of a possible line at 204 [59]. The observation of ^{194}Pt, ^{195}Pt and ^{196}Pt in the hyperfine spectra of platinum were reported [28] only three weeks before Dempster was able to identify them in his mass spectrograph [19]. Similarly, Dempster measured the two iridium isotopes ^{191}Ir and ^{193}Ir [112] only three months after Venkatesachar and Sibaiya identified these isotopes in the hyperfine spectra of iridium [110].

Finally there were four stable isotopes that were not discovered by either mass or molecular spectroscopy but in nuclear reactions. As early as 1919 had Rutherford shown that it was possible to transform nuclides by irradiating them with α-particles [113]. When he irradiated nitrogen gas with α-particles he observed energetic protons and concluded that they must originate from the disintegration of the nitrogen. In 1924 Blackett repeated the experiment in a Wilson chamber where he could observe the tracks of the incoming α particles as well as any outgoing reaction products. He recorded eight tracks that indicated a collision and, against expectations, the α

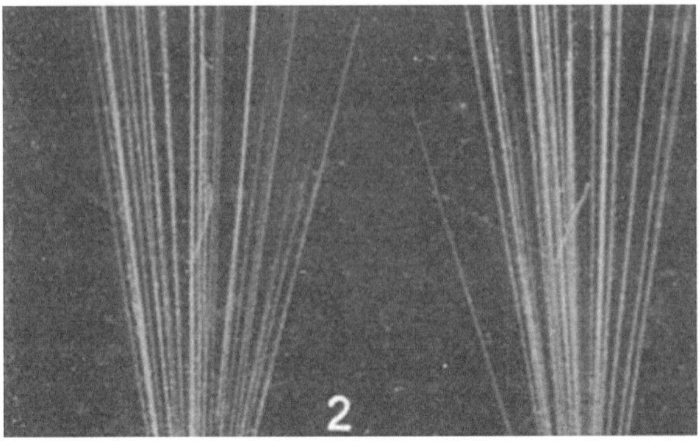

Fig. 3.6 Wilson chamber photograph of α-particle tracks in nitrogen gas. The *left* and *right* side of the picture shows the same tracks photographed from perpendicular angles. One track clearly shows a nuclear reaction, where the incoming α-particle disappears and only a proton (*thin track* to the *left*) and a residual nucleus (*thick track* to the *right*) emerge [100]

track did not break up into three branches but only into two and the α-particle was absorbed in the nucleus: "In ejecting a proton from a nitrogen nucleus the alpha-particle is therefore itself bound to the nitrogen nucleus. The resulting new nucleus must have a mass 17, and, provided no electrons are gained or lost in the process, an atomic number of 8... It ought therefore to be an isotope of oxygen" [100]. Figure 3.6 shows the second track of such a collision.

For the discovery of the next stable isotope in a nuclear reaction, the incoming particle did not originate from a source but from an accelerator. In 1934, Oliphant, Harteck, and Rutherford irradiated various deuteriated substances with deuterons accelerated by 20–100 kV. From the observation of a large number of neutrons they proposed the reaction "$D_1^2 + D_1^2 \rightarrow He_2^3 + n_0^1$ leading to the production of a helium isotope of mass 3 and a neutron" [107].

Marsh and Sugden discovered ^{176}Lu in 1935 during the irradiation of lutetium oxides with neutrons from a 400 mCi radon source in contact with powdered beryllium. In the chemically separated sample of lutetium after the irradiation they measured strong activity with the half-life of 4.0 ± 0.1 h which they attributed to the reaction ^{175}Lu(n,γ)^{176}Lu [109]. This half-life is close to the accepted half-life of 3.664 ± 0.019 h for an isomeric state of ^{176}Lu. The ground-state can be considered stable with a half-life of 38.5 ± 0.7 Gy and it was identified in a mass spectrometer in 1939 by Mattauch and Lichtblau [114].

^{180}Ta was discovered in 1937 by irradiating tantalum with fast neutrons produced by irradiating lithium with 5.5 MeV deuterons from the Berkeley cyclotron. An 8.2 h activity was proposed to originate from ^{180}Ta following the capture of one neutron and the ejection of two neutrons [111]. This half-life agrees with the presently adopted value of 8.152 ± 0.006 h. This state corresponds to the ground-state of ^{180}Ta. Nevertheless ^{180}Ta is considered to be stable because a state at an excitation energy of 77.1 keV has a half-life of larger than 1.2×10^{15} y. It has been measured with a mass spectrograph only in 1955 [115]. Pool, Cork, and Thornton reported a half-life of 9.1 h from a similar experiment, however, they did not assign the activity to a specific isotope [116]. Even earlier Bothe and Gentner measured a 20 min half-life from a photo disintegration experiment of tantalum which they assigned to ^{180}Ta [117]. However, such a decay has not been observed since.

References

1. E. Rutherford, H. Geiger, Proc. Roy. Soc. A **81**, 162 (1908)
2. J.J. Thomson, Nature **91**, 333 (1913)
3. J.J. Thomson, Proc. Roy. Soc. A **89**, 1 (1913)
4. J.J. Thomson, *Rays of Positive Electricity and their Application to Chemical Analyses* (Longmans, Green Co, London, 1913)
5. J. Hughes, Physics World, September, p. 31 (2003)
6. W.H. Brock, *From Protyle to Proton: William Prout and the Nature of Matter 1785–1985* (Adam Hilger, Bristol, 1985)
7. H. Kragh, Stud. Hist. Philos. Sci. Part B: Stud. Hist. Philos. Mod. Phys. **43**, 176 (2012)
8. F.W. Aston, Rep. Br. Assoc. Adv. Sci. **83**, 403 (1913)
9. J.J. Thomson, Proc. Roy. Soc. A **697**, 87 (1921)
10. A.J. Dempster, Phys. Rev. **11**, 316 (1918)
11. F.W. Aston, Philos. Mag. **38**, 707 (1919)
12. F.W. Aston, *Mass Spectra and Isotopes*, 2nd edn. (Longmans Green & Co, New York, 1942)
13. K.T. Bainbridge, A.O. Nier, *Relative Abundances of the Elements*, Preliminary Report No. 9, Nuclear Science Series, National Research Council (1950)
14. M. Mladenovic, *The History of Early Nuclear Physics (1896–1931)* (World Scientific, Singapore, 1992)
15. A.J. Dempster, Nature **136**, 180 (1935)
16. F. Soddy, Sci. Mon. **5**, 509 (1917)
17. A.J. Dempster, Science **52**, 559 (1920)
18. A.J. Dempster, Phys. Rev. **20**, 631 (1922)
19. A.J. Dempster, Nature **135**, 993 (1935)
20. A.J. Dempster, Nature **136**, 65 (1935)
21. A.J. Dempster, Phys. Rev. **49**, 947 (1936)
22. A.J. Dempster, Phys. Rev. **50**, 186 (1936)
23. A.J. Dempster, Phys. Rev. **51**, 289 (1937)
24. A.J. Dempster, Phys. Rev. **52**, 1074 (1937)
25. A.J. Dempster, Phys. Rev. **53**, 727 (1938)
26. A.J. Dempster, Phys. Rev. **55**, 794 (1939)
27. A.J. Dempster, Proc. Nat. Acad. Sci. **7**, 45 (1921)
28. B. Fuchs, H. Kopfermann, Naturwissenschaften **23**, 372 (1935)
29. F.W. Aston, Philos. Mag. **39**, 611 (1920)
30. F.W. Aston, Proc. Roy. Soc. A **115**, 487 (1927)
31. F.W. Aston, Nature **104**, 334 (1919)
32. F.W. Aston, Philos. Mag. **39**, 449 (1920)
33. F.W. Aston, Nature **104**, 393 (1919)
34. F.W. Aston, Nature **105**, 8 (1920)
35. F.W. Aston, Nature **105**, 547 (1920)
36. F.W. Aston, Nature **106**, 468 (1920)
37. F.W. Aston, G.P. Thomson, Nature **106**, 827 (1921)
38. F.W. Aston, Nature **107**, 72 (1921)
39. F.W. Aston, Nature **107**, 520 (1921)

40. F.W. Aston, Nature **109**, 813 (1922)
41. F.W. Aston, Nature **110**, 312 (1922)
42. F.W. Aston, Nature **110**, 664 (1922)
43. F.W. Aston, Nature **110**, 732 (1922)
44. F.W. Aston, Nature **111**, 771 (1923)
45. F.W. Aston, Nature **112**, 162 (1923)
46. F.W. Aston, Nature **112**, 449 (1923)
47. F.W. Aston, Nature **113**, 192 (1924)
48. F.W. Aston, Nature **113**, 856 (1924)
49. F.W. Aston, Nature **114**, 273 (1924)
50. F.W. Aston, Nature **114**, 717 (1924)
51. F.W. Aston, Nature **105**, 617 (1920)
52. T.R. Hogness, H.M. Kvalnes, Phys. Rev. **32**, 942 (1928)
53. R.S. Mulliken, Nature **113**, 423 (1924)
54. F.W. Aston, Nature **132**, 930 (1933)
55. A.J. Dempster, Nature **112**, 7 (1923)
56. K.T. Bainbridge, E.B. Jordan, Phys. Rev. **50**, 282 (1936)
57. F.W. Aston, Nature **116**, 208 (1925)
58. F.W. Aston, Nature **117**, 893 (1926)
59. F.W. Aston, Nature **120**, 224 (1927)
60. F.W. Aston, Nature **120**, 956 (1927)
61. F.W. Aston, Nature **122**, 345 (1928)
62. F.W. Aston, Nature **126**, 200 (1930)
63. F.W. Aston, Nature **126**, 348 (1930)
64. F.W. Aston, Nature **126**, 913 (1930)
65. F.W. Aston, Nature **127**, 233 (1931)
66. F.W. Aston, Nature **127**, 591 (1931)
67. F.W. Aston, Proc. Roy. Soc. A **132**, 487 (1931)
68. F.W. Aston, Nature **128**, 221 (1931)
69. F.W. Aston, Nature **130**, 130 (1932)
70. F.W. Aston, Nature **133**, 327 (1934)
71. F.W. Aston, Nature **133**, 684 (1934)
72. F.W. Aston, Nature **134**, 178 (1934)
73. F.W. Aston, Proc. Roy. Soc. A **149**, 396 (1935)
74. F.W. Aston, Nature **129**, 649 (1932)
75. H. Schüler, E.G. Jones, Naturwissenschaften **20**, 171 (1932)
76. A.O. Nier, J. Am. Chem. Soc. **60**, 1571 (1938)
77. F.W. Aston, Nature **130**, 847 (1932)
78. A.O. Nier, Phys. Rev. **54**, 275 (1938)
79. A.O. Nier, Phys. Rev. **49**, 272 (1936)
80. F.W. Aston, Nature **122**, 167 (1928)
81. K.T. Bainbridge, J. Frank. Inst. **215**, 509 (1933)
82. K.T. Bainbridge, Phys. Rev. **43**, 1056 (1933)
83. H. Ewald, Z. Phys. **122**, 487 (1944)
84. J. deGier, P. Zeeman, Proc. Akad. Soc. Amst. **38**, 810 (1935)
85. E. Svensson, Nature **131**, 28 (1933)
86. G.P. Thomson, Nature **107**, 395 (1921)
87. K.T. Bainbridge, Phys. Rev. **39**, 1021 (1932)
88. P. Zeeman, J. deGier, Proc. Akad. Soc. Amst. **37**, 127 (1934)
89. A.O. Nier, Phys. Rev. **48**, 283 (1935)
90. J. deGier, P. Zeeman, Proc. Akad. Soc. Amst. **38**, 959 (1935)
91. J.P. Blewett, M.B. Sampson, Phys. Rev. **49**, 778 (1936)
92. A.O. Nier, Phys. Rev. **52**, 885 (1937)
93. A.O. Nier, Phys. Rev. **53**, 282 (1938)

94. M.G. Inghram, R.J. Hayden, D.C. Hess Jr., Phys. Rev. **72**, 967 (1947)
95. D.C. Hess Jr., M.G. Inghram, Phys. Rev. **74**, 1724 (1948)
96. H.E. Duckworth, R.E. Black, R.F. Woodcock, Phys. Rev. **75**, 1438 (1949)
97. D.C. Hess Jr., M.G. Inghram, Phys. Rev. **76**, 1717 (1949)
98. W.T. Leland, Phys. Rev. **76**, 1722 (1949)
99. F.W. Loomis, Nature **106**, 179 (1920)
100. P.M.S. Blackett, Proc. Roy. Soc. A **107**, 349 (1925)
101. W.F. Giauque, H.L. Johnston, J. Am. Chem. Soc. **51**, 1436 (1929)
102. A.S. King, R.T. Birge, Phys. Rev. **34**, 376 (1929)
103. A.S. King, R.T. Birge, Nature **124**, 127 (1929)
104. S.M. Naude, Phys. Rev. **34**, 1498 (1929)
105. H. Schüler, J.E. Keyston, Naturwissenschaften **19**, 320 (1931)
106. H.C. Urey, F.G. Brickwedde, G.M. Murphy, Phys. Rev. **39**, 164 (1932)
107. M.L. Oliphant, P. Harteck, E. Rutherford, Nature **133**, 413 (1934)
108. M. Wehrli, Naturwissenschaften **22**, 289 (1934)
109. J.K. Marsh, S. Sugden, Nature **136**, 102 (1935)
110. B. Venkatesachar, L. Sibaiya, Nature **136**, 437 (1935)
111. O. Oldenberg, Phys. Rev. **53**, 35 (1938)
112. A.J. Dempster, Nature **136**, 909 (1935)
113. E. Rutherford, Philos. Mag. **37**, 581 (1919)
114. J. Mattauch, H. Lichtblau, Z. Phys. **111**, 514 (1939)
115. F.A. White, T.L. Collins, F.M. Rourke, Phys. Rev. **97**, 566 (1955)
116. M.L. Pool, J.M. Cork, R.L. Thornton, Phys. Rev. **52**, 239 (1937)
117. W. Bothe, W. Gentner, Naturwissenschaften **25**, 191 (1937)

Chapter 4
First Nuclear Reactions

4.1 Artificially Produced Isotopes

Rutherford had already demonstrated in 1919 that nuclear reactions could be used to break up nuclides [1] and Blackett formed ^{17}O in the reaction ^{14}N(α,p)^{17}O in 1925 [2]. Before nuclear reactions could be used to explain the various possible transformations, it was necessary to understand the composition of the nuclides in terms of its constituents. The last missing piece was found by Chadwick in 1932 when he introduced the neutron in order to explain the reaction ^{9}Be(α,n)^{12}C [3].

Soon after, this reaction was used as a source for neutrons to induce nuclear reactions. At the same time the first accelerators were developed which made it possible to accelerate protons and deuterons with larger intensities and higher energies than α-particles available from radioactive sources.

In the following sections, the discoveries of isotopes produced with α and neutron sources (Sect. 4.2) and with accelerators (Sects. 4.3 and 4.4) between 1932 and 1945 are described. The last two sections cover isotopes discovered in γ-ray induced reactions (Sect. 4.5) and in fission (Sect. 4.6).

4.2 Alpha-Particle and Neutron Sources

The first nuclear reactions were induced by using α-particles. The sources were extracted from the radio-elements of the thorium or uranium decay chains. They were typically radon or polonium sources. Neutrons were produced by mixing α sources with beryllium using the reaction ^{9}Be(α,n)^{12}C. Most of the neutron sources were radon-beryllium or radium-beryllium sources. Table 4.1 lists the isotopes produced with α (top) and neutron (bottom) sources.

In 1932, it was thought that the nucleus consisted of the particles known at the time, protons and electrons. When Chadwick irradiated beryllium with α-particles in 1932 he could not explain the kinematics of the reaction with just these two particles and

M. Thoennessen, *The Discovery of Isotopes*,
DOI 10.1007/978-3-319-31763-2_4

Table 4.1 Isotopes discovered with α-particle or neutron sources: The submission date, first author and reference of the publication as well as the intensity of the sources in mCi (1 Ci = 3.7×10^{10} Bq, 1 Bq = 1 decay/s), the targets, and the reactions are listed

Date	First author	Ref.	Int.	Target	Reaction	Isotope
02/17/1932	Chadwick	[3]	–	^{9}Be	(α,n)	^{1}n
01/15/1934[a]	Curie	[4]	100	^{10}B, ^{27}Al	(α,n)	^{13}N, ^{30}P
03/17/1934	Wertenstein	[5]	15	^{14}N	(α,n)	^{17}F
03/20/1934	Curie	[6]	60	^{25}Mg	(α,p)	^{28}Al
05/04/1934	Frisch	[7]	1	^{23}Na, ^{31}P	(α,n)	^{26}Al, ^{34}Cl
07/03/1935	Frisch	[8]	600	^{19}F, ^{40}Ca	(α,n), (α,p)	^{22}Na, ^{43}Sc
11/10/1933	Harkins	[9]	–	^{19}F	(n,α)	^{16}N
07/12/1934	Amaldi	[10]	800	^{51}V, ^{55}Mn, ^{75}As	(n,γ)	^{52}V, ^{56}Mn, ^{76}As
07/25/1934	Fermi	[11]	800	^{24}Mg, ^{27}Al, ^{31}P	(n,p)	^{24}Na, ^{27}Mg, ^{31}Si
				^{32}S, ^{127}I	(n,p), (n,γ)	^{32}P, ^{128}I
12/23/1934	Hevesy	[12]	–	^{45}Sc	(n,α)	^{42}K
03/15/1935	Hevesy	[13]	–	^{180}Hf	(n,γ)	^{181}Hf
05/02/1935	Hahn	[14]	–	^{232}Th	(n,γ)	^{233}Th
07/20/1935[a]	Hevesy	[15]	–	^{164}Dy	(n,γ)	^{165}Dy[b]
07/20/1935[a]	Marsh	[16]	400	^{139}La, ^{141}Pr, ^{164}Dy	(n,γ)	^{140}La, ^{142}Pr, ^{165}Dy[b]
03/07/1936	Andersen	[17]	100	^{35}Cl	(n,p)	^{35}S
07/06/1936[b]	Nahmias	[18]	400	^{19}F, ^{23}Na	(n,p)	^{19}O, ^{23}Ne
1936[c]	Neuninger	[19]	–	^{169}Tm	(n,γ)	^{170}Tm
07/29/1936	Bjerge	[20]	–	^{9}Be	(n,α)	^{6}He
10/29/1936	Hevesy	[21]	300	^{45}Sc, ^{165}Ho	(n,γ)	^{46}Sc, ^{166}Ho
05/14/1937	Meitner	[22]	–	^{238}U	(n,γ)	^{239}U
05/16/1938	Meitner	[23]	–	^{232}Th	(n,γ)	^{233}Pa[d]
05/12/1940	Zingg	[24]	–	^{190}Os, ^{192}Os	(n,γ)	^{191}Os, ^{193}Os

[a] Date of presentation
[b] Simultaneously discovered
[c] No date, not before July 1936
[d] Populated by β^--decay

he proposed the existence of the neutron: "[The] results, and others I have obtained in the course of the work, are very difficult to explain on the assumption that the radiation from beryllium is a quantum radiation, if energy and momentum are to be conserved in the collisions. The difficulties disappear, however, if it be assumed that the radiation consists of particles of mass 1 and charge 0, or neutrons. The capture of the α-particle by the Be^9 nucleus may be supposed to result in the formation of a C^{12} nucleus and the emission of the neutron [3]."

Only a year later, Harkins, Gans, and Newson already used the new particles as a source to induce the reaction ^{19}F(n,α)^{16}N at the University of Chicago. They irradiated a mixture of 30 % difluor-dichlor-methane and 70 % helium in a Wilson cloud chamber and took 3200 pairs of photographs. Ten photographs exhibited nuclear disintegrations as the one shown in Fig. 4.1. They could not determine whether ^{16}N was a stable or a radioactive nuclide because the reaction kinematics does not depend on the particle subsequently decaying or not [9].

The first demonstration of artificially produced radioactive isotopes was presented by Curie and Joliot in Paris about two months later in early 1934. They irradiated boron, magnesium, and aluminum samples with polonium α-particles and measured their activities with a Geiger Müller counter as a function of time. They assigned the observed half-lives to ^{13}N, ^{27}Si, and ^{30}P produced in (α,n) reactions on ^{10}B, ^{24}Mg, and ^{27}Si, respectively [4]. While the assignments of ^{13}N and ^{30}P were correct, Curie and Joliot revised the assignment for the reaction on magnesium in a later

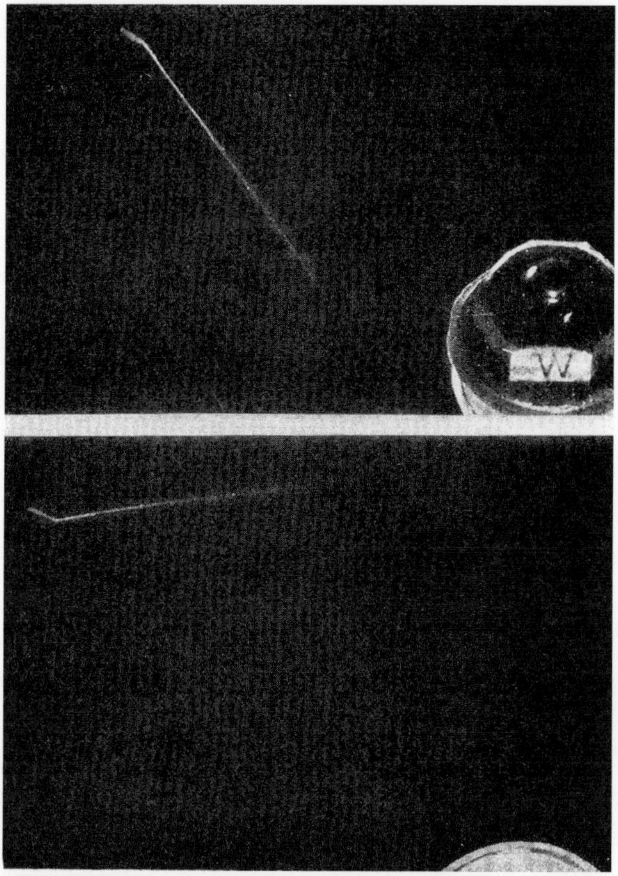

Fig. 4.1 Stereoscopic photographs of a neutron disintegrating fluorine into an α-particle and ^{16}N taken by Harkins et al. in 1933 [9] (Reprinted figure with permission from W.D. Harkins, D.M. Gans, and H.W. Newson, *Disintegration of fluorine nuclei by neutrons and the probable formation of a new isotope of nitrogen* (${}^{16}N$), Phys. Rev. 44 (1933) 945. Copyright 1933 by the American Physical Society.)

paper when they determined that in this reaction electrons instead of positrons were emitted. Thus they proposed the reaction ^{25}Mg(α,n)^{28}Al [6]. The decay curves for these three reactions are shown in Fig. 4.2.

It is interesting to note that the discovery of ^{17}F was reported by Wertenstein from the Radiological Institute in Warsaw, Poland, and although he acknowledges that the experiments were performed by M. Danysz and M. Zyw: "Messrs. M. Danysz and M. Zyw, working in this laboratory, have bombarded diverse substances with α-rays from a thin-walled glass tube containing some 15 millicuries of radon, and immediately afterwards have tested their activity with a Geiger Müller counter," they are not listed as coauthors [5].

Fermi had realized that nuclear reactions with charged-particles are limited to the lightest elements because of the Coulomb barrier. He thus started a program to irradiate a large number of elements from hydrogen to uranium with neutrons to search for artificial radioactivity. Most of the irradiated elements produced radioactivity and in many cases it was not possible to assign the observed activity to a specific isotope.

Fermi's group in Rome, Italy, performed chemical separations in order to determine the element of the radioactive isotope. They also assumed neutron capture reactions so that they could use the isotopic abundances of the irradiated substance as a guide to determine which isotope was initially formed. Possible exit channels were then the emission of γ-rays, protons, or α-particles.

In some cases, the same activity was observed following the irradiation of different elements. For example, a 150 min activity was found in chemically separated manganese after the irradiation of manganese, iron and cobalt. Thus it was determined that this activity was ^{56}Mn produced in the reactions ^{55}Mn(n,γ)^{56}Mn, ^{56}Fe(n,p)^{56}Mn, and ^{59}Co(n,α)^{56}Mn [10, 11]. Figure 4.3 shows the production and decay path of ^{56}Mn

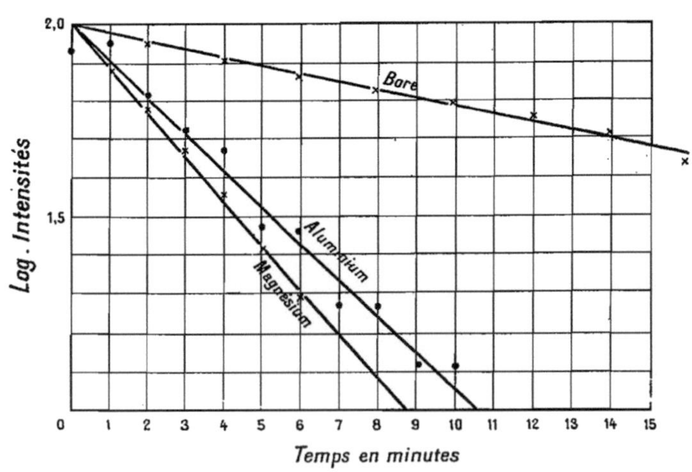

Fig. 4.2 Decay curves of the first artificially produced radioactive isotopes. The curves labeled Bore, Aluminium, and Magnesium correspond to the decays of ^{13}N, ^{30}P, and ^{28}Al, produced in the reactions ^{10}B(α,n)^{13}N, ^{27}Al(α,n)^{30}P, and ^{25}Mg(α,p)^{28}Al, respectively [6]

[11]. The figure also includes production and decay of ^{52}V. It was produced in the three different reactions: ^{51}V(n,γ)^{52}V, ^{52}Cr(n,p)^{52}V and ^{55}Mn(n,α)^{52}V and decays with a 4 min half-life [10, 11].

Fermi's interpretation of neutron capture on uranium as the production of new transuranic elements was incorrect (see Sects. 4.6 and 5.1). Fermi drew the conclusion from the chemical analysis which excluded the possibility that an observed activity with a half-life of 13 min was due to uranium or any other lighter neighboring element [25].

The identification of a new isotope based on the observed activity was quite often difficult. Frequently, activities were reported with no or ambiguous mass assignments. Also, inaccurate or incomplete measurements led to incorrect assignments. For example, half-lives of 35 h [15] and 33 h [26] were first reported in 1935 from neutron irradiation of holmium without a mass assignment. Later in 1935, a 2.6 h half-life [16] was assigned to ^{166}Ho. However, still in 1935, Hevesy and Levy in Copenhagen, Denmark, showed that this activity was due to the presence of dysprosium (^{165}Dy) in the samples and assigned the 35 h activity to ^{166}Ho [21]. The discovery of ^{165}Dy was reported earlier simultaneously by Hevesy and Levi [15] and Marsh and Sugden in Oxford, England [16]. The assignments for holmium and dysprosium are included in the section of the chart of nuclides presented in Fig. 4.4 taken from the paper by Hevesy and Levi [21].

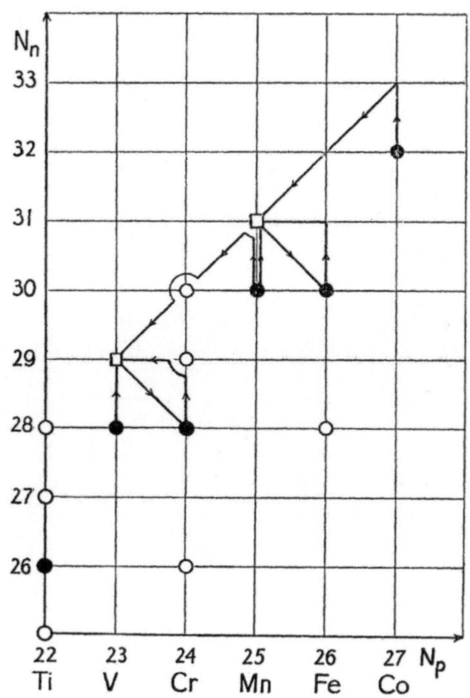

Fig. 4.3 Section of the chart of nuclides as presented by Fermi in 1934 where the number of neutrons and protons are plotted on the vertical and horizontal axis, respectively. It shows the production and decay paths of the radioactive isotopes ^{56}Mn and ^{52}V (*open squares*). *Solid circles* represent isotopes with more than 20 % abundance of an element and others are shown by *open circles* [11]

Marsh and Sugden also discovered ^{176}Lu which was later shown to be essentially stable as it has a half-life of 38.5±0.7 Gy (see Table 3.5). They measured a 4.0±0.1 h half-life which they identified by chemical separation as lutetium and assigned it to ^{176}Lu [16]. This activity corresponds to an excited isomeric state of ^{176}Lu which decays by β^- emission to ^{176}Hf. The ground-state of ^{176}Lu was first observed in 1939 by Libby at Berkeley [27].

At the Kaiser Wilhelm-Institut für Chemie in Berlin Meitner, Hahn, and Strassmann concentrated their work on the heavy elements. When irradiating uranium with neutrons they detected three different decay paths originating from ^{239}U: "In the third process we found only one active product, a β-emitting uranium isotope with a half-life of 28 min which must lead to a (probably long-lived) ekarhenium... The experiments to determine the nature of the neutron capture process demonstrated that none of the three processes exhibit α radiation. The magnitude of the cross-sections for all three processes seem to rule out reactions on uranium 235 and even more so on uranium 234. This result is very difficult to explain within the current understanding of the nucleus" [22]. At the time Meitner, Hahn, and Strassmann still assumed that these neutron capture processes led to transuranium elements. They proposed one path to decay by consecutive β^- emissions up to element 97 ("eka-gold"). The 28 min half-life corresponds to the β^--decay of ^{239}U to ^{239}Pu while the other decays were most likely due to fission products.

Hahn and Meitner also discovered the first heavy A = 4n + 1 radioactive isotope in 1935 by irradiating ^{232}Th with neutrons. They observed the β^--decay of ^{233}Th

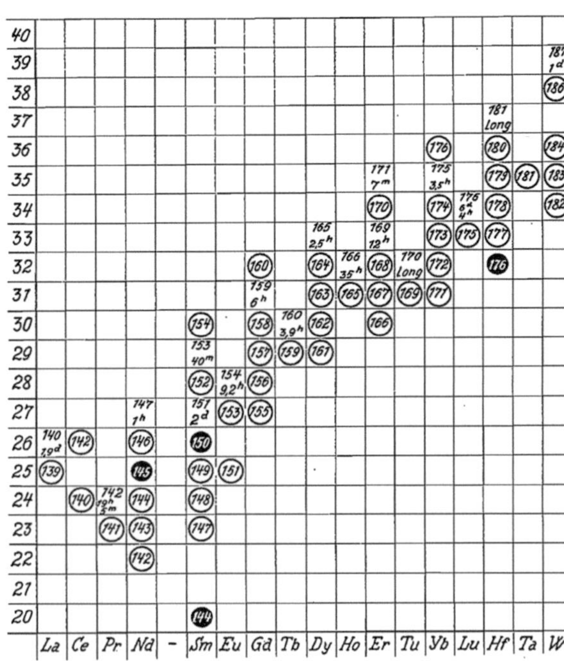

Fig. 4.4 Another early version of a section of the chart of nuclides by Hevesy and Levi. In this version the vertical axis displays the difference of the number of neutrons and twice the number of protons. Abundant isotopes are shown in *open circles*, rare isotopes (≤20 %) in *closed circles* and radioactive isotopes are indicated by their half-lives. Hevesy and Levi presented five sections covering all elements from hydrogen to uranium [21]

to ^{233}Pa [14]. Three years later, Meitner, Hahn, and Strassmann studied the same reaction further and detected the subsequent decay of ^{233}Pa to ^{233}U [23]. ^{233}Th is actually not part of the 4n + 1 decay chain, which populates ^{233}Pa by α-decay from ^{237}Np. This fourth decay chain which was the only chain not observed in nature was discovered almost simultaneously by Hagemann et al. [28] and English et al. [29] in 1947 (see Sect. 7.2). The chain is named after its member with the longest half-life, ^{237}Np, which was discovered in 1948 by Wahl and Seaborg [30] (Sect. 5.2).

4.3 First Accelerators

About two weeks before Chadwick proposed the existence of the neutron, Cockcroft and Walton succeeded with the construction of the first particle accelerator shown in Fig. 4.5 at the Cavendish Laboratory in Cambridge, England [31]. On February 2, 1932, they reported the acceleration of protons with high potentials up to energies of 600 keV [32]. Only two months later Lawrence accelerated protons to 1.22 MeV in the first cyclotron [33]. A schematic diagram of this cyclotron is shown in Fig. 4.6. Although some isotopes continued to be discovered using sources as described in the previous section, charged-particle accelerators began to dominate the production of new isotopes in the following years due to the availability of different probes (protons, deuterons, and α-particles) and higher bombarding energies.

Table 4.2 lists the isotopes discovered in charged-particle induced-reactions using accelerators until 1942. Most isotopes were discovered in the U.S. where the most productive institute was the Radiation Laboratory of the University of California at Berkeley. The isotopes discovered at Berkeley with charged-particle induced reactions between 1934 and 1941 are listed separately (see Table 4.3). In Europe, only the accelerators at the Cavendish Laboratory in Cambridge contributed to the discovery of isotopes produced in charged-particle reactions.

The identification techniques were similar to those described in the previous section. For example, Wilson cloud chambers were used to measure the tracks of the charged-particles and the kinematics of the reaction was determined from the track lengths and the angles between the tracks. In a different approach the produced activities were chemically separated to determine the element. The specific isotope was then deduced from the relative abundances of the isotopes of the target element and the characteristics of the emitted radiation. Quite often cross bombardments with different beams were used to narrow down the choices of isotopes that could be the source of the emitted radiation.

As the first application of their accelerator, Cockroft and Walton accelerated protons with a voltage between 125 and 400 kV in 1932, which then irradiated a lithium target. They observed the tracks of charged-particles with a Shimizu expansion chamber: "The brightness of the scintillations and the density of the tracks observed in the expansion chamber suggest that the particles are normal α-particles.

Fig. 4.5 Photograph of the Cockroft–Walton accelerator at the Cavendish Laboratory in Cambridge, England [31]

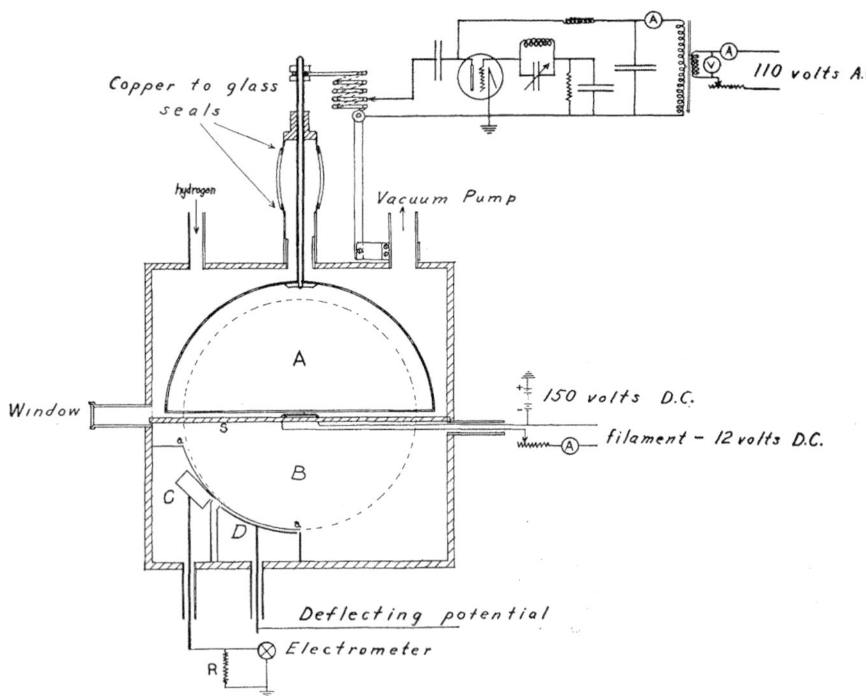

Fig. 4.6 Schematics of the first cyclotron [33] (Reprinted figure with permission from E.O. Lawrence and M.S. Livingston, *The production of high speed light ions without the use of high voltages*, Phys. Rev. 40 (1932) 19. Copyright 1932 by the American Physical Society.)

Table 4.2 Isotopes discovered in charged-particle induced-reactions using accelerators between 1932 and 1942: The submission date, first author and reference of the publication as well as the beam energy (in MeV), target, and reaction are listed. Isotopes discovered in Berkeley during this time are listed in Table 4.3

Date	First author	Ref.	Energy	Target	Reaction	Isotope
04/16/1932	Cockcroft	[34]	0.4	^{7}Li	(p,γ)	^{8}Be
03/01/1934	Crane	[35]	0.9	^{10}B	(d,n)	^{11}C
03/09/1934	Oliphant	[36]	0.02	^{2}H	(d,p)	^{3}H
03/19/1935	Oliphant	[37]	2	^{9}Be	(d,p)	^{10}Be
05/15/1935	Crane	[38]	0.5	^{11}B	(d,p)	^{12}B
05/31/1935	Crane	[39]	0.8	^{7}Li, ^{19}F	(d,p)	^{8}Li, ^{20}F
02/26/1937	Cork	[40]	6.3	^{116}Cd	(d,n)	^{117}In
05/07/1937	DuBridge	[41]	3.8	^{18}O	(p,n)	^{18}F
08/16/1937	Ridenour	[42]	9	^{58}Ni	(α,p)	^{61}Cu
08/23/1937	Kraus	[43]	6.3	^{108}Pd, ^{110}Pd	(d,p)	^{109}Pd, ^{111}Pd, ^{111}Ag[a]
09/08/1937	Stewart	[44]	6.3	^{88}Sr	(d,p)	^{89}Sr
05/27/1938	Roberts	[45]	1	^{6}Li	(d,n)	^{7}Be
05/03/1939	Fox	[46]	–	^{19}F	(p,n)	^{19}Ne
06/19/1939	Barnes	[47]	7.2	^{110}Cd	(p,n)	^{110}In
07/20/1939	White	[48]	6	^{23}Na	(p,n)	^{23}Mg
08/31/1939	Kuerti	[49]	6.8	^{27}Al	(p,n)	^{27}Si
09/11/1939	Enns	[50]	6.3	^{105}Pd	(p,n)	^{105}Ag
10/24/1939	Bethe	[51]	16	^{26}Mg	(α,p)	^{29}Al
11/01/1939	Davidson Jr.	[52]	3.1	^{40}Ca	(d,p)	^{41}Ca
04/22/1940	DuBridge	[53]	6.7	^{85}Rb, ^{87}Sr	(p,n)	^{85}Sr, ^{87}Y
05/15/1940	Pollard	[54]	2.6	^{20}Ne	(d,n)	^{21}Na
08/12/1940	Krishnan	[55]	9	^{204}Hg, ^{201}Hg	(d,p), (d,n)	^{205}Hg, ^{202}Tl
				^{208}Pb	(d,p)	^{209}Pb
08/12/1940	Turner	[56]	–	^{48}Ti	(d,n)	^{49}V[b]
10/10/1940	King	[57]	16	^{28}Si, ^{32}S	(α,n)	^{31}S, ^{35}Ar
11/14/1940	White	[58]	8.6	^{29}Si	(p,n)	^{29}P
01/31/1941	Elliott	[59]	8	^{40}Ca	(d,n)	^{41}Sc
07/24/1941	Allen	[60]	5	^{45}Sc	(p,n)	^{45}Ti
08/11/1941	Cork	[61]	9.5	^{133}Cs	(d,2n)	^{133}Ba
08/30/1941	Weimer	[62]	–	^{37}Cl	(d,2n)	^{37}Ar
07/24/1942	O'Connor	[63]	10, 20	^{46}Ti	(d,n), (α,n)	^{47}V, ^{49}Cr

[a]Populated by β^--decay
[b]Interpreted data from Walke et al. [64, 65]

If this point of view turns out to be correct, it seems not unlikely that the lithium isotope of mass 7 occasionally captures a proton and the resulting nucleus of mass 8 breaks into two α-particles, each of mass four and each with an energy of about eight million electron volts" [34]. This interpretation of the production of ^{8}Be subsequently breaking up into two α-particles was confirmed by a measurement of two back-to-back α-particles in coincidence only a few months later [66].

Also at the Cavendish Laboratory, Oliphant, Kempton, and Rutherford utilized electric and magnetic fields for the first time to identify and measure reaction products in 1935. They injected the produced charged-particles into electric and magnetic fields which were adjusted so that the particles were not deflected. The fields were calibrated with polonium α-particles and the velocity of the particles was then calculated from the ratio of the strength of the electric and magnetic field. With this method they discovered ^{10}Be by irradiating beryllium with deuterons in the reaction ^{9}Be(d,p)^{10}Be [37].

Krishnan and Nahum discovered ^{205}Hg, ^{202}Tl, and ^{209}Pb at Cambridge in 1940. They irradiated mercury and lead targets with 9 MeV deuterons but did not use the original Cockroft and Walton linear accelerator but the Cavendish cyclotron [55]. These were the last isotopes discovered at the Cavendish Laboratory.

Although Lawrence and Livingston built the first cyclotron already in 1932 at Berkeley, the first artificially produced isotopes discovered in the U.S. used a linear accelerator. Crane et al. observed ^{11}C [35], ^{12}B [38], and ^{8}Li and ^{20}F [39] by accelerating deuterons up to 900 keV with the 1 MV cascade transformer in the Kellogg Radiation Laboratory at the California Institute of Technology [67].

Subsequently most new isotopes observed in the 1930s were discovered at Berkeley as discussed below. Following the success of the Berkeley cyclotron several other cyclotrons were built in the U.S. and then used to search for new isotopes in nuclear reactions with protons, deuterons, and α-particles between 1937 and 1942. The first discoveries in each of the institutions were: ^{117}In at the University of Michigan [40], ^{18}F at the University of Rochester [41], ^{61}Cu at Princeton University [42], ^{29}Al at Purdue University [51], ^{41}Ca at Yale University [52], and ^{45}Ti at The Ohio State University [60].

The only isotope not produced from a cyclotron during this time was ^{7}Be. Roberts, Heydenburg, and Locher used the two-meter electrostatic generator of the Department of Terrestrial Magnetism of the Carnegie Institution of Washington to irradiate a LiF target with 1 MeV deuterons to form ^{7}Be in the (d,n) reaction [45].

The credit for the discovery of ^{49}V belongs to Turner from Princeton University who did, however, not perform the actual experiment [56]. In 1937, Walke et al. had proposed a 33 $\pm$ 1 min half-life for ^{49}V which was observed in the two reactions ^{48}Ti(d,n) and ^{46}Ti(α,p) [100]. Two years later Walke assigned a 600 $\pm$ 50 d half-life to ^{47}V from ^{46}Ti(d,n) [64]. This assignment seemed to have been confirmed in 1940 when the 600 d activity was not observed in the reaction ^{46}Ti(α,p) by Walke, Thompson, and Holt [65]. Turner reevaluated the data and argued that the latter experiment was not sufficiently sensitive to observe activities from the ^{46}Ti(α,p) reaction and assigned the 600 d half-life to ^{49}V. Turner did not reassign the 33 ± 1 min half-life proposing that both activities belong to ^{49}V [56]. However, in 1942 O'Connor et al. was not able to replicate the 33 ± 1 min half-life in the irradiation of titanium with α-particles and tentatively reassigned this activity to ^{47}V [63].

As mentioned earlier, the majority of new isotopes produced at accelerators during the 1930s were discovered in the Radiation Laboratory of the University of California at Berkeley and they are listed in Table 4.3.

The first new isotope discovered at Berkeley was ^{15}O by Livingston and McMillan in 1934. They irradiated a nitrogen gas target with a 1 μA, 2 MeV deuteron beam and measured the γ-ray emission from positron annihilation after the activated sample was chemically separated [68].

In 1938, Alvarez reported the first decay of an isotope by electron capture. He observed the decay of ^{67}Ga following the irradiation of zinc with deuterons from the Berkeley cyclotron. He interpreted the measured electron, γ-ray and X-ray spectra as ^{67}Ga transforming into ^{67}Zn by electron capture followed by the emission of a K X-ray [77]. Alvarez credits Mann with the discovery of the ^{67}Ga isotope, however, Mann had published the half-life of ^{67}Ga only in a conference abstract [101].

In addition to the almost 50 different isotopes discovered at Berkeley, two new elements were produced for the first time with the Berkeley cyclotron. In 1937 Lawrence sent a molybdenum sample which had been irradiated for several months with deuterons at the Berkeley cyclotron to Perrier and Segrè [102] at the University of Rome. Their chemical analysis and radiation measurements demonstrated that the sample contained isotopes of the yet unknown element 43. For ten years it was only referred to as "element 43" until in 1947 Perrier and Segrè suggested the name technetium [103]. It was accepted in 1949 at the 15th Conference of the International Union of Chemistry [104]. An earlier claim of discovery with the suggested name masurium [105] had been controversial for a long time. The history of the naming of technetium is described in a recent historical overview of the discovery of the element technetium [106]. The first unique identification of a technetium isotope (^{99}Tc) was reported a year later in 1938 by Segrè and Seaborg [87].

The element astatine was discovered by Corson, MacKenzie, and Segrè in 1940 by irradiating a bismuth target with 24 MeV α-particles from the Berkeley cyclotron [107]. In this initial publication, they did not specify the isotope, however, five months later in a follow-up paper [96] they described the formation reaction—^{209}Bi(α,2n)— and the subsequent two decay paths as shown in Fig. 4.7.

It is interesting to note, that already in the first paper Corson, MacKenzie, and Segrè immediately tested the new element for possible medical application as it was in the same chemical group as iodine: "At the suggestion of Dr. J.G. Hamilton and with his aid we have injected known amounts of the supposed eka-iodine into two hyperthyroid guinea pigs, on the chance that it might behave like iodine and be concentrated in the thyroid. The guinea pigs were killed about 4.5 h after administration

Table 4.3 Isotopes discovered in charged-particle reactions at Berkeley between 1934 and 1941: The submission date, first author and reference of the publication as well as the beam energy (in MeV), target, and reaction are listed

Date	First author	Ref.	Energy	Target	Reaction	Isotope
08/01/1934	Livingston	[68]	2	^{14}N	(d,n)	^{15}O
02/28/1936	Snell	[69]	3	^{40}Ar	(d,p)	^{41}Ar
04/07/1936	Cork	[70]	5	^{196}Pt	(d,p)	^{197}Pt
09/10/1936	Van Voorhis	[71]	5–6	^{63}Cu	(d,n)	^{64}Cu
01/27/1937	Walke	[72]	11	^{41}K	(α,n)	^{44}Sc
05/03/1937	Hurst	[73]	11	^{35}Cl	(α,n)	^{38}K
06/11/1937	Mann	[74]	11	^{63}Cu	(α,n)	^{66}Ga
08/03/1937	Snell	[75]	5.5	^{82}Se	(d,p), (d,2n)	^{83}Se, ^{83}Br[a], ^{82}Br
08/10/1937	Walke	[76]	11,5.5	^{45}Sc, ^{47}Ti	(α,p), (d,n)	^{48}V
03/15/1938	Alvarez	[77]	–	^{66}Zn	(d,n)	^{67}Ga
04/14/1938	Livingood	[78]	12.6,16	^{54}Fe	(α,n)	^{57}Ni
04/29/1938	Livingood	[79]	3.2	^{54}Fe	(p,γ)	^{55}Co
05/10/1938	Livingood	[80]	16,5.5	^{50}Cr, ^{58}Fe	(α,n), (d,p)	^{53}Fe, ^{59}Fe
05/28/1938	Sagane	[81]	5.5	^{73}Ge	(d,n)	^{74}As
07/11/1938	Livingood	[82]	5.5	^{50}Cr	(d,n)	^{51}Mn
				^{54}Fe, ^{56}Fe	(d,α)	^{52}Mn, ^{54}Mn
07/26/1938	Alvarez	[83]	–	^{64}Zn	(d,n)	^{65}Ga
08/23/1938	Mann	[84]	17	^{66}Zn	(α,n)	^{69}Ge
09/07/1938	Livingood	[85]	16,8	^{121}Sb, ^{130}Te	(α,n), (d,2n)	^{124}I, ^{130}I
10/03/1938	Sagane	[86]	5.5–7.6	^{70}Ge	(d,n)	^{71}As
10/14/1938	Segrè	[87]	–	^{98}Mo	(d,n)	^{99}Tc
12/15/1938	Livingood	[88]	8	^{64}Zn	(d,p)	^{65}Zn
01/30/1939	Livingood	[89]	5	^{122}Sn	(d,2n)	^{122}Sb
03/09/1939	Livingood	[90]	5	^{112}Sn, ^{124}Sn	(d,p)	^{113}Sn, ^{125}Sn
03/31/1939	Seaborg	[91]	8	^{121}Sb, ^{128}Te	(d,2n), (d,p)	^{121}Te, ^{129}Te
				^{130}Te, ^{130}Te	(d,p), (d,n)	^{131}Te, ^{131}I
05/30/1939	Livingood	[92]	5–8	^{54}Fe	(d,p)	^{55}Fe
11/20/1939	Walke	[93]	8	^{48}Ca	(d,n)	^{49}Sc
11/28/1939	Walke	[94]	8	^{44}Ca	(d,p)	^{45}Ca
11/28/1939	Walke	[65]	16	^{48}Ti	(α,n)	^{51}Cr
04/15/1940	Hoag	[95]	8	^{32}S	(d,n)	^{33}Cl
07/16/1940	Corson	[96]	32	^{209}Bi	(α,2n)	^{211}At
01/13/1941	Seaborg	[97]	16	^{71}Ga	(d,2n)	^{71}Ge
05/12/1941	Helmholz	[98]	8	^{88}Sr	(d,α)	^{86}Rb
12/01/1941	Livingood	[99]	–	^{56}Fe, ^{56}Fe	(d,2n), (d,n)	^{56}Co, ^{57}Co
				^{57}Fe, ^{59}Co	(d,n), (d,p)	^{58}Co, ^{60}Co

[a] Populated by β^--decay

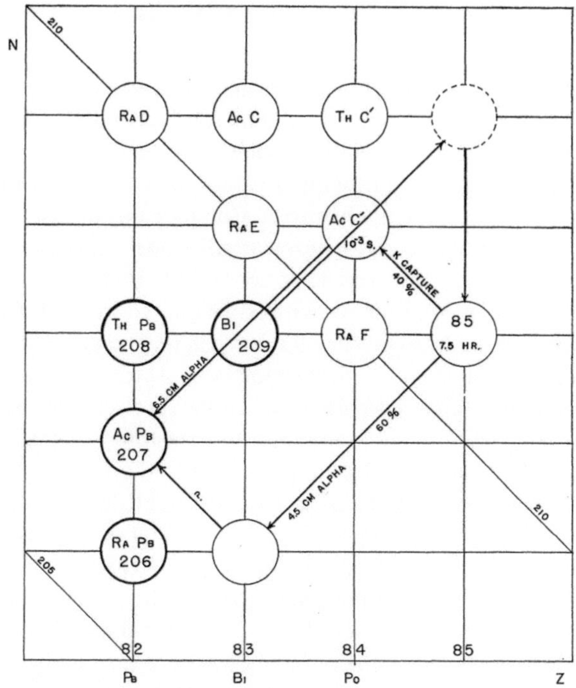

Fig. 4.7 Section of the chart of nuclides demonstrating the formation and decay of ^{211}At. ^{211}At was produced in the ^{209}Bi(α,2n) reaction and decays by either emitting an α-particle to ^{207}Bi or by electron capture to ^{211}Po and then by α-decay to ^{207}Pb. The β^+-decay of ^{207}Bi had not been observed. Stable isotopes are depicted by *thick solid circles* [96] (Reprinted figure with permission from D.R. Corson, K.R. MacKenzie, and E. Segrè, *Artificially radioactive element 85*, Phys. Rev. 58 (1940) 672. Copyright 1940 by the American Physical Society.)

of the radioactive material and various portions of the bodies were examined for activity. In one animal the thyroid contained roughly 100 times as much activity as equal masses of other portions of the body" [107].

As was discussed in Sect. 2.4, Thornton and Burdett [108] recently argued that Hulubei and Cauchois should receive at least partial credit for their observation of three X-rays which were close to the predicted values for eka-iodine (astatine) [109]. Several other claims for the discovery of element 85 were made. Already in 1931 Allison et al. had described the discovery of element 85 selecting the name alabamine [110]. However, this observation was later demonstrated to be incorrect [111]. Minder's 1940 report of element 85 naming it helvetium [112] and the 1942 claim by Leigh-Smith and Minder naming it anglohelvetium [113] were both proven to be wrong by Karlik and Bernert in 1942 [114] and 1944 [115], respectively. The name astatine was officially accepted at the 15th IUPAC conference in Amsterdam in 1949 [104].

4.4 Neutrons Produced with Accelerators

A few years after charged-particles had been used to induce a variety of reactions it was realized that certain deuteron induced reactions could be used as intensive neutron sources. Specifically, the reactions ^{9}Be(d,n) with a Q-value[1] of 4.3 MeV and ^{6}Li(d,n) and ^{7}Li(d,n) with Q-values of 14.2 and 8.6 MeV, respectively, were used to produce high energy neutrons. Comparisons of fast (on the order of MeV) and slow (on the order of eV) neutron irradiations yielded additional information about the source of the observed activities. While fast neutrons primarily induced (n,2n) reactions, slow neutrons were mostly captured by the target nuclides in (n,γ) reactions. The fast neutrons produced in nuclear reactions could be slowed down by moderating them in paraffin. Neutron irradiations were also combined with the several different options of charged-particle reactions discussed in the previous sections in order to assign measured new activities to a specific isotope.

The more than 60 isotopes discovered primarily by neutron-induced reactions at accelerators between 1935 and 1945 are listed in Tables 4.4 and 4.5.

The first isotope produced with accelerator neutrons was ^{14}C in 1935 at the Kellogg Radiation Laboratory of the California Institute of Technology. Bonner and Brubacker produced fast neutrons with Be+d and Li+d reactions and photographed several hundred tracks of reactions in a cloud chamber filled with nitrogen. From the angle between the outgoing tracks of the charged-particles and the track length, they deduced the particle energies and concluded that they observed the reaction ^{14}N(n,p)^{14}C [116].

In Europe, Heyn discovered ^{62}Cu [117] and ^{69}Zn [123] in 1936 and 1937, respectively, with neutrons from the X-ray Research Laboratory in Eindhoven, Netherlands, while Goldhaber and collaborators produced ^{66}Cu [118] as well as ^{115}Cd and ^{117}Cd [129] at the Cambridge Cavendish Laboratory using neutrons from the Li+d reaction.

However, most of the new isotopes produced with neutrons from accelerators were discovered at the University of Michigan and again at the University of California at Berkeley. In 1937, Pool et al. irradiated targets of almost all elements with fast and slow neutrons from the University of Michigan cyclotron [119, 124, 125]. Their results for the lower half of the rare earth elements together with other data known at the time are summarized in the chart of nuclides shown in Fig. 4.8. In this version of the chart the vertical axis shows the elements with increasing proton numbers towards the bottom as a function of mass (atomic) number. The element promethium had not been discovered and is labeled as illinium (see the discussion in Sect. 6.2). The tentative assignment of a 12.5 h half-life to ^{144}Il (which could not be replicated) was categorized "C" in the text [125], following the classification introduced a year earlier by Livingston and Bethe [154]. Category C corresponds to "One of few isotopes". Only categories A (Isotope certain) and B (Isotope probable) should qualify as an isotope discovery.

[1]The Q-Value corresponds to the amount of energy released in a reaction.

Table 4.4 Isotopes discovered by irradiations with accelerator neutrons between 1935 and 1939: The submission date, first author and reference of the publication as well as the target and the reaction are listed

Date	First author	Ref.	Target	Reaction	Isotope
11/25/1935	Bonner	[116]	^{14}N	(n,p)	^{14}C
09/30/1936	Heyn	[117]	^{63}Cu	(n,2n)	^{62}Cu
05/05/1937	Chang	[118]	^{69}Ga	(n,α)	^{66}Cu
07/03/1937	Pool	[119]	^{89}Y, ^{138}Ba, ^{197}Au	(n,γ)	^{90}Y, ^{139}Ba, ^{198}Au
07/19/1937	McMillan	[120]	^{191}Ir, ^{193}Ir	(n,γ)	^{192}Ir, ^{194}Ir
			^{198}Pt	(n,γ)	^{199}Pt, ^{199}Au[a]
			^{197}Au	(n,2n)	^{196}Au
07/19/1937	Lawson	[121]	^{115}In	(n,2n), (n,γ)	^{114}In, ^{116}In
08/03/1937	Snell	[75]	^{81}Br	(n,α)	^{78}As
08/26/1937	Walke	[100]	^{51}V, ^{48}Ti	(n,α), (n, p)	^{48}Sc
10/22/1937	Oldenberg	[122]	^{181}Ta	(n,γ)	^{182}Ta
11/17/1937	Heyn	[123]	^{68}Zn	(n,γ)	^{69}Zn
11/17/1937	Pool	[124]	^{115}In, ^{112}Cd	(n,α), (n, p)	^{112}Ag
01/15/1938	Pool	[125]	^{141}Pr, ^{150}Nd, ^{154}Sm	(n,2n)	^{140}Pr, ^{149}Nd, ^{153}Sm
			^{153}Eu, ^{165}Ho	(n,2n)	^{152}Eu, ^{164}Ho
			^{150}Nd, ^{170}Er	(n,γ)	^{151}Nd, ^{171}Er
06/01/1938	Livingood	[126]	^{127}I	(n,p), (n,2n)	^{127}Te, ^{126}I
08/22/1938	Sagane	[127]	^{90}Zr, ^{92}Mo, ^{100}Mo	(n,2n)	^{89}Zr, ^{91}Mo, ^{99}Mo
10/03/1938	Sagane	[86]	^{74}Ge, ^{76}Ge, ^{71}Ga	(n,γ)	^{75}Ge, ^{77}Ge, ^{72}Ga
10/22/1938	Sagane	[128]	^{93}Nb	(n,2n), (n,γ)	^{92}Nb, ^{94}Nb
11/05/1938	Goldhaber	[129]	^{116}Cd	(n,2n), (n,γ)	^{115}Cd, ^{117}Cd
12/29/1938	Sinma	[130]	^{185}Re, ^{187}Re	(n,γ)	^{186}Re, ^{188}Re
01/30/1939	Livingood	[89]	^{127}I	(n,α)	^{124}Sb
08/28/1939	Crittenden Jr.	[131]	^{103}Rh	(n,γ)	^{104}Rh

[a]Populated by β^--decay

In 1937 at Berkeley, McMillan et al. were the first to report the observation of a new isotope not produced directly in a nuclear reaction with either charged-particles or neutrons, but populated in the decay of an artificially produced radioactive isotope. They irradiated platinum with slow neutrons and assigned a 3.3 d activity in the chemically separated gold precipitate to ^{199}Au produced in the process ^{198}Pt(n,γ)^{199}Pt $\rightarrow$ ^{199}Au [120].

A rather curious method to produce a new isotope was reported in 1941 by Grahame and Walke. They discovered ^{36}Cl by neutron capture on ^{35}Cl: "Irradiation was carried out by allowing relatively large quantities (about a pound each) of sodium chlorate or of sodium perchlorate to stand in the neighborhood of the target holder of the Berkeley 37-in. cyclotron for periods of six months or more while the cyclotron was in use for other purposes" [142].

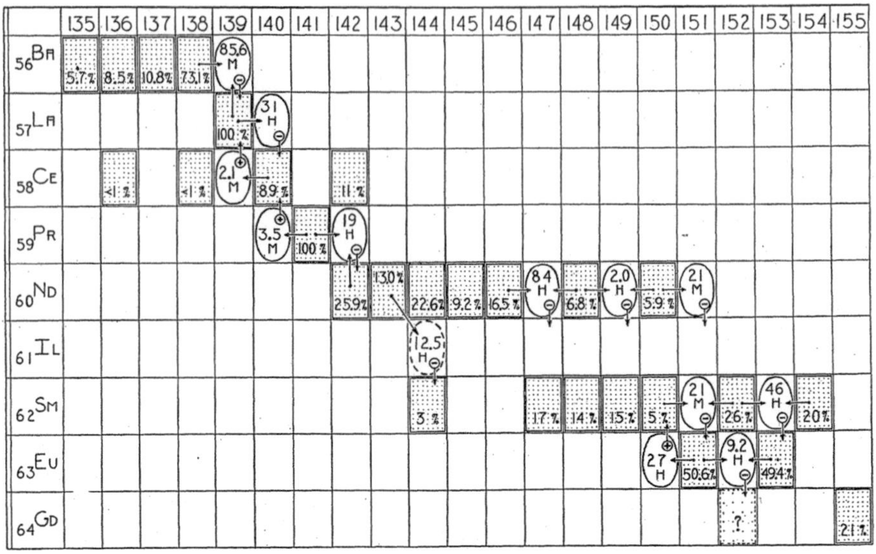

Fig. 4.8 Yet another version of the section of the chart of nuclides where the elements are plotted versus the mass or the atomic number for some of the rare earth elements. Stable isotopes are shown as *rectangles* with their relative abundances. *Question marks* indicate isotopes that were predicted to be stable at the time but not yet observed. *Ellipses* and *half-ellipses* correspond to radioactive isotopes and isomeric nuclides, respectively. The *arrows* indicate production and decay paths [125] (Reprinted figure with permission from M.L. Pool and L.L. Quill, *Radioactivity induced in the rare earth elements by fast neutrons*, Phys. Rev. 53 (1938) 437. Copyright 1938 by the American Physical Society.)

Also at Berkeley, Wu used neutron irradiations to identify the masses of activities that had been observed in fission fragments but which could not be assigned to specific isotopes. Two decay chains of thorium and uranium fission fragments had been observed in xenon, iodine, and tellurium but without mass assignments [155]. Wu irradiated barium and cesium targets with neutrons produced by irradiating beryllium with 16 MeV deuterons from the Berkeley cyclotron. Based on the analysis of Ba(n,α) and Cs(n,p) reactions, xenon activities of 5 d and 9.4 h were assigned to ^{133}Xe and ^{135}Xe, respectively [138]. These xenon half-lives then linked the previously measured iodine and tellurium activities to the isotopes ^{133}Te (60 min), ^{133}I (22 h), and ^{135}I (6.6 h). The 10 min half-life quoted for ^{135}Te could later not be replicated.

The Ivy League schools at Cornell and Harvard contributed the discoveries of ^{104}Rh in 1939 [131] and ^{197}Hg in 1941 [141], respectively.

The first isotopes discovered in Asia were produced by neutron irradiation at the cyclotron of the Institute of Physical and Chemical Research in Tokyo, Japan, where fifteen isotopes were discovered between 1938 and 1941. Sagane et al. discovered the first radioactive isotopes of niobium ^{92}Nb and ^{94}Nb still calling them columbium (Cb) [128].

Table 4.5 Isotopes discovered by irradiations with accelerator neutrons between 1940 and 1945: The submission date, first author and reference of the publication as well as the target and the reaction are listed

Date	First author	Ref.	Target	Reaction	Isotope
04/16/1940	Kennedy	[132]	^{37}Cl	(n,γ)	^{38}Cl
05/01/1940	Minakawa	[133]	^{184}W, ^{186}W	(n,γ)	^{185}W, ^{187}W
05/03/1940	Nishina	[134]	^{238}U	(n,2n)	^{237}U
05/09/1940	Sagane	[135]	^{92}Zr	(n,p)	^{92}Y
06/03/1940	Kalbfell	[136]	^{133}Cs	(n,γ)	^{134}Cs
07/02/1940	Fajans	[137]	^{185}Re, ^{187}Re	(n,2n)	^{184}Re
10/17/1940	Wu	[138]	^{136}Ba, ^{133}Cs	(n,α)	^{133}Xe, ^{133}I[a], ^{133}Te[a]
			^{138}Ba	(n,p)	^{135}Xe, ^{135}I[a]
05/30/1941	Maurer	[139]	^{100}Mo	(n,γ)	^{101}Mo, ^{101}Tc[b]
07/25/1941	Minikawa	[140]	^{103}Rh	(n,2n)	^{102}Rh
08/13/1941	Sherr	[141]	^{198}Hg	(n,2n)	^{197}Hg
11/29/1941	Grahame	[142]	^{35}Cl	(n,γ)	^{36}Cl
04/27/1942	Maurer	[143]	^{204}Pb	(n,2n)	^{203}Pb
12/18/1942	Friedlander	[144]	^{202}Hg, ^{204}Hg	(n,γ), (n,2n)	^{203}Hg
02/12/1943	Born	[145]	^{85}Rb	(n,p)	^{85}Kr, ^{85}Br[a]
			^{87}Rb	(n,p)	^{87}Kr, ^{87}Br[a]
			^{87}Rb	(n,α)	^{84}Br
04/25/1943	Riezler	[146]	^{136}Xe	(n,γ)	^{137}Xe, ^{137}I[a]
08/14/1943	Seelmann-E.	[147]	^{94}Zr	(n,α)	^{91}Sr, ^{91}Y[b]
08/25/1943	Seelmann-E.	[148]	^{138}Ba	(n,p)	^{138}Cs, ^{138}Xe[a]
09/29/1943	Bothe	[149]	^{159}Tb	(n,γ)	^{160}Tb
12/11/1944	Atterling	[150]	^{174}Yb, ^{176}Yb	(n,γ)	^{175}Yb, ^{177}Yb
			^{176}Lu	(n,γ)	^{177}Lu
04/25/1945	Bohr	[151]	^{102}Ru, ^{104}Ru	(n,γ)	^{103}Ru, ^{105}Ru, ^{105}Rh[b]
07/16/1945	Zünti	[152]	^{37}Cl	(n,α), (n,p)	^{34}P, ^{37}S
12/14/1945	Bothe	[153]	^{168}Yb	(n,γ)	^{169}Yb

[a]Not measured but linked by previously determined decay sequences of fission fragments
[b]Populated by β^--decay

The name of element 41 had been controversial for a long time. It was discovered in 1801 by Hatchett who named it columbium [156]. In 1844, Rose separated a supposedly new substance from tantalum naming it niobium [157]. However, columbium and niobium turned out to correspond to the same element. In a controversial decision a committee of the Council of the International Association of Chemical Societies endorsed the name niobium in 1913 [158]. Columbium was continued to be used in the U.S. while in Europe niobium was used. In 1949, at the 15th Conference of the International Union of Chemistry, niobium was voted to be the accepted name [104].

Researchers at the Kaiser Wilhelm Institut für Physik in Berlin, Germany, also used neutron irradiations to identify the masses of previously measured fission fragment activities which had not been assigned to specific isotopes. In early 1943, Seelmann-Eggebert and Born had observed the decay sequences Br(50 s) $\rightarrow$ Kr(75 min) $\rightarrow$ Rb and Br(3 min) $\rightarrow$ Kr(4.6 h) $\rightarrow$ Rb in the fission of uranium but had not been able to determine the corresponding masses [159]. Later in the year, Born and Seelmann-Eggebert reported the results of neutron irradiations of rubidium and strontium. They observed the 4.6 h half-life in the reactions Rb(n,p) and Sr(n,α) but the 75 min activity only in the Rb(n,p) reaction. From these results they assigned the above two decay sequences to masses 85 and 87, respectively [145].

Similarly, Strassmann and Hahn observed a new activity of iodine in neutron-induced fission of uranium with a half-life of 30 $\pm$ 6 s in 1940 [160]. Three years later Seelmann-Eggebert and Born established that this 30 s activity decays to a 3.8 min activity of Xenon [159]. Riezler used neutron irradiation of xenon to identify the origin of this 3.8 min half-life as ^{137}Xe [146]. The submission did not have a specific date, instead it stated "Easter 1943". Although Riezler did not measure or mention ^{137}I specifically the mass identification of ^{137}Xe directly implied the identification of ^{137}I as acknowledged later by Snell et al. [161] and Sugarman [162].

Riezler's assignment of the 3.8 min half-life to ^{137}Xe also had direct implications for the mass 138 fission fragments. In 1939, Hahn and Strassmann observed a $\sim$15 min half-life of xenon decaying to a 33 min cesium activity in neutron-induced fission of uranium [163]. A year later these results were confirmed by Glasoe and Steigman reporting a 17 $\pm$ 1 min xenon half-life decaying to a 32.0 $\pm$ 5 min cesium isotope [164]. Seelmann-Eggebert used fast neutron irradiation of barium to identify the origin of this cesium activity as ^{138}Cs: "In addition to the xenon isotopes produced in (n,α) processes with fast neutrons, a cesium isotope with a half-life of 33 min was produced in (n,p) processes also induced with fast neutrons. This isotope is according to the 33 min half-life as well as the β-ray absorption curve identical with the 33 min cesium isotope observed in uranium fission. W. Riezler was recently able to assign a 3.8 min xenon isotope to mass 137. The present cesium isotope must therefore be assigned to mass 138, because there is no isotope with a 33 min half-life in the decay chain of this xenon isotope" [148]. Seelmann-Eggebert did not measure or mention ^{138}Xe but the mass identification of ^{138}Cs links the previously measured half-life of 15–17 min to ^{138}Xe.

^{175}Yb, ^{177}Yb, and ^{177}Lu were the first isotopes discovered in Sweden. In 1945, Atterling, Bohr, and Sigurgeirsson irradiated ytterbium and lutetium targets with fast and slow neutrons and assigned the measured half-lives to the respective isotopes based on the isotopic abundances of the targets and the relative intensities of the activities produced in the different irradiations [150]. In the same year, Bohr and Hole irradiated ruthium samples with fast and slow neutrons produced with deuterons on LiOH or beryllium targets at the cyclotron of the Forskningsinstitutet för Fysik in Stockholm. They measured the decay of ^{103}Ru, and ^{105}Ru, and in the latter case the subsequent decay of ^{105}Rh [151].

Shortly before the end of the Second World War, Zünti and Bleuler discovered ^{34}P and ^{37}S with a tensator at the Eidgenössische Technische Hochschule Zürich [152].

The last isotope discovered in Europe during the War was ^{169}Yb. Bothe had submitted a manuscript already in April 1944 to the Zeitschrift für Physik, however, the journal had not been published between October 1944 and December 1946, thus he published the results in Zeitschrift für Naturforschung shortly after the end of the War [153].

4.5 Photonuclear Reactions

Early attempts to induce nuclear reactions with γ-rays did not succeed because the energy of the available γ-ray sources was limited to a few MeV which was not sufficient to induce nuclear reactions because the nucleon separation energies for all stable nuclides (with the exception of the deuteron and ^{9}Be) is about 8 MeV. In 1937, Delsasso, Fowler, and Lauritsen measured a 17.1 MeV γ-ray resonance in the decay of an excited state in ^{8}Be at Caltech. They irradiated a lithium target with protons and obtained stereoscopic pictures from a cloud chamber placed in a magnetic field. At the center the γ-rays impinged on a lead scatterer. An example of one of the 770 electron positron pair emissions is shown in Fig. 4.9 [165]. The resonance for this reaction had been measured to be 440 keV earlier by Hafstad, Heydenburg, and Tuve at the Carnegie Institution of Washington [166].

Delsasso, Fowler, and Lauritsen submitted their results on December 29, 1936 and already less than a month later (January 26, 1937) Bothe and Gentner used the 17 MeV ^{7}Li+p γ-rays to induce nuclear reactions [167]. The eleven isotopes discovered with high energy γ-rays are listed in Table 4.6.

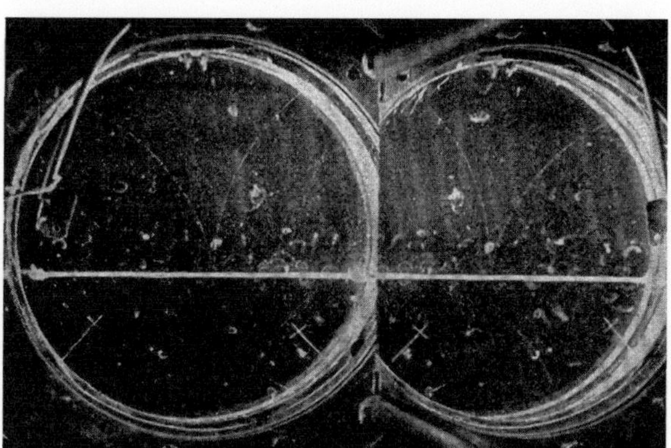

Fig. 4.9 Stereoscopic picture of a e^+/e^- pair emitted following the scattering of a 17 MeV γ-ray of lead inside a cloud chamber placed in a magnetic field [165] (Reprinted figure with permission from L.A. Delsasso, W.A. Fowler, and C.C. Lauritsen, *Energy and absorption of the gamma-radiation from* $Li^7 + H^1$, Phys. Rev. 51 (1937) 391. Copyright 1937 by the American Physical Society.)

Table 4.6 Isotopes discovered in photonuclear reactions between 1937 and 1943: The submission date, first author and reference of the publication are listed

Date	First author	Ref.	Target	Reaction	Isotope
01/26/1937	Bothe	[167]	^{81}Br	(γ,n)	^{80}Br
02/09/1937	Bothe	[169]	^{79}Br, ^{121}Sb	(γ,n)	^{78}Br, ^{120}Sb
			^{107}Ag, ^{109}Ag, ^{111}Ag	(γ,n)	^{106}Ag, ^{108}Ag, ^{110}Ag
03/08/1937	Bothe	[170]	^{64}Zn, ^{69}Ga, ^{71}Ga	(γ,n)	^{63}Zn, ^{68}Ga, ^{70}Ga
01/07/1943	Huber	[171]	^{40}Ca	(γ,n)	^{39}Ca
08/28/1943[a]	Huber	[172]	^{26}Mg	(γ,p)	^{25}Na

[a]Date of meeting where paper was presented

Bothe and Gentner performed their experiments with the Van de Graaff electrostatic generator of the Kaiser Wilhelm-Institut für medizinische Forschung in Heidelberg, Germany. In their first experiment they identified ^{80}Br by measuring an 18 min half-life in chemically separated bromine following the irradiation of bromine with the 17 MeV γ-rays. This half-life had been measured earlier in neutron irradiations of bromine by Amaldi et al. but without a mass assignment [168] because bromine has two stable isotopes (^{79}Br and ^{81}Br) with almost identical abundances. Bothe and Gentner were then able to assign the activity to ^{80}Br as it was produced in (n,γ) reactions from ^{79}Br as well as (γ,n) reactions from ^{81}Br [167].

In two subsequent papers, Bothe and Gentner were able to identify eight additional isotopes for the first time [169, 170]. In most cases half-lives had been previously measured in neutron capture reactions but no definite mass assignments had been possible. Similar to the case of ^{80}Br, Bothe and Gentner then combined their (γ,n) results with the previous data of (n,γ) reactions to determine the isotope emitting the measured radiation.

Huber et al. discovered ^{39}Ca in the reaction ^{40}Ca(γ,n) at the Van de Graaff electrostatic high voltage accelerator of the Eidgenössische Technische Hochschule in Zürich, Switzerland in 1943 [171]. The same group also observed for the first time a (γ,p) reaction a few months later when they irradiated a magnesium target with the 17 MeV γ-rays. The observed half-life of 62 ± 2 s was assigned to ^{25}Na formed in the reaction ^{26}Mg(γ,p) [172].

4.6 Identification of Fission Fragments

When Fermi irradiated a uranium sample with an 800 mCi beryllium-radon source he interpreted the observed activities as the potential discovery of elements heavier than uranium. He drew this conclusion from the inability to assign the activities to a known heavy element: "This negative evidence about the identity of the 13 min.-activity from a large number of heavy elements suggests the possibility that the atomic number of the element may be greater than 92... A careful search for such heavy particles has not yet been carried out, as they require for their observation that the active product should be in the form of a very thin layer. It seems therefore at present premature to form any definite hypothesis on the chain of disintegrations involved" [25].

Fermi's interpretation was questioned by Noddack shortly after: "Proof that the new radio-element has a Z of 93, has not been established yet, as Fermi deduced it from incomplete exclusion arguments" [173]. She continued by already suggesting the possibility of fission: "It would be conceivable that heavy nuclei which are bombarded with neutrons break up into several larger fragments which are isotopes of known elements but not neighbors of the irradiated elements" [173]. Although initial experiments by Hahn and Meitner seemed to confirm the production of transuranium elements [174, 175], some doubts were expressed by Meitner, Hahn, and Strassmann in 1937: "Therefore these processes must be capture processes of Uranium 238, which would lead to three isomeric nuclei of Uranium 239. This result is hard to understand within the current understanding of nuclei" [22].

Nevertheless Hahn, Meitner, and Strassmann continued to publish papers assuming the formation of transuranium elements [176, 177]. When Fermi in 1938 received the Nobel prize in part for the apparent discovery of transuranium elements he stated in his acceptance speech delivered on December 12, 1938: "We concluded that the carriers were one or more elements of atomic number larger than 92; we, in Rome, used to call the elements 93 and 94 Ausenium and Hesperium respectively. It is known that O. Hahn and L. Meitner have investigated very carefully and extensively the decay products of irradiated uranium and were able to trace among them elements up to the atomic number 96" [178]. However, in the writeup of the acceptance speech he acknowledged the potential misidentification in a footnote: "The discovery by Hahn and Strassmann of barium among the disintegration products of bombarded uranium, as a consequence of a process in which uranium splits into two approximately equal parts, makes it necessary to reexamine all the problems of the transuranic elements, as many of them might be found to be products of a splitting of uranium" [178].

Commonly, the paper by Hahn and Strassmann submitted on December 22, 1938 and published on January 6, 1939 [179] is credited with the discovery of fission. However, they did not dare to announce the discovery of fission: "As far as the 'Transuranium elements' are concerned, they are chemically similar to their homologues rhenium, osmium, iridium, platinum, but not equal. If they correspond to the even lower homologues technetium, ruthenium, rhodium, palladium has not been tested. This is not something that one could have thought of earlier. The sum of the Ba+Ma mass numbers ($138 + 101$) is 239!)." They go on to state: "As chemists we should rename Ra, Ac, Th to Ba, La, Ce based on the above described experiments. As 'nuclear chemists' close to physics, we cannot take this step, because it contradicts all present knowledge of nuclear physics. It still could be possible that a string of weird coincidences could have simulated these events" [179].

Table 4.7 Isotopes discovered in fission between 1939 and 1943: The submission date, first author and reference of the publication as well as the target, and neutron source are listed

Date	First author	Ref.	Target	Source	Isotope
01/28/1939	Hahn	[181]	U	Ra–Be	^{140}Ba
03/10/1939	Heyn	[182]	U	d+Li,Be	^{88}Rb, ^{139}Cs
04/11/1939	Abelson	[183]	U	d+Be	^{127}Sb, ^{129}Sb
06/14/1939	Langsdorf Jr.	[184]	Th	d+Be	^{88}Kr
05/13/1940	Glasoe	[164]	U	p+Be	^{89}Rb
08/08/1941	Götte	[185]	U	Ra–Be	^{92}Sr
04/16/1943	Hahn	[186]	U	d+Li	^{89}Kr

The first correct interpretation of the Hahn and Strassmann results were submitted on January 16, 1939 and published on February 11, 1939 by Meitner and Frisch: "On the basis, however, of present ideas about the behavior of heavy nuclei, an entirely different and essentially classical picture of these new disintegration processes suggest itself. On account of their close packing and strong energy exchange, the particles in a heavy nucleus would be expected to move in a collective way which has some resemblance to the movement of a liquid drop. If the movement is made sufficiently violent by adding energy, such a drop may divide itself into two smaller drops" [180].

Hahn and Strassmann submitted their proof that they had actually observed barium fragments less than two weeks after the Meitner and Frisch paper on January 28, 1939, and the paper was published one day earlier (February 10, 1939). This paper correctly identified the decay of ^{140}Ba as the first isotope discovered in fission [181]. The nine isotopes discovered in neutron-induced fission reactions between 1939 and 1943 are listed in Table 4.7.

For the discovery of ^{140}Ba Hahn and Strassmann used a Ra-Be source to induce fission of uranium and measured decay curves following chemical separations. They observed a 300 h activity which they had already observed and labeled earlier "Ra IV" [177, 179]. They changed the name to "Ba IV" and tentatively assigned it to ^{140}Ba: "Concerning the other from uranium produced barium isotopes, it is hypothesized that Ba IV may be the parent of the radioactive lanthanum which was reported with a half-life of 31–46 h with the probable atomic weight of 140" [181]. Hahn and Strassmann did not assign the 300 h activity to ^{140}Ba in subsequent papers [163, 187–189], although they confirmed the relationship of the activity to ^{140}La [189]. Subsequently this assignment was specifically made by other authors [190, 191] and it was generally adopted in the 1944 Table of Isotopes [192]. The final proof was given by mass-spectroscopic measurements in 1947 [193].

Less than two months later Heyn, Aten, and Bakker reported the identification of ^{88}Rb and ^{139}Cs as fission fragments from uranium at the Philips X-ray Laboratory in Eindhoven, The Netherlands [182].

Even before the Meitner and Frisch [180] and Hahn and Strassmann [181] papers were published, news of the discovery of fission had crossed the Atlantic. Abelson submitted his confirmation of the "cleavage" of uranium by detecting iodine K X-rays already on February 3, 1939 [194]. Later in the year, Abelson discovered the antimony isotopes ^{127}Sb and ^{129}Sb by irradiating uranium with neutrons produced by irradiating beryllium with 8 MeV deuterons from the Berkeley cyclotron [183]. He had published the observation of the 4.6 h half-life of ^{127}Sb earlier without a mass assignment [195].

A few months later, Langsdorf observed a 3 h krypton activity which decayed to a rubidium activity of 18 min at Berkeley [184]. As Heyn, Aten, and Bakker had assigned the rubidium activity to ^{88}Rb [182] Langsdorf concluded that this new krypton activity must belong to ^{88}Kr.

The credit for the discovery of ^{92}Sr is given to Götte who observed a strontium isotope with a half-life of 2.7 h decay via an yttrium isotope with a half-life of 3.5 h to stable zirconium in the fission of uranium [185]. Götte did not make a mass assignment, but the 3.5 h yttrium activity had been assigned to ^{92}Y a year earlier by Sagane et al. [135] (see Sect. 4.4) thus linking the 2.7 h strontium activity measured by Götte directly to ^{92}Sr.

The rubidium and krypton activities of the mass 89 chain were discovered by Glasoe and Steigmann at Columbia University [164] and Hahn and Strassmann [186] in 1940 and 1943, respectively.

References

1. E. Rutherford, Philos. Mag. **37**, 581 (1919)
2. P.M.S. Blackett, Proc. R. Soc. A **107**, 349 (1925)
3. J. Chadwick, Nature **129**, 312 (1932)
4. I. Curie, F. Joliot, Compt. Rend. Acad. Sci. **198**, 254 (1934)
5. L. Wertenstein, Nature **133**, 564 (1934)
6. I. Curie, F. Joliot, J. Phys. Radium **5**, 153 (1934)
7. O.R. Frisch, Nature **133**, 721 (1934)
8. O.R. Frisch, Nature **136**, 220 (1935)
9. W.D. Harkins, D.M. Gans, H.W. Newson, Phys. Rev. **44**, 945 (1933)
10. E. Amaldi et al., Ric. Scient. **5**, 21 (1934)
11. E. Fermi et al., Proc. R. Soc. A **146**, 483 (1934)
12. G. Hevesy, Nature **135**, 96 (1935)
13. G. Hevesy, Nature **135**, 580 (1935)
14. O. Hahn, L. Meitner, Naturwissenschaften **23**, 320 (1935)
15. G. Hevesy, H. Levi, Nature **136**, 103 (1935)
16. J.K. Marsh, S. Sugden, Nature **136**, 102 (1935)
17. E.B. Andersen, Z. Phys. Chem. **32B**, 237 (1936)
18. M.E. Nahmias, R.J. Walen, Compt. Rend. Acad. Sci. **203**, 71 (1936)
19. E. Neuninger, E. Rona, Wiener Akad. Anz. **73**, 159 (1936)
20. T. Bjerge, K.J. Broström, Nature **138**, 400 (1936)
21. G. Hevesy, H. Levi, Det Kgl. Danske Vid. Selskab., Math.-fys. Med. **XIV**(5) (1936)
22. L. Meitner, O. Hahn, F. Strassmann, Z. Phys. **106**, 249 (1937)
23. L. Meitner, F. Strassmann, O. Hahn, Z. Phys. **109**, 538 (1938)
24. E. Zingg, Helv. Phys. Acta **13**, 219 (1940)
25. E. Fermi, Nature **113**, 898 (1934)

26. E. Neuninger, E. Rona, Wiener Akad. Anz. **72**, 275 (1935)
27. W.F. Libby, Phys. Rev. **56**, 21 (1939)
28. F. Hagemann et al., Phys. Rev. **72**, 252 (1947)
29. A.C. English et al., Phys. Rev. **72**, 253 (1947)
30. A.C. Wahl, G.T. Seaborg, Phys. Rev. **73**, 940 (1948)
31. J.D. Cockcroft, E.T.S. Walton, Proc. R. Soc. A **136**, 619 (1932)
32. J.D. Cockroft, E.T.S. Walton, Nature **129**, 242 (1932)
33. E.O. Lawrence, M.S. Livingston, Phys. Rev. **40**, 19 (1932)
34. J.D. Cockcroft, E.T.S. Walton, Nature **129**, 649 (1932)
35. H.R. Crane, C.C. Lauritsen, Phys. Rev. **45**, 430 (1934)
36. M.L. Oliphant, P. Harteck, E. Rutherford, Nature **133**, 413 (1934)
37. M.L. Oliphant, A.E. Kempton, E. Rutherford, Proc. R. Soc. A **150**, 241 (1935)
38. H.R. Crane et al., Phys. Rev. **47**, 887 (1935)
39. H.R. Crane et al., Phys. Rev. **47**, 971 (1935)
40. J.M. Cork, R.L. Thornton, Phys. Rev. **51**, 608 (1937)
41. L.A. DuBridge, S.W. Barnes, J.H. Buck, Phys. Rev. **51**, 995 (1937)
42. L.N. Ridenour, W.J. Henderson, Phys. Rev. **52**, 889 (1937)
43. J.D. Kraus, J.M. Cork, Phys. Rev. **52**, 763 (1937)
44. D.W. Stewart, J.L. Lawson, J.M. Cork, Phys. Rev. **52**, 901 (1937)
45. R.B. Roberts, N.P. Heydenburg, G.L. Locher, Phys. Rev. **53**, 1016 (1938)
46. J.G. Fox et al., Phys. Rev. **55**, 1106 (1939)
47. S.W. Barnes, Phys. Rev. **56**, 414 (1939)
48. M.G. White et al., Phys. Rev. **56**, 512 (1939)
49. G. Kuerti, S.N. van Voorhis, Phys. Rev. **56**, 614 (1939)
50. T. Enns, Phys. Rev. **56**, 872 (1939)
51. H.A. Bethe, W.J. Henderson, Phys. Rev. **56**, 1060 (1939)
52. W.L. Davidson Jr., Phys. Rev. **56**, 1061 (1939)
53. L.A. DuBridge, J. Marshall, Phys. Rev. **58**, 7 (1940)
54. E. Pollard, W.W. Watson, Phys. Rev. **58**, 12 (1940)
55. R.S. Krishnan, E.A. Nahum, Proc. Camb. Philos. Soc. **36**, 490 (1940)
56. L.A. Turner, Phys. Rev. **58**, 679 (1940)
57. L.D.P. King, D.R. Elliott, Phys. Rev. **58**, 846 (1940)
58. M.G. White et al., Phys. Rev. **59**, 63 (1941)
59. D.R. Elliott, L.D.P. King, Phys. Rev. **59**, 403 (1941)
60. J.S.V. Allen et al., Phys. Rev. **60**, 425 (1941)
61. J.M. Cork, G.P. Smith, Phys. Rev. **60**, 480 (1941)
62. P.K. Weimer, J.D. Kurbatov, M.L. Pool, Phys. Rev. **60**, 469 (1941)
63. J.J. O'Connor, M.L. Pool, J.D. Kurbatov, Phys. Rev. **62**, 413 (1942)
64. H. Walke, E.J. Williams, G.R. Evans, Proc. R. Soc. A **171**, 360 (1939)
65. H. Walke, F.C. Thompson, J. Holt, Phys. Rev. **57**, 171 (1940)
66. J.D. Cockcroft, E.T.S. Walton, Proc. R. Soc. A **137**, 229 (1932)
67. H.R. Crane, C.C. Lauritsen, A. Soltan, Phys. Rev. **44**, 514 (1933)
68. M.S. Livingston, E. McMillan, Phys. Rev. **46**, 437 (1934)
69. A.H. Snell, Phys. Rev. **49**, 555 (1936)
70. J.M. Cork, E.O. Lawrence, Phys. Rev. **49**, 788 (1936)
71. S.N. Van Voorhis, Phys. Rev. **50**, 895 (1936)
72. H. Walke, Phys. Rev. **51**, 439 (1937)
73. D.G. Hurst, H. Walke, Phys. Rev. **51**, 1033 (1937)
74. W.B. Mann, Phys. Rev. **52**, 405 (1937)
75. A.H. Snell, Phys. Rev. **52**, 1007 (1937)
76. H. Walke, Phys. Rev. **52**, 669 (1937)
77. L.W. Alvarez, Phys. Rev. **53**, 606 (1938)
78. J.J. Livingood, G.T. Seaborg, Phys. Rev. **53**, 765 (1938)
79. J.J. Livingood, G.T. Seaborg, Phys. Rev. **53**, 847 (1938)

80. J.J. Livingood, G.T. Seaborg, Phys. Rev. **54**, 51 (1938)
81. R. Sagane, S. Kojima, M. Ikawa, Phys. Rev. **54**, 149 (1938)
82. J.J. Livingood, G.T. Seaborg, Phys. Rev. **54**, 391 (1938)
83. L.W. Alvarez, Phys. Rev. **54**, 486 (1938)
84. W.B. Mann, Phys. Rev. **54**, 649 (1938)
85. J.J. Livingood, G.T. Seaborg, Phys. Rev. **54**, 775 (1938)
86. R. Sagane, Phys. Rev. **55**, 31 (1939)
87. E. Segrè, G.T. Seaborg, Phys. Rev. **54**, 772 (1938)
88. J.J. Livingood, G.T. Seaborg, Phys. Rev. **55**, 457 (1939)
89. J.J. Livingood, G.T. Seaborg, Phys. Rev. **55**, 414 (1939)
90. J.J. Livingood, G.T. Seaborg, Phys. Rev. **55**, 667 (1939)
91. G.T. Seaborg, J.J. Livingood, J.W. Kennedy, Phys. Rev. **55**, 794 (1939)
92. J.J. Livingood, G.T. Seaborg, Phys. Rev. **55**, 1268 (1939)
93. H. Walke, Phys. Rev. **57**, 163 (1940)
94. H. Walke, F.C. Thompson, J. Holt, Phys. Rev. **57**, 177 (1940)
95. J.B. Hoag, Phys. Rev. **57**, 937 (1940)
96. D.R. Corson, K.R. MacKenzie, E. Segrè, Phys. Rev. **58**, 672 (1940)
97. G.T. Seaborg, J.J. Livingood, G. Friedlander, Phys. Rev. **59**, 320 (1941)
98. A.C. Helmholz, C. Pecher, P.R. Stout, Phys. Rev. **59**, 902 (1941)
99. J.J. Livingood, G.T. Seaborg, Phys. Rev. **60**, 913 (1941)
100. H. Walke, Phys. Rev. **52**, 777 (1937)
101. W.B. Mann, Phys. Rev. **53**, 212 (1938)
102. C. Perrier, E. Segrè, J. Chem. Phys. **5**, 712 (1937)
103. C. Perrier, E. Segrè, Nature **159**, 24 (1947)
104. International Union of Chemistry, Chem. Eng. News **27**, 2996 (1949)
105. O. Berg, I. Tacke, Naturwissenschaften **13**, 571 (1925)
106. R. Zingales, J. Chem. Educt. **82**, 221 (2005)
107. D.R. Corson, K.R. MacKenzie, E. Segrè, Phys. Rev. **57**, 459 (1940)
108. B.F. Thornton, S.C. Burdette, Bull. Hist. Chem. **35**(2), 86 (2010)
109. H. Hulubei, Y. Cauchois, Compt. Rend. Acad. Sci. **209**, 39 (1939)
110. F. Allison et al., Phys. Rev. **37**, 1178 (1931)
111. H.G. MacPherson, Phys. Rev. **47**, 310 (1935)
112. W. Minder, Helv. Phys. Acta **13**, 144 (1940)
113. A. Leigh-Smith, W. Minder, Nature **150**, 767 (1942)
114. B. Karlik, T. Bernert, Naturwissenschaften **30**, 685 (1942)
115. B. Karlik, T. Bernert, Z. Phys. **123**, 51 (1944)
116. T.W. Bonner, W.M. Brubaker, Phys. Rev. **49**, 223 (1936)
117. F.A. Heyn, Nature **138**, 723 (1936)
118. W.Y. Chang, M. Goldhaber, R. Sagane, Nature **139**, 962 (1937)
119. M.L. Pool, J.M. Cork, R.L. Thornton, Phys. Rev. **52**, 239 (1937)
120. E. McMillan, M. Kamen, S. Ruben, Phys. Rev. **52**, 375 (1937)
121. J.L. Lawson, J.M. Cork, Phys. Rev. **52**, 531 (1937)
122. O. Oldenberg, Phys. Rev. **53**, 35 (1938)
123. F.A. Heyn, Physica **4**, 1224 (1937)
124. M.L. Pool, Phys. Rev. **53**, 116 (1938)
125. M.L. Pool, L.L. Quill, Phys. Rev. **53**, 437 (1938)
126. J.J. Livingood, G.T. Seaborg, Phys. Rev. **53**, 1015 (1938)
127. R. Sagane et al., Phys. Rev. **54**, 542 (1938)
128. R. Sagane et al., Phys. Rev. **54**, 970 (1938)
129. M. Goldhaber, R.D. Hill, L. Szilard, Phys. Rev. **55**, 47 (1939)
130. K. Sinma, F. Yamasaki, Phys. Rev. **55**, 320 (1939)
131. E.C. Crittenden Jr., Phys. Rev. **56**, 709 (1939)
132. J.W. Kennedy, G.T. Seaborg, Phys. Rev. **57**, 843 (1940)
133. O. Minakawa, Phys. Rev. **57**, 1189 (1940)

134. Y. Nishina et al., Phys. Rev. **57**, 1182 (1940)
135. R. Sagane et al., Phys. Rev. **57**, 1179 (1940)
136. D.C. Kalbfell, R.A. Cooley, Phys. Rev. **58**, 91 (1940)
137. K. Fajans, W.H. Sullivan, Phys. Rev. **58**, 276 (1940)
138. C.S. Wu, Phys. Rev. **58**, 926 (1940)
139. W. Maurer, W. Ramm, Naturwissenschaften **29**, 368 (1941)
140. O. Minikawa, Phys. Rev. **60**, 689 (1941)
141. R. Sherr, K.T. Bainbridge, H.H. Anderson, Phys. Rev. **60**, 473 (1941)
142. D.C. Grahame, H.W. Walker, Phys. Rev. **60**, 909 (1941)
143. W. Maurer, W. Ramm, Z. Phys. **119**, 602 (1942)
144. G. Friedlander, C. Wu, Phys. Rev. **63**, 227 (1943)
145. H.J. Born, W. Seelmann-Eggebert, Naturwissenschaften **31**, 86 (1943)
146. W. Riezler, Naturwissenschaften **31**, 326 (1943)
147. W. Seelmann-Eggebert, Naturwissenschaften **31**, 510 (1943)
148. W. Seelmann-Eggebert, Naturwissenschaften **31**, 491 (1943)
149. W. Bothe, Naturwissenschaften **31**, 551 (1943)
150. H. Atterling, E. Bohr, T. Sigurgeirsson, Arkiv Mat. Astron. Fysik **32A**(2) (1945)
151. E. Bohr, N. Hole, Arkiv Mat. Astron. Fysik **32A**(15) (1945)
152. W. Zünti, E. Bleuler, Helv. Phys. Acta **18**, 263 (1945)
153. W. Bothe, Z. Naturforsch. **1**, 173 (1946)
154. M.S. Livingston, H.A. Bethe, Rev. Mod. Phys. **9**, 245 (1937)
155. E. Segrè, C.S. Wu, Phys. Rev. **57**, 552 (1940)
156. C. Hatchett, Philos. Trans. R. Soc. Lond. **92**, 49 (1802)
157. H. Rose, Ann. Physik **139**, 317 (1844)
158. F.W. Clarke, Science **39**, 139 (1914)
159. W. Seelmann-Eggebert, H.J. Born, Naturwissenschaften **31**, 59 (1943)
160. F. Strassmann, O. Hahn, Naturwissenschaften **28**, 817 (1940)
161. A.H. Snell et al., Phys. Rev. **72**, 545 (1947)
162. N. Sugarman, J. Chem. Phys. **17**, 11 (1949)
163. O. Hahn, F. Strassmann, Naturwissenschaften **27**, 529 (1939)
164. G.N. Glasoe, J. Steigman, Phys. Rev. **58**, 1 (1940)
165. L.A. Delsasso, W.A. Fowler, C.C. Lauritsen, Phys. Rev. **51**, 391 (1937)
166. L.R. Hafstad, N.P. Heydenburg, M.A. Tuve, Phys. Rev. **50**, 504 (1936)
167. W. Bothe, W. Gentner, Naturwissenschaften **25**, 90 (1937)
168. E. Amaldi et al., Proc. R. Soc. A **149**, 522 (1935)
169. W. Bothe, W. Gentner, Naturwissenschaften **25**, 126 (1937)
170. W. Bothe, W. Gentner, Naturwissenschaften **25**, 191 (1937)
171. O. Huber et al., Helv. Phys. Acta **16**, 33 (1943)
172. O. Huber et al., Helv. Phys. Acta **16**, 431 (1943)
173. I. Noddack, Angew. Chem. **47**, 653 (1934)
174. O. Hahn, L. Meitner, Naturwissenschaften **23**, 37 (1935)
175. O. Hahn, L. Meitner, Naturwissenschaften **23**, 230 (1935)
176. O. Hahn, L. Meitner, F. Strassmann, Naturwissenschaften **26**, 475 (1938)
177. O. Hahn, F. Strassmann, Naturwissenschaften **26**, 755 (1938)
178. E. Fermi, *Nobel Lectures, Physics 1922–1941* (Elsevier Publishing Company, Amsterdam, 1965)
179. O. Hahn, F. Strassmann, Naturwissenschaften **27**, 11 (1939)
180. L. Meitner, O.R. Frisch, Nature **143**, 239 (1939)
181. O. Hahn, F. Strassmann, Naturwissenschaften **27**, 89 (1939)
182. F.A. Heyn, A.H.W. Aten, C.J. Bakker, Nature **143**, 516 (1939)
183. P. Abelson, Phys. Rev. **55**, 876 (1939)
184. A. Langsdorf Jr., Phys. Rev. **56**, 205 (1939)
185. H. Götte, Naturwissenschaften **29**, 496 (1941)
186. O. Hahn, F. Strassmann, Naturwissenschaften **31**, 249 (1943)

187. O. Hahn, F. Strassmann, S. Flügge, Naturwissenschaften **27**, 544 (1939)
188. O. Hahn, F. Strassmann, Naturwissenschaften **28**, 54 (1940)
189. O. Hahn, F. Strassmann, Naturwissenschaften **28**, 543 (1940)
190. H.J. Born, W. Seelmann-Eggebert, Naturwissenschaften **31**, 201 (1943)
191. K.R. Weimer, M.L. Pool, J.D. Kurbatov, Phys. Rev. **63**, 67 (1943)
192. G.T. Seaborg, Rev. Mod. Phys. **16**, 1 (1944)
193. R.J. Hayden, Phys. Rev. **74**, 650 (1948)
194. P. Abelson, Phys. Rev. **55**, 418 (1939)
195. P. Abelson, Phys. Rev. **55**, 670 (1939)

Chapter 5
Transuranium Elements

5.1 Creating New Elements

The feasibility to change elements artificially was first demonstrated in 1919 by Rutherford when he observed the disintegration of nitrogen in the irradiation with α-particles [1]. Later on Blackett showed that in this reaction nitrogen was actually transformed into oxygen in the reaction $^{14}N + \alpha \rightarrow {}^{17}O + p$ [2].

In 1934, Curie and Joliot created the first artificial radioactive isotopes ^{13}N and ^{30}P in α-particle induced reactions [3]. These first experiments with charged-particles were limited to the lightest elements because the available beam energies were not sufficient to overcome the Coulomb barrier in heavier elements. Also, these reactions primarily produced nuclides on the proton-rich side of the stable isotopes.

In contrast, neutron capture reactions could be used across the whole periodic table. The newly formed nuclides were more neutron-rich and decayed by the emission of β^--particles leading to an isotope of an element with a larger atomic number. Thus, when Fermi in 1934 observed an activity after neutron irradiation of uranium ($Z = 82$), the heaviest element at the time, he suspected that he had formed a new element with $Z = 93$ [4]. The subsequent experimental and theoretical efforts leading to the correct interpretation by Meitner and Frisch in 1939 [5] that Fermi's activity was due to fission are discussed in Sect. 4.6 in the previous chapter [4].

It should be mentioned that shortly after Fermi's initial announcement in 1934 Koblic claimed to have extracted element 93 from a pitchblende sample. He called the element bohemium [6] but his claim was very quickly discredited by Speter in an article entitled "Bohemium—An Orbituary" [7]. Also a report by Hulubei and Cauchois of natural occurring element 93 in 1939 named sequanium [8] was considered not credible because it had been accepted that element 93 does not occur in nature [9–11]. Nevertheless it has been suggested that Hulubei and Cauchois might have observed neptunium [12] after minuscule amounts of neptunium have been observed in nature [13, 14].

Element 93 was eventually discovered by McMillan and Abelson in 1940. They irradiated uranium samples with neutrons produced by the Berkeley cyclotron and measured β-decay curves of successive fluoride extractions as shown in Fig. 5.1: "When a thin layer of uranium is bombarded with and without cadmium around it, the fission product intensity is changed by a large factor, while the 2.3-day period and

M. Thoennessen, *The Discovery of Isotopes*,
DOI 10.1007/978-3-319-31763-2_5

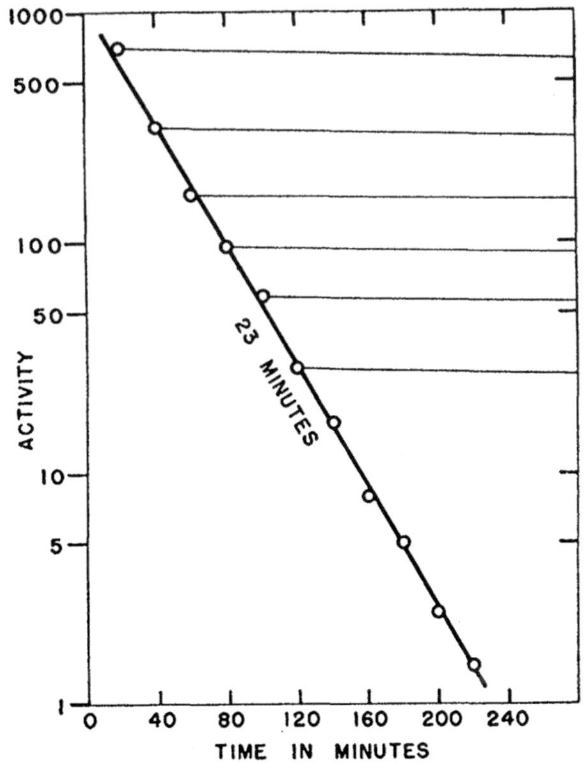

Fig. 5.1 Discovery of ^{239}Np. The *thick line* represents the decay of the 23 min ^{239}U and the *thin* (almost) *horizontal lines* correspond to the decay of a 2.3 d half-life extracted at the time of the circles attributed to ^{239}Np [15] (Reprinted figure with permission from E. McMillan and P. Abelson, *Radioactive element 93*, Phys. Rev. 57 (1940) 1185. Copyright 1940 by the American Physical Society.)

the 23-min uranium period are only slightly changed, and their ratio remains constant. Also absorption of resonance neutrons by uranium changes these two periods in the same ratio, suggesting a genetic relation between them, and the consequent identification of the longer period with element 93" [15].

The name neptunium was officially accepted by the International Union of Pure and Applied Chemistry (IUPAC) at the 15th IUPAC conference in Amsterdam in 1949 [16, 17].

The discovery of fission and neptunium coincided with the beginning of the Second World War and most of the subsequent research was classified. Following the war, research related to nuclear weapons testing continued to be classified. Transuranium isotopes discovered during this time are described in the next section (Sect. 5.2). In the early 1950s three different experimental approaches contributed to the discovery of isotopes of transuranium elements: neutron-capture reactions in reactors (Sect. 5.3), light-particle-induced reactions (Sect. 5.4), and fusion-evaporation reactions (Sect. 5.5). Transuranium elements up to element 101 (mendelevium) are covered in the present chapter while heavier elements will be discussed in Chap. 12 on superheavy elements.

5.2 Plutonium Project and Thermonuclear Tests

The objective of the plutonium project within the Manhattan project was to develop methods to produce plutonium in sufficient quantities for a nuclear weapon. The technical and scientific results of the project were declassified only in 1949 and published in the Plutonium Project Records as Division IV of the National Nuclear Energy Series. In two parts of Volume 14 B, Seaborg, Katz, and Manning edited the research papers describing the transuranium elements [18].

The isotopes discovered during the plutonium project and in the subsequent nuclear weapons tests are listed in Table 5.1.

Table 5.1 Transuranium isotopes discovered as part of the plutonium project and nuclear weapons tests. The 1940 discovery of neptunium is included for completeness. The submission date, first author and reference of the publication as well as the target and reaction are listed. For the weapons test, the name of the shots are included instead of the target. The number of consecutive (n,γ) reactions listed for the weapons tests assume neutron capture on ^{238}U

Date	First author	Ref.	Target	Reaction	Isotope
05/27/1940	McMillan	[15]	^{238}U	(n,γ)	^{239}Np
05/29/1941	Kennedy	[19]	^{238}U	(n,γ)	^{239}Pu[a]
07/27/1942[b]	Kennedy	[20]	^{238}U	(d,2n)	^{238}Np, ^{238}Pu[a]
03/15/1946[b]	Manning	[21]	^{241}Am	(n,γ)	^{242}Am
01/19/1948	Wahl	[22]	^{238}U	(n,2n)	^{237}Np[a]
August 1949[c]	Hyde	[23]	^{233}U	(d,n), (α,3n)	^{234}Np, ^{234}Pu
August 1949[c]	James	[24]	^{235}U	(d,2n), (d,n)	^{235}Np, ^{236}Np, ^{236}Pu[a]
			^{235}U	(α,2n)	^{237}Pu, ^{240}Pu
August 1949[c]	Seaborg	[25]	^{238}U	(α,n)	^{241}Pu, ^{241}Am[a]
			^{237}Np	(α,2n), (α,n)	^{239}Am, ^{240}Am
August 1949[c]	Seaborg	[26]	^{239}Pu	(α,3n), (α,n)	^{240}Cm, ^{242}Cm
06/30/1955	Engelkemeir	[27]	Mike	8(n,γ)	^{246}Pu, ^{246}Am
12/05/1955	Fields	[28]	Mike	10(n,γ)	^{248}Cm, ^{249}Cm[d]
03/28/1966	Combined radio-chemistry group	[29]	Par	12(n,γ)	^{250}Cm
12/29/1970	Hulet	[30]	Hutch	^{257}Fm(d,p)	^{258}Fm[e]
08/05/1977	Lougheed	[31]	Hutch	12(n,γ)	^{251}Cm[d]
11/10/1980	Lougheed	[32]	Hutch	16(n,γ)	^{255}Cf[d]

[a]Populated by subsequent β^--decay
[b]Published in 1949 as part of the plutonium project records, date of internal report submission
[c]Published in 1949 as part of the plutonium project records, date of editor's preface
[d]Produced by neutron capture on targets from debris in a reactor
[e]Produced by deuteron irradiation of a target from debris

With the exception of ^{242}Am all isotopes discovered within the plutonium project were produced at the radiation laboratory of the University of California at Berkeley. Deuterons or α-particles were accelerated with the 60-in. cyclotron and neutrons were produced by irradiating beryllium targets. The half-life of the produced activities were measured after chemical separation of the elements of interest. ^{242}Am was produced by neutron irradiation in the thimble of the Argonne heavy-water pile [21].

The element plutonium was discovered by Seaborg et al. in December 1940 and submitted for publication in Physical Review on January 28, 1941. However, it was only published in 1946 as explained in a footnote of the paper: "This letter was received for publication on the date indicated but was voluntarily withheld from publication until the end of the war" [33]. Although the paper reported the discovery of the element plutonium, it did not identify any specific isotope. The observed β-decay activity was assigned to either mass 238, 236 or 235 [33]. The name plutonium was officially accepted at the 15th IUPAC conference in Amsterdam in 1949 [16, 17].

The first identified isotope of plutonium was ^{239}Pu and was published in 1946 by Kennedy et al. with the same footnote as the discovery of the element. They identified fission as well as α-decay from this plutonium isotope [19]. The paper was submitted on May 29, 1941 and published in the October 1946 issue of Physical Review. It is interesting to note that about seven months later, on December 4, 1941, Kennedy and Wahl submitted the results of an unsuccessful search for spontaneous fission of ^{239}Pu. This manuscript was then published prior to the discovery paper by Kennedy et al. in the April 1946 issue [34].

The only other isotope published in a refereed journal prior to the publication of the whole declassified work of the plutonium project was ^{237}Np. The article was submitted to Physical Review on January 19, 1948, but a footnote stated: "This article was mailed, as a secret report, from Berkeley, California to the Uranium Committee in Washington, D.C., on April 14, 1942. The experimental work was done during 1941 and the early part of 1942" [22].

Seaborg later reminisced about the classified research: "These first two transuranium elements were referred to simply as 'element 93' and 'element 94' or by code names,... Throughout 1941, element 94 was referred to by the code name of 'copper', which was satisfactory until it was necessary to introduce the element copper into some of the experiments. This posed the problem of distinguishing between the two... For a while, the plutonium was referred to as 'copper' and the real copper as 'honest-to-God copper'" [35].

The discoveries of the elements americium ($Z = 95$) and curium ($Z = 96$) were announced in 1945 [36] and first published in 1949 by Seaborg et al. as part of the Plutonium Project Record identifying the isotopes ^{239}Am, ^{240}Am, and ^{241}Am [25] and ^{240}Cm and ^{242}Cm [26], respectively. The names americium and curium were also officially accepted at the 15th IUPAC conference in Amsterdam in 1949 [16, 17].

After the War, classified research continued during the nuclear weapons testing. Two new elements ($Z = 99$ and $Z = 100$) and several new isotopes were discovered in the debris of the nuclear test explosion Mike on November 1, 1952 as part of the Ivy test series. The explosion took place on the Enewetak Atoll of the Marshall Islands (see Fig. 5.2 [37, 38]).

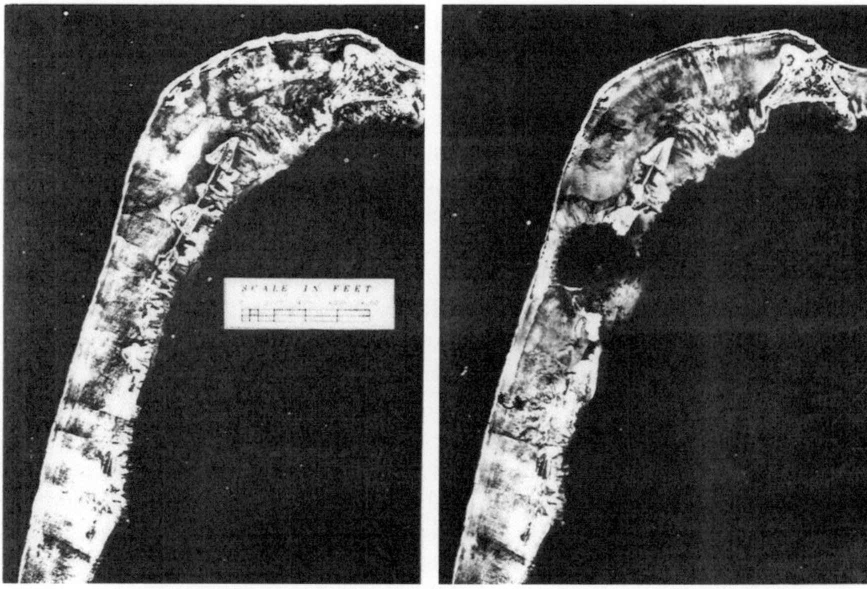

Fig. 5.2 Aerial pictures of the Enewetak Atoll before (*left*) and after (*right*) the Mike explosion [37, 38]

Uranium was subjected to the high instantaneous neutron flux of the explosion and samples were collected on filters which were flown through the cloud of the explosion. The new elements were identified by chemical separation performed at the University of California Radiation Laboratory, Argonne National Laboratory and Los Alamos Scientific Laboratory [39]. It is interesting to note that the loss of life during the collection of these samples was only mentioned in the more recent accounts [35, 40] of the discovery of these transuranium elements: "These samples cost the life of First Lieutenant Jimmy Robinson, who waited too long before he went home, tried to land on Eniwetok, and ditched about a mile short of the runway" [40].

The data from the Mike explosion were only declassified in 1955 and in the meantime several of the isotopes had been produced by neutron capture in reactors.

The researchers involved in the classified work also performed unclassified research and were aware of the competition from other research groups: "At this juncture we began to worry that other laboratories might discover lighter isotopes of the elements 99 and 100 by the use of reactions with cyclotron-produced heavy ions. They would be able to publish that work without any problem and would feel that they should be able to name these elements. This might well happen before we could declassify the Mike work and it would make it difficult for us to claim priority in discovery. (Traditionally, the right to name a new element goes to the first to find it, but it is not clear that the world would accept that premise if the work is done secretly.)" [41].

In 1953, Ghiorso et al. had succeeded in forming ^{246}Es in the fusion-evaporation reaction $^{238}U(^{14}N, 6n)$ and they submitted the results on November 8, 1953 (see Sect. 5.5). However, they did not want this observation to be regarded as the discovery of einsteinium and added the note: "There is unpublished information relevant to element 99 at the University of California, Argonne National Laboratory, and Los Alamos Scientific Laboratory. Until this information is published the question of the first preparation should not be prejudged on the basis of this paper" [42].

Less than two months later, Harvey et al. reported the identification of the first isotope of element Z = 100 (^{254}Fm) following the neutron irradiation of heavy californium isotopes in the Materials Testing Reactor. Again, a statement regarding the discovery was added: "Because of the existence of unpublished information on element 100 the question of its first preparation should not be prejudged on the basis of this paper" [44]. A few other new einsteinium and fermium isotopes were reported during the next few months (see Sects. 5.3 and 5.5).

The official announcement of the discovery of the two new elements was made in the summer of 1955 by Ghiorso et al. [39]. They suggested the names einsteinium and fermium for the elements with Z = 99 and Z = 100, respectively. These names were accepted at the 19th IUPAC Conference in Paris 1957 [17, 45]. However, they changed the originally suggested symbol for einsteinium from E to Es.

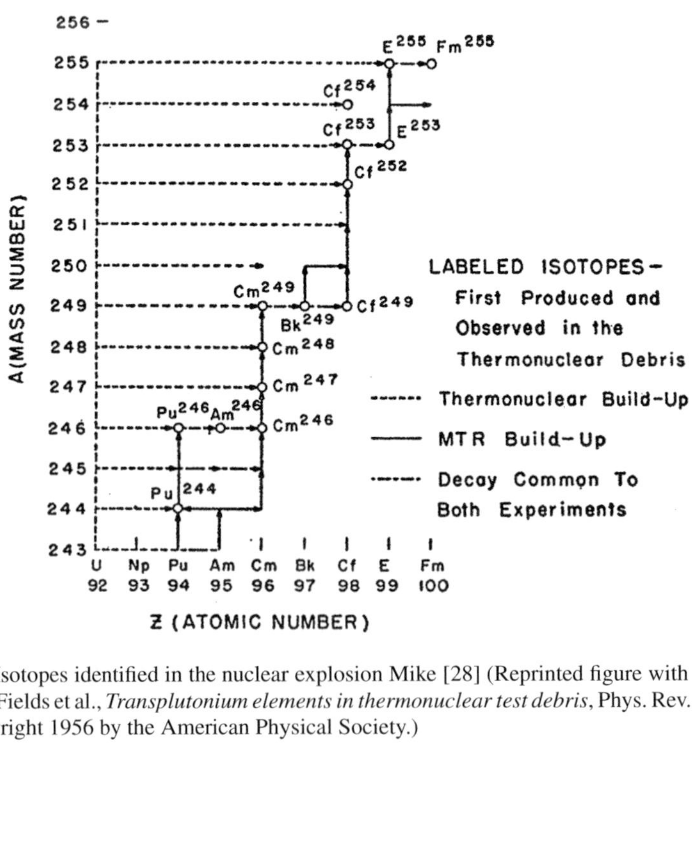

Fig. 5.3 Isotopes identified in the nuclear explosion Mike [28] (Reprinted figure with permission from P.R. Fields et al., *Transplutonium elements in thermonuclear test debris*, Phys. Rev. 102 (1956) 180. Copyright 1956 by the American Physical Society.)

In addition to the discovery of the two new elements einsteinium and fermium, fifteen new isotopes of elements from plutonium to fermium were identified in the samples from the filters collected in the Mike explosion. Most of the isotopes had also been measured by neutron irradiation in reactors and the results had been published before the Mike data were declassified. Figure 5.3 shows the isotopes produced during the Mike explosion and indicates the different neutron capture paths in the explosion and in reactors [28]. Some of the papers describing new isotopes produced in reactors did acknowledge the presence of unpublished results without mentioning the origin of the data: "These isotopes of curium [^{246}Cm and ^{247}Cm] have previously been discovered in other work at Argonne National Laboratory, not yet published" [46], and "Nuclear properties of some of these isotopes were measured in other work at Argonne National Laboratory, University of California Radiation Laboratory, and at Los Alamos Scientific Laboratory, not yet published" (^{249}Bk) [47].

By the time the Mike data were declassified, three isotopes had not been published elsewhere, ^{246}Pu, ^{246}Am [27], and ^{248}Cm [28]. The elements were chemically separated from the debris and the mass of the isotopes were determined with the

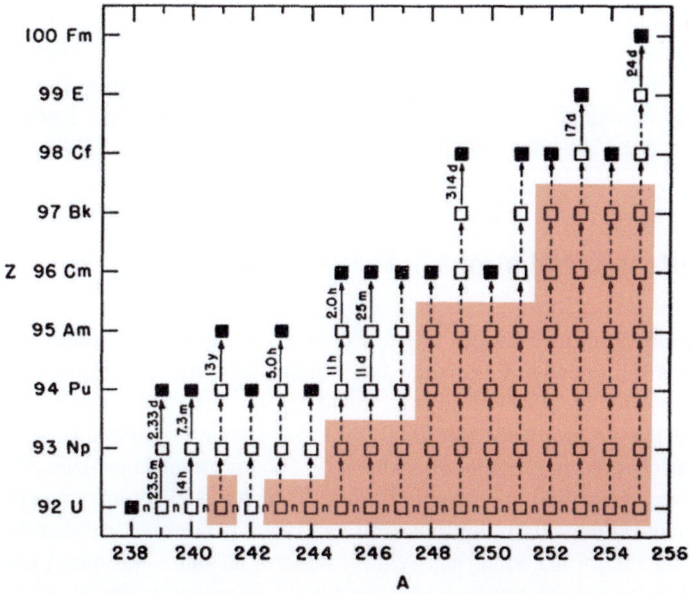

Fig. 5.4 Isotopes produced in the nuclear explosion Mike. From the observation and identification of the β-stable isotopes (*black squares*) it can be deduced that the over 70 β-unstable isotopes (*white squares*) had been produced by successive neutron capture on ^{238}U during the explosion (adapted from [43]). The *red shaded area* includes the approximately 50 isotopes that are still unknown today. Note that the axes in this figure are reversed relative to the chart shown in the previous figure (Reprinted figure with permission from H. Diamond et al., *Heavy isotope abundances in Mike thermonuclear device*, Phys. Rev. 119 (1960) 2000. Copyright 1960 by the American Physical Society.)

Argonne 12-in., 60° mass spectrometer. In addition, a curium sample was irradiated in the Materials Testing Reactor (MTR) to produce ^{249}Cm [28].

The new isotopes observed in the Mike explosion were produced by successive neutron capture reactions on ^{238}U and the subsequent β-decay towards heavier elements as shown in Fig. 5.4. This directly implies that the over 70 β-unstable isotopes (white squares) along the paths must have been produced and exist. Approximately 50 of these isotopes (shown within the red area) remain unknown today. With the exception of ^{251}Cf and ^{250}Cm all β-stable isotopes indicated in the figure by black squares were discovered in the Mike data.

^{251}Cf was discovered by Diamond et al. by neutron irradiation of plutonium in the Materials Testing Reactor in Arco, Idaho (see Table 5.2) in 1954 [47]. ^{250}Cm was identified in 1966 in the debris of the nuclear explosion Par on October 9, 1964 as part of the Whetstone nuclear test series 1964–1965. The results were published by the Combined Radiochemistry Group which consisted of 60 authors from the four national laboratories at Livermore, Los Alamos, Berkeley, and Argonne [29].

The last three isotopes discovered within the weapons program were ^{258}Fm [30], ^{251}Cm [31] and ^{255}Cf [32]. They were formed in reactions on targets extracted from the debris of the Hutch thermonuclear explosion (July, 6, 1969, as part of the Mandrell test series). Hulet et al. irradiated a chemically separated ^{257}Fm target with deuterons from the Berkeley heavy-ion linear accelerator HILAC to populate ^{258}Fm [30]. ^{251}Cm [31] and ^{255}Cf [32] were identified after chemically separated curium and californium targets were irradiated with neutrons in the Livermore Pool Type Reactor and the General Electric Vallecitos Test Reactor, respectively.

5.3 Neutron Irradiations in Reactors

About 30 transuranium isotopes were discovered by neutron irradiation in reactors using targets that were not extracted from debris of a nuclear explosion. The first isotope discovered by this method was ^{242}Am in 1946 as part of the plutonium project as listed in the previous section [21]. Table 5.2 lists all isotopes of transuranium elements discovered in reactors after the end of the plutonium project. The isotopes that had been identified in classified work from the Mike nuclear explosion are indicated in the table. All irradiations until 1955 were performed in either the Argonne Heavy-Water Pile or the Materials Testing Reactor at Idaho National Laboratory.

Most of the isotopes were then identified by mass spectroscopy after the elements of interest were chemically separated. As an example, Fig. 5.5 shows the spectrum of the curium fraction of irradiated ^{241}Am from the Berkeley 60° focusing mass spectrograph. The new isotopes ^{243}Cm and ^{244}Cm only show up at mass 259 and 260, respectively, corresponding to the oxides (^{243}CmO$^+$ and ^{244}CmO$^+$) [49].

The observation of ^{254}Fm in the neutron irradiation of ^{252}Cf represented the first identification of a fermium isotope [44]. However, as mentioned in the previous section it was not supposed to be considered as the discovery of the element fermium.

The isotopes ^{245}Pu and ^{245}Am were discovered simultaneously by Browne et al. at Los Alamos [61] and Fields et al. at Argonne [60], respectively.

Some of the targets were irradiated for a long time. For example ^{257}Fm was produced from a target consisting of a mixture of ^{242}Pu,^{243}Am, and ^{244}Cm which was irradiated in the Idaho Materials Testing Reactor for four years [62].

The last new transuranium isotopes produced in a reactor in the U.S. were ^{255}Cf and ^{256}Es irradiated in the core of the Livermore Pool-Type Reactor in 1980 [32].

Later in the 1980s two further isotopes were discovered at Dimitrovgrad in Russia. ^{247}Pu [64] and ^{257}Es [65] were produced by irradiating plutonium and californium targets in the central channel of the high-flux SM-2 reactor, respectively.

Table 5.2 Isotopes of transuranium elements discovered after the plutonium project by neutron irradiation in reactors: The submission date, first author and reference of the publication as well as the target and the number of neutron captures involved are listed

Date	First author	Ref.	Target	# of (n,γ)	Isotope
06/12/1950	Street	[48]	^{241}Am	2, 3	^{243}Am, ^{244}Am
09/07/1950	Reynolds	[49]	^{241}Am	2, 3	^{243}Cm, ^{244}Cm
10/20/1950	Thompson	[50]	^{239}Pu	3	^{242}Pu
07/23/1951	Sullivan	[51]	$^{239-242}$Pu	1–4	^{243}Pu
04/21/1953	Knight	[52]	^{238}U	2	^{240}Np
12/21/1953	Thompson	[53]	^{239}Pu	14	^{253}Es^{a}
01/14/1954	Harvey	[44]	^{252}Cf	2	^{254}Fm
02/01/1954	Studier	[54]	^{239}Pu	5	^{244}Pu^{a}
02/11/1954	Fields	[55]	^{253}Es	1	^{254}Es
02/15/1954	Stevens	[46]	Pu	$\geq$3	^{245}Cm, ^{246}Cm^{a}, ^{247}Cm^{a}
03/19/1954	Choppin	[56]	Pu	$\geq$11	^{253}Cf^{a}, ^{255}Es^{a}, ^{255}Fm^{a}
03/19/1954	Ghiorso	[57]	^{249}Bk	1, 1, 3	^{250}Bk, ^{250}Cf^{a}, ^{252}Cf^{a}
			^{239}Pu	10	^{249}Cf^{a}
03/24/1954	Diamond	[47]	Pu	$\geq$7	^{249}Bk^{a}, ^{251}Cf
04/18/1955	Choppin	[58]	^{255}Es	1	^{256}Fm
05/09/1955	Harvey	[59]	^{253}Es	1	^{254}Cf
05/20/1955	Fields	[60]	^{244}Pu	1	^{245}Pu^{a}, ^{245}Am
05/20/1955	Browne	[61]	Pa-Pu	$\geq$3	^{245}Pu^{a}, ^{245}Am
08/10/1964	Hulet	[62]	^{242}Pu, ^{243}Am, ^{244}Cm	$\geq$13	^{257}Fm
08/31/1966	Diamond	[63]	Es	$\geq$1	^{251}Bk^{b}
11/10/1980	Lougheed	[32]	^{255}Es	1	^{256}Es
08/28/1982	Popov	[64]	Pu	$\geq$5	^{247}Pu
07/29/1985	Popov	[65]	^{252}Cf	5	^{257}Es

[a] Also identified in the Mike nuclear explosion during classified research
[b] Populated by α-decay of ^{255}Es

Fig. 5.5 Curium mass spectra after irradiation of ^{241}Am. The new isotopes ^{243}Cm and ^{244}Cm only show up at mass 259 and 260, respectively, corresponding to the oxides ($^{243}CmO^+$ and $^{244}CmO^+$) [49] (Reprinted figure with permission from F.L. Reynolds, E.K. Hulet, and K. Street, *Mass-spectrographic identification of Cm*^{243} *and Cm*^{244}, Phys. Rev. 80 (1950) 467. Copyright 1950 by the American Physical Society.)

5.4 Light-Charged-Particle Reactions

Over 50 new transuranium isotopes were produced with light-charged-particles from accelerators. They are listed in Table 5.3. All isotopes until 1958 with the exception of ^{252}Fm, which was first observed at Argonne National Laboratory, were discovered at the University of California Radiation Laboratory at Berkeley with deuterons and α-particles accelerated by either the 60-in. or the 184-in. cyclotron. Most of the transuranium targets used for the reactions were made by lengthy neutron irradiations in reactors. In one case (discovery of ^{258}Fm) it was specifically mentioned that the target was extracted from the Hutch nuclear explosion so this discovery was included in Sect. 5.2.

The first new transuranium element discovered after the Second World War was berkelium in 1950 by Thompson et al. identifying ^{243}Bk. The isotope was produced by irradiating ^{241}Am with helium ions from the Berkeley 60-in. cyclotron. They suggested the name berkelium with the symbol Bk: "It is suggested that element 97 be given the name berkelium (symbol Bk), after the city of Berkeley, in a manner similar to that used in naming its chemical homologue terbium (atomic number 65) whose name was derived from the town of Ytterby, Sweden, where the rare earth minerals were first found" [66].

It is interesting to note that the tentative assignment of ^{243}Bk ("The particular isotope discovered is thought to be 97^{243}, or possibly 97^{244} decaying with a 4.8-h half-life by electron capture with approximately 0.1 percent alpha-decay branching.") [66]) is up to now still the only half-life measurement of ^{243}Bk published in a refereed journal. Two later measurements were only reported in Ph.D. theses [100, 101]. Although this assignment was very tentative it is currently being accepted as correct [102].

Table 5.3 Isotopes of transuranium elements with Z < 102 discovered with light-charged-particles: The submission date, first author and reference of the publication as well as the target and reaction are listed

Date	First author	Ref.	Target	Reaction	Isotope
01/23/1950	Thompson	[66]	^{241}Am	(α,2n)	^{243}Bk
02/07/1950	Magnusson	[67]	^{233}U	(d,xn)	^{231}Np, ^{232}Np, ^{233}Np
06/12/1950	Street	[48]	^{239}Pu	(d,3n)	^{238}Am
09/05/1951	Hulet	[68]	^{244}Cm	(d,n)	^{245}Bk
02/25/1952	Higgins	[69]	^{239}Pu	(α,2n)	^{241}Cm
07/23/1954	Hulet	[70]	^{243}Am	(α,n)	^{246}Bk
04/18/1955	Ghiorso	[71]	^{253}Es	(α,n)	^{256}Md
11/14/1955	Hulet	[72]	^{246}Cm	(α,pn)	^{248}Bk
01/16/1956	Chetham-Strode	[73]	^{244}Cm	(α,4n), (α,3n)	^{244}Cf, ^{245}Cf
02/20/1956	Friedman	[74]	^{249}Cf	(α,n)	^{252}Fm
08/07/1956	Chetham-Strode	[75]	^{249}Cf	(d,3n)	^{248}Es
08/13/1956	Harvey	[76]	^{249}Bk	(α,xn)	^{249}Es, ^{250}Es, ^{251}Es, ^{252}Es
12/21/1956	Amiel	[77]	^{252}Cf	(α,3n)	^{253}Fm
01/22/1957	Amiel	[78]	^{249}Cf	(α,2n)	^{251}Fm
03/07/1957	Thomas	[79]	^{233}U	(α,4n, (α,2n)	^{233}Pu, ^{235}Pu
08/21/1958	Phillips	[80]	^{253}Es	(α,2n)	^{255}Md
09/08/1958	Vandenbosch	[81]	^{238}U	(α,p)	^{241}Np
11/30/1964	Milsted	[82]	^{244}Cm	(α,p)	^{247}Bk
03/13/1967	Fields	[83]	^{242}Cm	(^{3}He,3n), (^{3}He,2n)	^{242}Cf, ^{243}Cf[a]
06/02/1967	Orth	[84]	^{244}Pu	(α,p)	^{247}Am
03/01/1968	Hahn	[85]	^{233}U	(p,5n), (p,4n)	^{229}Np, ^{230}Np
04/06/1970	Fields	[86]	^{253}Es, ^{255}Es	(α,3n), (α,n)	^{254}Md, ^{258}Md
04/07/1970	Polikanov	[87]	^{238}Pu	(p,2n)	^{237}Am
02/01/1972	Wolf	[88]	^{241}Am, ^{243}Am	(α,3n)	^{242}Bk, ^{244}Bk
12/04/1972	Jäger	[89]	^{233}U	(α,5n)	^{232}Pu
05/15/1978	Flynn	[90]	^{244}Pu	(t,α)	^{243}Np
01/09/1979	Haustein	[91]	^{244}Pu	(n,p2n)[b]	^{242}Np
07/17/1979	Hoffman	[92]	^{254}Cf	(t,p)	^{256}Cf
07/17/1979	Hulet	[93]	^{257}Fm	(t,p)	^{259}Fm
07/26/1993	Kreek	[94]	^{233}U	(p,6n)	^{228}Np
08/09/1993	Kreek	[95]	^{241}Am	(α,7n)	^{238}Cm, ^{238}Bk[c]
02/28/1996	Guo	[96]	^{238}Pu	(p,4n)	^{235}Am
12/24/1998	Laue	[97]	^{233}U	(^{3}He,5n)	^{231}Pu

[a] Simultaneously discovered by Sikkeland et al. in fusion-evaporation reactions [98, 99]
[b] Neutrons were produced by 200 MeV protons from the Brookhaven AGS
[c] Populated by subsequent β^{-}-decay

After californium was discovered in a fusion-evaporation reaction later in 1950 (see Sect. 5.5) and einsteinium and fermium were observed in the Mike explosion in 1952, the identification of the next element (Z = 101) was reported in 1955 by Ghiorso et al. They measured the decay of ^{256}Md and suggested the name mendelevium: "We would like to suggest the name mendelevium, symbol Mv, for the new element in recognition of the pioneering role of the great Russian chemist, Dmitri Mendeleev, who was the first to use the periodic system of the elements to predict the chemical properties of undiscovered elements, a principle which has been the key to the discovery of the last seven transuranium (actinide) elements" [71]. IUPAC adopted the name mendelevium but changed the symbol from "Mv" to "Md" at the 19th IUPAC Conference in Paris 1957 [17, 45].

In 1986, the IUPAC in conjunction with the International Union of Pure and Applied Physics (IUPAP) established a Transfermium Working Group (TWG) with the task to establish criteria for the discovery of new elements [103, 104] and assign priorities for discoveries of the transfermium elements. In the final report of the TWG the credit for the discovery of mendelevium was given to Berkeley: "Element 101 was discovered by the Berkeley group—with certainty in 1958 [80] following strong indications in 1955 [71]" [104, 105].

^{242}Np was not directly produced in a charged-particle reaction but by neutron irradiation of ^{244}Pu, where the neutrons were produced by irradiating a copper beam stop with 200 MeV protons from the Alternating Gradient Synchrotron (AGS) at the Brookhaven Medium Energy Intense Neutron (MEIN) facility [91].

Only three isotopes were discovered at laboratories outside of the U.S. Polikanov et al. produced ^{237}Am with the tandem Van de Graaff accelerator of the Niels Bohr Institute in Copenhagen, Denmark [87], ^{232}Pu was identified by Jäger et al. with the Karlsruhe Isochronous Cyclotron in Germany [89], and Guo et al. discovered ^{235}Am with the proton linear accelerator in the Beijing Institute of High Energy Physics in China [96].

5.5 Heavy-Ion Fusion-Evaporation Reactions

Although Alvarez had already demonstrated in 1940 with the Berkeley 37-in. cyclotron that it was possible to accelerate ions heavier than helium [106] it took another decade before Miller et al. successfully accelerated detectable intensities of completely stripped carbon nuclei in the Berkeley 60-in. cyclotron [107]. The possibility of accelerating heavy ions to energies where they would fuse with the target nuclei forming a compound nuclide with the combined Z of the beam and the target immediately opened up the opportunity to extend the nuclear chart to even heavier transuranium elements. Table 5.4 lists all isotopes of transuranium elements discovered with heavy ions up to Z = 101.

Fusion evaporation turned out to be the only viable method to produce elements with even larger Z and these discoveries are discussed in Chap. 12 on superheavy elements.

Table 5.4 Isotopes of transuranium elements with $Z < 102$ discovered with fusion-evaporation reactions: The submission date, first author and reference of the publication as well as the beam, target, and evaporation channel (xn) are listed. A few isotopes were discovered in transfer (TR), incomplete fusion (ICF) or deep inelastic (DI) reactions

Date	First author	Ref.	Beam	Target	Decay	Isotope
11/08/1950	Ghiorso	[108]	^{12}C	^{238}U	4n	^{246}Cf
11/25/1953	Ghiorso	[42]	^{14}N	^{238}U	6n	^{246}Es
					5n, 4n	^{247}Cf^{a}, ^{248}Cf^{a}
05/18/1954	Atterling	[109]	^{16}O	^{238}U	4n	^{250}Fm
06/06/1958	Ghiorso	[110]	^{12}C	^{240}Pu	4n	^{248}Fm
07/04/1959	Perelygin	[111]	^{16}O	^{238}U	5n	^{249}Fm
06/03/1965	Sikkeland	[112]	^{11}B, ^{12,13}C	^{250,252}Cf	ICF	^{257}Md
05/15/1966	Mikheev	[113]	^{14}N	^{238}U	7n, 5n	^{245}Es, ^{247}Es
05/17/1966	Akapev	[114]	^{16}O	^{235}U	5n	^{246}Fm
08/11/1966	Kuznetsov	[115]	^{10,11}B	^{230}Th	6, 7n	^{234}Am
11/14/1966	Kuznetsov	[116]	^{10}B	^{230}Th	8n	^{232}Am
11/29/1966	Flerov	[117]	^{12}C	^{239}Pu	4n	^{247}Fm
03/13/1967	Sikkeland	[98]	^{12}C	U	3–8n	^{242}Cf^{b}
03/13/1967	Sikkeland	[99]	^{12}C	U	2–7n	^{243}Cf^{b}
11/21/1967	Nurmia	[118]	^{16}O	^{233}U	5n, 4n	^{244}Fm, ^{245}Fm
07/20/1970	Silva	[119]	^{12}C	$^{233-235}$U	4–7n	^{240}Cf, ^{241}Cf
06/09/1972	Eskola	[120]	^{12,13}C	^{243}Am	3–6n	^{250}Md, ^{251}Md, ^{252}Md
			^{12}C	^{241}Am	5, 4n	^{248}Md, ^{249}Md
10/05/1973	Eskola	[121]	^{15}N	^{233}U	5n, 4n	^{243}Es, ^{244}Es
08/11/1975	Ter-Akopian	[122]	^{40}Ar	^{204,206}Pb	2, 4n	^{242}Fm
07/21/1979	Gangrskii	[123]	^{14}N	^{232}Th	6n	^{240}Bk
05/12/1981	Münzenberg	[124]	^{40}Ar	^{209}Bi	2n	^{247}Md
				^{206}Pb	3n	^{243}Fm, ^{239}Cf^{b}
04/12/1982	Wild	[125]	^{16}O	^{248}Cm	α3n	^{259}Md^{a}
07/28/1987	Moody	[126]	^{136}Xe	^{244}Pu	TR	^{244}Np
12/28/1988	Hulet	[127]	^{18}O	^{254}Es	ICF	^{260}Md
12/20/1989	Andreyev	[128]	^{22}Ne	^{209}Bi	4n	^{227}Np
05/14/1990	Ninov	[129]	^{22}Ne	^{209}Bi	5n	^{226}Np
06/11/1990	Andreyev	[130]	^{26}Mg	^{208}Pb	4n	^{230}Pu
04/19/1991	Kadkhodayan	[131]	^{13}C	^{243}Am	3n	^{253}Md
11/15/1993	Andreyev	[132]	^{24,26}Mg	^{207,208}Pb	4n, 5n	^{228}Pu, ^{229}Pu
03/08/1994	Lazarev	[133]	^{40}Ar	^{209}Bi	α3n	^{242}Es
03/28/1994	Yeremin	[134]	^{22}Ne	^{209}Bi	6n	^{225}Np
08/22/1994	Lazarev	[135]	^{34,36}S	$^{206-208}$Pb	3–6n	^{237}Cf, ^{238}Cf
07/11/1996	Ninov	[136]	^{40}Ar	^{209}Bi	3n, 4n	^{246}Md, ^{245}Md, ^{241}Es^{c}

(continued)

Table 5.4 (continued)

Date	First author	Ref.	Beam	Target	Decay	Isotope
11/21/1997	Tsukada	[137]	^{6}Li	^{235}U	5n	^{236}Am
09/05/2000	Sakama	[138]	^{6}Li	^{233}U	6n	^{233}Am
07/12/2001	Ichikawa	[139]	^{6}Li	^{237}Np	6n	^{237}Cm
10/14/2002	Asai	[140]	^{6}Li	^{239}Pu	4n	^{241}Bk
11/27/2007	Qin	[141]	^{12}C	^{232}Th	5n	^{239}Cm
04/24/2008	Khuyagbaatar	[142]	^{40}Ar	^{204}Pb	3n	^{241}Fm
07/05/2010	Khuyagbaatar	[143]	^{36}S	^{204,207}Pb	3n	^{233}Cm^{c}, ^{236}Cm^{c}
05/05/2015	Devaraja	[144]	^{48}Ca	^{248}Cm	DI	^{233}Bk, ^{229}Am^{a}, ^{223}Am

^{a}Populated by subsequent β^{+}-decay/electron capture
^{b}Simultaneously discovered by Fields et al. in ^{3}He induced reactions [83]
^{c}Populated by subsequent α-decay

Less than two months (11/8/1950) after sufficiently intense heavy-ion beams became available Ghiorso et al. reported the discovery of ^{246}Cf in the heavy-ion fusion-evaporation reaction ^{238}U(^{12}C, 4n) [108]. This represented the first correct identification of a californium nuclide because the isotope assignment in the paper reporting the discovery of the element californium a few months earlier (8/7/1950) was incorrect.

Thompson et al. had assigned a 45 min half-life tentatively to ^{244}Cf in the irradiation of ^{242}Cm with helium ions [145], however, this assignment was later changed to ^{245}Cf [73]. Thus, the paper by Thompson et al. constitutes the discovery of the element californium but not a californium isotope. Thompson et al. suggested the name californium with the symbol Cf: "It is suggested that element 98 be given the name californium (symbol Cf) after the university and state where the work was done. This name, chosen for the reason given, does not reflect the observed chemical homology of element 98 to dysprosium (No. 66) as the names americium (No. 95), curium (No. 96), and berkelium (No. 97) signify that these elements are the chemical homologs of europium (No. 63), gadolinium (No. 64), and terbium (No. 65), respectively; the best we can do is point out, in recognition of the fact that dysprosium is named on the basis of a Greek word meaning 'difficult to get at,' that the searchers for another element a century ago found it difficult to get to California" [145].

Three years later, Ghiorso et al. identified the first isotope of the element 99 in the reaction ^{238}U(^{14}N,6n)^{246}Es [42]. The authors were also involved in the classified nuclear weapons research where they had already evidence for the discovery of the elements 99 and 100. Thus, they did not want the observation of ^{246}Es to be considered the discovery of the element 99 (see Sect. 5.2).

The concern by the Berkeley group that other groups might discover elements 99 and 100 with heavy-ions was justified. Only four months after Harvey et al. reported the observation of ^{254}Fm in neutron capture reactions on ^{252}Cf [44], Atterling et al. discovered ^{250}Fm by irradiating ^{238}U with ^{16}O from the 225-cm cyclotron of the Nobel Institute in Stockholm, Sweden [109].

Another major breakthrough was achieved in 1958 when Ghiorso et al. succeeded in separating the fusion products from the high-intensity beam. The recoils were slowed down in helium gas and removed from the beam with an electric field (see Fig. 12.1 of Sect. 12.2). The first isotope discovered with this new method was ^{248}Fm [110]. Essentially all subsequent transuranium isotope discoveries until 1980 used some variation of this method.

Three new isotopes were not produced in fusion-evaporation reactions. Two mendelevium isotopes—^{257}Md [112] and ^{260}Md [127]—were populated in

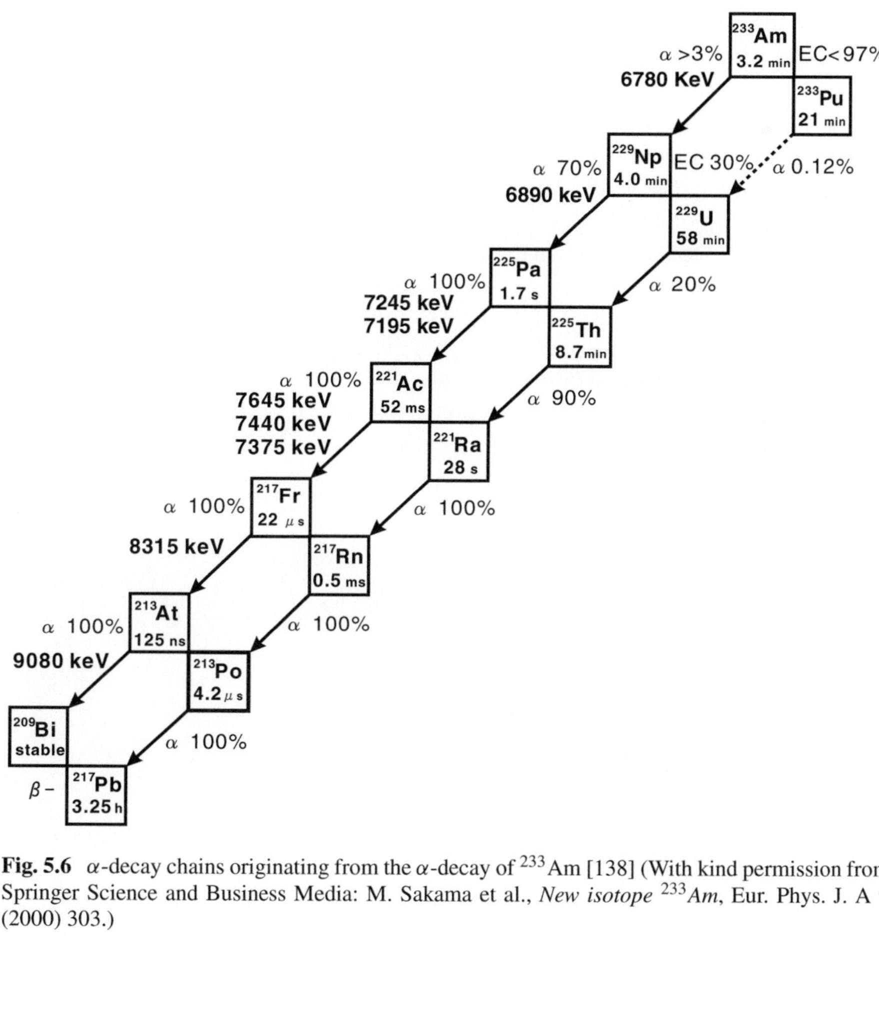

Fig. 5.6 α-decay chains originating from the α-decay of ^{233}Am [138] (With kind permission from Springer Science and Business Media: M. Sakama et al., *New isotope* ^{233}*Am*, Eur. Phys. J. A 9 (2000) 303.)

incomplete fusion reactions, and ^{244}Np was identified following a charge-exchange transfer-type reaction [126].

Between 1974 and 1976, Münzenberg et al. built the velocity filter SHIP (Separator for Heavy Ion reaction Products) at the Gesellschaft für Schwerionenforschung (GSI) in Darmstadt, Germany [146]. At the time, fusion-evaporation reactions had pushed the limit of the nuclear chart up to element 106, seaborgium. These new types of recoil separators, where the residues were implanted into position-sensitive detector arrays which also could measure the subsequent decays and determine their spatial and temporal correlation [147] were used to discover essentially all of the superheavy elements above rutherfordium (see Sects. 12.3 and 12.4).

These devices were also used to separate and identify a significant number of new isotopes of the elements from neptunium through mendelevium. The first three transuranium isotopes discovered with this method were ^{247}Md, ^{243}Fm and ^{239}Cf by Münzenberg et al. using beams from the GSI UNILAC accelerator and SHIP [124]. Nine years later Andreyev et al. reported the first transuranium isotope identified with the VASSILISSA kinematic separator. ^{227}Np was produced by irradiating a ^{209}Bi target with ^{22}Ne beam from the U400 cyclotron of the Laboratory of Nuclear Reactions, JINR at Dubna [128].

A different efficient technique to produce, separate and identify new isotopes, the ISOL method (see Sect. 6.5), was used to produce four new transuranium isotopes between 1998 and 2003 at the Japan Atomic Energy Research Institute (JAERI) in Tokai. Fusion-evaporation residues produced with ^{6}Li beams from the JAERI tandem accelerator were separated and analyzed with the online isotope separator (JAERI-ISOL) where the subsequent decay was measured. Figure 5.6 shows the α-decay chain measured from the decay of the new isotope ^{233}Am [138].

The seven most recent isotopes were again discovered at GSI. Qin et al. collected recoil fragments on a rotating target wheel and identified ^{239}Cm by subsequent chemical separation [141] while Khuyagbaatar et al. identified spontaneous fission (^{241}Fm [142]), α-decay (^{233}Cm and ^{236}Cm [143]) events with SHIP. In 2015, Devaraja et al. populated very neutron-deficient berkelium and americium isotopes in deep inelastic multinucleon transfer reactions [144].

References

1. E. Rutherford, Phil. Mag. **37**, 581 (1919)
2. P.M.S. Blackett, Proc. R. Soc. A **107**, 349 (1925)
3. I. Curie, F. Joliot, Compt. Rend. Acad. Sci. **198**, 254 (1934)
4. E. Fermi, Nature **113**, 898 (1934)
5. L. Meitner, O.R. Frisch, Nature **143**, 239 (1939)
6. O. Koblic, Nature **134**, 55 (1934)
7. M. Speter, Science **80**, 588 (1934)
8. H. Hulubei, Y. Cauchois, Compt. Rend. Acad. Sci. **209**, 476 (1939)
9. J. Emsley, *Nature's Building Blocks: an A-Z Guide To The Elements* (Oxford University Press, Oxford, 2002)

10. M. Fontani, The twilight of the naturally-occurring elements: Moldavium (Ml), Sequanium (Sq) and Dor (Do), in *Proceedings of the International Conference on the History of Chemistry*, Lisbon (2005), p. 1
11. M. Fontani, M. Costa, M.V. Orna, *The Lost Elemens* (Oxford University Press, Oxford, 2015)
12. AnVi Open Source Knowledge Trust, *Actinides: A Complete Guide*, https://en.wikipedia.org/wiki/Neptunium
13. D.F. Peppard et al., J. Am. Chem. Soc. **74**, 6081 (1952)
14. W.A. Myers, M. Lindner, J. Inorg. Nucl. Chem. **33**, 3233 (1971)
15. E. McMillan, P. Abelson, Phys. Rev. **57**, 1185 (1940)
16. IUPAC: Commission de Nomenclature de Chimie Inorganique, in *Comptes Rendus de la Quinzième Conférence*, Amsterdam (1949)
17. W.H. Koppenol, Helv. Phys. Acta **88**, 95 (2005)
18. G.T. Seaborg, J.J. Katz, W.M. Manning, *The Transuranium Elements: Research Papers, Part I and II*, vol. 14b, National Nuclear Energy Series IV (McGraw-Hill, New York, 1949)
19. J.W. Kennedy et al., Phys. Rev. **70**, 555 (1946)
20. J.W. Kennedy et al., *The Transuranium Elements: Research Papers, Paper 1.9*, vol. 14b, National Nuclear Energy Series IV (McGraw-Hill, New York, 1949), p. 79
21. W.M. Manning, L.B. Asprey, *The Transuranium Elements: Research Papers, Paper 22.7*, vol. 14b, National Nuclear Energy Series IV (McGraw-Hill, New York, 1949), p. 1595
22. A.C. Wahl, G.T. Seaborg, Phys. Rev. **73**, 940 (1948)
23. E.K. Hyde, M.H. Studier, A. Ghiorso, *The Transuranium Elements: Research Papers, Paper 22.15*, vol. 14b, National Nuclear Energy Series IV (McGraw-Hill, New York, 1949), p. 1622
24. R.A. James et al., *The Transuranium Elements: Research Papers, Paper 22.8*, vol. 14b, National Nuclear Energy Series IV (McGraw-Hill, New York, 1949), p. 1604
25. G.T. Seaborg, R.A. James, L.O. Morgan, *The Transuranium Elements: Research Papers, Paper 22.1*, vol. 14b, National Nuclear Energy Series IV (McGraw-Hill, New York, 1949), p. 1525
26. G.T. Seaborg, R.A. James, A. Ghiorso, *The Transuranium Elements: Research Papers, Paper 22.2*, vol. 14b, National Nuclear Energy Series IV (McGraw-Hill, New York, 1949), p. 1554
27. D. Engelkemeir et al., J. Inorg. Nucl. Chem. **1**, 345 (1955)
28. P.R. Fields et al., Phys. Rev. **102**, 180 (1956)
29. Combined Radiochemistry Group, Phys. Rev. **148**, 1192 (1966)
30. E.K. Hulet et al., Phys. Rev. Lett. **26**, 523 (1971)
31. R.W. Lougheed et al., J. Inorg. Nucl. Chem. **40**, 1865 (1978)
32. R.W. Lougheed et al., J. Inorg. Nucl. Chem. **43**, 2239 (1981)
33. G.T. Seaborg et al., Phys. Rev. **69**, 366 (1946)
34. J.W. Kennedy, A.C. Wahl, Phys. Rev. **69**, 367 (1946)
35. G.T. Seaborg, W.D. Loveland, *The elements beyond uranium* (Wiley, New York, 1990)
36. G.T. Seaborg, Chem. Eng. News **23**, 2190 (1945)
37. D.W. Dorn, Heavy isotope production by nuclear explosive devices, in *Proceedings of the of the Third Plowshare Symposium Engineering with Nuclear Explosives*, TID 7695 (1964), p. 11
38. S.A. Becker, Approximating the r-Process on Earth with Thermonuclear, Talk at Monash University Australia, LA-UR-12-25146 (2012)
39. A. Ghiorso et al., Phys. Rev. **99**, 1048 (1955)
40. G.T. Seaborg (ed.), *Proceedings of the Symposium Commemorating the 25th Anniversary of Elements 99 and 100*, 23 January 1978, LBL-7701, UC-34C, April 1979
41. D.C. Hoffman, A. Ghiorso, G.T. Seaborg, *The transuranium people* (Imperial College Press, London, 2000)
42. A. Ghiorso et al., Phys. Rev. **93**, 257 (1954)
43. H. Diamond et al., Phys. Rev. **119**, 2000 (1960)
44. B.G. Harvey et al., Phys. Rev. **93**, 1129 (1954)
45. IUPAC: Commission de Nomenclature de Chimie Inorganique, in *Comptes Rendus de la Dix-Neuvième Conférence*, Paris (1957), p. 93

46. C.M. Stevens et al., Phys. Rev. **94**, 974 (1954)
47. H. Diamond et al., Phys. Rev. **94**, 1083 (1954)
48. K. Street, A. Ghiorso, G.T. Seaborg, Phys. Rev. **79**, 530 (1950)
49. F.L. Reynolds, E.K. Hulet, K. Street, Phys. Rev. **80**, 467 (1950)
50. S.G. Thompson et al., Phys. Rev. **80**, 1108 (1950)
51. J.C. Sullivan et al., Phys. Rev. **83**, 1267 (1951)
52. J.D. Knight et al., Phys. Rev. **91**, 889 (1953)
53. S.G. Thompson et al., Phys. Rev. **93**, 908 (1954)
54. M.H. Studier et al., Phys. Rev. **93**, 1433 (1954)
55. P.R. Fields et al., Phys. Rev. **94**, 209 (1954)
56. G.R. Choppin et al., Phys. Rev. **94**, 1080 (1954)
57. A. Ghiorso et al., Phys. Rev. **94**, 1081 (1954)
58. G.R. Choppin et al., Phys. Rev. **98**, 1519 (1955)
59. B.G. Harvey et al., Phys. Rev. **99**, 337 (1955)
60. P.R. Fields et al., J. Inorg. Nucl. Chem. **1**, 262 (1955)
61. C.I. Browne et al., J. Inorg. Nucl. Chem. **1**, 254 (1955)
62. E.K. Hulet et al., Phys. Rev. Lett. **13**, 343 (1964)
63. H. Diamond et al., J. Inorg. Nucl. Chem. **29**, 601 (1967)
64. Yu.S. Popov et al., Sov. Radiochem. **25**, 458 (1983)
65. Yu.S. Popov et al., Sov. Radiochem. **29**, 431 (1987)
66. S.G. Thompson, A. Ghiorso, G.T. Seaborg, Phys. Rev. **77**, 838 (1950)
67. L.B. Magnusson, S.G. Thompson, G.T. Seaborg, Phys. Rev. **78**, 363 (1950)
68. E.K. Hulet et al., Phys. Rev. **84**, 366 (1951)
69. G.H. Higgins, K. Street, Phys. Rev. **86**, 252 (1952)
70. E.K. Hulet, S.G. Thompson, A. Ghiorso, Phys. Rev. **95**, 1703 (1954)
71. A. Ghiorso et al., Phys. Rev. **98**, 1518 (1955)
72. E.K. Hulet, Phys. Rev. **102**, 182 (1956)
73. A. Chetham-Strode, G.R. Choppin, B.G. Harvey, Phys. Rev. **102**, 747 (1956)
74. A.M. Friedman et al., Phys. Rev. **102**, 585 (1956)
75. A. Chetham-Strode, L.W. Holm, Phys. Rev. **104**, 1314 (1956)
76. B.G. Harvey et al., Phys. Rev. **104**, 1315 (1956)
77. S. Amiel, Phys. Rev. **105**, 1412 (1957)
78. S. Amiel et al., Phys. Rev. **106**, 553 (1957)
79. T.D. Thomas et al., Phys. Rev. **106**, 1228 (1957)
80. L. Phillips et al., Phys. Rev. Lett. **1**, 215 (1958)
81. R. Vandenbosch, Phys. Rev. **113**, 259 (1959)
82. J. Milsted, A.M. Friedman, C.M. Stevens, Nucl. Phys. **71**, 299 (1965)
83. P.R. Fields et al., Phys. Lett. B **24**, 340 (1967)
84. C.J. Orth et al., Phys. Rev. Lett. **19**, 128 (1967)
85. R.L. Hahn, M.F. Roche, K.S. Toth, Nucl. Phys. A **113**, 206 (1968)
86. P.R. Fields et al., Nucl. Phys. A **154**, 407 (1970)
87. S.M. Polikanov, G. Sletten, Nucl. Phys. A **151**, 656 (1970)
88. K.L. Wolf, K.P. Unik, Phys. Lett. B **38**, 405 (1972)
89. U. Jäger, H. Münzel, G. Pfennig, Z. Phys. **258**, 337 (1973)
90. E.R. Flynn, D.L. Hanson, R.A. Hardekopf, Phys. Rev. C **19**, 355 (1979)
91. P.E. Haustein et al., Phys. Rev. C **19**, 2332 (1979)
92. D.C. Hoffman et al., Phys. Rev. C **21**, 972 (1980)
93. E.K. Hulet et al., Phys. Rev. C **21**, 966 (1980)
94. S.A. Kreek et al., Phys. Rev. C **50**, 2288 (1994)
95. S.A. Kreek et al., Phys. Rev. C **49**, 1859 (1994)
96. J. Guo et al., Z. Phys. A **355**, 111 (1996)
97. C.A. Laue et al., Phys. Rev. C **59**, 3086 (1999)
98. T. Sikkeland, A. Ghiorso, Phys. Lett. B **24**, 331 (1967)
99. T. Sikkeland et al., Phys. Lett. B **24**, 333 (1967)

100. E.K. Hulet, Thesis, Univ. California (1953); UCRL-2283 (1953)
101. A. Chetham-Strode Jr., Ph.D. Thesis, University of California, June 1956, also published as UCRL-3322
102. Y.A. Akovali, Nucl. Data Sheets **103**, 515 (2004)
103. IUPAC Transfermium Working Group, Pure Appl. Chem. **63**, 879 (1991)
104. R.C. Barber et al., Prog. Part. Nucl. Phys. **29**, 453 (1992)
105. Transfermium Working Group, Pure Appl. Chem. **65**, 1757 (1993)
106. L.W. Alvarez, Phys. Rev. **58**, 192 (1940)
107. J.F. Miller et al., Phys. Rev. **80**, 486 (1950)
108. A. Ghiorso et al., Phys. Rev. **81**, 154 (1951)
109. H. Atterling et al., Phys. Rev. **95**, 585 (1954)
110. A. Ghiorso et al., Phys. Rev. Lett. **1**, 18 (1958)
111. V.P. Perelygin, E.D. Donets, G.N. Flerov, Sov. Phys. JETP **10**, 1106 (1960)
112. T. Sikkeland et al., Phys. Rev. **140**, B277 (1965)
113. V.L. Mikheev, V.I. Ilyushchenko, M.B. Miller, Sov. J. Nucl. Phys. **5**, 35 (1967)
114. G.N. Akapev et al., Sov. J. At. Energy **21**, 908 (1966)
115. V.I. Kuznetsov, N.K. Skobelev, G.N. Flerov, Sov. J. Nucl. Phys. **5**, 191 (1967)
116. V.I. Kuznetsov, N.K. Skobelev, Sov. J. Nucl. Phys. **5**, 810 (1967)
117. G.N. Flerov et al., Sov. At. Energy **22**, 434 (1967)
118. M.J. Nurmia et al., Phys. Lett. B **26**, 78 (1967)
119. R.J. Silva et al., Phys. Rev. C **2**, 1948 (1970)
120. P. Eskola, Phys. Rev. C **7**, 280 (1973)
121. P. Eskola et al., Physica Fennica **8**, 357 (1973)
122. G.M. Ter-Akopian et al., Nucl. Phys. A **255**, 509 (1975)
123. YuP Gangrskii et al., Sov. J. Nucl. Phys. **31**, 162 (1980)
124. G. Münzenberg et al., Z. Phys. A **302**, 7 (1981)
125. J.F. Wild et al., Phys. Rev. C **26**, 1531 (1982)
126. K.J. Moody et al., Z. Phys. A **328**, 417 (1987)
127. E.K. Hulet et al., Phys. Rev. C **40**, 770 (1989)
128. A.N. Andreyev et al., Z. Phys. A **337**, 229 (1990)
129. V. Ninov et al., Z. Phys. A **336**, 473 (1990)
130. A.N. Andreyev et al., Z. Phys. A **337**, 231 (1990)
131. B. Kadkhodayan et al., Radiochim. Acta **56**, 1 (1992)
132. A.N. Andreyev et al., Z. Phys. A **347**, 225 (1994)
133. YuA Lazarev et al., Nucl. Phys. A **580**, 113 (1994)
134. A.V. Yeremin et al., Nucl. Instrum. Meth. A **350**, 608 (1994)
135. YuA Lazarev et al., Nucl. Phys. A **588**, 501 (1995)
136. V. Ninov et al., Z. Phys. A **356**, 11 (1996)
137. K. Tsukada et al., Phys. Rev. C **57**, 2057 (1998)
138. M. Sakama et al., Eur. Phys. J. A **9**, 303 (2000)
139. S. Ichikawa et al., Nucl. Instrum. Meth. B **187**, 548 (2002)
140. M. Asai et al., Eur. Phys. J. A **16**, 17 (2003)
141. Z. Qin et al., Radiochim. Acta **96**, 455 (2008)
142. J. Khuyagbaatar et al., Eur. Phys. J. A **37**, 177 (2008)
143. J. Khuyagbaatar et al., Eur. Phys. J. A **46**, 59 (2010)
144. H.M. Devaraja et al., Phys. Lett. B **748**, 199 (2015)
145. S.G. Thompson et al., Phys. Rev. **78**, 298 (1950)
146. G. Münzenberg et al., Nucl. Instrum. Meth. **161**, 65 (1979)
147. S. Hofmann et al., Nucl. Instrum. Meth. **223**, 312 (1984)

Chapter 6
Neutron-Induced Fission

6.1 Identification of Fission Fragments

Soon after Meitner and Frisch [1] interpreted the data from neutron irradiation of uranium by Hahn and Strassmann [2] as fission, it was realized that the neutrons released in this process could be used to sustain a chain reaction. The implications for the energy release in a controlled or explosive way were immediately obvious. The events coincided with the outbreak of the Second World War and thus the research related to fission was mostly classified.

The first chain reaction was achieved by Fermi et al. on December 2, 1942 in the first nuclear reactor which was called the Chicago Pile 1 (CP-1). The first successful atomic bomb detonated less than three years later, in July 1945 in New Mexico.

One important aspect in the understanding of the fission process was the study of the fission fragments produced in the process. Figure 6.1 shows the fission fragment distribution populated in the fission of ^{235}U by thermal neutrons. The chart also shows the stable isotopes (black squares) and the radioactive isotopes known prior to 1940 (red squares). The figure demonstrates the large number of new isotopes that were accessible with fission.

For the next twenty years fission was the only mechanism available to produce neutron-rich isotopes which were not close to the valley of stability. At the beginning most isotopes were discovered utilizing the reactors built within the Manhattan project (Sect. 6.2). Only a few isotopes were first identified using neutrons produced with accelerators (Sect. 6.3). Reactors had the advantage that the available neutron flux was larger and that several targets could be irradiated simultaneously over a long period of time. At accelerators every experiment required dedicated beam time to produce the neutrons for the irradiations.

In the 1950s, new isotopes were produced at many different research reactors around the world (Sect. 6.4). The main challenge was the separation and identification of the specific isotopes. Following the irradiations, the samples were highly radioactive containing a wide range of elements and the isotopes decayed with different half-lives successively towards the stable isotopes. Chemical separations were

M. Thoennessen, *The Discovery of Isotopes*,
DOI 10.1007/978-3-319-31763-2_6

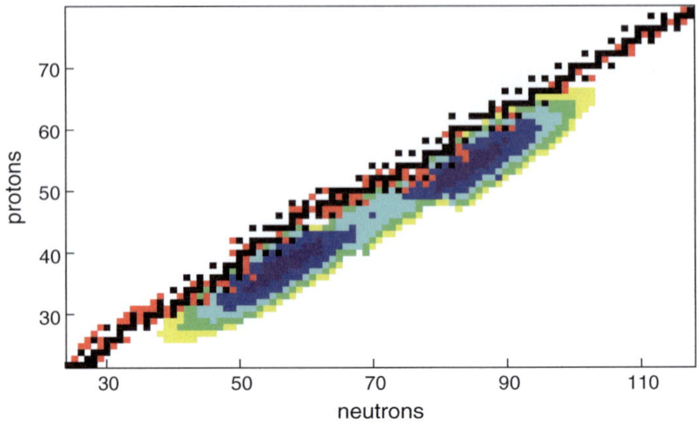

Fig. 6.1 Section of the chart of nuclides including stable (*black squares*) and radioactive isotopes known before 1940 (*red squares*). Isotopes that are produced in fission of ^{235}U with thermal neutrons are shown in the different colors corresponding to their calculated relative yields [3] (increasing from *yellow to black*)

essential where ultimately the observed activities had to be correlated with known half-lives of a given mass chain.

A major breakthrough was the development of online physical separation techniques. The ISOL (Isotope Separation On-Line) method originally designed for neutron-induced fission at the Copenhagen cyclotron in 1951 [4] did not contribute to isotope discoveries at reactors until 1970 (Sect. 6.5).

6.2 Fission Fragments Discovered During the Plutonium Project

The fission research within the Manhattan project was performed as part of the plutonium project. The main goal of the plutonium project was the development of methods to produce plutonium in sufficient quantities for an explosive device. The work was classified at the time and it took several years after the War to compile and edit the research papers on the fission fragments. In order to publish the main results sooner and give appropriate credit to the researchers involved in classified work, the plutonium project issued a survey of the properties of the fission fragments in 1946. It was published simultaneously in the Journal of the American Chemical Society and Reviews of Modern Physics [5]. The compilation of the papers was published in 1951 in the Plutonium Project Records as Division IV of the National Nuclear Energy Series (NNES). Coryell and Sugarman edited three books in Volume 9 entitled “Radiochemical studies: the fission products” [6]. Book 2 contains most of the papers discussing the properties of the fission fragments themselves.

The isotopes discovered in neutron-induced fission within the plutonium project are listed in Table 6.1.

The isotopes discovered within the project and published in the NNES are shown in the top half of the table. The dates listed correspond to the dates of the last internal report contributing to the corresponding paper. Isotope discoveries reported in regular refereed journals are listed in the bottom half of the table.

The 1946 compilation of the plutonium project results classified the available data based on the degree of certainty in the element and mass assignment. The classification was similar to the ones introduced earlier by Livingston and Bethe [34] and Seaborg [35]. The four categories were: (A) element certain, isotope certain, (B) element certain, isotope probably, (C) element certain, isotope uncertain, and (D) insufficient evidence [5].

The credit for discovery is given to the classified work if (i) a half-life was measured which agreed within about a factor of two with the currently accepted values, (ii) the results were rated A, B or C, and (iii) the quoted internal report was included in the 1951 unclassified compilation of the plutonium project fission fragment work.

The first fission fragment identified as part of the plutonium project was produced with neutrons from a cyclotron and not a reactor. Nottorf discovered ^{90}Sr in 1943 at Iowa State College. Strontium samples that had been separated from slow neutron irradiated uranyl nitrate seven months earlier were measured for a period of 20 months and a half-life of 23 ± 3 y was extracted [7]. Powers and Voigt continued the measurements for a total of seven years until 1950 and revised the half-life to 19.9 ± 0.3 y. They stated that the sample had been irradiated at a cyclotron, but they did not specify the location [36]. It is assumed to be the Chicago cyclotron as researchers from Iowa State College had collaborated with Argonne, however, the samples could also have been irradiated at the Washington University cyclotron in St. Louis.

The only other two isotopes produced with neutrons from an accelerator were the mother-daughter nuclides ^{141}Ba–^{141}La. Katcoff irradiated uranyl nitrate with slow neutrons at the Chicago cyclotron for 10 min [12]. As was the case with many of the other fission fragment activities the half-lives had previously been measured without being able to assign them to a specific isotope. In 1938, even prior to the discovery of fission, Curie and Savitch observed a radio-element with a half-life of 3.5 h which had chemical properties that resembled lanthanum. They assumed it was a transuranium element but did not assign it to a specific atomic number [37]. After the discovery of fission Hahn and Strassmann established in 1942 the relationship between this lanthanum activity and an 18 min barium activity [38]. Katcoff was then able to link these two isotopes to the known 28 d half-life of ^{141}Ce [39].

The first new isotopes discovered with a reactor were ^{107}Ru and its daughter nuclide ^{107}Rh. In 1944, Glendenin irradiated uranyl nitrate in the Clinton pile in Oak Ridge, Tennessee, and determined a relationship between a 4 min ruthenium (^{107}Ru) and a 24 min rhodium (^{107}Rh) activity [8]. The Clinton pile, later called the X-10 Graphite Reactor, was the second nuclear reactor built after the Chicago pile where the first chain reaction was achieved. A schematic of the reactor is shown in Fig. 6.2.

Table 6.1 Isotopes discovered in fission during the time of the plutonium project: For the papers published within the NNES the date corresponds to the date of the last internal report on which the paper was based on. For the papers published in refereed journals the submission date is listed. In addition, first author and reference of the publication, as well as the pile of the irradiations are included

Date	First author	Ref.	Pile	Isotope
Published within the National Nuclear Energy Series (NNES)				
06/15/1943	Nottorf	[7]	Chicago[a]	^{90}Sr
09/30/1944	Glendenin	[8]	Clinton	^{107}Ru, ^{107}Rh[b]
01/01/1945	Steinberg	[9]	CP-3	^{77}As, ^{78}Ge
01/01/1945	Katcoff	[10]	CP-3	^{97}Zr, ^{97}Nb
01/01/1945	Seiler	[11]	CP-3	^{112}Pd
01/01/1945	Katcoff	[12]	Chicago[a]	^{141}Ba, ^{141}La[b]
01/01/1945	Winsberg	[13]	CP-3	^{157}Eu, ^{158}Eu
01/01/1945	Winsberg	[14]	Clinton	^{155}Eu
09/04/1945	Dillard	[15]	CP-2	^{90}Kr, ^{91}Kr, ^{92}Kr, ^{93}Kr, ^{139}Xe, ^{140}Xe, ^{141}Xe, ^{143}Xe
April 1945	Gest	[16]	Clinton	^{143}La
04/15/1945	Winsberg	[17]	CP-3	^{156}Sm, ^{156}Eu
06/15/1945	Siegel	[18]	Clinton	^{72}Zn, ^{73}Ga
03/01/1946	Stanley	[19]	Clinton	^{125}Sb
04/02/1946	Winsberg	[20]	Water boiler	^{155}Sm
no date	Glendenin	[21]	Clinton	^{136}Cs
Published in refereed journals				
06/01/1946	Grummitt	[22]	ZEEP	^{95}Zr, ^{95}Nb, ^{106}Ru, ^{106}Rh, ^{144}Ce, ^{144}Pr
07/16/1947	Marinsky	[23]	Clinton	^{147}Nd, ^{147}Pm, ^{149}Pm
12/06/1947	Grummitt	[24]	ZEEP	^{123}Sn
04/23/1948	Sugarman	[25]	CP-3	^{88}Br, ^{138}I, ^{139}I
05/28/1948	Katcoff[c]	[26]	Water boiler	^{93}Y, ^{94}Y, ^{132}Te, ^{134}Te, ^{132}I, ^{134}I
06/03/1948	Hayden	[27]	Clinton	^{88}Y
12/02/1948	Stanley[c]	[28]	Water boiler	^{136}I
02/02/1949	Sugarman	[29]	Water boiler	^{135}Cs
05/31/1949	Sugarman	[30]	CP-3	^{140}Cs
04/10/1950	Steinberg	[31]	CP-2 or CP-3	^{93}Zr
02/09/1951	Kofoed-Hansen	[4]	Copenhagen[d]	^{90}Rb, ^{91}Rb
02/15/1951	Katcoff	[32]	Clinton	^{129}I
08/21/1951	Barnes	[33]	Water boiler	^{127}Sn

[a] Produced at the Chicago cyclotron
[b] Populated by β^--decay
[c] Work performed as part of the plutonium project
[d] Produced at the Copenhagen cyclotron

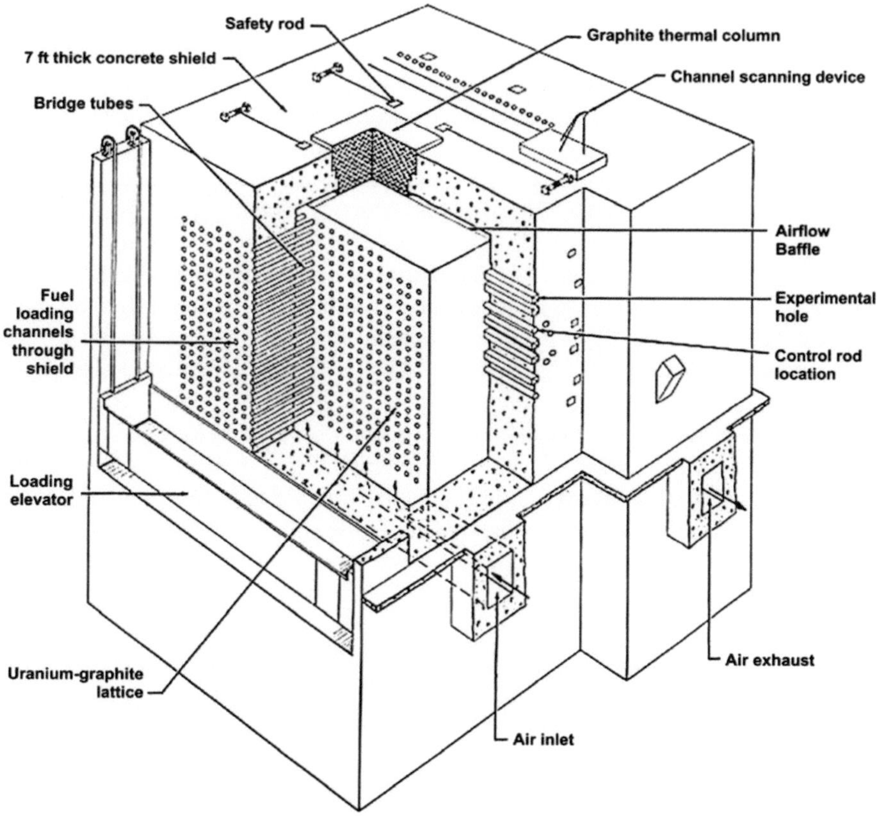

Fig. 6.2 Schematic of the X-10 graphite reactor in Oak Ridge, Tennessee [40]

Subsequently another eleven isotopes were discovered by irradiations in the Clinton pile including some very long-lived activities. For example, ^{155}Eu was identified after the irradiation of a uranium slug for about 10 days and a waiting time of seven months before the beginning of the measurements of the chemically separated europium sample. Winsberg measured the activity for nearly 1.5 years and extracted a half-life of 2 y [14]. The large yield and long half-life of ^{88}Y (108 d) allowed Hayden to determine the isotope using a mass spectrometer [27]. Katcoff also used a mass spectrometer to extract the very long half-life of ^{129}I ($1.72 \pm 0.09 \times 10^7$ y) by measuring the absolute decay rates and the isotopic ratio of ^{129}I relative to ^{127}I. A uranium slug had been irradiated in the Clinton pile for four years and had cooled for 21 months [32].

While the long-lived activities required long waiting times for the shorter (more intense) activities to decay away, the identification of shorter activities relied on fast chemical separations in order to start the half-life measurements as soon as possible. For not too short half-lives most experiments also measured absorption plots in order to determine the energy of the emitting β-radiation. Feather had developed a method to relate the half-thickness of the exponentially decreasing intensity through increasing thicknesses of aluminum absorbers to the decay energy [41].

Figure 6.3 shows as an example the identification of ^{77}As by Steinberg and Engelkemeir. They irradiated uranyl nitrate in the thimble of the Argonne heavy water pile (CP-3) [9]. The decay curve (left) of the arsenic fractions exhibits two half-lives of 90 min and 40 h. The first half-life was tentatively assigned to ^{78}As with a previously reported half-life of 65 min [42]. The identification of ^{77}As was deduced from the Feather analysis. The half-thickness of the 40 h arsenic components was the same as the half-thickness observed in the germanium fraction with a similar half-life. Periodic extraction of arsenic from germanium established the genetic relationship of the 40 h arsenic to a 12 h germanium which was known to be ^{77}Ge [43].

Most of the other isotopes discovered at Argonne were also extracted from samples irradiated in the CP-3 heavy water pile. The graphite pile (CP-2) was used in the identification of eight neutron-rich isotopes of the noble gases krypton and xenon [15]. CP-2 was essentially Fermi's first reactor (CP-1) after it was moved and rebuilt on the Argonne site. For the irradiation of the long-lived zirconium isotope ^{93}Zr

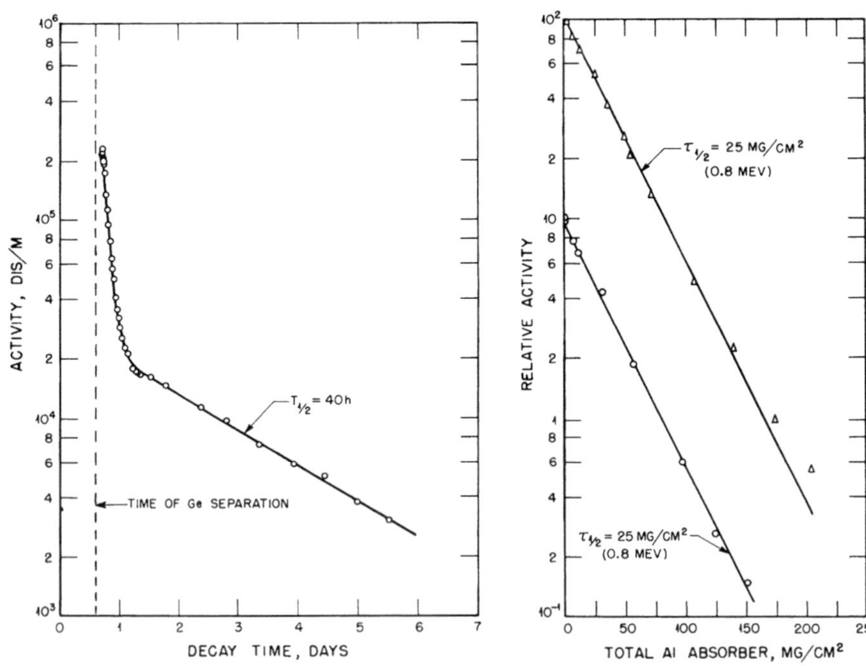

Fig. 6.3 Decay curve (*left*) and absorption plot (*right*) of arsenic activities. The decay curve shows in addition to the 40 h half-life assigned to ^{77}As a 90 min component from ^{78}As. In the absorption plot the *top curve* was measured in the arsenic fraction while the *bottom curve* exhibits the same half-thickness in the parent germanium fraction [9]

Steinberg and Glendenin did not specify which reactor was used for the ten-month irradiation. They determined a half-life of 5.2×10^6 y for ^{93}Zr from the ratio of ^{93}Zr–^{95}Zr measured more than four years after the irradiation [31].

Another ten isotopes were discovered with the Los Alamos homogeneous pile, also called the water boiler. Katcoff, Miskel, and Stanley measured the range of fission fragments in air after irradiation of a plutonium foil for 0.5−15 h in the water boiler. From the range-mass curve determined for well-known masses they were able to make mass assignments for the previously six unknown isotopes ^{93}Y, ^{94}Y, ^{132}Te, ^{134}Te, ^{132}I and ^{134}I [26].

The only country outside the U.S. where isotopes were discovered in a reactor during this time period was Canada. Grummitt and Wilkinson used the Zero Energy Experimental Pile (ZEEP) at the Chalk River Laboratory to identify a total of seven new isotopes. They discovered ^{95}Zr, ^{95}Nb, ^{106}Ru, ^{106}Rh, ^{144}Ce, and ^{144}Pr [22] and ^{123}Sn [24] by irradiating ^{235}U in 1946 and ^{233}U in 1948, respectively.

In addition, ^{90}Rb and ^{91}Rb were first observed in Denmark at the cyclotron of the University of Copenhagen in 1951. Kofoed-Hansen and Nielsen used neutrons produced in the cyclotron to irradiate uranium and then pumped gaseous fission fragments through a 10-m long tube directly into the ion source of an isotope separator. With this method they discovered ^{90}Rb and ^{91}Rb as the daughters populated in the decay of the noble gas isotopes of ^{90}Kr and ^{91}Kr, respectively [4]. This experiment is generally accepted as the first application of the Isotope Separation On-Line (ISOL) method which later was very successfully used in the discovery of many isotopes at reactors (Sect. 6.5), and in spallation (Sect. 10.4), charged-particle-induced-fission (Sect. 10.5) and fusion evaporation (Sect. 11.2) reactions. Interestingly enough Kofoed-Hansen and Nielsen were members of the Institute for Theoretical Physics at the University of Copenhagen.

6.3 Fission Induced by Neutrons Produced with Accelerators

In spite of the larger neutron fluxes available at reactors and easier access to extended irradiation times, some additional isotopes were discovered using neutrons produced at accelerators using (d,n) reactions after the plutonium project fission fragment data were declassified as part of the NNES in 1951. The thirteen isotopes discovered between 1953 and 1973 are listed in Table 6.2.

^{142}La and ^{102}Mo were discovered at the Instituut voor Kernphysisch Onderzoek in Amsterdam, Netherlands, in 1953 [44] and at the Laboratory for Nuclear Science of the Massachusetts Institute of Technology in 1954 [45], respectively.

Between 1954 and 1956 several isotopes were discovered at the Comisión Nacional de la Energia Atómica (CNEA) in Buenos Aires, Argentina. Researchers at the CNEA used charged-particle reactions in cross bombardments to verify the isotope identifications from neutron-induced reactions. For example, in 1956 Fränz,

Table 6.2 Isotopes discovered in fission induced by neutrons from accelerators: The submission date, first author and reference of the publication, as well as the accelerator type and the deuteron energy (in MeV) are listed

Date	First author	Ref.	Accelerator	E_d	Isotope
02/27/1953	Vanden Bosch	[44]	Cyclotron	–	^{142}La
07/20/1954	Wiles	[45]	Cyclotron	15	^{102}Mo
08/14/1954	Flegenheimer	[46]	Philips Kaskaden Generator	1.4	^{102}Tc
01/10/1955	Baró	[47]	Synchrocyclotron	28	^{108}Ru, ^{108}Rh
07/12/1955	Flegenheimer	[48]	Synchrocyclotron[a]	28	^{105}Tc
11/17/1956	Fränz	[49]	Synchrocyclotron	28	^{128}Sn, ^{128}Sb
02/17/1959	Wolfsberg	[50]	Cyclotron	10	^{142}Xe
08/24/1961	Wahl	[51]	Cyclotron	10	^{141}Cs, ^{143}Ba
07/18/1973	Aronsson	[52]	Philips Neutron Generator	14	^{145}La, ^{146}La

[a]Experiments were also performed with 1.4 MeV deuterons from the Philips Kaskaden Generator

Rodriguez, and Radicella identified ^{128}Sb and ^{128}Sn in uranium fission using neutrons produced in the (d,n) reaction with 28 MeV deuterons from the synchrocyclotron of the institute. In addition, they observed ^{128}Sb in the reactions ^{130}Te(d,α) and ^{128}Te(n,p) [49].

The 1961 paper "Nuclear-charge distribution in low-energy fission" by Wahl et al. is one of many examples where the authors assumed that the isotopes they studied had previously been observed and identified [51]. The goal of the paper was to measure the fractional cumulative yield for thermal-neutron fission of ^{235}U. The experiment was performed at the Washington University cyclotron at St. Louis. In an appendix the measured half-lives are discussed and compared to previously measured values. The observation of a 25 $\pm$ 3s half-life for ^{141}Cs was considered to be in good agreement with the value of 24 $\pm$ 2s reported by Fritze and Kennett. However, the reference was listed only as an unpublished private communication and the 1958 Table of Isotopes had listed the ^{141}Cs half-life only as "short" [53]. Similarly, prior to the work by Wahl et al. the half-life of ^{143}Ba was only known to be smaller than 0.5 min [53]. Wahl et al. determined a half-life of 12.0 $\pm$ 1.2s without quoting any earlier measurements.

Twelve years later, Aronsson et al. discovered ^{145}La and ^{146}La at the Chalmers University of Technology in Göteborg, Sweden, using an online chemical separation technique [52]. In the SISAK (Short-lived Isotopes Studied by the AKufve) technique continuous multistage chemical separations of liquid phases are accomplished by H-centrifuges [54].

6.4 Fission Induced by Reactor Neutrons After 1951

Almost 80 new isotopes were discovered in many different research reactors around the world between 1952 and 1975. They are listed in Tables 6.3 (1952–1965) and 6.4 (1966–1975).

From the reactors built as part of the plutonium project where new fission fragments were identified, only the Los Alamos homogeneous pile (water boiler) continued to be used for the discovery of new isotopes. Between 1959 and 1964 another eleven new nuclides were identified in neutron-induced fission of uranium.

The Materials Testing Reactor (MTR) at Idaho National Laboratory had been instrumental in the discovery of isotopes of transuranium elements in the early 1950s (see Sect. 5.3). The observation of ^{142}Ba by Schuman, Turk, and Heath in 1959 was one of the first experiments to measure high-resolution γ-spectra as part of the discovery. Enriched uranium was irradiated for one or two minutes in the MTR and γ-ray spectra were measured with a 3-in. × 3-in. NaI(Tl) crystal following chemical

Table 6.3 Isotopes discovered in fission at reactors between 1952 and 1965: The submission date, first author and reference of the publication, as well as the reactor and the neutron flux (in 1/cm^{2}s) are listed

Date	First author	Ref.	Pile	Flux	Isotope
11/25/1952	Caretto Jr.	[55]	Brookhaven	4×10^{12}	^{146}Ce, ^{146}Pr
11/21/1953	Markowitz	[56]	Brookhaven	4×10^{12}	^{145}Ce, ^{145}Pr
06/09/1955	Pappas	[57]	JEEP	–	^{131}Sb, ^{132}Sb
08/08/1958	Perlow	[58]	CP-5	10^{10}	^{89}Br, ^{90}Br
12/04/1958	Knight	[59]	Water boiler	–	^{93}Sr, ^{94}Sr, ^{95}Y
02/13/1959	Schuman	[60]	MTR	–	^{142}Ba
02/17/1959	Sattizahn	[61]	Water boiler	–	^{84}Se, ^{85}Se
10/30/1959	Orth	[62]	Water boiler	10^{13}	^{98}Nb
09/02/1960	Fritze	[63]	MNR	–	^{92}Rb, ^{93}Rb
110/3/1960	Fritze	[64]	MNR	3×10^{12}	^{94}Rb, ^{95}Sr
01/24/1962	Dropesky	[65]	Water boiler	5×10^{12}	^{126}Sn
03/20/1962	Kienle	[66]	FRM	–	^{104}Mo
03/20/1962	Kienle	[67]	FRM	–	^{105}Mo
03/27/1962	Fritze	[68]	MNR	3×10^{13}	^{142}Cs, ^{143}Cs
10/30/1962	Kienle	[69]	FRM	–	^{103}Mo
05/16/1963	Bemis	[70]	MIT	10^{13}	^{135}Sb
05/20/1963	Greendale	[71]	Stanford[a]	–	^{131}Sn, ^{132}Sn
02/27/1964	Hoffman	[72]	Water boiler	–	^{147}Ce, ^{148}Ce, ^{147}Pr, ^{148}Pr
01/29/1965	von Baeckmann	[73]	FR-2	3×10^{13}	^{106}Tc, ^{107}Tc
03/05/1965	Weiss	[74]	NTR	10^{12}	^{121}Cd
12/06/1965	Strom	[75]	Stanford	10^{10}	^{133}Sb

[a]It is not obvious from the paper where the irradiations were performed

Table 6.4 Isotopes discovered in fission at reactors after 1965: The submission date, first author and reference of the publication, as well as the reactor and the neutron flux (in 1/cm^{2}s) are listed

Date	First author	Ref.	Pile	Flux	Isotope
07/14/1966	del Marmol	[76]	BR-2	10^9	^{85}As
09/03/1966	Fritze	[77]	MNR	5×10^{12}	^{125}In
11/12/1966	Griffiths	[78]	MNR	–	^{109}Ru
05/03/1967	Hübenthal	[79]	Siloé	4×10^{13}	^{98}Zr
10/12/1967	Tomlinson	[80]	LIDO	–	^{134}Sb
12/18/1967	Tomlinson	[81]	LIDO	–	^{87}Se
03/25/1968	Wish	[82]	NTR	5×10^{11}	^{77}Ga
03/25/1968	Weiss	[83]	NTR	10^{12}	^{117}Pd
03/25/1968	Weiss	[84]	NTR	10^{12}	^{118}Ag
04/09/1968	del Marmol	[85]	BR-2	10^9	^{83}As, ^{84}As
05/07/1968	Hastings	[86]	Oak Ridge	6×10^{13}	^{106}Mo
12/12/1968	Denschlag	[87]	TRIGA	10^{14a}	^{135}Te
07/01/1969	Weiss	[88]	NTR	5×10^{11}	^{118}Pd
07/28/1969	del Marmol	[89]	BR-1	10^{12}	^{88}Se
11/26/1969	Wakat	[90]	Ford	2×10^{12}	^{152}Nd
05/18/1970	Kratz	[91]	TRIGA	5×10^{13a}	^{87}As
03/01/1971	Tomlinson	[92]	LIDO	–	^{89}Se
07/14/1971	Izak	[93]	IRR-1	5×10^{12}	^{130}Sn
01/26/1972	Schussler	[94]	TRIGA	10^{14a}	^{140}I
04/11/1972	Trautmann	[95]	TRIGA	10^{14a}	^{102}Nb, ^{107}Mo, ^{108}Mo
06/02/1972	del Marmol	[96]	BR-1	3×10^{11}	^{78}Ga, ^{80}Ge, ^{81}Ge, ^{82}Ge, ^{83}Ge, ^{84}Ge
06/19/1972	Kratz	[97]	TRIGA	10^{14a}	^{86}As
06/18/1973	Tamai	[98]	KUR	2×10^{13}	^{86}Se
03/06/1974	Kratz	[99]	TRIGA	10^{15a}	^{91}Br, ^{92}Br, ^{141}I
12/19/1974	Kratz	[100]	TRIGA	10^{15a}	^{142}I
06/30/1975	Ahrens	[101]	TRIGA	–	^{106}Nb
07/08/1975	Franz	[102]	TRIGA	5×10^{13}	^{111}Rh
09/02/1975	Trautmann	[103]	TRIGA	3×10^{13a}	^{109}Tc, ^{110}Tc

[a]Pulsed operation, flux given in cm^{-2}

separation. Figure 6.4 shows γ-ray spectra of a barium sample [60]. The γ-transitions occur in the daughter nucleus (^{142}La) following β-decay of ^{142}Ba.

The other two National Laboratories which evolved from the plutonium project, Argonne and Oak Ridge, built new reactors at which new fission fragments were discovered. Perlow and Stehney irradiated uranium in the Argonne research reactor CP-5 and identified ^{89}Br and ^{90}Br in 1958 [58]. ^{106}Mo was discovered by Hastings and Troutner in 1968 using the Oak Ridge Research Reactor [86].

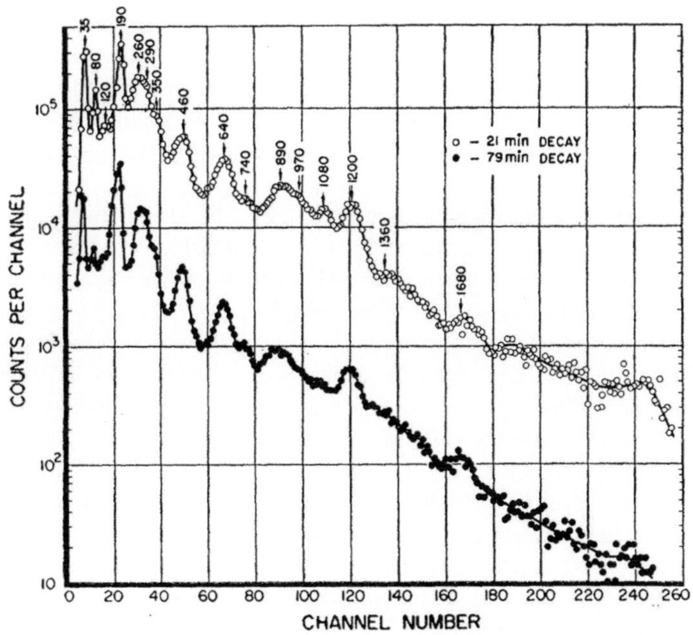

Fig. 6.4 Gamma-ray spectra of a barium sample from neutron irradiated uranium approximately eight minutes after chemical separation from lanthanum measured with a 3-in × 3-in. NaI(Tl) crystal. The γ-ray energies are in keV and correspond to transitions following the decay of at the time known ^{141}Ba and the new isotope ^{142}Ba. The *upper* and *lower* spectra were taken after a decay period of 21 min and 79 min, respectively, before the chemical separation [60] (Reprinted figure with permission from R.P. Schuman, E.H. Turk, and R.L. Heath, *Decay of short-lived barium and lanthanum fission products*, Phys. Rev. 115 (1959) 185. Copyright 1959 by the American Physical Society.)

Brookhaven National Laboratory was established after the War and completed the construction of a research reactor, the Brookhaven pile, in 1949. Two fission fragment cerium-praseodymium mass chains were identified after irradiation of uranyl nitrate in the pile. Caretto and Katcoff discovered ^{146}Ce/^{146}Pr in 1952 [55] and a year later Markowitz, Bernstein, and Katcoff resolved an outstanding puzzle regarding the properties of ^{145}Ce and ^{145}Pr [56].

Between 1963 and 1969, researchers from the U.S. Naval Radiological Defense Laboratory in San Francisco used the Stanford University research reactor and the Vallecitos nuclear test reactor (NTR) to discover a total of eight new fission fragments primarily in the mass region between palladium and antimony. The NTR was designed and constructed by General Electric.

The Stanford University research reactor was not the first reactor on a university campus that was used in the discovery of fission fragments. Only four days prior to the submission of the discovery of ^{131}Sn and ^{132}Sn by Greendale and Love who may have irradiated their samples at Stanford [71], Bemis, Gordon, and Coryell submitted their observation of ^{135}Sb following the irradiation of $UO_2(NO_3)_2 \times 6H_2O$ for 30 s in the MIT reactor [70].

The last fission fragment isotope identified in the U.S. using chemical separation techniques at a reactor was ^{152}Nd in 1969. Wakat and Griffin irradiated ^{235}U for ten minutes in the Ford research reactor of the University of Michigan and assigned a half-life of 11.3 ± 0.4 min decay to ^{152}Nd [90].

The first university-based research reactor in Canada was the McMaster Nuclear Reactor (MNR) in Hamilton, Ontario, which began operating in 1959. Fritze and collaborators discovered eight new fission fragments between 1960 and 1967.

The first reactor in Europe which was used in the discovery of new fission fragments was built at the Joint Establishment for Nuclear Energy Research in Kjeller, Norway. The JEEP (Joint Establishment Experimental Pile) went critical for the first time in 1951. In 1955, Pappas and Wiles developed a fast chemical separation method which allowed them to separate tin within 45 s and antimony within another 30 s after irradiation of $(PyH)_2UCl_6$ in JEEP for one minute. The presence of several long-lived tellurium isomers as shown in Fig. 6.5 makes the assignment of specific antimony masses to the observed activities in this mass region especially challenging. Pappas

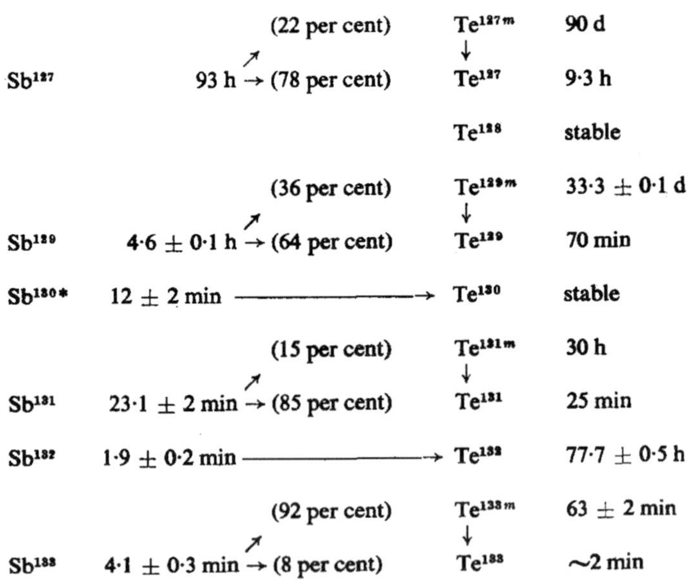

Fig. 6.5 Various decay paths of antimony isotopes populated in fission of uranium [57] (Reprinted from A.C. Pappas and D.R. Wiles, *New short-lived isotopes of tin found in fission*, J. Inorg. Nucl. Chem. 2 (1956) 69, Copyright 1956, with permission from Elsevier.)

and Wiles measured half-lives of 2.0 min and 23.1 min and assigned them to ^{131}Sb and ^{132}Sb, respectively [57]. These assignments had previously only been published in internal reports [104, 105] and a conference proceeding [106].

In 1957, the first research reactor in Germany began operation. The Forschungsreaktor München (FRM) was located in Garching near Munich. In a series of three papers, Kienle et al. reported the discovery of ^{104}Mo [66], ^{105}Mo [67], and ^{103}Mo [69] in 1962. The first two papers were submitted on the same day and published sequentially in the same issue of the journal Naturwissenschaften.

At the same time when the first reactor began operation, construction was started on the second research reactor in Germany. The Forschungsreaktor 2 (FR2) was built at the Kernforschungszentrum Karlsruhe and began operation in 1962. Three years later, von Baeckmann and Feuerstein discovered two technetium isotopes by irradiating ^{239}Pu with thermal neutrons for 10−30 s [73]. Again, the identification relied on fast and clean chemical separation which was capable of measuring half-lives as short as 10 s, sufficient for the measured half-lives of 37 s and 29 s for ^{106}Tc and ^{107}Tc , respectively.

In Belgium the BR-1 and BR-2—Belgian reactors one and two—became operational in 1956 and 1961, respectively. Both reactors were used to identify new fission fragments. Del Marmol and collaborators measured the half-lives of ten new fission fragments between 1967 and 1972. ^{85}As [76] and ^{88}Se [89] were identified by β-delayed neutron emission.

Three additional β-delayed neutron emitters were discovered between 1967 and 1971 at the LIDO reactor of the Atomic Energy Research Establishment in Harwell, England, which had been in operation since 1956. Tomlinson and Hurdus identified ^{134}Sb [80], ^{87}Se [81], and ^{89}Se [92] by obtaining fast separation times of about 5 s and counting the neutrons emitted from the daughter nuclei.

In 1967, Hübenthal irradiated ^{235}U for 30 s in the Siloé reactor at the Centre d'Etude Nucléaires de Grenoble in France, which had been in operation since 1963. A half-life of 31 ± 3s was measured in the zirconium fraction and assigned to ^{98}Zr based on auxiliary studies of the daughter ^{98}Nb in the reaction ^{98}Mo(n,p) [79].

Even shorter half-lives became accessible with the pulsed operation of the TRIGA reactor of the Institut für Kernchemie der Johannes Gutenberg-Universität in Mainz, Germany. This type of small reactor was designed and manufactured by General Atomics, and the acronym stands for Training, Research, Isotopes, General Atomic. The Mainz TRIGA reactor, also called the Forschungsreaktor Mainz (FRMZ) started operation in 1965. Denschlag irradiated uranyl nitrate three years later with pulses of 0.05 s and was able to perform the chemical separation of iodine in 2 s which was sufficient to extract the 18 ± 2s half-life of the ^{135}Te, the precursor of ^{135}I [87].

Subsequently, the pulsed operation of the Mainz TRIGA reactor enabled the discovery of fifteen additional short-lived fission fragments, including several delayed neutron emitters. Figure 6.6 demonstrates the need for the capability to measure shorter and shorter half-lives as more neutron-rich fission fragments are explored. It shows a section of the chart of nuclides indicating the half-lives of the radioactive isotopes. The new neodymium and molybdenum isotopes discovered by Trautmann et al. are shown in red.

	^{100}Ru	^{101}Ru	^{102}Ru	^{103}Ru 39.5d	^{104}Ru	^{105}Ru 4.4h	^{106}Ru 1.0a	^{107}Ru 4.2m	^{108}Ru 4.5m		
	^{99}Tc $2.1 \cdot 10^5$a	^{100}Tc 17s	^{101}Tc 14m	^{102}Tc 5.3s	^{103}Tc 50s	^{104}Tc 18m	^{105}Tc 7.7m	^{106}Tc 36s	^{107}Tc 21s	^{108}Tc 5.0s	
^{97}Mo	^{98}Mo	^{99}Mo 66.7h	^{100}Mo	^{101}Mo 14.6m	^{102}Mo 11.5m	^{103}Mo 62s	^{104}Mo 64s	^{105}Mo 48s	^{106}Mo 11s	^{107}Mo ~5s	^{108}Mo 1.5s
^{96}Nb 23.4h	^{97}Nb 74m	^{98}Nb 2.7s	^{99}Nb 15s	^{100}Nb 2.2s	^{101}Nb 7.1s	^{102}Nb 2.9s					
^{95}Zr 65.6d	^{96}Zr	^{97}Zr 16.8h	^{98}Zr 30.7s	^{99}Zr 2.0s	^{100}Zr 1.3s	^{101}Zr 2.0s	^{102}Zr 0.8s				
^{94}Y 20m	^{95}Y 10.9m	^{96}Y 2.3m	^{97}Y 1.9s								

Fig. 6.6 Section of the chart of nuclides. The half-lives decrease for increasingly neutron-rich isotopes. Stable isotopes are shown in *black* and the isotopes discovered by Trautmann et al. using the pulsed operation of the TRIGA reactor and fast chemical separation techniques [95] are shown in *red*

Two fission fragments were observed for the first time in Asia. Izak and Amiel discovered ^{130}Sn in 1971 at the Israel Research Reactor (IRR-1) located at the Soreq Nuclear Research Centre in Yavne, Israel. The reactor reached criticality for the first time in 1960. A $UO_2(NO_3)_2$ solution was irradiated for ten minutes and the tin fraction was separated within 10 s. Izak and Amiel assigned a half-life of 3.69 ± 0.10 min to ^{130}Sn on the basis of half-life measurements of four γ-ray lines [93].

Finally, ^{86}Se was discovered in 1973 at the Kyoto University Reactor (KUR) in Japan which had been operational since 1964. Tamai et al. irradiated uranyl nitrate for 10 s and γ-ray spectra of the selenium fraction were measured with a Ge(Li) detector. The half-life of seven γ-rays attributed to ^{86}Se were measured resulting in an average half-life of 16 s [98].

6.5 Online Separation Facilities at Reactors

As mentioned in Sect. 6.2 Kofoed-Hansen and Nielsen had developed the first isotope separation on-line (ISOL) system at the Copenhagen cyclotron in 1951 which they used in the discovery of ^{90}Rb and ^{91}Rb. The method was subsequently further developed at the Gustaf Werner Institute for Nuclear Chemistry in Uppsala, Sweden, before the fully dedicated ISOL facility ISOLDE (Isotope Separation On-Line DEtector) was installed at CERN in 1964 [144, 145] (see Sect. 10.4).

ISOL has two tremendous advantages over pure chemical separations. The separation is very fast and the mass separation simplifies the identification of new isotopes. Isotopes discovered with the ISOL method at reactors are listed in Table 6.5.

Table 6.5 Isotopes discovered in fission at reactors with ISOL: The submission date, first author and reference of the publication, as well as the name of the separator facility are listed

Date	First author	Ref.	Separator	Isotope
08/09/1971	Fogelberg	[107]	OSIRIS	^{120}Ag
09/16/1971	Amiel	[108]	SOLIS	^{94}Kr
02/05/1973	Borg	[109]	OSIRIS	^{133}Sn
03/08/1973	Kerek	[110]	OSIRIS	^{132}In
04/06/1973	Kerek	[111]	OSIRIS	^{130}In
01/10/1974	Grapengiesser	[112]	OSIRIS	^{75}Zn, ^{76}Zn, ^{79}Ga, ^{80}Ga, ^{126}In, ^{136}Te , ^{145}Ba
02/18/1974	Shalev	[113]	OSIRIS	^{134}Sn
06/21/1974	Fogelberg	[114]	OSIRIS	^{124}Cd
07/10/1974	Gunther	[115]	(OSTIS)[b]	^{96}Y[a]
08/27/1974	Kawase	[116]	OSIRIS	^{119}Ag
10/28/1974	Aleklett	[117]	OSIRIS	^{127}In, ^{128}In, ^{129}In
01/06/1975	Rudstam	[118]	OSIRIS	^{81}Ga, ^{82}Ga, ^{83}Ga
10/21/1975	Lund	[119]	OSIRIS	^{123}Ag, ^{131}In, ^{136}Sb
03/07/1977	Aleklett	[120]	OSIRIS	^{77}Zn, ^{78}Zn
12/12/1977	Shih	[121]	TRISTAN	^{122}Ag
01/03/1978	Wohn	[122]	OSTIS	^{97}Sr[a], ^{147}Cs, ^{147}Ba[a]
05/17/1978	Gartner	[123]	TRISTAN	^{126}Cd
08/14/1978	Koglin	[124]	OSTIS	^{100}Rb, ^{100}Sr[a], ^{148}Cs
08/23/1978	Engler	[125]	SOLIS	^{148}Ba, ^{147}La, ^{149}La
06/23/1982	Fogelberg	[126]	OSIRIS	^{121}Ag
08/05/1982	Gabelmann	[127]	OSTIS	^{148}La[a]
01/11/1983	Hill	[128]	TRISTAN	^{152}Pr
02/10/1983	Reeder	[129]	TRISTAN	^{123}Cd
06/23/1983	Wohn	[130]	TRISTAN	^{101}Sr, ^{101}Y
07/01/1983	Hill	[131]	TRISTAN	^{124}Ag
10/22/1984	Reeder	[132]	TRISTAN	^{75}Cu
11/20/1985	Mach	[133]	TRISTAN	^{156}Pm, ^{159}Sm, ^{160}Sm, ^{161}Eu
12/16/1985	Hill	[134]	TRISTAN	^{102}Sr
02/12/1986	Gill	[135]	TRISTAN	^{80}Zn
04/01/1986	Ekstrom	[136]	OSIRIS	^{79}Zn
04/21/1986	Gokturk	[137]	OSIRIS	^{125}Cd, ^{127}Cd, ^{128}Cd, ^{129}Cd
08/08/1986	Okano	[138]	KUR-ISOL	^{155}Nd
03/21/1988	Kratz	[139]	OSTIS[c]	^{93}Br, ^{94}Br
04/26/1988	Kawase	[140]	KUR-ISOL	^{154}Pr
12/22/1989	Tago	[141]	KUR-ISOL	^{152}Ce
11/16/1990	Omtvedt	[142]	OSIRIS	^{85}Ge
01/01/1993	Rudstam	[143]	OSIRIS	^{149}Ba, ^{150}La

[a]Populated by β^--decay
[b]At the time, the online separator was installed at the FRM in Munich before it was transferred to the high-flux reactor at Grenoble where it was named OSTIS
[c]Also measured at ISOLDE

The first discovery of a fission fragment at an ISOL facility installed at a reactor was ^{120}Ag in 1971 by Fogelberg, Bäcklin, and Nagarajan [107] at the R2-0 reactor of the Swedish Research Councils' Laboratory in Studsvik, Sweden. The layout of the On-line Separation of Isotopes at a Reactor In Studsvik (OSIRIS) is shown in Fig. 6.7. The name extended the tradition of acronyms for isotope separators from medieval (TRISTAN and ISOLDE) to Egyptian mythology. Between 1971 and 1993 a total of 34 new isotopes were identified at OSIRIS.

Isotopes were identified either directly by their β-decay (for example ^{133}Sn [109]) or by β-delayed neutron emission (e.g. ^{134}Sn [113]). Half-lives as short as 300 ms could be measured (e.g. ^{129}Cd with 270 ± 40 ms) [137]).

Seven new isotopes were first reported in the refereed literature by Grapengiesser, Lund, and Rudstam in 1974 in a survey article including the half-life measurements of 85 fission fragments with half-lives between 1 s and 10 min [112]. This article quotes several previous references, however, they were all either conference proceedings or internal reports.

In another review article later in the year Aleklett, Nyman, and Rudstam described the discovery of ^{127}In, ^{128}In, and ^{129}In among the observation of 50 neutron-rich fission products with overall somewhat shorter half-lives (between 0.8 and 30 s) [117].

The observation of ^{149}Ba and ^{150}La included in a compilation of delayed-neutron branching ratios of fission fragments published in Atomic Data and Nuclear Data Tables in 1993 by Rudstam, Aleklett, and Sihver [143] corresponded to the first

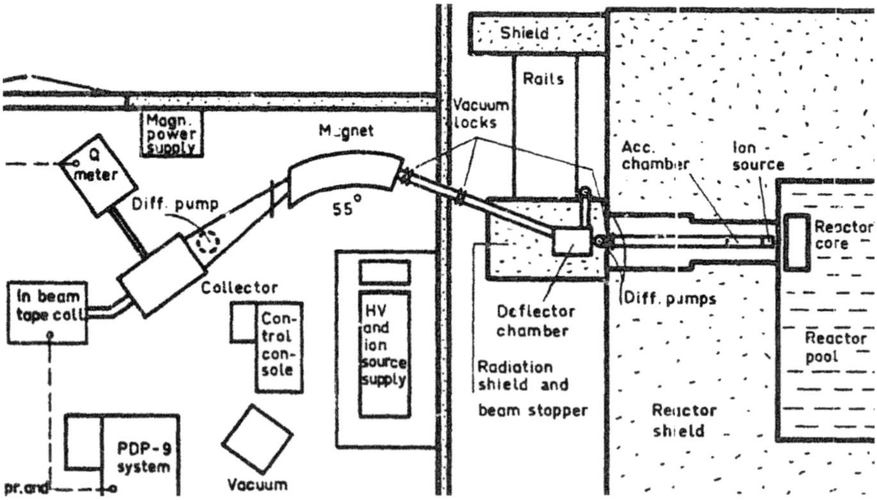

Fig. 6.7 Layout of the OSIRIS facility at the R2-0 reactor of the Swedish Research Councils' Laboratory in Studsvik, Sweden [146] (Reprinted from S. Borg et al., *Alpha decay of neutron-deficient osmium isotopes*, Nucl. Instrum. Meth. 91 (1971) 109. Copyright 1971, with permission from Elsevier.)

identification of these isotopes. They are the most neutron-rich isotopes discovered with OSIRIS.

The first on-line separator in Asia was installed at the Soreq Nuclear Research Centre in Yavne, Israel in 1969. Amiel et al. discovered ^{94}Kr in the thermal neutron fission of ^{235}U at the Israel Research Reactor 1 (IRR-1). Gaseous fission fragments were transferred by helium flow into the plasmatron ion source of the Soreq On-Line Isotope Separator and a 200 $\pm$ 10 ms half-life was measured for ^{94}Kr in 1971 [108]. The acronym SOLIS for the separator was only used later. Engler et al. discovered another three isotopes—^{148}Ba, ^{147}La, and ^{149}La—with SOLIS seven years later [125].

In 1974, Gunther et al. discovered ^{96}Y by separating rubidium fission fragments with an online separator at the Munich research reactor (FRM) [115]. After the development and testing of the separator, OSTIS (Online Separator für Thermisch Ionisierbare Spaltprodukte) was installed at the high flux reactor in Grenoble. Between 1978 and 1982 seven isotopes were discovered with OSTIS at Grenoble. In the three experiments, fission fragments of the alkali elements rubidium and cesium were analyzed and their decay and subsequent growth of their respective daughter nuclei was measured. As an example, Fig. 6.8 shows the measured growth and decay curve for mass 147 isotopes. ^{147}Cs and its daughter ^{147}Ba were discovered in this experiment by Wohn et al. in 1978 [122].

In 1988, Kratz et al. first identified ^{93}Br and ^{94}Br using spectroscopic information from experiments at OSTIS as well as experiments at CERN-ISOLDE [139].

The ISOL system which started the use of acronyms taken from mythology for separators was the Terrific Reactor Isotope Separator to Analyze Nuclei TRISTAN

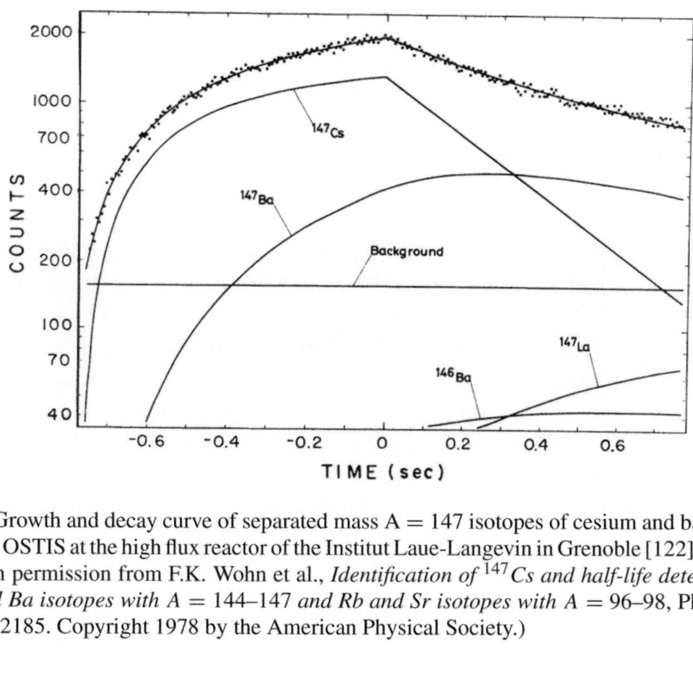

Fig. 6.8 Growth and decay curve of separated mass A = 147 isotopes of cesium and barium measured with OSTIS at the high flux reactor of the Institut Laue-Langevin in Grenoble [122] (Reprinted figure with permission from F.K. Wohn et al., *Identification of* ^{147}*Cs and half-life determinations for Cs and Ba isotopes with* $A = 144$–147 *and Rb and Sr isotopes with* $A = 96$–98, Phys. Rev. C 17 (1978) 2185. Copyright 1978 by the American Physical Society.)

(in response to ISOLDE). TRISTAN was actually the first on-line isotope separator facility to become operational at a reactor. It was initially located at the Ames Laboratory Research Reactor (ALRR) in Ames, Iowa, which became operational in 1965. TRISTAN produced the first short-lived gaseous fission fragments in 1966 [147].

However, it took eleven years before the first isotope was discovered with TRISTAN. Shih, Hill, and Williams separated and identified ^{122}Ag in 1977 [121]. Only one additional fission fragment (^{126}Cd [123]) was discovered with TRISTAN at the ALRR before the reactor was shut down at the end of 1977.

TRISTAN was then transported and reinstalled at the High Flux Beam Reactor (HFBR) at Brookhaven National Laboratory which had been in operation since 1965. Between 1983 and 1986 another twelve isotopes were discovered with TRISTAN at the HFBR. Isotopes were identified by their β-decay and subsequent γ-ray transitions in the corresponding daughter nuclei (e.g. ^{152}Pr [128]) and β-delayed neutron emission (e.g. ^{123}Cd [129]). The shortest half-life of a new isotope measured at TRISTAN was 68 ± 8 ms for ^{102}Sr [134].

Three neutron-rich isotopes of the rare earth elements neodymium, praseodymium, and cerium were discovered at the ISOL facility at the Kyoto University Reactor (KUR-ISOL). The ion source of the separator was coupled to a He-jet transport system which delivered the neutron-rich fission fragments from the irradiation position in the reactor. In three separate experiments between 1986 and 1990 the half-lives of ^{155}Nd [138], ^{154}Pr [140], and ^{152}Ce [141] were measured.

At the same time as ISOL systems were developed and installed at reactors, an alternative method was also pursued. Rather than thermalizing the fission fragments, online systems were designed where the fast fragment recoils from the reaction were directly guided into a mass separator. This method allowed for even faster separation times. While the thermalization times in ISOL systems is on the order of 100 ms, the separation in the recoil separators is on the order of 1 μs. The disadvantage of the recoil separators are the smaller acceptances of the devices. Isotopes discovered at reactors with recoil separators are listed in Table 6.6.

The first online recoil separator did not have specific acronyms. Eidens, Roeckl, and Armbruster discovered ^{97}Y, ^{99}Zr and ^{101}Nb in 1969 at the Forschungsreaktor Jülich 2 (FRJ-2 or DIDO) which was built in 1962. They used a modified version of the gas-filled online separator which had previously demonstrated the feasibility to separate short-lived fission fragments at the FRM reactor in Munich from 1961 through 1963 [148]. The extremely fast separation time allowed Grüter et al. a year later to identify an isomeric state of ^{98}Y with a a half-life of 0.83 ± 0.01 μs [149].

In yet another reference to mythology, the recoil mass separator built at the high-flux reactor of the Institut Laue-Langevin in Grenoble, France, was called LOHENGRIN. The high-flux reactor began operation in 1971 and LOHENGRIN was installed in 1975. Asghar et al. identified ^{91}Se, ^{99}Sr, ^{99}Y, ^{137}Te, and ^{138}Te by measuring the β-decay of the separated fragments as well as their delayed neutron emission [150]. Twelve years later, Armbruster et al. identified new neutron-rich nickel and copper isotopes event-by-event using the ΔE-E techniques which had been pioneered by Poskanzer et al. in 1966 in the spallation of uranium by protons (see Sect. 10.3) [157]. Figure 6.9 shows the scatter plot of particles identified in the

Table 6.6 Isotopes discovered in fission at reactors with recoil separators: The submission date, first author and reference of the publication, as well as the separator name are listed

Date	First author	Ref.	Separator	Isotope
03/20/1969	Eidens	[148]	Gas-filled separator	^{97}Y, ^{99}Zr, ^{101}Nb
10/23/1970	Grüter	[149]	Gas-filled separator	^{98}Y
02/17/1975	Asghar	[150]	LOHENGRIN	^{91}Se, ^{99}Sr, ^{99}Y, ^{137}Te, ^{138}Te
05/17/1976	Pfeiffer	[151]	LOHENGRIN	^{100}Y
01/17/1983	Shizuma	[152]	JOSEF	^{102}Y
07/20/1983	Shizuma	[153]	LOHENGRIN/JOSEF	^{105}Nb
04/03/1987	Graefenstedt	[154]	LOHENGRIN	^{103}Zr
04/27/1987	Armbruster	[155]	LOHENGRIN	^{70}Ni, ^{71}Ni, ^{72}Ni, ^{73}Ni, ^{74}Ni, ^{74}Cu, ^{76}Cu, ^{77}Cu
10/02/1989	Graefenstedt	[156]	LOHENGRIN	^{151}Pr

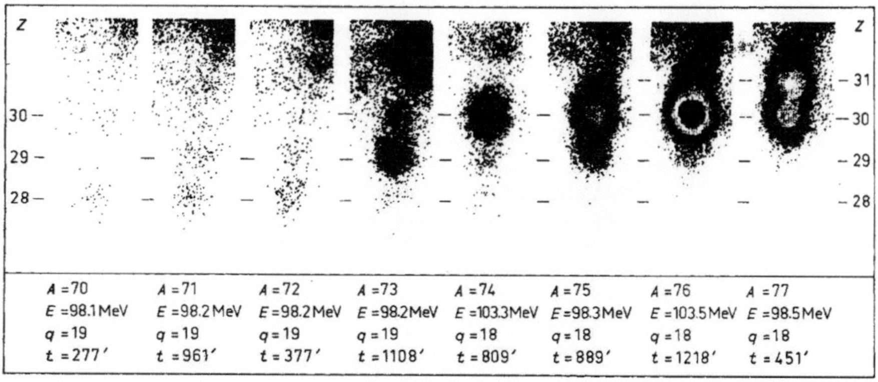

Fig. 6.9 Scatter plot of particles in the focal plane of LOHENGRIN identified by energy loss (*vertical axis*) and total energy (*horizontal axis*). The panel below the plot lists the mass number (A), energy (E), charge state (q), and data collection time (t) for the different setting. Events for the new nickel (^{70}Ni, ^{71}Ni, ^{72}Ni, ^{73}Ni, and ^{74}Ni) and copper (^{74}Cu, ^{76}Cu and ^{77}Cu) isotopes are visible. [155]

focal plane of LOHENGRIN [155] where vents for the new nickel (^{70}Ni, ^{71}Ni, ^{72}Ni, ^{73}Ni, and ^{74}Ni) and copper (^{74}Cu, ^{76}Cu and ^{77}Cu) isotopes are visible.

In 1983, Shizuma et al. discovered ^{102}Y at the Jülich DIDO reactor with the gas-filled separator JOSEF (Jülich On-line SEparator for Fission products) by measuring β-ray decay curves ($T_{1/2}$ = 360 ± 40 ms) and γ-ray spectra of the daughter nucleus ^{102}Zr [152]. Later in the same year, Shizuma et al. reported the observation of ^{105}Nb from experiments with JOSEF and LOHENGRIN [153].

References

1. L. Meitner, O.R. Frisch, Nature **143**, 239 (1939)
2. O. Hahn, F. Strassmann, Naturwissenschaften **27**, 11 (1939)
3. A.C. Wahl, At. Data Nucl. Data Tables **39**, 1 (1988)
4. O. Kofoed-Hansen, K.O. Nielsen, Phys. Rev. **82**, 96 (1951)
5. The Plutonium Project, J. Am. Chem. Soc. 68 (1946) 2411; Rev. Mod. Phys. 18 (1946) 513
6. C.D. Coryell, N. Sugarman (eds.), in *Radiochemical Studies: The Fission Products*. National Nuclear Energy Series IV, vol. 9 (McGraw-Hill, New York, 1951)
7. R.W. Nottorf, in *Radiochemical Studies: The Fission Products, Paper 77*. National Nuclear Energy Series IV, vol. 9 (McGraw-Hill, New York, 1951), p. 682
8. L.E. Glendenin, *Radiochemical Studies: The Fission Products, Paper 115*. National Nuclear Energy Series IV, vol. 9 (McGraw-Hill, New York, 1951), p. 849
9. E.P. Steinberg, D. Engelkemeir, in *Radiochemical Studies: The Fission Products, Paper 54*. National Nuclear Energy Series IV, vol. 9 (McGraw-Hill, New York, 1951), p. 566
10. S. Katcoff, B. Finkle, in *Radiochemical Studies: The Fission Products, Paper 83*. National Nuclear Energy Series IV, vol. 9 (McGraw-Hill, New York, 1951), p. 705
11. J.A. Seiler, in *Radiochemical Studies: The Fission Products, Paper 119*. National Nuclear Energy Series IV, vol. 9 (McGraw-Hill, New York, 1951), p. 860
12. S. Katcoff, in *Radiochemical Studies: The Fission Products, Paper 172*. National Nuclear Energy Series IV, vol. 9 (McGraw-Hill, New York, 1951), p. 1147
13. L. Winsberg, in *Radiochemical Studies: The Fission Products, Paper 197*. National Nuclear Energy Series IV, vol. 9 (McGraw-Hill, New York, 1951), p. 1292
14. L. Winsberg, in *Radiochemical Studies: The Fission Products, Paper 198*. National Nuclear Energy Series IV, vol. 9 (McGraw-Hill, New York, 1951), p. 1311
15. C.R. Dillard et al., in *Radiochemical Studies: The Fission Products, Paper 68*. National Nuclear Energy Series IV, vol. 9 (McGraw-Hill, New York, 1951), p. 624
16. H. Gest, R.R. Edwards, in *Radiochemical Studies: The Fission Products, Paper 171*. National Nuclear Energy Series IV, vol. 9 (McGraw-Hill, New York, 1951), p. 1144
17. L. Winsberg, in *Radiochemical Studies: The Fission Products, Paper 198*. National Nuclear Energy Series IV, vol. 9 (McGraw-Hill, New York, 1951), p. 1302
18. J.M. Siegel, L.E. Glendenin, in *Radiochemical Studies: The Fission Products, Paper 53*. National Nuclear Energy Series IV, vol. 9 (McGraw-Hill, New York, 1951), p. 549
19. C.W. Stanley, L.E. Glendenin, in *Radiochemical Studies: The Fission Products, Paper 134*. National Nuclear Energy Series IV, vol. 9 (McGraw-Hill, New York, 1951), p. 947
20. L. Winsberg, in *Radiochemical Studies: The Fission Products, Paper 196*. National Nuclear Energy Series IV, vol. 9 (McGraw-Hill, New York, 1951), p. 1284
21. L.E. Glendenin, in *Radiochemical Studies: The Fission Products, Paper 159*. National Nuclear Energy Series IV, vol. 9 (McGraw-Hill, New York, 1951), p. 1092
22. W.E. Grummitt, G. Wilkinson, Nature **158**, 163 (1946)
23. J.A. Marinsky, L.E. Glendenin, C.D. Coryell, J. Am. Chem. Soc. **69**, 2781 (1947)
24. W.E. Grummitt, G. Wilkinson, Nature **161**, 520 (1948)
25. N. Sugarman, J. Chem. Phys. **17**, 11 (1949)
26. S. Katcoff, J.A. Miskel, C.W. Stanley, Phys. Rev. **74**, 631 (1948)
27. R.J. Hayden, Phys. Rev. **74**, 650 (1948)
28. C.W. Stanley, S. Katcoff, J. Chem. Phys. **17**, 653 (1949)
29. N. Sugarman, Phys. Rev. **75**, 1473 (1949)
30. N. Sugarman, H. Richter, J. Chem. Phys. **18**, 174 (1950)
31. E.P. Steinberg, L.E. Glendenin, Phys. Rev. **78**, 624 (1950)
32. S. Katcoff, O.A. Schaeffer, J.M. Hastings, Phys. Rev. **82**, 688 (1951)
33. J.W. Barnes, A.J. Freedman, Phys. Rev. **84**, 365 (1951)
34. M.S. Livingston, H.A. Bethe, Rev. Mod. Phys. **9**, 245 (1937)
35. G.T. Seaborg, Rev. Mod. Phys. **16**, 1 (1944)

36. R.I. Powers, A.F. Voigt, Phys. Rev. **79**, 175 (1950)
37. I. Curie, P. Savitch, J. Phys. Radium **9**, 355 (1938)
38. O. Hahn, F. Strassmann, Naturwissenschaften **30**, 324 (1942)
39. N.E. Ballou, W.H. Burgus, in *Radiochemical Studies: The Fission Products, Paper 177*. National Nuclear Energy Series IV, vol. 9 (McGraw-Hill, New York, 1951), p. 1168
40. M.W. Rosenthal, *An Account of Oak Ridge National Laboratory's Thirteen Nuclear Reactors*, ORNL/TM-2009/181, August 2009 (revised March 2010)
41. N. Feather, Math. Proc. Camb. Philos. Soc. **34**, 599 (1938)
42. A.H. Snell, Phys. Rev. **52**, 1007 (1937)
43. G.T. Seaborg, J.J. Livingood, G. Friedlander, Phys. Rev. **59**, 320 (1941)
44. A. Vanden, Bosch. Physica **19**, 374 (1953)
45. D.R. Wiles, C.D. Coryell, Phys. Rev. **96**, 696 (1954)
46. J. Flegenheimer, W. Seelmann-Eggebert, Z. Naturforsch. **9a**, 806 (1954)
47. G.B. Baró, P. Rey, W. Seelmann-Eggebert, Z. Naturforsch. **10a**, 81 (1955)
48. J. Flegenheimer, G.B. Baró, A. Medina, Z. Naturforsch. **10a**, 798 (1955)
49. I. Fränz, J. Rodriguez, R. Radicella, Z. Naturforsch. **11a**, 1037 (1956)
50. D.R. Wolfsberg et al., J. Inorg. Nucl. Chem. **12**, 201 (1960)
51. A.C. Wahl et al., Phys. Rev. **126**, 1112 (1962)
52. P.O. Aronsson, G. Skarnemark, M. Skarestad, J. Inorg. Nucl. Chem. **36**, 1689 (1974)
53. D. Strominger, J.M. Hollander, G.T. Seaborg, Rev. Mod. Phys. **30**, 585 (1958)
54. P.O. Aronsson et al., J. Inorg. Nucl. Chem. **36**, 2397 (1974)
55. A.A. Caretto Jr., S. Katcoff, Phys. Rev. **89**, 1267 (1953)
56. S.S. Markowitz, W. Bernstein, S. Katcoff, Phys. Rev. **93**, 178 (1954)
57. A.C. Pappas, D.R. Wiles, J. Inorg. Nucl. Chem. **2**, 69 (1956)
58. G.J. Perlow, A.F. Stehney, Phys. Rev. **113**, 1269 (1959)
59. J.D. Knight et al., J. Inorg. Nucl. Chem. **10**, 183 (1959)
60. R.P. Schuman, E.H. Turk, R.L. Heath, Phys. Rev. **115**, 185 (1959)
61. J.E. Sattizahn, J.D. Knight, J. Inorg. Nucl. Chem. **12**, 206 (1960)
62. C.J. Orth, R.K. Smith, J. Inorg. Nucl. Chem. **15**, 4 (1960)
63. K. Fritze, T.J. Kennett, Can. J. Phys. **38**, 1614 (1960)
64. K. Fritze, T.J. Kennett, W.V. Prestwich, Can. J. Phys. **39**, 67 (1961)
65. B.J. Dropesky, C.J. Orth, J. Inorg. Nucl. Chem. **24**, 1301 (1962)
66. P. Kienle et al., Naturwissenschaften **49**, 294 (1962)
67. P. Kienle et al., Naturwissenschaften **49**, 295 (1962)
68. K. Fritze, Can. J. Chem. **40**, 1344 (1962)
69. P. Kienle et al., Radiochim. Acta **1**, 84 (1963)
70. C.E. Bemis, G.E. Gordon, C.D. Coryell, J. Inorg. Nucl. Chem. **26**, 213 (1964)
71. A.E. Greendale, D.L. Love, Anal. Chem. **35**, 1712 (1963)
72. D.C. Hoffman, W.R. Daniels, J. Inorg. Nucl. Chem. **26**, 1769 (1964)
73. A. von Baeckmann, H. Feuerstein, Radiochim. Acta **4**, 111 (1965)
74. H.V. Weiss, Phys. Rev. **139**, B304 (1965)
75. P.O. Strom et al., Phys. Rev. **144**, 984 (1966)
76. P. del Marmol, M. Nève de Mévergnies, J. Inorg. Nucl. Chem. **29**, 273 (1967)
77. K. Fritze, K. Griffiths, Radiochim. Acta **7**, 59 (1967)
78. K. Griffiths, K. Fritze, Z. Anal, Chem. **226**, 122 (1967)
79. K. Hübenthal, C. R. Acad. Sci. **264**, 1468 (1967)
80. L. Tomlinson, M.H. Hurdus, Phys. Lett. B **25**, 545 (1967)
81. L. Tomlinson, M.H. Hurdus, J. Inorg. Nucl. Chem. **30**, 1995 (1968)
82. L. Wish, Phys. Rev. **172**, 1262 (1968)
83. H.V. Weiss, J.L. Elzie, J.M. Fresco, Phys. Rev. **172**, 1269 (1968)
84. H.V. Weiss, J.M. Fresco, W.L. Reichert, Phys. Rev. **172**, 1266 (1968)
85. P. del Marmol, J. Inorg. Nucl. Chem. **30**, 2873 (1968)
86. J.D. Hastings, D.E. Troutner, R.L. Ferguson, Radiochim. Acta **11**, 51 (1969)
87. H.O. Denschlag, J. Inorg. Nucl. Chem. **31**, 1873 (1969)

88. H.V. Weiss et al., Phys. Rev. **188**, 1893 (1969)
89. P. del Marmol, D.C. Perricos, J. Inorg. Nucl. Chem. **32**, 705 (1970)
90. A. Wakat, C. Griffin, Radiochem. Radioanal. Lett. **2**, 351 (1969)
91. J.V. Kratz, G. Herrmann, J. Inorg. Nucl. Chem. **32**, 3713 (1970)
92. L. Tomlinson, M.H. Hurdus, J. Inorg. Nucl. Chem. **33**, 3609 (1971)
93. T. Izak, S. Amiel, J. Inorg. Nucl. Chem. **34**, 1469 (1972)
94. H.-D. Schussler, G. Herrmann, Radiochim. Acta **18**, 123 (1972)
95. N. Trautmann et al., Radiochim. Acta **18**, 86 (1972)
96. P. del Marmol, P. Fettweis, Nucl. Phys. A **194**, 140 (1972)
97. J.V. Kratz, H. Franz, G. Herrmann, J. Inorg. Nucl. Chem. **35**, 1407 (1973)
98. T. Tamai et al., Inorg. Nucl. Chem. Lett. **9**, 1145 (1973)
99. K.-L. Kratz, G. Herrmann, Nucl. Phys. A **229**, 179 (1974)
100. K.-L. Kratz, W. Lauppe, G. Herrmann, Inorg. Nucl. Chem. Lett. **11**, 331 (1975)
101. H. Ahrens et al., Phys. Rev. C **14**, 211 (1976)
102. G. Franz, G. Herrmann, Inorg. Nucl. Chem. Lett. **11**, 857 (1975)
103. N. Trautmann et al., Phys. Rev. C **13**, 872 (1976)
104. G.B. Cook, British Atomic Energy Report AERE C/R-729 June, 1951 (unpublished)
105. A.C. Pappas, Laboratory for Nuclear Science, Technical Report No. 63, Massachusetts Institute of Technology (September 1953)
106. A.C. Pappas, Z. Elektrochem. **58**, 620 (1954)
107. B. Fogelberg, A. Bäcklin, T. Nagarajan, Phys. Lett. B **36**, 334 (1971)
108. S. Amiel et al., Phys. Rev. C **5**, 270 (1972)
109. S. Borg, G.B. Holm, B. Rydberg, Nucl. Phys. A **212**, 197 (1973)
110. A. Kerek et al., Phys. Lett. B **44**, 252 (1973)
111. A. Kerek et al., Nucl. Phys. A **209**, 520 (1973)
112. B. Grapengiesser, E. Lund, G. Rudstam, J. Inorg. Nucl. Chem. **36**, 2409 (1974)
113. S. Shalev, G. Rudstam, Nucl. Phys. A **230**, 153 (1974)
114. B. Fogelberg, T. Nagarajan, B. Grapengiesser, Nucl. Phys. A **230**, 214 (1974)
115. H. Gunther et al., Nucl. Phys. A **242**, 56 (1975)
116. Y. Kawase et al., Nucl. Phys. A **241**, 237 (1975)
117. K. Aleklett, G. Nyman, G. Rudstam, Nucl. Phys. A **246**, 425 (1975)
118. G. Rudstam, E. Lund, Phys. Rev. C **13**, 321 (1976)
119. E. Lund, G. Rudstam, Phys. Rev. C **13**, 1544 (1976)
120. K. Aleklett et al., Nucl. Phys. A **285**, 1 (1977)
121. L.L. Shih, J.C. Hill, S.A. Williams, Phys. Rev. C **17**, 1163 (1978)
122. F.K. Wohn et al., Phys. Rev. C **17**, 2185 (1978)
123. M.L. Gartner, J.C. Hill, Phys. Rev. C **18**, 1463 (1978)
124. E. Koglin et al., Z. Phys. A **288**, 319 (1978)
125. G. Engler et al., Phys. Rev. C **19**, 1948 (1979)
126. B. Fogelberg, P. Hoff, Nucl. Phys. A **391**, 445 (1982)
127. H. Gabelmann et al., Z. Phys. A **308**, 359 (1982)
128. J.C. Hill, H. Yamamoto, A. Wolf, Phys. Rev. C **27**, 2857 (1983)
129. P.L. Reeder, R.A. Warner, R.L. Gill, Phys. Rev. C **27**, 3002 (1983)
130. F.K. Wohn et al., Phys. Rev. Lett. **51**, 873 (1983)
131. J.C. Hill et al., Phys. Rev. C **29**, 1078 (1984)
132. P.L. Reeder et al., Phys. Rev. C **31**, 1029 (1985)
133. H. Mach et al., Phys. Rev. Lett. **56**, 1547 (1986)
134. J.C. Hill et al., Phys. Rev. C **33**, 1727 (1986)
135. R.L. Gill et al., Phys. Rev. Lett. **56**, 1874 (1986)
136. B. Ekström et al., Phys. Scr. **34**, 614 (1986)
137. H. Gokturk et al., Z. Phys. A **324**, 117 (1986)
138. K. Okano, Y. Kawase, K. Aoki, Radiochim. Acta **40**, 57 (1986)
139. K.-L. Kratz et al., Z. Phys. A **330**, 229 (1988)
140. Y. Kawase, K. Okano, Z. Phys. A **330**, 231 (1988)

141. I. Tago, Y. Kawase, K. Okano, Z. Phys. A **335**, 477 (1990)
142. J.P. Omtvedt et al., Z. Phys. A **338**, 241 (1991)
143. G. Rudstam, K. Aleklett, L. Sihver, At. Data Nucl. Data Tables **53**, 1 (1993)
144. G. Andersson, G. Rudstam, Nucl. Instrum. Meth. **29**, 93 (1964)
145. G. Rudstam, Nucl. Instrum. Meth. **38**, 282 (1965)
146. S. Borg et al., Nucl. Instrum. Meth. **91**, 109 (1971)
147. J.R. McConnell, W.L. Talbert, Nucl. Instrum. Meth. **128**, 227 (1975)
148. J. Eidens, E. Roeckl, P. Armbruster, Nucl. Phys. A **141**, 289 (1970)
149. J.W. Grüter et al., Phys. Lett. B **33**, 474 (1970)
150. M. Asghar et al., Nucl. Phys. A **247**, 359 (1975)
151. B. Pfeiffer et al., J. Phys. (Paris) **38**, 9 (1977)
152. K. Shizuma et al., Phys. Rev. C **27**, 2869 (1983)
153. K. Shizuma et al., Z. Phys. A **315**, 65 (1984)
154. M. Graefenstedt et al., Z. Phys. A **327**, 383 (1987)
155. P. Armbruster et al., Europhys. Lett. **4**, 793 (1987)
156. M. Graefenstedt et al., Z. Phys. A **336**, 247 (1990)
157. A.M. Poskanzer et al., Phys. Rev. Lett. **17**, 1271 (1966)

Chapter 7
Neutron-Induced Reactions

7.1 Thermal and Fast Neutrons

Neutron interactions with all nuclides—with the exception of nuclides located at the neutron drip-line—are exothermic, i.e. during the reaction the nuclides absorb the neutron and gain energy. In addition, as uncharged particles, neutrons do not have to overcome a Coulomb barrier, thus capture reactions can proceed at the lowest energies. This is actually the process by which all elements (and most isotopes) beyond iron were formed in the universe by either the slow neutron (s-process) or the rapid neutron (r-process) capture process.

Neutron capture reactions are predominantly performed at reactors where thermal neutrons ($\sim$0.025 eV) are readily available (Sect. 7.2). Fast neutrons with energies above about 1 MeV can interact with individual neutrons and protons in the target and knock out a single or sometimes multiple nucleons. These knock out or nucleon removal reactions together with charge exchange (n,p) reactions were primarily performed with fast neutrons in order to discover new isotopes as described in Sect. 7.3.

The present chapter discusses the discovery of isotopes up to uranium which were produced by neutron capture, neutron removal or charge exchange reactions. These reactions were used to discover many neutron-rich isotopes close to the valley of stability. Isotopes of transuranium elements and isotopes discovered by neutron-induced fission have already been discussed in Chap. 5, Sects. 5.2 and 5.3, and Chap. 6, respectively.

7.2 Reactor Neutrons

Thermal neutrons are naturally available in nuclear reactors. The neutrons emitted by nuclear fuel (uranium or plutonium) have to be moderated to thermal energies to maintain the chain reactions. Samples can be placed inside the reactor core for the irradiation with thermal neutrons. Activated isotopes are next to stable isotopes

M. Thoennessen, *The Discovery of Isotopes*,
DOI 10.1007/978-3-319-31763-2_7

as most are produced in single neutron capture reactions and thus the half-lives are typically long and samples can be measured offline. Also chemical separations are not critical as neutron capture reactions are the most dominant reactions and do not change the atomic number (Z) so the element is known. The number of available isotopes to be discovered is of course limited and the last isotope discovered by neutron capture in a reactor was produced in 1966. Isotopes discovered by neutron irradiations in reactors as part of the plutonium project are listed in Table 7.1 while other discoveries are listed in Table 7.2.

The research within the plutonium project was classified during Second World War so the results were only published after the War. Several reactors built as part of the plutonium project were used to irradiate various materials with neutrons to measure capture cross-sections.

Turkevich et al. assigned the mass of ^{137}Cs by irradiating natural xenon in the Argonne heavy water pile CP-3 and analyzed the subsequent decay of cesium isotopes. The 1945 classified report of this work was included in the fission fragment volume of the National Nuclear Energy Series (NNES) which was only published in 1951 [1]. One of only two charge exchange reactions used to produce new isotopes in a reactor was performed by Swartout et al. in 1946. They irradiated enriched ^{65}Cu

Table 7.1 Isotopes discovered by neutron irradiations in reactors as part of the plutonium project: The submission date, first author and reference of the publication, as well as the reactor name and reaction are listed

Date	First author	Ref.	Reactor	Reaction	Isotope
01/01/1945[a]	Turkevich	[1]	CP-3	(n,γ)	^{137}Cs[b]
07/31/1946	Swartout	[2]	Clinton	(n,p)	^{65}Ni
07/31/1946	Motta	[3]	Clinton	(n,γ)	^{97}Tc[c]
12/06/1946	Inghram	[4]	CP-2[d]	(n,γ)	^{154}Eu
02/11/1947	Seaborg	[5]	Clinton[d]	(n,γ)	^{233}U[b]
02/20/1947	Seren	[6]	CP-2	(n,γ)	^{51}Ti
03/08/1947	Friedlander	[7]	CP-2[d]	(n,γ)	^{75}Se
03/27/1947	Inghram	[8]	Hanford	(n,γ)	^{145}Sm, ^{151}Sm, ^{153}Gd
06/19/1947	Hagemann	[9]	CP-2	(n,γ)	^{229}Th[e], ^{225}Ra[e], ^{225}Ac[b], ^{221}Fr[e], ^{217}At[e], ^{213}Bi[e], ^{213}Po[b]
08/11/1947	Katcoff	[10]	Clinton	(n,γ)	^{131}Ba
04/02/1951	Ghiorso	[11]	Clinton[d]	(n,γ)	^{236}U

[a] Date of classified plutonium project report
[b] Populated by β^--decay
[c] Populated by β^+-decay
[d] Probably reactor used for irradiation but not specified in publication
[e] Populated by α-decay

targets at the Clinton pile at Oak Ridge and observed the decay of ^{65}Ni in chemically separated nickel [2].

The long half-lives of the isotopes populated in neutron capture reactions allowed the identification by mass spectroscopy. Inghram and Hayden identified ^{154}Eu with their mass spectrograph after irradiation of europium in one of the piles at Argonne National Laboratory (CP-2 or CP-3) in 1946 [4]. A few months later Inghram, Hayden, and Hess Jr. analyzed samarium samples which had been irradiated for a long time in one of the piles at the Hanford site near Richland, Washington, and identified mass lines for ^{145}Sm, ^{151}Sm, and ^{153}Gd in the spectrometer [8]. These isotopes were the only isotopes discovered by irradiation at Hanford.

The discovery of ^{233}U was probably one of the shortest papers reporting the observation of a new isotope in the refereed literature.[1] The paper entitled "Nuclear Properties of U^{233}: A New Fissionable Isotope of Uranium" was published in 1947 in Physical Review: "The bombardment of thorium with slow neutrons produces Th^{233} (by the reaction $Th^{232}(n,\gamma)Th^{233}$), which emits beta-particles and has a half-life of 23.5 min. The daughter of Th^{233} is the 27.4-day beta-emitting Pa^{233} which in turn decays to U^{233}. We have measured the radioactive and the fission properties of U^{233}. Our measurements on a sample of U^{233} weighing 3.8 μg show that this isotope undergoes fission with neutrons. The same result was obtained in a check experiment with another sample of U^{233} weighing 0.8 microgram" [5]. That was the whole paper! The footnote stated: "This paper was mailed from Berkeley, California, to the "Uranium Committee" in Washington, D.C. on April 14, 1942. The experimental work was done during 1941 and the early part of 1942. It is being published in the open literature now for historical purposes, with the original text somewhat changed, by omissions, in order to conform to present declassification standards. The information covered in this document will appear in Division IV of the MPTS, as part of a contribution of the University of California."

Hagemann et al. from Argonne National Laboratory followed the decay of ^{233}U as they realized that it corresponded to the still missing $A = 4n + 1$ radioactive decay chain which does not occur in nature. They identified five α-emitters and two β-decays of the chain in 1947 [9]. Essentially simultaneously, English et al. measured the same chain at Chalk River, Canada [12]. They submitted their results only a day later and it was published in the same issue of Physical Review immediately following the paper by Hagemann et al. The decay scheme reported by English et al. is shown in Fig. 7.1.

^{236}U was the last isotope that was discovered as part of the plutonium project. Ghiorso et al. observed the α-decay of ^{235}U samples that had been irradiated strongly with neutrons and from the intensity deduced a half-life of 2×10^7 y [11].

The first isotope (^{131}Cs) discovered at one of the plutonium project reactors after the Second World War was published by Yu, Gideon, and Kurbatov from Ohio State University in 1947. They observed a strong intensity of the decay of ^{131}Ba after irradiation of barium in the Clinton pile while also measuring a weak intensity following the (d,p) reaction induced by deuterons from a cyclotron [13]. A few months

[1] See also the article on the discovery of spontaneous fission in Chap. 13.

Table 7.2 Isotopes discovered by neutron irradiations in reactors outside of the plutonium project: The submission date, first author and reference of the publication, as well as the reactor name and reaction are listed

Date	First author	Ref.	Reactor	Reaction	Isotope
02/23/1947	Yu	[13]	Clinton	$(n,\gamma)^a$	$^{131}Cs^b$
06/06/1947	Parker	[14]	Clinton	(n,γ)	^{148}Pm
06/24/1948	DeBenedetti	[15]	Clinton	(n,γ)	$^{171}Tm^c$
07/09/1948	Inghram	[16]	CP-2	(n,γ)	$^{137}La^b$
03/03/1949	Butement	[17]	GLEEP[d]	(n,γ)	^{159}Gd, ^{161}Gd, $^{161}Tb^c$
09/06/1949	Ketelle	[18]	Clinton	$(2n,\gamma)$	^{166}Dy
03/13/1950	Hagemann	[19]	CP-2	(n,γ)	$^{209}Tl^c$
05/05/1950	der Mateosian	[20]	CP-3	(n,γ)	^{49}Ca
06/29/1950	Reynolds	[21]	CP-2	(n,γ)	^{81}Kr
07/12/1950	Brosi	[22]	Clinton	(n,p)	^{39}Ar
07/27/1950	Lindner	[23]	Clinton	$(2n,\gamma)$	^{188}W
09/28/1950	Brosi	[24]	Clinton	(n,γ)	^{59}Ni, ^{63}Ni
11/07/1950	Butement	[25]	GLEEP[d]	(n,γ)	$^{145}Pm^b$
01/23/1951	Butement	[26]	GLEEP[d]	(n,γ)	^{159}Dy
07/02/1951	Lindner	[27]	Clinton	$(2n,\gamma)$	^{194}Os
12/06/1951	Depocas	[28]	NRX	(n,γ)	$^{229}Ac^c$
02/04/1952	Rutledge	[29]	CP-3	(n,γ)	$^{151}Pm^c$
04/24/1952	Boyd	[30]	Clinton	(n,γ)	^{100}Tc
05/16/1952	Katcoff	[31]	BGRR	$(2n,\gamma)$	^{42}Ar
04/21/1953	Knight	[32]	Water boiler[d]	$(2n,\gamma)$	^{240}U
05/04/1953	Butler	[33]	NRX	(n,γ)	^{227}Ra
05/26/1953	Harbottle	[34]	Clinton	(n,γ)	^{204}Tl
10/22/1954	LeBlanc	[35]	CP-2[d]	(n,γ)	^{71}Zn
03/27/1956	Nethaway	[36]	MTR	$(2n,\gamma)$	^{172}Er, $^{172}Tm^c$
06/25/1956	Bisi	[37]	GLEEP[d]	(n,γ)	^{169}Er
11/14/1956	Roy	[38]	NRX	$(2n,\gamma)$	^{200}Pt
07/12/1960	Naumann	[39]	MTR	(n,γ)	$^{157}Tb^b$
10/11/1960	Hutchin	[40]	MTR	$(2n,\gamma)$	^{182}Hf
04/16/1962	Facetti	[41]	BONUS	(n,α)	^{201}Pt
07/26/1966	Wahlgren	[42]	Savannah River	$(2n,\gamma)$	^{162}Gd

[a] Also observed in (d,p) reactions in a cyclotron
[b] Populated by β^+-decay
[c] Populated by β^--decay
[d] Probably reactor used for irradiation but not specified in publication

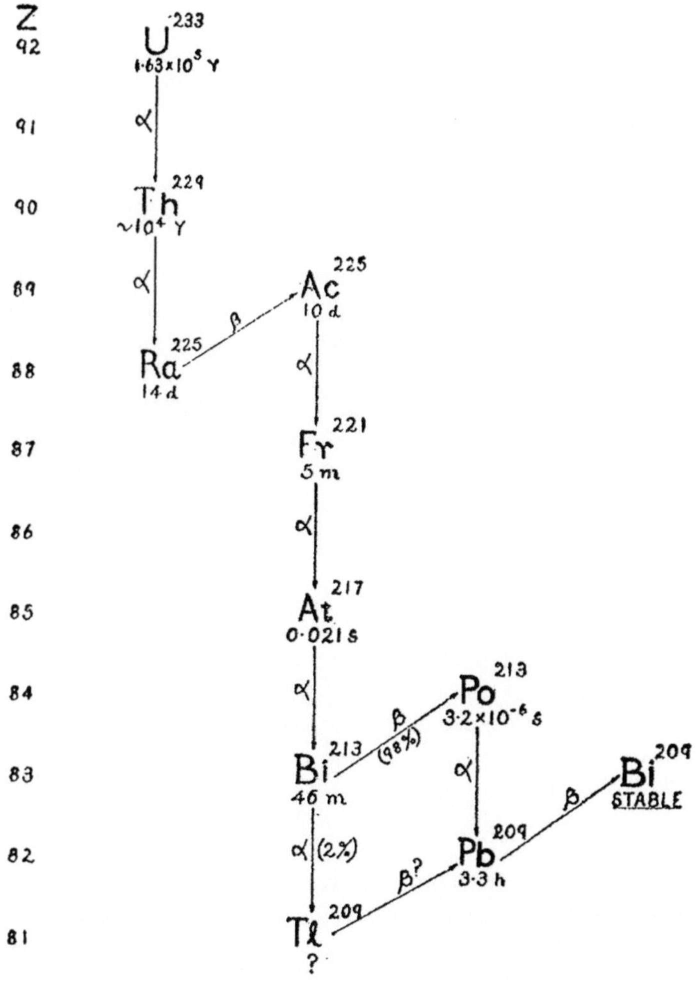

Fig. 7.1 Decay path of the neptunium (A = 4n+1) chain measured by English et al. [12] (Reprinted figure with permission from A.C. English et al., *The* (4*n* + 1) *Radioactive Series*, Phys. Rev. 72 (1947) 253. Copyright 1947 by the American Physical Society.)

later, Parker et al. irradiated a promethium sample—which itself was produced by fission—in the Clinton pile and identified ^{148}Pm in a mass spectrograph [14].

In 1948, DeBenedetti and McGowan developed a new coincidence detection technique to measure short-lived isomeric states. Radioactive sources of ^{171}Er were produced by neutron capture in the Clinton Pile and delayed coincidences between the β-particles and γ-rays were measured with two Geiger counters. They observed a 2.5 μs decay in the daughter nucleus ^{171}Tm which corresponded to the first measurement of a property of this isotope [15].

The only isotopes discovered at a reactor in Europe were produced at the research reactors of the Atomic Energy Research Establishment in Harwell, England, between 1949 and 1956. At the time two reactors were operating, the Graphite Low Energy Experimental Pile (GLEEP) and the British Experimental Pile (BEPO). In 1949, Butement irradiated gadolinium oxide and identified ^{159}Gd, ^{161}Gd, and ^{161}Tb [17]. One year later Butement irradiated samarium for 28 days and after a waiting period of 206 days, he determined a half-life of approximately 30 y for ^{145}Pm populated by the decay of ^{145}Sm ($T_{1/2} = 410$ d) [25]. Shortly after Butement also reported the observation of ^{159}Dy. In addition to the neutron irradiation of dysprosium the isotope was also populated in the reaction ^{159}Tb(d,2n) [26]. In 1956, Bisi, Terrani, and Zappa identified ^{169}Er measuring the β-decay following irradiation of spectroscopically pure erbium [37].

The first sequential two-neutron capture reaction populating a new isotope was observed in 1949 at Oak Ridge National Laboratory. Ketelle attributed an 80 h half-life to the decay of ^{166}Dy from ^{164}Dy by irradiating a highly purified dysprosium target in the Clinton pile [18]. The next two new isotopes populated by two-neutron capture reactions (^{188}W [23] and ^{194}Os [27]) were also irradiated in the Clinton pile in 1950 and 1951, respectively.

In 1950, Hagemann et al. discovered the β-decay of ^{209}Tl which corresponded to a missing branch $^{213}\text{Bi} \rightarrow {}^{209}\text{Tl} \rightarrow {}^{209}\text{Pb}$ of the $A = 4n + 1$ radioactive decay chain [19]. In the 1947 decay path published by English et al. shown in Fig. 7.1 ^{209}Tl and it's decay were labeled with question marks [12].

The second and only other charge-exchange reaction producing a new isotope in a reactor was reported in 1950 by Brosi, Zeldes, and Ketelle also in the irradiation of samples in the Clinton pile. They observed a long-lived activity (>15 y) in the argon fraction following irradiation of potassium which was assigned to the reaction ^{39}K(n,p)^{39}Ar [22].

Three heavy nuclides were identified between 1951 and 1957 after neutron irradiations in the National Research eXperimental (NRX) pile of the National Research Council of Canada at Chalk River. Depocas and Harvey irradiated ^{228}Ra and measured a 66 ± 5 min half-life of ^{229}Ac following the decay of ^{229}Ra [28], and Butler and Adam identified the 41.2 ± 0.2 β-decay of ^{227}Ra by irradiating ^{226}Ra [33]. ^{200}Pt was discovered by L.P. Roy, L.-C. Roy, and Merritt by sequential two-neutron capture of ^{198}Pt [38].

Katcoff irradiated ^{40}Ar at the Brookhaven Graphite Research Reactor (BGRR) for twelve days in 1952 and observed ^{42}Ar produced by two-neutron capture. He identified ^{42}Ar by measuring the decay of ^{42}K daughter activity after successive extractions over a period of 400 days [31].

In 1953, Knight et al. measured the half-life and energy of the β-decay of ^{240}U populated by two-neutron capture on ^{238}U at Los Alamos National Laboratory [32]. Most of the fission fragment measurements discussed in Sect. 6.2 were performed at the water boiler reactor (also called Los Alamos homogeneous pile). This reactor was probably also used in the irradiation of ^{238}U. With their measured β-decay energies of ^{240}U and ^{240}Np and the network of known β- and α-decay energies as well as

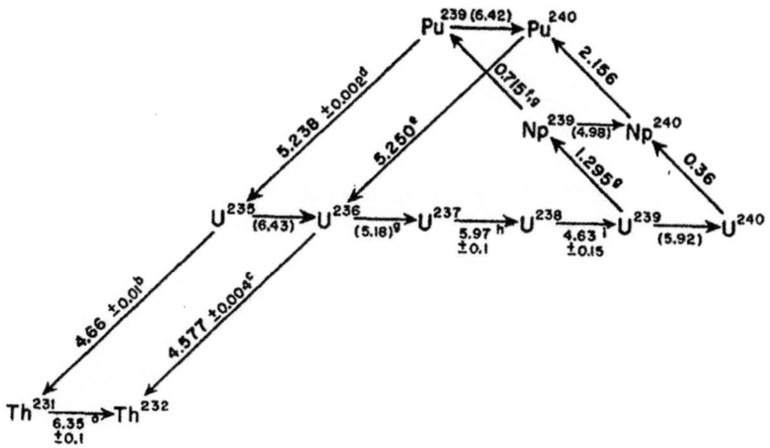

Fig. 7.2 Network relationships between α-decay, β-decay and neutron separation energies for thorium, uranium and plutonium isotopes. The numbers in brackets were calculated. The neutron separation energies of the new isotopes ^{240}U and ^{240}Np were deduced from known α- and β-decay energies [32] (Reprinted figure with permission from J.D. Knight et al., *The Radiations of* U^{240} *and* Np^{240}, Phys. Rev. 91 (1953) 889. Copyright 1953 by the American Physical Society.)

neutron separation energies in neighboring nuclides they were able to determine the neutron separation energies of ^{240}U and ^{240}Np as shown in Fig. 7.2.

The Materials Test Reactor (MTR) at Idaho National Laboratory had been used extensively in the discovery of isotopes of transuranium elements during the plutonium project (see Sects. 5.2 and 5.3). It was also used to identify four new lighter isotopes in neutron capture reactions. In 1956, Nethaway, Michel, and Nervik discovered ^{172}Er and ^{172}Tm by two-neutron capture on ^{170}Er [36]. Two long-lived isotopes of terbium and hafnium were produced in several months long irradiations and identified by mass spectroscopy. Naumann, Michel, and Power discovered ^{157}Tb in 1960 [39] and later in the year Hutchin and Lindner observed ^{182}Hf after a cooling period of four years [40].

^{201}Pt was discovered by Facetti et al. in 1962 at the BOiling NUclear Superheater (BONUS) Reactor Facility at the Puerto Rico Nuclear Center. They irradiated mercury samples for 30s to 3 min and observed the β-decay of chemically separated platinum activities. A 2.3 ± 0.2 min activity was assigned to ^{201}Pt formed in the (n,α) reaction [41].

The last isotope discovered with a reactor by neutron capture was ^{162}Gd. Wahlgren et al. irradiated isotopically enriched ^{160}Gd targets in the Savannah River high flux reactor with a thermal neutron flux of $\sim 2 \times 10^{15}$ cm^{-2} s^{-1} and ^{162}Gd was populated by two-neutron capture [42].

7.3 Neutrons Produced at Accelerators

Two experiments within the plutonium project utilized neutrons from an accelerator to identify new isotopes. Ballou and Burgus irradiated beryllium with deuterons from the Chicago cyclotron to produce fast and slow (moderated by several inches of paraffin) neutrons. They observed two separate half-lives in the chemically separated cerium fractions. From the intensity ratios for fast and slow neutrons they could assign one half-life to ^{141}Ce and the other to ^{143}Ce because the latter could not be produced in (n,2n) reactions induced by fast neutrons. In addition, they identified a praseodymium β-decay daughter of ^{143}Ce. These results were only published in 1951 as part of the NNES [43].

The second isotope discovered within the plutonium project (^{107}Cd) was measured at the Berkeley 60-in. cyclotron by Helmholz and published in the refereed literature already in 1946 [44].

After the Second World War nuclear reactions induced by neutrons were probably the most economical way to study new nuclei. Intense fast neutron beams could be produced, for example, by deuterons of a few MeV on lithium and beryllium in the reactions ^{7}Li(d,n) and ^{9}Be(d,n). This did not require large accelerators. In addition, similar to neutron capture reactions in reactors discussed in the previous section, the nuclides produced by fast neutrons (the most common reaction was (n,p) charge exchange) were still close to the valley of stability and had long half-lives. Therefore, very fast chemical separations were typically not required. Also, the element of the isotope of interest had to be extracted from only a few other elements that could have been produced by neutron irradiations.

Thus, it is not surprising that many different laboratories contributed to the discovery of isotopes produced by neutron irradiations. The over 75 isotopes discovered between 1946 and 2003 with neutrons produced with accelerators were identified at 27 different laboratories in 13 different countries on 4 continents and are listed in Tables 7.3 and 7.4. These tables include isotopes that were only or predominantly identified by neutron irradiations. Isotopes that were discovered also by complementary charged-particle cross irradiations are included in Chap. 9.

The discoveries of the first two isotopes outside of the plutonium project by Parmley and Moyer (^{61}Co [45]) in 1947 and Parmley, Moyer, and Lilly (^{62}Co [46]) in 1948 utilized charge exchange reactions with fast neutrons produced at the Berkeley 60-in. cyclotron.

In Europe, the first isotopes were discovered by fast neutron irradiations about five years after the end of the War. Flammersfeld and Herr used deuterons from the high voltage (HV) installation of the Max-Planck-Institut für Chemie in Mainz, Germany, to identify ^{79}Se in 1950 [47] and ^{55}Cr in 1952 [48] for the first time. In 1953, Flammersfeld also discovered ^{68}Cu [49]. Four years later Stribel used the HV facility in Hechingen, Germany, to observe ^{178}Lu [59].

Several isotopes were discovered using cyclotrons to accelerate protons or deuterons to generate neutrons for the subsequent irradiations. In 1953, Cohen and coworkers irradiated a beryllium target with 22 MeV protons from the 86-in. cyclotron at Oak

Table 7.3 Isotopes discovered by neutrons from accelerators between 1943 and 1966: The submission date, first author and reference of the publication, as well as the accelerator and the reaction are listed

Date	First author	Ref.	Accelerator	Reaction	Isotope
5/20/1943[a]	Ballou	[43]	Cyclotron	(n,γ), (n,2n)	^{141}Ce, ^{143}Ce, ^{143}Pr[b]
07/15/1946	Helmholz	[44]	60-in. cyclotron	(n,γ)	^{107}Cd
05/17/1947	Parmley	[45]	60-in. cyclotron	(n,p)	^{61}Co
10/28/1948	Parmley	[46]	60-in. cyclotron	(n,p)	^{62}Co
10/06/1950	Flammersfeld	[47]	HV facility	(n,p)	^{79}Se
07/19/1952	Flammersfeld	[48]	HV facility	(n,γ), (n,p)	^{55}Cr
03/05/1953	Flammersfeld	[49]	HV facility	(n,p)	^{68}Cu
09/18/1953	Cohen	[50]	86-in. cyclotron	(n,p)	^{44}K
02/15/1954	Cohen	[51]	86-in. cyclotron	(n,p)	^{57}Mn
01/13/1955	Butement	[52]	110-in. cyclotron	(n,p)	^{184}Ta
08/10/1955	Poe	[53]	110-in. cyclotron	(n,p)	^{186}Ta
11/18/1955	Flegenheimer	[54]	Synchrocyclotron	(n,p)	^{104}Tc
05/25/1956	Morinaga	[55]	37-in. cyclotron	(n,p)	^{40}Cl, ^{74}Ga
06/12/1956	Gatti	[56]	Synchrocyclotron	(n,α)	^{183}Hf
02/25/1957	Flegenheimer	[57]	Synchrocyclotron	(n,np)	^{103}Tc
05/20/1957	Baró	[58]	Synchrocyclotron	(n,α)	^{195}Os
05/23/1957	Stribel	[59]	HV facility	(n,α)	^{178}Lu
06/26/1957	McGinnis	[60]	60-in. cyclotron	(n,p)	^{120}In
09/23/1957	Schindewolf	[61]	42-in. cyclotron	(n,γ), (n,2n)	^{107}Pd
05/16/1958	Nurmia	[62]	400 kV C-W	(n,p)	^{11}Be
06/16/1958	Nurmia	[63]	400 kV C-W	(n,p)	^{26}Na
08/04/1958	Wille	[64]	400 kV C-W	(n,p)	^{152}Pm, ^{154}Pm
11/18/1958	Poularikas	[65]	400 kV C-W	(n,p)	^{50}Sc
05/25/1959	Meads	[66]	1 MV C-W[c]	(n,p)	^{80}As
09/21/1959	Wille	[67]	400 kV C-W	(n,α)	^{167}Dy
				(n,p)	^{168}Ho, ^{170}Ho, ^{174}Tm
10/07/1960	Kumabe	[68]	400 kV C-W	(n,^{3}He)	^{53}V
10/10/1960	Chittenden II	[69]	400 kV C-W	(n,p)	^{58}Mn
04/06/1961	Robinson	[70]	37-in. cyclotron	(n,p)	^{30}Al
04/18/1961	Takahashi	[71]	160 cm cyclotron	(n,p)	^{76}Ga, ^{176}Tm
07/24/1961	Hagebo	[72]	150 cm cyclotron	(n,p)	^{130}Sb
01/12/1962	Stehney	[73]	60-in. cyclotron	(n,p)	^{86}Br
07/23/1962	Kantele	[74]	400 kV C-W	(n,p)	^{122}In
03/25/1963	Flegenheimer	[75]	Synchrocyclotron	(n,α)	^{189}W
04/03/1963	Crasemann	[76]	60-in. cyclotron	(n,p), (n,np)	^{189}Re, ^{191}Re
07/23/1963	Karras	[77]	400 kV C-W	(n,np)	^{110}Rh

(continued)

Table 7.3 (continued)

Date	First author	Ref.	Accelerator	Reaction	Isotope
01/27/1964	Karras	[78]	400 kV C-W	(n,p)	^{124}In
11/10/1964	Chase Jr.	[79]	3.5 MeV VdG	(n,p)	^{18}N
04/27/1965	Schneider	[80]	Synchrocyclotron	(n,p)	^{162}Tb
08/02/1965[d]	Blachot	[81]	SAMES T400	(n,p)	^{192}Re
08/26/1965	Vaughn	[82]	3.5 MeV VdG	(n,p)	^{22}F
04/06/1966	Vonach	[83]	C-W	(n,p)	^{196}Ir

[a]Date of last internal report on which the paper was based
[b]Populated by β^--decay
[c]Not explicitly mentioned
[d]Date of presentation

Ridge National Laboratory to produce fast neutrons which then irradiated enriched ^{44}Ca and ^{57}Fe targets. They identified the new isotopes ^{44}K [50] and ^{57}Mn [51] produced in (n,p) reactions by measuring the β- and γ-radiation following chemical separations. Similarly, two years later, 20 MeV protons from the cyclotron of the Atomic Energy Research Establishment in Harwell, England, produced fast neutrons to populate ^{184}Ta [52] and ^{186}Ta [53].

In a series of experiments between 1955 and 1963 at the synchrocyclotron of the Laboratorios de Radioquímica de la Comisión Nacional de la Energía Atómica in Buenos Aires, Argentina, the research group of Flegenheimer discovered four new isotopes. The neutrons were produced by irradiating beryllium with deuterons of energies up to 28 MeV. The (n,α) and (n,np) reactions were used for the first time to identify ^{183}Hf [56] and ^{103}Tc [57], respectively. In addition, Baró and Rey discovered ^{195}Os with an (n,α) reaction on ^{198}Pt [58].

During this time, cyclotrons were also used in the discovery of isotopes by fast neutron irradiation at Purdue University (^{40}Cl and ^{74}Ga [55] and ^{30}Al [70]), the Carnegie Institution of Washington (^{120}In [60]), Massachusetts Institute of Technology (^{107}Pd [61]), the Institute for Nuclear Studies of the University of Tokyo (^{76}Ga and ^{176}Tm [71]), the Institute for Nuclear Research in Amsterdam (^{130}Sb [72]), Argonne National Laboratory (^{86}Br [73]), Brookhaven National Laboratory (^{189}Re and ^{191}Re [76]), and the Institut für Radiochemie of the Kernforschungszentrum Karlsruhe (^{162}Tb [80]).

By far the most isotopes discovered by fast neutron irradiation at a single institution were observed at the University of Arkansas in Fayetteville, Arkansas. Between 1958 and 1970 nineteen isotopes over a wide mass range from ^{11}Be [62] to ^{202}Au [84] were identified using 14.8 MeV neutrons produced in ^{3}H(d,n)^{4}He reactions with the 400 kV Cockcroft-Walton accelerator. The neutron flux was up to 10^{10} cm^{-2}s^{-1} and most isotopes were produced in charge exchange reactions. One exception was ^{53}V which is the only isotope discovered in the reaction (n,^{3}He) on ^{55}Mn [68].

Cockcroft-Walton (C-W) type accelerators were also used between 1959 and 1970 to produce new isotopes at the Clarendon Laboratory in Oxford (^{80}As [66]), the Institut für Radiumforschung der Österreichischen Akademie der Wissenschaften

Table 7.4 Isotopes discovered by neutrons from accelerators between 1967 and 2003: The submission date, first author and reference of the publication, as well as the accelerator and the reaction are listed

Date	First author	Ref.	Accelerator	Reaction	Isotope
06/12/1967	Ward	[84]	400 kV C-W	(n,p)	^{202}Au
07/31/1967[a]	Hübenthal	[85]	SAMES T400	(n,p)	^{100}Nb
03/04/1968	Monnand	[86]	SAMES T400	(n,p)	^{164}Tb
10/08/1968	Mathew	[87]	3 MV VdG[b]	(n,p)	^{82}As
10/15/1968	Trautmann	[88]	C-W	(n,p)	^{238}Pa
02/13/1969	Ward	[89]	400 kV C-W	(n,p)	^{64}Co
03/18/1969	Trautmann	[90]	C-W	(n,α)	^{235}Th
09/19/1969	Karras	[91]	400 kV C-W	(n,α)	^{79}Ge
01/12/1970	Ward	[92]	400 kV C-W	(n,p)	^{54}V
06/30/1970	Gujrathi	[93]	TNC Model 9400	(n,p)	^{180}Lu
07/24/1970	Ward	[94]	400 kV C-W	(n,p)	^{150}Pr
12/21/1970	Taff	[95]	C-W	(n,p)	^{70}Cu
10/11/1971	Pakkanen	[96]	SAMES J150	(n,p)	^{204}Au
03/03/1972	Pursiheimo	[97]	SAMES T400	(n,α)	^{173}Er
12/01/1972	D'Auria	[98]	TNC Model 9400	(n,α), (n,p)	^{157}Sm, ^{160}Eu
05/10/1973	Szalay	[99]	T + d generator	(n,p)	^{198}Ir
04/09/1975	Franz	[100]	AGS Injector Linac	(n,2pn)	^{62}Fe
05/19/1976	Haustein	[101]	AGS Injector Linac	(n,2pn)	^{190}W
04/18/1977	Gilat	[102]	AGS Injector Linac	(n,2pn)	^{230}Ra
06/06/1977	Haustein	[103]	AGS Injector Linac	(n,2pn)	^{196}Os
01/09/1979	Haustein	[104]	AGS Injector Linac	(n,2pn)	^{242}U
02/03/1982	Mirzadeh	[105]	AGS Injector Linac	(n,2p)	^{207}Hg
01/06/1992	Shi	[106]	K = 40 Cyclotron	(n,2pn)	^{202}Pt
08/10/1992	Yuan	[107]	600 kV C-W	(n,2p)	^{185}Hf
03/01/1993	Yuan	[108]	600 kV C-W	(n,2p)	^{237}Th
10/25/1995	Zhang	[109]	600 kV C-W	(n,2p)	^{175}Er
05/08/2003	Xu	[110]	600 kV C-W	(n,2p)	^{197}Os

[a]Date of presentation
[b]Not explicitly mentioned

in Vienna (^{196}Ir [83]), the Institut für Anorganische Chemie und Kernchemie der Universität Mainz (^{238}Pa [88] and ^{235}Th [90]), and the Natuurkundig Laboratorium der Rijksuniversiteit, Groningen (^{70}Cu [95]).

A few isotopes were discovered at Van de Graaff (VdG) laboratories. Two of the lightest isotopes were identified for the first time at the research laboratories of the Lockheed Missiles and Space Company in Palo Alto, California, in 1964 and 1965. Chase et al. and Vaughn et al. discovered ^{18}N [79] and ^{22}F [82], respectively, using neutrons from the $^{3}H(d,n)^{4}He$ reaction at the Lockheed 3.5 MV Van de Graaff accelerator.

The only isotope ever discovered in New Zealand was produced at the Institute of Nuclear Science in Lower Hutt using a 3 MV Van de Graaff accelerator. Mathew, McCallum, and Freeman produced neutrons by irradiating a tritium target with 0.8 MeV deuterons and irradiated enriched ^{82}Se targets with a neutron flux of about 10^9 cm^{-2}s^{-1} in the discovery of ^{82}As [87].

Several institutes bought commercial neutron generators for their research with neutrons. The French company SAMES began manufacturing electrostatic generators developed at the French National Center for Scientific Research (CNRS) in 1947. The 400 kV SAMES T400 generator was used at the Centre d'Études Nucléaire de Grenoble to discover ^{192}Re in 1965 [81], ^{100}Nb in 1967 [85], and ^{164}Tb in 1968 [86]. In 1972, this generator was also the neutron source in the discovery of ^{173}Er at the University of Helsinki [97]. Also in Finland, the smaller 150 kV SAMES J150 was used at the University of Jyväskylä to produce ^{204}Au [96] in 1971.

^{180}Lu [93] and ^{157}Sm and ^{160}Eu [98] were discovered at Simon Fraser University in Burnaby, Canada, in 1970 and 1972, respectively, using a TNC Model 9400 neutron generator. The generator was manufactured by Texas Nuclear Corporation (TNC) which was formed in 1956 by researchers from the University of Texas in Austin.

The only isotope ever discovered in Hungary was ^{198}Ir, which was identified by Szalay and Uray in 1973. They used a T + d generator with a neutron flux of 5×10^9 s^{-1} at the Institute of Nuclear Research of the Hungarian Academy of Sciences in Debrecen to irradiate enriched ^{198}Pt targets [99].

Between 1975 and 1982, neutrons between 25 and 200 MeV were produced at the Brookhaven Medium Energy Intense Neutron (MEIN) Facility by stopping 200 MeV protons from the linac injector of the alternating gradient synchrotron in a copper beam stop. Five isotopes were discovered in (n,2pn) reactions and ^{207}Hg was populated with a (n,2p) reaction [105]. The two heaviest isotopes, ^{230}Ra [102] and

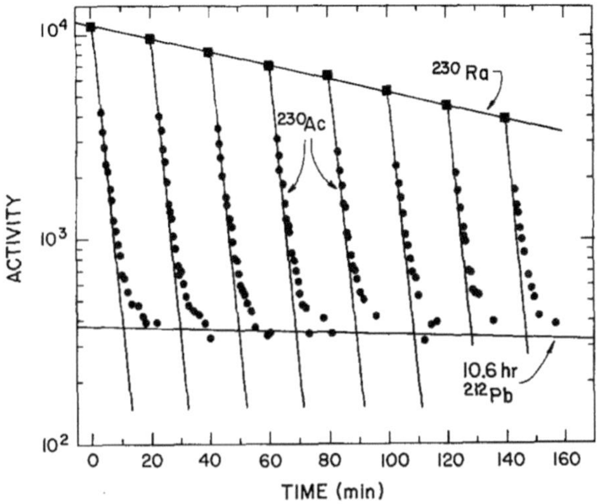

Fig. 7.3 Decay of actinium fractions (*circles*) milked from a radium parent at twenty minute intervals. The times of the separation from radium are shown by the square points [102] (Reprinted from J. Gilat and S. Katcoff, *Decay of* ^{230}*Ra and* ^{230}*Ac*, J. Inorg. Nucl. Chem. 40 (1978) 369. Copyright 1978, with permission from Elsevier.)

^{242}U [104] were identified by half-life measurements of the daughter nuclides from successive extractions. Figure 7.3 shows the decay curve for ^{230}Ra extracted from the decay curves of ^{230}Ac [102].

Similarly, Shi et al. used the (n,2pn) reaction to discover ^{202}Pt with high energy neutrons in 1992. They irradiated a beryllium target with 250 MeV protons from the K = 40 cyclotron of the Institute of Nuclear Research in Shanghai, China, and the neutrons then irradiated an enriched ^{204}Hg target [106]. The last four isotopes discovered with neutrons produced at accelerators were identified between 1992 and 2003 at the Institute of Modern Physics in Lanzhou, China. ^{185}Hf [107], ^{237}Th [108], ^{175}Er [109], and ^{197}Os [110] were populated with (n,2p) reactions with 14 MeV neutrons produced by irradiating a TiT target with deuterons from a 600-kV Cockroft-Walton accelerator.

References

1. A. Turkevich et al., in *Radiochemical Studies: The Fission Products, Paper 153*. The National Nuclear Energy Series IV, vol. 9 (McGraw-Hill, New York, 1951), p. 1070
2. J.A. Swartout, G.E. Boyd, Phys. Rev. **70**, 232 (1946)
3. E.E. Motta, G.E. Boyd, A.R. Brosi, Phys. Rev. **71**, 210 (1946)
4. M.G. Inghram, R.J. Hayden, Phys. Rev. **71**, 130 (1947)
5. G.T. Seaborg, J.W. Gofman, R.W. Stoughton, Phys. Rev. **71**, 378 (1947)
6. L. Seren, H.N. Friedlander, S.H. Turkel, Phys. Rev. **72**, 888 (1947)
7. H.N. Friedlander, L. Seren, S.H. Turkel, Phys. Rev. **72**, 23 (1947)
8. M.G. Inghram, R.J. Hayden, D.C. Hess Jr., Phys. Rev. **71**, 643 (1947)
9. F. Hagemann et al., Phys. Rev. **72**, 252 (1947)
10. S. Katcoff, Phys. Rev. **72**, 1160 (1947)
11. A. Ghiorso et al., Phys. Rev. **82**, 558 (1951)
12. A.C. English et al., Phys. Rev. **72**, 253 (1947)
13. F.C. Yu, D. Gideon, J.D. Kurbatov, Phys. Rev. **71**, 382 (1947)
14. G.W. Parker et al., Phys. Rev. **72**, 85 (1947)
15. S. DeBenedetti, F.K. McGowan, Phys. Rev. **74**, 728 (1948)
16. M.G. Inghram, D.C. Hess Jr., Phys. Rev. **74**, 627 (1948)
17. F.D.S. Butement, Phys. Rev. **75**, 1276 (1949)
18. B.H. Ketelle, Phys. Rev. **76**, 1256 (1949)
19. F. Hagemann, Phys. Rev. **79**, 534 (1950)
20. E. der Mateosian, M. Goldhaber, Phys. Rev. **79**, 192 (1950)
21. J.H. Reynolds, Phys. Rev. **79**, 886 (1950)
22. A.R. Brosi, H. Zeldes, B.H. Ketelle, Phys. Rev. **79**, 902 (1950)
23. M. Lindner, J.S. Coleman, J. Am. Chem. Soc. **73**, 1610 (1951)
24. A.R. Brosi et al., Phys. Rev. **81**, 391 (1951)
25. F.D.S. Butement, Nature **167**, 400 (1951)
26. F.D.S. Butement, Proc. Phys. Soc. A **64**, 428 (1951)
27. M. Lindner, Phys. Rev. **84**, 240 (1951)
28. F. Depocas, B.G. Harvey, Phys. Rev. **85**, 499 (1952)
29. W.C. Rutledge, J.M. Cork, S.B. Burson, Phys. Rev. **86**, 775 (1952)
30. G.E. Boyd, Q.V. Larson, G.W. Parker, Phys. Rev. **86**, 1051 (1952)
31. S. Katcoff, Phys. Rev. **87**, 886 (1952)
32. J.D. Knight et al., Phys. Rev. **91**, 889 (1953)
33. J.P. Butler, J.S. Adam, Phys. Rev. **91**, 1219 (1953)

34. G. Harbottle, Phys. Rev. **91**, 1234 (1953)
35. J.M. LeBlanc, J.M. Cork, S.B. Burson, Phys. Rev. **97**, 750 (1955)
36. D.R. Nethaway, M.C. Michel, W.E. Nervik, Phys. Rev. **103**, 147 (1956)
37. A. Bisi, S. Terrani, L. Zappa, Nuovo Cimento **4**, 758 (1956)
38. L.P. Roy, J.-C. Roy, J.S. Merritt, Phys. Rev. **105**, 1337 (1957)
39. R.A. Naumann, M.C. Michel, J.L. Power, J. Inorg. Nucl. Chem. **16**, 163 (1960)
40. W.H. Hutchin, M. Lindner, J. Inorg. Nucl. Chem. **16**, 369 (1961)
41. J. Facetti et al., Phys. Rev. **127**, 1690 (1962)
42. M.A. Wahlgren et al., Phys. Rev. **153**, 1310 (1967)
43. N.E. Ballou, W.H. Burgus, in *Radiochemical Studies: The Fission Products, Paper 177*. The National Nuclear Energy Series IV, vol. 9 (McGraw-Hill, New York, 1951), p. 1168
44. A.C. Helmholz, Phys. Rev. **70**, 982 (1946)
45. T.J. Parmley, B.J. Moyer, Phys. Rev. **72**, 82 (1947)
46. T.J. Parmley, B.J. Moyer, R.C. Lilly, Phys. Rev. **75**, 619 (1949)
47. A. Flammersfeld, W. Herr, Z. Naturforsch. **5a**, 569 (1950)
48. A. Flammersfeld, W. Herr, Z. Naturforsch. **7a**, 649 (1952)
49. A. Flammersfeld, Z. Naturforsch. **8a**, 274 (1953)
50. B.L. Cohen, Phys. Rev. **94**, 117 (1954)
51. B.L. Cohen et al., Phys. Rev. **94**, 953 (1954)
52. F.D.S. Butement, A.J. Poe, Philos. Mag. **46**, 482 (1955)
53. A.J. Poe, Philos. Mag. **46**, 1165 (1955)
54. J. Flegenheimer, W. Seelmann-Eggebert, Z. Naturforsch. **11a**, 678 (1956)
55. H. Morinaga, Phys. Rev. **103**, 504 (1956)
56. O.O. Gatti, J. Flegenheimer, Z. Naturforsch. **11a**, 679 (1956)
57. J. Flegenheimer, D. Geithoff, Z. Naturforsch. **12a**, 351 (1957)
58. G.B. Baró, P. Rey, Z. Naturforsch. **129**, 520 (1957)
59. T. Stribel, Z. Naturforsch. **12a**, 519 (1957)
60. C.L. McGinnis, Phys. Rev. **109**, 888 (1958)
61. U. Schindewolf, Phys. Rev. **109**, 1280 (1958)
62. M.J. Nurmia, R.W. Fink, Phys. Rev. Lett. **1**, 23 (1958)
63. M.J. Nurmia, R.W. Fink, Nucl. Phys. **8**, 139 (1958)
64. R.G. Wille, R.W. Fink, Phys. Rev. **112**, 1950 (1958)
65. A. Poularikas, R.W. Fink, Phys. Rev. **115**, 989 (1959)
66. R.E. Meads, J.E.G. McIldowie, Proc. Phys. Soc. **74**, 693 (1959)
67. R.G. Wille, R.W. Fink, Phys. Rev. **118**, 242 (1960)
68. I. Kumabe et al., Phys. Rev. **117**, 1568 (1960)
69. D.M. Chittenden II, D.G. Gardner, R.W. Fink, Phys. Rev. **122**, 860 (1961)
70. E.L. Robinson, O.E. Johnson, Phys. Rev. **123**, 1349 (1961)
71. K. Takahashi et al., J. Phys. Soc. Jpn. **16**, 1664 (1961)
72. E. Hagebo, A. Kjelberg, A.C. Pappas, J. Inorg. Nucl. Chem. **24**, 117 (1962)
73. A.F. Stehney, E.P. Steinberg, Phys. Rev. **127**, 563 (1962)
74. J. Kantele, M. Karras, Phys. Rev. **129**, 270 (1963)
75. J. Flegenheimer, G.B. Baró, M. Viirsoo, Radiochim. Acta **2**, 7 (1963)
76. B. Crasemann et al., Phys. Rev. **132**, 1681 (1963)
77. M. Karras, J. Kantele, Phys. Lett. **6**, 98 (1963)
78. M. Karras, Phys. Rev. **135**, B1301 (1964)
79. L.F. Chase Jr. et al., Phys. Rev. Lett. **13**, 665 (1964)
80. V.T. Schneider, H. Münzel, Radiochim. Acta **4**, 171 (1965)
81. J. Blachot, E. Monnand, A. Moussa, C. R. Acad. Sci. **261**, 1835 (1965)
82. F.J. Vaughn et al., Phys. Rev. Lett. **15**, 555 (1965)
83. H. Vonach, H. Munzer, P. Hille, Z. Phys. **195**, 343 (1966)
84. T.E. Ward et al., Phys. Rev. **164**, 1545 (1967)
85. K. Hübenthal, E. Monnand, A. Moussa, C. R. Acad. Sci. **265**, 579 (1967)
86. E. Monnand, A. Moussa, J. Phys. (Paris) **29**, 545 (1968)

87. P.J. Mathew, G.J. McCallum, R.M. Freeman, Phys. Lett. B **28**, 106 (1968)
88. N. Trautmann et al., Z. Naturforsch. **23a**, 2127 (1968)
89. T.E. Ward, P.H. Pile, P.K. Kuroda, J. Inorg. Nucl. Chem. **31**, 2679 (1969)
90. N. Trautmann, R. Denig, G. Herrmann, Radiochim. Acta **11**, 168 (1969)
91. M. Karras, T.E. Ward, H. Ihochi, Nucl. Phys. A **147**, 120 (1970)
92. T.W. Ward, P.H. Pile, P.K. Kuroda, Nucl. Phys. A **148**, 225 (1970)
93. S.C. Gujrathi, J.M. D'Auria, Nucl. Phys. A **161**, 410 (1971)
94. T.E. Ward, N.A. Morcos, P.K. Kuroda, Phys. Rev. C **2**, 2410 (1970)
95. L.M. Taff, B.K.S. Koene, J. van Klinken, Nucl. Phys. A **164**, 565 (1971)
96. A. Pakkanen, T. Komppa, H. Helppi, Nucl. Phys. A **184**, 157 (1972)
97. V. Pursiheimo, T. Tuurnala, T. Raunemaa, Z. Phys. **252**, 283 (1972)
98. J.M. D'Auria, R.D. Guy, S.C. Gujrathi, Can. J. Phys. **51**, 686 (1973)
99. A. Szalay, S. Uray, Radiochem. Radioanal. Lett. **14**, 135 (1973)
100. E.M. Franz et al., Phys. Rev. C **12**, 616 (1975)
101. P.E. Haustein et al., Phys. Rev. C **14**, 645 (1976)
102. J. Gilat, S. Katcoff, J. Inorg. Nucl. Chem. **40**, 369 (1978)
103. P.E. Haustein et al., Phys. Rev. C **16**, 1559 (1977)
104. P.E. Haustein et al., Phys. Rev. C **19**, 2332 (1979)
105. S. Mirzadeh, Y.Y. Chu, Radiochem. Radioanal. Lett. **51**, 279 (1982)
106. S. Shi et al., Z. Phys. A **342**, 369 (1992)
107. S. Yuan et al., Z. Phys. A **344**, 355 (1993)
108. S. Yuan et al., Z. Phys. A **346**, 187 (1993)
109. X. Zhang et al., Z. Phys. A **353**, 353 (1996)
110. Y. Xu et al., J. Radioanal. Nucl. Chem. **258**, 439 (2003)

Chapter 8
Photon and Pion Induced Reactions

8.1 Secondary Photon and Meson Beams

Although the term secondary beams is typically associated with beams of radioactive nuclides produced either by projectile fragmentation reactions or with the ISOL method, neutrons produced in nuclear reactions at charged-particle accelerators also represent a secondary beam. Similarly, photons, produced in nuclear reactions or by bremsstrahlung as well as pions are secondary beams.

In contrast to neutron induced reactions (n,γ) which are exothermic, the inverse reactions, photon induced-emission of nucleons (γ,n) or (γ,p) are endothermic. They require a significant amount of energy (~8 MeV) to break up a nucleus. Thus early attempts with γ-ray sources did not lead to any new radioactive isotopes. Only with the availability of ~17 MeV γ-rays produced in the reaction ^{7}Li(p,γ)^{8}Be first observed by Delsasso et al. in 1937 [1] it became feasible to explore (γ,n) and (γ,p) reactions. Isotopes discovered by photon induced reactions after the Second World War are described in the next section. New isotopes produced prior to the end of the Second World War were covered in Sect. 4.5.

The first meson beam facilities were constructed in the mid 1960s. They required intense high-energy proton beams with currents between 100 μA and 1 mA and energies between 500 and 1000 MeV [2]. The three major facilities in the world were the Los Alamos Meson Physics Facility (LAMPF) in the U.S., the TRI-Universities Meson Facility (TRIUMF) in Vancouver, Canada, and the Swiss Institute for Nuclear Research (SIN), later renamed the Paul Scherrer Institute (PSI) in Villigen, Switzerland.

The production of new isotopes was not the main research focus of these facilities. Only in a few special cases, pion induced reactions could reach new isotopes which were not accessible with other nuclear reactions at the time. The particle-bound isotopes discovered at meson facilities are described in Sect. 8.3. Particle-unbound nuclides will be discussed in Chap. 16.

M. Thoennessen, *The Discovery of Isotopes*,
DOI 10.1007/978-3-319-31763-2_8

8.2 Photo-Nuclear Reactions

Isotopes discovered in photo-nuclear reactions after the end of the Second World War are listed in Table 8.1. Only the first isotope ^{81}Se was discovered with 17.5 MeV γ-rays produced in the ^{7}Li(p,γ)^{8}Be reaction. Wäffler from the Eidgenössische Technische Hochschule Zürich continued his work performed during the War and together with Hirzel reported the first identification of ^{81}Se populated with the (γ,n) reaction in 1948 [3].

A year later Duffield and Knight used 23 MeV bremsstrahlung from the betatron at the University of Illinois to produce ^{118}In and ^{119}In in (γ,p) reactions on enriched ^{119}Sn and ^{120}Sn targets, respectively [4]. Later in 1949, they also reported the observation of ^{113}Ag and ^{115}Ag [6] and in 1950, Duffield, Hsiao, and Sloth discovered ^{99}Nb [8].

^{39}Cl was discovered by Haslam et al. in 1949 by irradiating argon gas at the betatron of the University of Saskatchewan in Saskatoon, Canada, with an energy of 23 MeV. They assigned a half-life of 55.5 ± 0.2 min to ^{39}Cl by measuring the activity of a chemically separated glass wool filter which had absorbed the chlorine from the argon gas [5].

At the Atomic Energy Research Establishment in Harwell, England, Butement used 23 MeV bremsstrahlung from an electron synchrotron to produce ^{79}As, ^{183}Ta, ^{185}Ta [7] and ^{200}Au [9] in (γ,p) and ^{150}Eu [7] in (γ,n) reactions in 1949 and 1950. Two years later Butement and Shillito identified ^{201}Au and ^{203}Au using 28 MeV bremsstrahlung from the synchrotron and cross bombardments with 18 MeV neutrons produced in a cyclotron [12].

Sheline, Holtzman, and Fan used several different photonuclear reactions to populate ^{33}P for the first time. In 1951, they irradiated sulfur and chlorine samples with 48 MeV γ-rays from the University of Chicago betatron. ^{33}P was produced in the three reactions ^{34}S(γ,p), ^{35}Cl(γ,2p), and ^{37}Cl(γ,α) [10]. Two years later, Sheline and Johnson used the γ-ray spectrum of the 100 MeV Chicago betatron to populate ^{28}Mg in the (γ,2p) reaction on ^{30}Si. They also produced ^{28}Mg in the reaction ^{26}Mg(α,2p) with 39 MeV α-particles from the Berkeley cyclotron [13].

Between 1952 and 1966, five isotopes were discovered using bremsstrahlung with energies of up to 75 MeV from the Iowa State College synchrotron in Ames, Iowa. Christian, Mitchell, and Martin discovered ^{195}Ir and ^{197}Ir in 1952 [11], Hammer and Stewart identified ^{158}Tb and ^{163}Ho in 1956 [15], and Van Klinken et al. observed the decay of ^{69}Cu in 1966 [24]. The discovery of ^{163}Ho corresponds to the first two-neutron removal reaction induced by γ-rays following the irradiation of ^{165}Ho. Hammer and Stewart observed a 305 keV γ-ray decay of a 1 s isomeric state and measured the excitation function by varying the bremsstrahlung energies. The assignment was partially based on the high activation threshold for the state as shown in Fig. 8.1 [15].

Table 8.1 Isotopes discovered in photo-nuclear reactions from accelerators between 1948 and 1973: The submission date, first author and reference of the publication, as well as the maximum photon energy (in MeV), the target and the reaction are listed

Date	First author	Ref.	Energy	Target	Reaction	Isotope
05/08/1948[a]	Wäffler	[3]	17.5	^{82}Se	(γ,n)	^{81}Se
04/28/1949	Duffield	[4]	23	^{119}Sn, ^{120}Sn	(γ,p)	^{118}In, ^{119}In
07/15/1949	Haslam	[5]	23	^{40}Ar	(γ,p)	^{39}Cl
10/03/1949	Duffield	[6]	21	^{114}Cd, ^{116}Cd	(γ,p)	^{113}Ag, ^{115}Ag
11/24/1949	Butement	[7]	23	^{80}Se, ^{151}Eu	(γ,p), (γ,n)	^{79}As, ^{150}Eu
				^{184}W, ^{186}W	(γ,p)	^{183}Ta, ^{185}Ta
07/14/1950	Duffield	[8]	23	^{100}Mo	(γ,p)	^{99}Nb
08/01/1950	Butement	[9]	23	^{201}Hg	(γ,p)	^{200}Au
05/21/1951	Sheline	[10]	48	^{34}S, ^{35}Cl, ^{37}Cl	(γ,p), (γ,2p), (γ,α)	^{33}P
02/21/1952	Christian	[11]	70	^{196}Pt, ^{198}Pt	(γ,p)	^{195}Ir, ^{197}Ir
05/26/1952	Butement	[12]	28	^{202}Hg, ^{204}Hg	(γ,p)	^{201}Au, ^{203}Au
12/01/1952	Sheline	[13]	100	^{30}Si	(γ,2p)[b]	^{28}Mg
09/26/1955	Silva	[14]	22	^{144}Sm	(γ,n)	^{143}Sm
12/28/1956	Hammer	[15]	75	^{159}Tb, ^{165}Ho	(γ,n), (γ,2n)	^{158}Tb, ^{163}Ho
11/05/1959	Morinaga	[16]	25	^{64}Ni, ^{76}Ge	(γ,p)	^{63}Co, ^{75}Ga
				^{82}Se	(γ,p)	^{81}As
12/28/1959	Yuta	[17]	25	^{122}Sn, ^{124}Sn	(γ,p)	^{121}In, ^{123}In
08/18/1961	Kuroyanagi	[18]	25	^{160}Gd, ^{174}Yb	(γ,p)	^{159}Eu, ^{173}Tm
				^{176}Yb, ^{180}Hf	(γ,p)	^{175}Tm, ^{179}Lu
07/13/1962	Kotajima	[19]	20	^{154}Sm	(γ,p)	^{153}Pm
07/25/1963	Miyano	[20]	21	^{170}Er	(γ,p)	^{169}Ho
08/05/1963	Kuroyanagi	[21]	23	^{48}Ca	(γ,p)	^{47}K
02/27/1964	Hoffman	[22]	24	^{150}Nd	(γ,p)	^{149}Pr
02/18/1966	Funke	[23]	27	^{164}Dy	(γ,p)	^{163}Tb
03/30/1966	van Klinken	[24]	70	^{70}Zn	(γ,p)	^{69}Cu
05/09/1969	Larson	[25]	55	^{48}Ca	(γ,α)	^{44}Ar
11/17/1972	Chayawattanangkur	[26]	150	^{232}Th	(γ,pn), (γ,p)	^{230}Ac, ^{231}Ac
02/05/1973	Kaffrell	[27]	140	^{238}U	(γ,2p)	^{236}Th

[a]Date of presentation
[b]Also observed in the reaction ^{26}Mg(α,2p) at 39 MeV

The first and only isotope discovered in Brazil was ^{143}Sm in 1955. Silva and Goldemberg used bremsstrahlung from the betatron of the Laboratório de Física Nuclear at the Universidade de São Paulo to irradiate Sm_2O_3. They measured decay curves, absorption spectra and excitation functions of the photodisintegration of ^{144}Sm and determined a half-life of 9.03 min for ^{143}Sm [14].

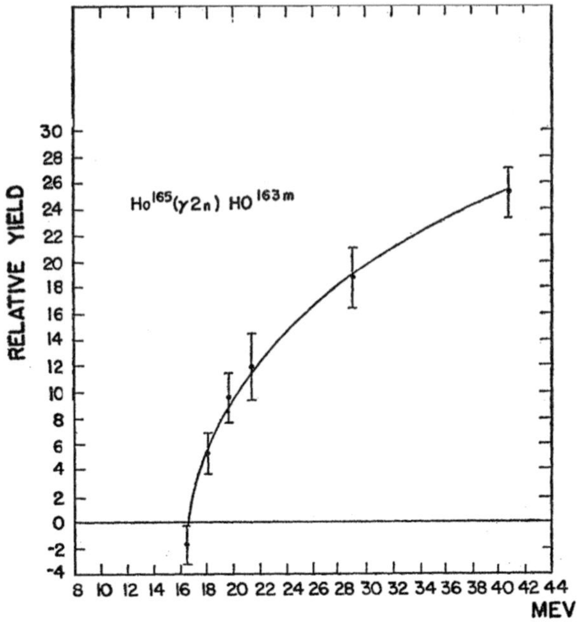

Fig. 8.1 Excitation function of the 305 keV isomeric state in ^{163}Ho produced in a (γ,2n) reaction [15] (Reprinted figure with permission from C.L. Hammer and M.G. Stewart, *Isomeric transitions in the rare-earth elements*, Phys. Rev. 106 (1957) 1001. Copyright 1957 by the American Physical Society.)

The research group of Morinaga discovered nine isotopes at Tôhoku University in Sendai, Japan, between 1959 and 1961 [16–18]. They irradiated samples of various elements with bremsstrahlung inside the 25 MeV betatron. The internal target set-up is shown in Fig. 8.2 [28]. The isotopes were identified by subsequent β- and γ-decay measurements without chemical separations.

In the subsequent three years 1962–1964, three additional isotopes were identified in Japan. Kotajima [19], Miyano and Kuroyanagi [20], and Kuroyanagi et al. [21] used the bremsstrahlung from the electron linear accelerator of the Japan Atomic Energy Research Institute (JAERI) in Tokai to induce (γ,p) reactions and discover ^{153}Pm, ^{169}Ho, and ^{47}K, respectively.

Hoffman and Daniels irradiated enriched ^{150}Nd samples with 22–24 MeV betatron bremsstrahlung at Los Alamos National Laboratory producing ^{149}Pr in 1964. They identified the isotopes by measuring the resulting radiation with γ- and β-scintillation spectrometers [22].

The first and only isotope discovered in the former East Germany (Deutsche Demokratische Republik, DDR) is ^{163}Tb in 1966. Funke et al. irradiated enriched ^{164}Dy samples with 27 MeV bremsstrahlung of the betatron of the Technisch-Physikalisches Institut der Universität Jena and produced ^{163}Tb in (γ,p) reactions [23].

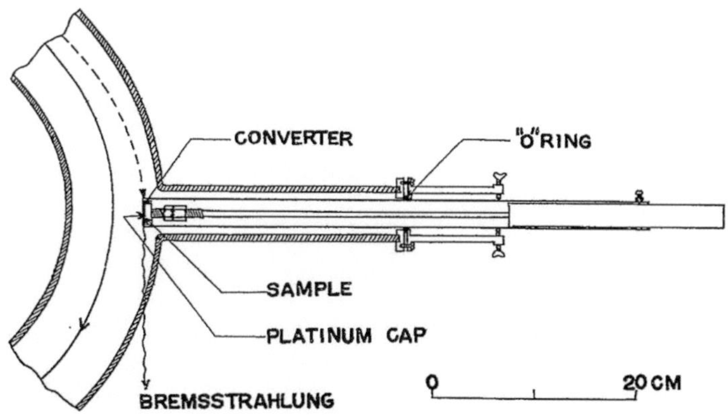

Fig. 8.2 Internal target of the betatron at Tôhoku University [28] (Reprinted from H. Morinaga and T. Kuroyanagi, *Three new isotopes, Co^{63}, Ga^{75}, and As^{81}*, Nucl. Instrum. Meth. 6 (1960) 66. Copyright 1960, with permission from Elsevier.)

The last three discovery experiments using photodisintegration reactions populated the isotopes with three new exit channels. Larson and Gordon discovered ^{44}Ar by irradiating enriched ^{48}Ca targets with bremsstrahlung from the Linac of the Naval Research Laboratory (NRL) in Washington, D.C., in 1969 [25]. By comparing irradiations on calcium targets enriched in ^{44}Ca and ^{46}Ca they assigned a half-life of 14 min to ^{44}Ar produced in the (γ,α) reaction. Bremsstrahlungs-energies of up to 150 MeV from the linear electron accelerator in Mainz, Germany, were utilized to irradiate thorium and uranium targets in 1972. Chayawattanangkur et al. discovered ^{230}Ac and ^{231}Ac in the ^{232}Th(γ,pn) reaction [26], while Kaffrell and Trautmann identified ^{236}Th in the ^{238}U(γ,2p) reaction [27].

8.3 Pion-Induced Reactions

Only three particle-bound isotopes were discovered in pion-induced reactions as listed in Table 8.2. All three were observed at the Los Alamos Meson Physics Facility (LAMPF) with the Energetic PIon Channel and Spectrometer (EPICS) in double-charge exchange reactions. In 1982, Morris et al. irradiated natural silicon and calcium targets with 164 MeV positively charged pions and discovered ^{28}S and ^{40}Ti,

Table 8.2 Particle-bound isotopes discovered by pion induced reactions: The submission date, first author and reference of the publication, as well as target, pion energy (in MeV) and the reaction are listed

Date	First author	Ref.	Target	Energy	Reaction	Isotope
01/25/1982	Morris	[29]	^{28}Si, ^{40}Ca	164	(π^+, π^-)	^{28}S, ^{40}Ti
04/01/1986	Seth	[30]	^{58}Ni	292	(π^+, π^-)	^{58}Zn

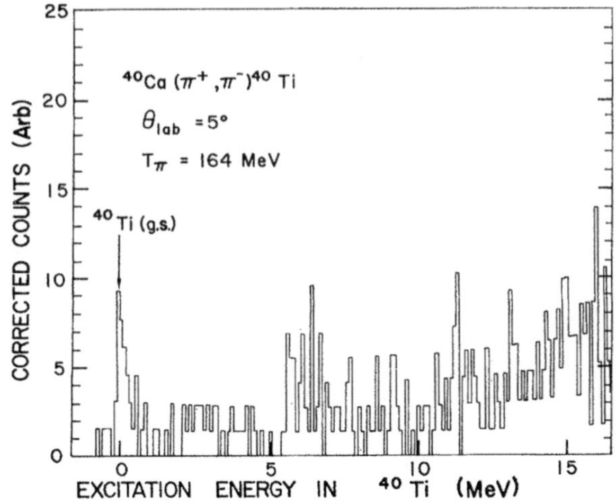

Fig. 8.3 Excitation energy spectrum of ^{40}Ti produced in the reaction ^{40}Ti(π^+, π^-) [29] (Reprinted figure with permission from C.L. Morris et al., *Target mass dependence of isotensor double charge exchange: Evidence for deltas in nuclei*, Phys. Rev. C 25 (1982) 3218. Copyright 1982 by the American Physical Society.)

respectively. They deduced the excitation energy (missing mass) spectra of the isotopes from the energy of the negatively charged pions measured in the spectrometer [29].

Figure 8.3 shows the excitation energy spectrum of ^{40}Ti [29]. Seth et al. used the same set-up to irradiate an enriched ^{58}Ni with 292 MeV π^+ to identify ^{58}Zn in [30].

In addition to these three isotopes four particle-unbound nuclides were discovered using pion beams. The scattering length of the di-neutron was extracted from the reaction ^{2}H(π^-,γ) at the 184-in. cyclotron at Berkeley [31] and particle unstable ^{4}H was identified with the reaction ^{7}Li(π^-,t) at PSI [32]. In addition, double charge exchange reactions with negative and positive pions at LAMPF populated ^{9}He [33] and ^{16}Ne [34], respectively. These discoveries are discussed further in Sects. 16.2 and 16.3, respectively.

References

1. L.A. Delsasso, W.A. Fowler, C.C. Lauritsen, Phys. Rev. **51**, 391 (1937)
2. T.E.O. Ericson, V.W. Hughes, D.E. Nagle, *The Meson Factories* (University of California Press, Berkeley, 1991)
3. H. Wäffler, O. Hirzel, Helv. Phys. Acta **21**, 200 (1948)
4. R.B. Duffield, J.D. Knight, Phys. Rev. **75**, 1967 (1949)
5. R.N.H. Haslam et al., Phys. Rev. **76**, 704 (1949)
6. R.B. Duffield, J.D. Knight, Phys. Rev. **75**, 1613 (1949)
7. F.D.S. Butement, Nature **165**, 149 (1950)
8. R.B. Duffield, L. Hsiao, E.N. Sloth, Phys. Rev. **79**, 1011 (1950)
9. F.D.S. Butement, Proc. Phys. Soc. A **64**, 395 (1951)
10. R.K. Sheline, R.B. Holtzman, C.-Y. Fan, Phys. Rev. **83**, 919 (1951)
11. D. Christian, R.F. Mitchell, D.S. Martin Jr., Phys. Rev. **86**, 946 (1952)
12. F.D.S. Butement, R. Shillito, Proc. Phys. Soc. A **65**, 945 (1952)

13. R.K. Sheline, N.R. Johnson, Phys. Rev. **89**, 520 (1953)
14. E. Silva, J. Goldemberg, Nuovo Cimento **3**, 12 (1956)
15. C.L. Hammer, M.G. Stewart, Phys. Rev. **106**, 1001 (1957)
16. H. Morinaga et al., J. Phys. Soc. Jpn. **15**, 213 (1960)
17. H. Yuta, H. Morinaga, Nucl. Phys. **16**, 119 (1960)
18. T. Kuroyanagi et al., J. Phys. Soc. Jpn. **16**, 2393 (1961)
19. K. Kotajima, Nucl. Phys. **39**, 89 (1962)
20. K. Miyano, T. Kuroyanagi, Nucl. Phys. **49**, 315 (1963)
21. T. Kuroyanagi et al., Nucl. Phys. **50**, 417 (1964)
22. D.C. Hoffman, W.R. Daniels, J. Inorg. Nucl. Chem. **26**, 1769 (1964)
23. L. Funke et al., Nucl. Phys. **84**, 424 (1966)
24. J. van Klinken et al., Phys. Rev. **150**, 886 (1966)
25. R.E. Larson, C.M. Gordon, Nucl. Phys. A **133**, 237 (1969)
26. K. Chayawattanangkur, G. Herrmann, N. Trautmann, J. Inorg. Nucl. Chem. **35**, 3061 (1973)
27. N. Kaffrell and N. Trautmann, Z. Naturforsch. **28a**, 541 (1973)
28. H. Morinaga, T. Kuroyanagi, Nucl. Instrum. Meth. **6**, 66 (1960)
29. C.L. Morris et al., Phys. Rev. C **25**, 3218 (1982)
30. K.K. Seth et al., Phys. Lett. B **173**, 397 (1986)
31. R.P. Haddock et al., Phys. Rev. Lett. **14**, 318 (1965)
32. U. Sennhauser et al., Phys. Lett. B **103**, 409 (1981)
33. K.K. Seth et al., Phys. Rev. Lett. **58**, 1930 (1987)
34. R.J. Holt et al., Phys. Lett. B **69**, 55 (1977)

Chapter 9
Light-Charged-Particle Reactions

9.1 Discoveries with Charged-Particles After 1942

Charged-particle induced reactions had been utilized to discover new isotopes immediately after Cockcroft-Walton completed the construction of the first accelerator in 1932 (see Sect. 4.3). The beginning of the Second World War essentially stopped these investigations. Only two papers reporting the discoveries of new isotopes in charged-particle reactions were submitted shortly after the beginning of the War. In Europe, Krishnan and Nahum reported the observation of ^{202}Tl, ^{205}Hg, and ^{209}Pb at Cambridge on 8/12/1940 [1]. About seven months after the U.S. entered the War on 12/8/1941, O'Connor, Pool, and Kurbatov submitted their results for ^{47}V and ^{49}Cr with the cyclotron at Ohio State University (7/24/1942) [2]. These discoveries have been discussed in Sect. 4.3. Only one additional new isotope was reported before the end of the War. Hibdon and Pool discovered ^{47}Sc also at Ohio State in a paper submitted on 3/12/1945 [3]. Another seven isotopes were discovered in the U.S. during the War as part of the classified research of the plutonium project and were published within the National Nuclear Energy Series on the transuranium elements and the fission fragments in 1949 and 1951.

Soon after the War, charged-particle accelerators were instrumental in the discoveries of many new isotopes. Between 1945 and 1950, almost 100 new isotopes were discovered in the U.S. at several different accelerators. Researchers at the Radiation Laboratory of the University of California at Berkeley continued their predominant position from before the War with the discovery of over 60 of these isotopes.

Light-charged particle-induced reactions can populate isotopes of several different elements near the target element. For example, deuterons can produce a more neutron-rich isotope of the element of the target nucleus in (d,p) reactions or it can form an isotone of the next heavier element in (d,n) reactions. Thus chemical separation following the irradiation can select a specific element and eliminate contaminant radiation from isotopes of other elements. The specific isotope can then be identified by the type of radiation emitted (electron or positron emission, or known γ-ray transitions in the daughter nucleus).

M. Thoennessen, *The Discovery of Isotopes*,
DOI 10.1007/978-3-319-31763-2_9

For the most dominant reaction channels chemical separations were not always necessary and the measured radiation could be assigned to a specific radioactive isotope based on other arguments. For example, by changing the incident energy one can determine the threshold energy at which the isotope is produced or measure the cross-section for the production as a function of beam energy. Cross bombardments with different probes including fast and slow neutrons can also help with the identification. For non-monoisotopic targets, irradiation of targets with different enrichments of the various isotopes is another method to assign a newly measured half-life to a specific isotope.

Also, chemical separations take some time and are not feasible for radioactive isotopes with very short half-lives. Special fast counting techniques and fast transport systems to move the target with the produced isotopes from the irradiation position to a counting station were developed for the identification of these isotopes.

Most of the isotopes produced in light-charged-particle reactions were identified using one or more of these methods without (Sect. 9.2) or in addition to chemical separation (Sect. 9.3). In some cases chemical and physical separation were combined or other special physical identification methods (β-delayed particle emission and missing mass technique) were employed (Sect. 9.4).

9.2 Isotope Identification Without Chemical Separation

In contrast to charged-particles, neutrons do not have to overcome the Coulomb barrier so they can be emitted as long as the excitation energy imparted by the reaction exceeds the neutron separation energy. At beam energies below the Coulomb barrier, neutron emission is the only open decay channel to produce new isotopes. In these cases, the element of the produced isotope is known so that no chemical separations are necessary. Isotopes identified without chemical separation using various techniques and methods for the mass assignment are listed in Table 9.1.

Wiedenbeck used the "stacked foil" technique to identify ^{93}Mo at the University of Michigan in 1946. He irradiated a stack of niobium (which he still called columbium, see discussion in Sect. 4.4) metal foils with 10 MeV deuterons [4]. As the deuterons traverse the foils they lose a known amount of energy, so that each subsequent foil is irradiated with a lower energy. By measuring the produced activity of interest for each foil separately the excitation function can be extracted from a single measurement. Two years later Kelly and Segrè applied the method in the identification of ^{209}Po and ^{210}At by irradiating a stack of bismuth foils (aluminum foils with evaporated bismuth) with deuterons and α-particles from the Berkeley 60-in. cyclotron, respectively. The measured excitation function for the assignment of ^{210}At is shown in Fig. 9.1 [6].

Also in 1948, Schelberg, Sampson, and Mitchell developed a rather tedious method to measure activities with short half-lives. They irradiated a metallic calcium target with 23 MeV α-particles from the Indiana University 45-in. cyclotron and transported the targets in a pneumatic tube within 200 ms in front of a Geiger counter. The pulses of the Geiger counter were displayed on an oscilloscope which

Table 9.1 Isotopes discovered by light-charged-particle reactions without chemical separation: The submission date, first author and reference of the publication, as well as the beam energy (in MeV), target and reaction are listed

Date	First author	Ref.	Energy	Target	Reaction	Isotope
09/03/1946	Wiedenbeck	[4]	10	^{93}Nb	(p,n)	^{93}Mo
04/05/1948	Schelberg	[5]	23	^{40}Ca	(α,n)	^{43}Ti
12/17/1948	Kelly	[6]	15, 40	^{209}Bi	(d,2n), (α,3n)	^{209}Po, ^{210}At
02/14/1949	Alvarez	[7]	32	^{12}C	(p,n)	^{12}N
01/19/1950	Hudspeth	[8]	2.8	^{14}C	(d,p)	^{15}C
07/11/1952	Martin	[9]	15–22	^{46}Ti, ^{50}Cr	(p,n)	^{46}V, ^{50}Mn
				^{54}Fe, ^{58}Ni	(p,n)	^{54}Co, ^{58}Cu
03/02/1953	Glass	[10]	20	^{24}Mg, ^{28}Si, ^{32}S	(p,n)	^{24}Al, ^{28}P, ^{32}Cl
07/28/1953	Churchill	[11]	0.418	^{24}Mg	(p,γ)	^{25}Al
01/08/1954	Gow	[12]	30	^{19}F	(p,2n)	^{18}Ne
05/29/1954	Winn	[13]	28	^{209}Bi	(α,n)	^{212}At
01/18/1955	Glass	[14]	20	^{40}Ca	(p,n)	^{40}Sc
04/25/1955	Kistiakowsky	[15]	31.5	^{182}W	(p,3n)	^{180}Re
08/12/1955	Morinaga	[16]	18	^{39}K	(α,n)	^{42}Sc
08/05/1957	Sun	[17]	12.8	^{40}Ca	(p,α)	^{37}K
10/26/1959	Wilson	[18]	6	^{168}Yb	(p,n)	^{168}Lu
01/22/1960	Ames	[19]	18.5	^{102}Pd	(p,n)	^{102}Ag
04/08/1960	Robinson	[20]	8	^{24}Mg	(^{3}He,n)	^{26}Si
12/19/1960	Robinson	[21]	8	^{28}Si	(^{3}He,n)	^{30}S
04/18/1962	Tomlinson	[22]	30	^{197}Au	(p,4n)	^{194}Hg
10/05/1964	Shida	[23]	3.2	^{48}Ca	(t,p)	^{50}Ca
09/07/1965	Messlinger	[24]	10	^{144}Sm	(p,n)	^{144}Eu
07/21/1966	Miller	[25]	10	^{32}S	(^{3}He,n)	^{34}Ar
09/12/1966	Berg	[26]	23	^{36}Ar	(p,n)	^{36}K
04/10/1968	Yamazaki	[27]	28–50	^{106}Cd, ^{108}Cd	(α,2n), (α,4n)	^{108}Sn
06/06/1971	Doron	[28]	16.5	^{92}Mo	(^{3}He,2n)	^{93}Ru
06/08/1971	Paradellis	[29]	40	^{70}Ge	(p,3n)	^{68}As
04/14/1972	Miyano	[30]	52	^{112}Sn	(p,3n)	^{110}Sb
06/12/1972	Singh	[31]	30	^{112}Sn	(p,2n)	^{111}Sb
09/14/1973	Iafigliola	[32]	30, 40	^{92}Mo	(p,3n), (p,2n)	^{90}Tc, ^{91}Tc
01/18/1974	Charvet	[33]	48	^{112}Sn	(α,3n)	^{113}Te
10/15/1974	Kosanke	[34]	70	^{144}Sm	(t,4n)	^{143}Gd
06/20/1975	Burminskii	[35]	21–59	^{106}Cd	(^{3}He,3n)	^{106}Sn
03/11/1976	Hseuh	[36]	30	^{105}Cd	(^{3}He,n)	^{107}Sn
08/27/1976	Oxorn	[37]	45–65	^{112}Sn	(p,5n), (p,4n)	^{108}Sb, ^{109}Sb
05/12/1977	Deslauriers	[38]	55	^{144}Sm	(p,4n)	^{141}Eu
01/13/1978	Chiba	[39]	36–52	^{64}Zn	(p,3n)	^{62}Ga
03/28/1978	Zarifov	[40]	34–59	^{96}Ru	(^{3}He,p4n)	^{94}Rh
06/30/1980	Aras	[41]	60	^{96}Ru	(α,4n)	^{96}Pd

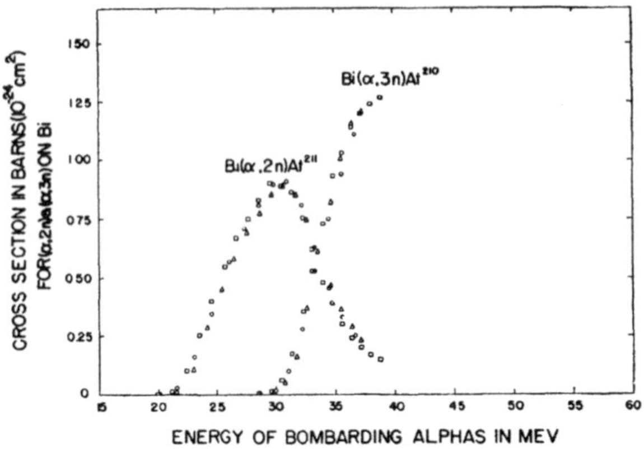

Fig. 9.1 Absolute cross-section for the ^{209}Bi(α,2n) reaction and the ^{209}Bi(α,3n) reaction as a function of beam energy [6]. The latter reaction populated the previously unknown isotope ^{210}At (Reprinted figure with permission from E.L. Kelly and E. Segrè, *Some Excitation Functions of Bismuth*, Phys. Rev. 75 (1949) 999. Copyright 1949 by the American Physical Society.)

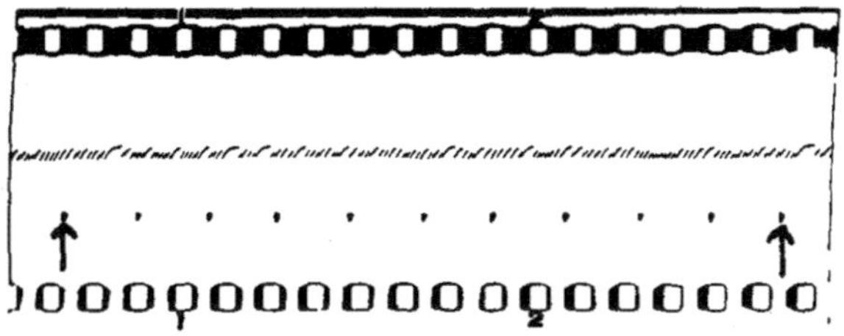

Fig. 9.2 Photograph of an oscilloscope with an oscillograph recorder. The time interval between the arrows is 0.05 s during which 153 pulses from the decay of ^{43}Ti were counted [5] (Reprinted with permission from A.D. Schelberg, M.B. Sampson, and A.C.G. Mitchell, *A method for measuring short period activities*, Rev. Sci. Instrum. 19 (1948) 458. Copyright 1948, AIP Publishing LLC.)

was photographed with an oscillograph recorder as shown in Fig. 9.2. The half-life of ^{43}Ti was then extracted by counting the pulses in each tenth second interval for about 8 s [5].

Similar systems were used by Martin and Breckon at the McGill University cyclotron in the discovery of ^{46}V, ^{50}Mn, ^{54}Co, and ^{58}Cu [9] and by Glass and Richardson [14] and Sun and Wright [17] who discovered ^{40}Sc and ^{37}K, respectively, with the UCLA 20 MeV cyclotron.

In 1949, Alvarez was able to measure the shortest β-decay half-life at the time (12.5 ± 0.1 ms) in the discovery of ^{12}N. He measured the produced activity at the position of the irradiation at the Berkeley pulsed linear accelerator with two Geiger counters in coincidence. The lifetime was determined with an electronic gate circuit where the number of decays within the gate were counted as a function of delay after the accelerator pulse [7]. It is interesting to note that ^{12}N is located at the proton drip-line and before the experiment it was not clear if ^{12}N was bound: "According

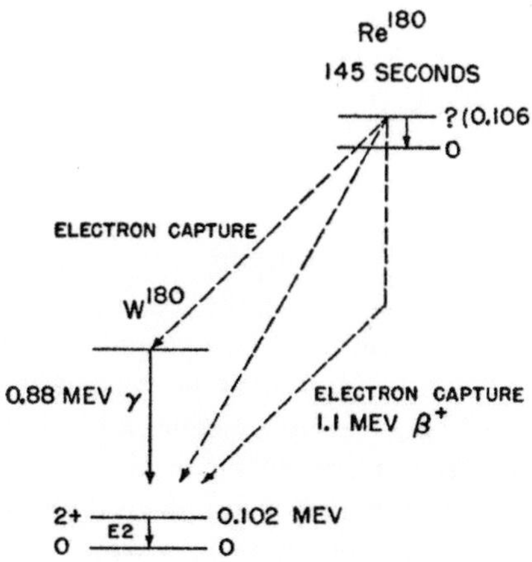

Fig. 9.3 Proposed decay scheme of the new isotope ^{180}Re identified by a γ-ray transition in its daughter nuclide ^{180}W [15] (Reprinted figure with permission from V. Kistiakowsky Fischer, *Metastable States of Re*^{180}, *Ir*^{191}, *Au*^{193}, *Pb*^{201}, *and Pb*^{203}, Phys. Rev. 99 (1955) 764. Copyright 1955 by the American Physical Society.)

to Professor E. Teller, the existence of N^{12} may be of some importance in theories of the evolution of the light elements" [7].

Rather simple counting techniques were also utilized in 1950 by Hudspeth, Swann, and Heydenburg. They irradiated an enriched ^{14}C target with 2.8 MeV deuterons from the large Van de Graaff generator of the Carnegie Institution of Washington and identified ^{15}C produced in a (d,p) reaction: "We were able to measure the half-life by simply following the activity of the target with a stop-watch; counts were recorded after the irradiating beam had been shut off for varying lengths of time" [8]. Three years later Churchill, Jones, and Hunt recorded the counts from a Geiger–Müller tube for 2 s periods with a nine-channel scaler in the discovery of ^{25}Al at the Research Laboratory of Associated Electrical Industries in Aldermaston, England [11].

A different transport system to move the target out of the beam was developed in 1954 by Winn at the Nuffield cyclotron of the University of Birmingham, England. He built an electromechanical device that moved the target away from the irradiation position in front of a detector to measure the α-decays. Winn discovered ^{212}At with a half-life of only 220 ± 30 ms with the device [13]. A similar device was employed by Kistiakowsky Fischer at Berkeley in the discovery of ^{180}Re. At the end of the irradiation a catch holding the target was solenoidally released and the target dropped between a NaI (Tl) scintillation and a gas counter. ^{180}Re was identified by observing a known γ-ray transition in the ^{180}W daughter nucleus. The decay scheme is shown in Fig. 9.3 [15].

With the increasing availability of larger NaI(Tl) detectors and later high resolution germanium detectors this identification method was widely used. Essentially all isotopes discovered between 1964 and 1974 listed in Table 9.1 were identified by measuring known γ-rays in the daughter or even grand-daughter from the target at the irradiation position after switching off the beam.

In the meantime Wilson and Pool discovered ^{168}Lu by irradiating enriched ^{168}Yb with 6 MeV protons from the Ohio State University cyclotron by identifying ytterbium K X-rays and comparing activities produced with similar proton irradiations of other enriched ytterbium isotopes [18]. Also, a pneumatic extraction system was again used by Robinson and collaborators in the discovery of ^{26}Si in 1960 [20] and ^{30}S in 1960 [21] at the Purdue University 37-in. cyclotron. Other pneumatic systems were used about 15 years later at the isochronous cyclotron of the Nuclear Physics Institute of the Kazakh Academy of Sciences in Almaty, Kaszakhstan, to discover ^{106}Sn [35] and ^{94}Rh [40] and at McGill University to discover ^{108}Sb and ^{109}Sb [37] and ^{141}Eu [38]. In 1974, Kosanke et al. developed an improved helium-jet recoil transport system at the sector-focused cyclotron at Michigan State University discovering ^{143}Gd [34]. Two years later, Hseuh and Macias identified ^{107}Sn with a similar helium jet system at the Washington University at St. Louis cyclotron [36].

Table 9.1 contains two special cases—^{194}Hg and ^{108}Sn—where the authors did not consider their measurements of properties of these nuclei a new discovery because half-life measurements had previously been reported. However, later on these measurements turned out to be incorrect.

When Tomlinson and Strokes measured nuclear moments and isotope shifts of ^{194}Hg in 1962 at the Harvard University cyclotron, they quoted a known half-life of 120 d [22]. Brunner et al. in 1955 and Merz in 1961 had reported half-lives of $\sim$130 d [42] and 146 $\pm$ 6 d [43], respectively. Only in 1973, Orth et al. determined the half-life of ^{194}Hg to be longer than 90 y and shorter than 540 y [44]. The currently accepted value is 444 $\pm$ 77 y [45].

In 1968, Yamazaki, Ewan, and Prussin determined the first three excited rotational states of ^{108}Sn by measuring γ-ray spectra with Ge (Li) detectors at the Berkeley 88-in. cyclotron [27]. They assumed ^{108}Sn to be well known as a half-life of 4.5 h had been already reported in 1949 by Mallary and Pool [46]. The correct half-life of 10.5 $\pm$ 0.4 min for ^{108}Sn was only measured for the first time in 1970 by Kiselev and Burmistrov [47].

9.3 Radioactive Decays Following Chemical Separations

Most of the methods described in the previous sections were also applied for the discovery of isotopes where the corresponding element was first selected by chemical separation. Table 9.2 lists the isotopes discovered as part of the plutonium project, while Table 9.3 lists all other isotopes discovered in light charged particle induced reactions at Berkeley after the War (1947–1960). Isotopes discovered at other laboratories are shown in Tables 9.4 (1945–1954) and 9.5 (1955–1973). With the exception of ^{61}Fe and ^{126}Sb which were discovered in Argentina, all isotopes were first observed in the U.S. or Europe. The reactions in the tables are only examples for the production mechanisms and in many cases the isotopes were also observed in other reactions.

The research within the plutonium project concentrated on the understanding of the fission process and the majority of the effort was concentrated on neutron induced reactions to study and produce fissionable material. Charged-particle induced reactions were mostly used to measure the properties of fission fragments or isotopes

Table 9.2 Isotopes discovered by light-charged-particle induced reactions as part of the plutonium project: The submission date, first author and reference of the publication, as well as the beam energy (in MeV), target and reaction are listed

Date	First author	Ref.	Energy	Target	Reaction	Isotope
10/20/1942[a]	Gofman	[48]	14	^{232}Th	(d,2n)	^{232}Pa, ^{232}U[b]
12/02/1944[a]	Jacobson	[49]	16	^{92}Mo, ^{90}Zr	(d,α), (d,n)	^{90}Nb, ^{91}Nb
11/01/1946	Sullivan	[50]		^{96}Ru	(d,p)	^{97}Ru
03/08/1947	Howland	[51]	40	^{203}Tl, ^{204}Pb	(α,3n), (α,2n)	^{204}Bi, ^{206}Po, ^{206}Bi[c]
				^{206}Pb, ^{207}Pb	(α,3n)	^{207}Po, ^{208}Po
05/20/1948	Studier	[52]	19	^{232}Th	(d,4n)	^{230}Pa, ^{230}U[c] ^{226}Th[d], ^{222}Ra[d], ^{218}Rn[d]
Aug. 1949[e]	Hyde	[53]	22	^{230}Th	(d,3n)	^{229}Pa
Aug. 1949[e]	Osborne	[54]	21	^{231}Pa	(d,2n)	^{231}U

[a]Date of preparation of classified report
[b]Populated by β^--decay
[c]Populated by β^+-decay
[d]Populated by α-decay
[e]Month of publication of the plutonium project volume

of heavy elements close to uranium. With the exception of ^{97}Ru all isotopes were produced by deuteron or α-particle irradiation from the Berkeley 60-in. cyclotron.

Although technically not transuranium isotopes, the discovery of ^{232}Pa and its β-decay daughter ^{232}U [48] as well as ^{229}Pa [53] and ^{231}U [54] were first reported in the volume entitled "The Transuranium Elements: Research Papers" of the National Nuclear Energy Series (NNES).

Sullivan, Sleight, and Gladrow reported ^{97}Ru in Physical Review already in 1946 quoting two plutonium project reports from 1943 to 1944 and stating: "This document is based on work performed under contract No. W–7405-ENG–82 for the Manhattan Project and the information covered in this document will appear in Division IV of the Manhattan Project Technical Series, as part of the contribution of the Iowa State College" [50]. The paper was included in the NNES published in 1951 [169].The irradiation probably occurred at the Chicago cyclotron as stated in another paper in the NNES on ruthenium isotopes by the same authors which refers to the same two internal reports [170]. However, another paper by these authors also mentions irradiations at the St. Louis cyclotron [171].

The only other fission fragments identified in charged-particle reactions were ^{90}Nb and ^{91}Nb. Jacobson and Overstreet observed the activities corresponding to these isotopes in an experiment designed to assign mass values to a zirconium-niobium chain previously observed in fission [49].

Ten isotopes were discovered within the Manhattan project after the end of the War and before the project was terminated at the end of 1946. While the experiments were performed in 1946, the results on bismuth and polonium isotopes by Howland,

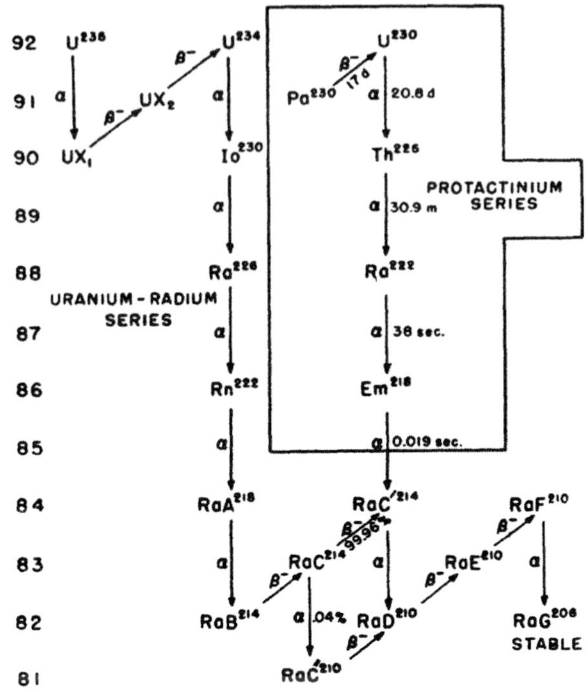

Fig. 9.4 Relation of the newly observed protactinium series to the uranium-radium series [52] (Reprinted figure with permission from M.H. Studier and E.K. Hyde, *A new radioactive series—The protactinium series*, Phys. Rev. 74 (1948) 591. Copyright 1948 by the American Physical Society.)

Templeton, and Perlman [51] and on a new radioactive decay chain originating at ^{230}Pa by Studier and Hyde [52] were published in 1947 and 1948, respectively.

^{230}Pa was populated in the reaction (d,4n) on a ^{232}Th target by measuring the subsequent β^- and α-decays of the daughters. Studier and Hyde were able to link the chain to the naturally occurring uranium-radium (A = 4n + 2) decay chain as shown in Fig. 9.4. They proposed to call it the protactinium series after its first member ^{230}Pa [52]. Six months later, Walen reported a weak β-decay branch of ^{218}At linking ^{218}Rn to the uranium decay chain. He was probably not aware of the results by Studier and Hyde and claimed the discovery of ^{218}Rn [172] (see Sect. 2.4).

The more than 80 isotopes discovered at Berkeley between 1947 and 1960 (Table 9.3) were produced with several different accelerators. While most of them utilized the 60-in. cyclotron others were produced with the 134-in. cyclotron and the proton linear accelerator. ^{59}Cu and ^{60}Cu were the only isotopes that were populated with protons from the 37-in. cyclotron [56].

One of the most productive discoverer of new isotopes was Geoffrey Wilkinson, who has the most first author publications of any Berkeley researcher. He identified a total of 33 neutron-deficient isotopes between neodymium and gold all with light-charged-particle induced reactions within four years between 1947 and 1950. Wilkinson was the sole author of three papers, while the other seven papers were coauthored with Harry Hicks. After he left Berkeley, Wilkinson went on to study transition metal complexes and won the 1973 Nobel Prize in chemistry for his work on the chemistry of organometallic compounds.

Table 9.3 Isotopes discovered by light-charged-particle induced reactions at Berkeley: The submission date, first author and reference of the publication, as well as the beam energy (in MeV), target and reaction are listed

Date	First author	Ref.	Ener.	Target	Reaction	Isotope
08/13/1947	Barber	[55]	18	^{86}Sr	(d,α)	^{84}Rb
09/25/1947	Leith	[56]	5–15	^{60}Ni	(p,2n), (p,n)	^{59}Cu, ^{60}Cu
12/20/1947	Wilkinson, G.	[57]	20	^{181}Ta	(d,2n)	^{181}W
12/22/1947	Wilkinson, G.	[58]	18	^{191}Ir, ^{193}Ir	(d,2n)	^{191}Pt, ^{193}Pt
			38	^{191}Ir, ^{193}Ir	(α,3n), (α,2n)	^{192}Au, ^{194}Au, ^{193}Au, ^{195}Au
03/10/1948	Lindner	[59]	18, 40	^{120}Sn, ^{121}Sb	(d,p), (d,xn)	^{121}Sn, ^{118}Te, ^{119}Te
08/30/1948	Chubbuck	[60]	30, 60	^{133}Cs, ^{139}La	(α,2n), (d,xn)	^{135}La, ^{135}Ce, ^{137}Ce
09/13/1948	Overstreet	[61]	40	^{40}Ar	(α,p)	^{43}K
10/13/1948	Wilkinson, G.	[62]	38	^{165}Ho, ^{175}Lu	(α,3n), (α,2n)	^{166}Tm, ^{167}Tm, ^{176}Ta, ^{177}Ta
01/03/1949	Wilkinson, G.	[63]	10, 19	^{175}Lu	(p,n), (d,2n)	^{175}Hf
01/19/1949	Wilkinson, G.	[64]	19	^{165}Ho	(α,n)	^{168}Tm
02/14/1949	Orth	[65]	38	^{197}Au	(α,xn)	^{198}Tl, ^{199}Tl, ^{200}Tl
02/23/1949	Wilkinson, G.	[66]	9, 19	^{141}Pr	(d,3n), (d,2n)	^{140}Nd, ^{141}Nd
08/16/1949	Wilkinson, G.	[67]	38	^{181}Ta	(α,3n)	^{182}Re
01/13/1950	Marquez	[68]	60, 45	^{121}Sb	(α,4n), (α,3n)	^{121}I, ^{122}I
03/31/1950	Meinke	[69]	60, 18	^{232}Th, ^{238}U	(p,3n), (d,αn)	^{226}Ac^{a}, ^{235}Pa
04/24/1950	Chu	[70]	38	^{185}Re, ^{187}Re	(α,n), (α,3n)	^{188}Ir
05/16/1950	Wilkinson, G.	[71]	19	^{151}Eu, ^{153}Eu	(α,n)	^{154}Tb, ^{156}Tb
			38	^{159}Tb	(α,3n)	^{160}Ho
06/12/1950	Wilkinson, G.	[72]	25–70	^{181}Ta	(p,xn)	^{176}W, ^{177}W, ^{178}W, ^{179}W
			10	^{178}Hf, ^{179}Hf	(p,n)	^{178}Ta, ^{179}Ta
07/28/1950	Germain	[73]	30	^{209}Bi	(α,2n)	^{207}Bi^{a}
08/10/1950	Stover	[74]	40, 25	^{185}Re	(p,4n), (p,3n)	^{182}Os, ^{183}Os, ^{183}Re^{b}
08/18/1950	Stover	[75]	32, 40	^{140}Ce, ^{141}Pr	(p,xn), (p,3n)	^{138}Pr, ^{139}Pr, ^{139}Nd
10/26/1950	Wilkinson, G.	[76]	15–38	^{169}Tm, ^{175}Lu	(α,xn), (p,p2n)	^{170}Lu, ^{171}Lu, ^{172}Lu, ^{173}Lu
			19,38	^{176}Hf, Yb	(d,α), α,xn)	^{174}Lu, ^{171}Hf, ^{172}Hf, ^{173}Hf

(continued)

Table 9.3 (continued)

Date	First author	Ref.	Ener.	Target	Reaction	Isotope
07/12/1951	Hoff	[77]	9–50	^{nat}Sm, ^{nat}Eu	(p,xn)	^{145}Eu, ^{147}Eu, ^{148}Eu, ^{149}Gd
03/25/1952	Kistiakowski	[78]	18–35	^{142}Nd, ^{141}Pr	(p,2n), (α,xn)	^{141}Pm, ^{143}Pm, ^{144}Pm
08/25/1952	Castner	[79]	25–60	^{85}Rb	(p,xn)	^{81}Sr, ^{82}Sr, ^{83}Sr
09/15/1952	Rasmussen	[80]	36, 19	^{147}Sm, ^{151}Eu	(α,3n), (d,3n)	^{148}Gd, ^{150}Gd
			45	^{151}Eu	(α,4n)	^{151}Tb
08/17/1953	Dunlavey	[81]	40	^{nat}Nd	(α,xn)	^{146}Sm
07/30/1954	Crane	[82]	190	^{238}U	(d,2pn)	^{237}Pa
04/18/1955	Wilkinson, J.R.	[83]	9.5	^{53}Cr	(p,n)	^{53}Mn
04/23/1957	Shirley	[84]	24–48	^{147}Sm	(α,4n)	^{147}Gd
06/11/1957	Gallagher	[85]	48	^{181}Ta	(α,4n)	^{181}Re
09/13/1957	Toth	[86]	48	^{152}Gd, ^{155}Gd	(α,xn)	^{152}Dy, ^{153}Dy, ^{155}Dy
09/30/1957	Nethaway	[87]	48	^{37}Cl	(α,3p)	^{38}S
11/21/1957	Toth	[88]	48	^{159}Tb	(α,4n)	^{159}Ho
05/07/1958	Diamond	[89]	25–45	^{185}Re	(α,xn)	^{185}Ir, ^{186}Ir, ^{187}Ir
11/06/1958	Toth	[90]	37	^{151}Eu	(α,3n)	^{152}Tb
09/16/1960	Macfarlane	[91]	48	^{154}Gd	(α,4n)	^{154}Dy

[a]Populated by α-decay
[b]Populated by β^+-decay

The research at Berkeley covered a wide range of masses from ^{38}S by Nethaway and Caretto [87] to ^{237}Pa by Crane and Iddings [82] including a large number of rare earth isotopes. Dunlavey and Seaborg discovered the quasi-stable ^{146}Sm with a half-life of 1.03×10^8 y in 1953 by irradiating a neodymium target with 40 MeV α-particles. They took emulsion photographs of 72 h intervals following chemical separation of samarium. Examination with a microscope revealed ten α-particle tracks corresponding to an α-energy of 2.55 ± 0.05 MeV [81].

Figure 9.5 shows an example of how excitation functions were used to identify new isotopes. Shirley, Smith, and Rasmussen irradiated enriched $^{147}Sm_2O_3$ targets with 24–48 MeV α-particles from the 60-in. cyclotron. They could assign an unknown activity starting at a beam energy of 44 MeV to the new isotope ^{147}Gd produced in the (α, 4n) reaction by measuring the cross-section peaks for the known activities of ^{145}Gd and ^{146}Gd produced at lower energies in (α,2n) and (α,3n) reactions, respectively [84].

Next to Berkeley, the group of Pool at Ohio State University was most productive utilizing light-charged particles to discover new isotopes. Pool had discovered several isotopes in neutron irradiation at the University of Michigan prior to the War and then initiated a program at the Ohio State 42-in. cyclotron at the beginning of the War.

Table 9.4 Isotopes discovered by light-charged-particle induced reactions from 1945 to 1954: The submission date, first author and reference of the publication, as well as the beam energy E (in MeV), target and reaction are listed

Date	First author	Ref.	Ener.	Target	Reaction	Isotope
03/12/1945	Hibdon	[3]	20	^{44}Ca	(α,p)	^{47}Sc
11/27/1946	Goodman	[92]	10	^{185}Re, ^{189}Os	(d,2n), (d,n)	^{185}Os, ^{190}Ir
04/09/1947	Glendenin	[93]	14	^{125}Te, ^{124}Te	(d,2n), (d,n)	^{125}I
05/01/1947	Mitchell	[94]	23	^{69}Ga	(α,n)	^{72}As
05/19/1947	Edwards	[95]	10	^{94}Mo, ^{95}Mo	(d,n)	^{95}Tc, ^{96}Tc
08/08/1947	Coleman	[96]	10	^{116}Sn, ^{118}Sn	(d,n)	^{117}Sb, ^{119}Sb
			20	^{115}In	(α,n)	^{118}Sb
10/23/1947	Tendam	[97]	15–20	^{109}Ag	(α,2n), (α,n)	^{111}In, ^{112}In
01/21/1948	Pool	[98]	10	^{139}La	(d,2n)	^{139}Ce
03/08/1948	Cowart	[99]	20	^{70}Ge	(α,n)	^{73}Se
03/16/1948	Eggen	[100]	20	^{92}Mo	(α,n)	^{95}Ru
05/26/1948	Motta	[101]	16	^{94}Mo	(d,2n)	^{94}Tc
06/21/1948	Woodward	[102]	20	^{74}Se, ^{76}Se	(α,n)	^{77}Kr, ^{79}Kr
07/02/1948	Woodward	[103]	10	^{74}Se, ^{76}Se	(d,n)	^{75}Br, ^{77}Br
08/04/1948	McCown	[104]	10	^{72}Ge	(d,n)	^{73}As
09/03/1948	Kundu	[105]	5	^{92}Mo	(p,γ)	^{93}Tc
10/04/1948	Sherr	[106]	17	^{10}B, ^{14}N	(p,n)	^{10}C, ^{14}O
02/16/1949	Kundu	[107]	5	^{96}Zr	(p,n)	^{96}Nb
04/13/1949	Hinshaw	[108]	20	^{108}Cd	(α,n)	^{111}Sn
08/01/1949	Mallary	[46]	10	^{106}Cd, ^{108}Cd	(d,n), (d,2n)	^{107}In, ^{108}In
08/15/1949	Mitchell	[109]	23	^{121}Sb	(α,2n)	^{123}I
08/24/1949	Robertson	[110]	20	^{84}Zr	(α,n)	^{87}Zr
11/17/1949	Anderson	[111]	20	^{122}Te, ^{124}Te	(α,n)	^{125}Xe, ^{127}Xe
04/24/1950	Hein	[112]	20	^{151}Eu	(d,2n)	^{151}Gd
04/26/1950	Mei	[113]	23	^{103}Rh	(α,p3n)	^{103}Pd
05/07/1950	Butement	[114]	10	^{165}Ho	(p,n)	^{165}Er
06/20/1950	Gum	[115]	20	^{102}Pd	(α,n)	^{105}Cd
11/21/1951	Long	[116]	6	^{150}Nd	(p,n)	^{150}Pm
12/31/1951	Fultz	[117]	7.3	^{76}Se	(p,n)	^{76}Br
01/09/1952	Scoville	[118]	6.3	^{99}Ru	(p,n)	^{99}Rh
04/16/1952	Smith	[119]	23	^{127}I	(α,n)	^{130}Cs
04/18/1952	van der Wiel	[120]	52	^{92}Mo	(α,2n)	^{94}Ru
12/20/1952	Wapstra	[121]	26	^{133}Cs	(d,p2n)	^{132}Cs

(continued)

Table 9.4 (continued)

Date	First author	Ref.	Ener.	Target	Reaction	Isotope
01/13/1953	Diamond	[122]	55–60	^{93}Nb	(p,4n)	^{90}Mo
03/18/1953	Handley	[123]	24	^{159}Tb	(p,3n)	^{157}Dy
03/27/1953	Cohen	[124]	13	^{64}Zn	(p,n)	^{64}Ga
07/20/1953	Handley	[125]	24	^{165}Ho, ^{166}Er	(p,3n), (p,2n)	^{163}Er, ^{165}Tm
10/05/1953	Handley	[126]	24	^{162}Er	(p,2n)	^{161}Er^{a}, ^{161}Ho^{a}
11/30/1953	Sharp	[127]	30–45	^{45}Sc	(p,2n)	^{44}Ti
01/27/1954	Maeder	[128]	26	^{203}Tl	(d,3n)	^{202}Pb
01/27/1954	Handley	[129]	24	^{169}Tm	(p,3n)	^{167}Yb
04/01/1954	Diamond	[130]	40–45	^{90}Zr	(p,2n)	^{89}Nb
05/12/1954	Handley	[131]	22.4	^{170}Er	(p,α)	^{167}Ho
06/28/1954	Naumann	[132]	50	^{191}Ir	(p,4n)	^{188}Pt
08/12/1954	Handley	[133]	22.4	^{136}Ce	(p,2n)	^{135}Pr

^{a}Populated by β^{+}-decay

Between 1945 and 1952 Pool and his collaborators published 20 papers describing the discovery of 28 isotopes. As mentioned in the introduction, the first paper by Hibdon and Pool describing the discovery of ^{47}Sc was submitted before the end of the War [3].

Most of the papers by Pool and collaborators included an "inverted" chart of nuclei where the elements (with the lighter elements at the top) were displayed as a function of mass to show the relationships between the various production mechanisms and the decay of the radioactive isotopes. Figure 9.6 shows as an example the mass region around scandium [3]. These charts were quite often included in the papers at that time and they became increasingly more complicated because the higher available energies from the accelerators opened up more reaction channels.

Additional isotopes were discovered at other U.S. universities during this time with the cyclotrons at MIT (^{125}I [93]), Indiana University (^{72}As [94], ^{123}I [109], and ^{103}Pd [113]), Purdue University (^{111}In and ^{112}In [97]), Princeton University (^{10}C and ^{14}O [106]), Iowa State College (^{151}Gd [112]) and Harvard University (^{90}Mo [122], ^{44}Ti [127], ^{89}Nb [130], and ^{188}Pt [132]).

For the discovery of the short-lived isotopes ^{10}C and ^{14}O at Princeton, Sherr, Muether, and White developed a gas transport system shown in Fig. 9.7 to quickly remove the produced activities from the cyclotron to a counting station. The activity of ^{151}Gd was measured by Hein and Voigt at Iowa State College for over 625 days after it was produced by irradiating a europium oxide target with 20 MeV deuterons at the Berkeley 60-in. cyclotron [112]. Naumann was at Princeton University when he reported the discovery of ^{188}Pt using the Columbia (Nevis Laboratory) and Harvard University cyclotrons [132].

Although Motta, as the first author for the discovery of ^{94}Tc in 1948 [101] was at Oak Ridge National Laboratory (ORNL), it is more likely that the experiment was performed at the Carnegie Institution (home institution of the second author, Boyd),

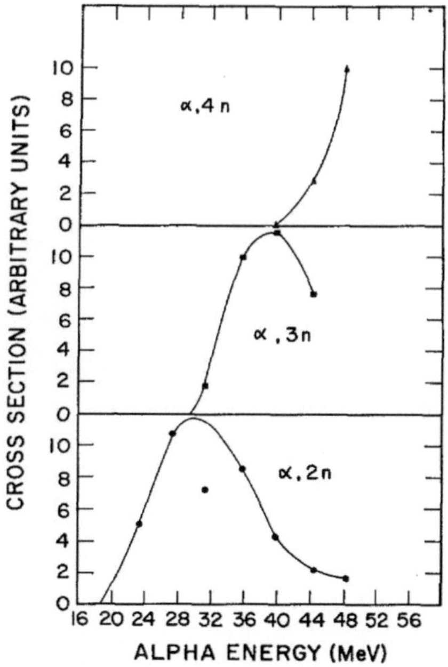

Fig. 9.5 Excitation functions for producing ^{149}Gd (*bottom*), ^{148}Gd (*middle*) and the newly identified isotope ^{147}Gd (*top*) in the reactions ^{147}Sm(α,2n), ^{147}Sm(α,3n), and ^{147}Sm(α,4n), respectively [84] (Reprinted from V.S. Shirley, W.G. Smith, and J.O. Rasmussen, *Conversion-electron and photon spectra of Gd*^{147} *and Gd*^{149}, Nucl. Phys. 4 (1957) 395. Copyright 1957, with permission from Elsevier.)

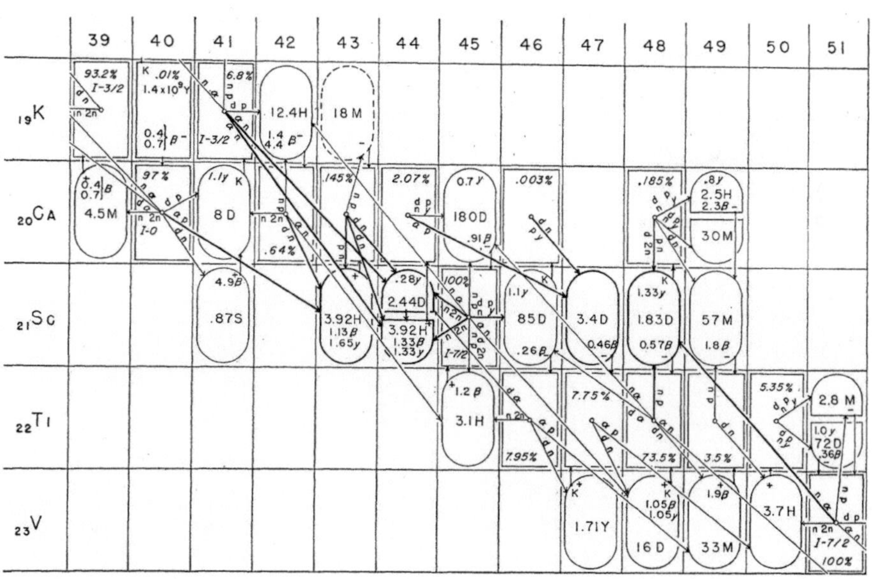

Fig. 9.6 Transmutation chart showing the scandium region and reactions leading to radioactive scandium isotopes [3] (Reprinted figure with permission from C.T. Hibdon, M.L. Pool, *Radioactive scandium. II*, Phys. Rev. 67 (1945) 313. Copyright 1945 by the American Physical Society.)

Table 9.5 Isotopes discovered by light-charged-particle induced reactions from 1955 to 1973: The submission date, first author and reference of the publication, as well as the beam energy (in MeV), target and reaction are listed

Date	First author	Ref.	Ener.	Target	Reaction	Isotope
01/13/1955	Butement	[134]	30	^{70}Ge	(p,2n)	^{69}As
05/05/1955	Aten Jr.	[135]	17–52	^{nat}Ru	(d,xn), (α,xn)	^{97}Rh, ^{98}Pd, ^{98}Rh^{a}, ^{99}Pd
05/06/1955	Aten Jr.	[136]	26	^{192}Os	(d,α)	^{190}Re
07/15/1955	Knight	[137]	20	^{196}Hg	(d,3n)	^{195}Tl
08/01/1955	Lindner	[138]	52	^{58}Ni	(α,2n), (α,n)	^{60}Zn, ^{61}Zn
08/04/1955	Johnson	[139]	50	^{107}Ag	(p,4n)	^{104}Cd, ^{104}Ag^{a}
01/12/1956	Dropesky	[140]	1.5	^{22}Ne	(t,p)	^{24}Ne
11/17/1956	Fränz	[141]	28	^{128}Te	(d,α)	^{126}Sb
01/01/1957	Ricci	[142]	28	^{64}Ni	(d,pα)	^{61}Fe
06/21/1957	Mihelich	[143]	22	^{nat}Gd, ^{nat}Dy	(p,xn)	^{153}Tb, ^{155}Tb, ^{156}Ho, ^{162}Ho
09/04/1958	Selinov	[144]	10	^{112}Sn, ^{114}Sn	(d,n)	^{113}Sb, ^{115}Sb
11/17/1958	Harmatz	[145]	12–22	^{162}Er, ^{164}Er	(p,2n)	^{161}Tm, ^{163}Tm
02/07/1959	Selinov	[146]	15	^{112}Sn, ^{114}Sn	(p,n)	^{112}Sb, ^{114}Sb
05/21/1959	Grover	[147]	40	^{144}Sm	(α,3n)	^{145}Gd
06/11/1959	Gratot	[148]	11, 44	^{142}Nd	(p,n), (α,4n)	^{142}Pm, ^{142}Sm
12/21/1959	Dropesky	[149]	2.9	^{54}Cr	(t,p)	^{56}Cr
02/17/1960	Harmatz	[150]	22	^{180}W	(p,2n)	^{179}Re
03/29/1960	Selinov	[151]	21	^{112}Sn	(α,n)	^{115}Te
07/18/1960	Funk Jr.	[152]		^{148}Nd	(p,3n)	^{146}Pm
12/11/1961	Catura	[153]	14	^{106}Cd	(p,n)	^{106}In
04/20/1963	Wolzak	[154]	26	^{238}U	(d,α)	^{236}Pa
03/02/1964	van Lieshout	[155]	17	^{92}Mo	(d,2n)	^{92}Tc
06/20/1964	Morinaga	[156]	52	^{48}Ca	(α,3p4n)	^{45}K
06/29/1964	Kotajima	[157]	26	^{144}Sm	(d,3n)	^{143}Eu
01/25/1965	Khulelidze	[158]	21	^{106}Cd	(α,n)	^{109}Sn
07/09/1965	Malan	[159]	40	^{144}Sm	(d,4n)	^{142}Eu
03/14/1966	Norris	[160]	20	^{130}Ba	(^{3}He,2n)	^{131}Ce
04/15/1966	Aten Jr.	[161]	50	^{142}Nd	(p,3n)	^{140}Pm
05/09/1966	Hofstetter	[162]	32	^{182}W	(^{3}He,4n)	^{181}Os
11/10/1966	Aten Jr.	[163]	17	^{96}Ru	(p,2n), (p,n)	^{95}Rh, ^{96}Rh
10/18/1967	Keller	[164]	52	^{144}Sm	(α,4n)	^{144}Gd
04/03/1968	Toeset	[165]	60	^{79}Br	(^{3}He,4n)	^{78}Rb
06/12/1969	Aten Jr.	[166]	15–22	^{96}Ru	(^{3}He,2n)	^{97}Pd
05/15/1972	Robertson	[167]	50	^{64}Zn	(^{3}He,xn)	^{64}Ge, ^{65}Ge
03/23/1973	Orth	[44]	12	^{176}Yb	(t,p)	^{178}Yb
06/14/1973	Rivier	[168]	25	^{106}Cd	(p,2n)	^{105}In

[a]Populated by β^{+}-decay

as neither the 63-in. nor the 86-in. cyclotrons were operational at ORNL at the time. After the completion of the 86-in. cyclotron in 1950, Handley and Olson [123, 125, 126, 129, 133] and Handley et al. [131] discovered several isotopes of rare earth elements and Cohen identified ^{64}Ga [124].

The first isotope discovered in a light-charged-particle induced reaction in Europe was ^{165}Er which was produced at the cyclotron of the Atomic Energy Research Establishment in Harwell, England, by Butement in 1950 [114]. Five years later Butement and Prout discovered ^{69}As [134]. The most active laboratory in Europe identifying new isotopes was the Instituut voor Kernphysisch Onderzoek in Amsterdam, The Netherlands. Led by Aten Jr., nineteen new isotopes were produced with the Philips synchrocyclotron between 1952 and 1969.

After 1955 the number of new isotopes produced in light-charged-particle reactions and only separated by chemical means dropped significantly. While between 1947 and 1955 more than seven isotopes were discovered every year, that number decreased to less than two between 1956 and 1973.

Cyclotrons continued to be the dominant accelerators. Only ^{24}Ne [140], ^{56}Cr [149], and ^{178}Yb [44] were discovered at a Van de Graaff accelerator. They were produced in (t,p) reactions at Los Alamos National Laboratory. Dropesky and Schardt used a fast separation technique similar to the set-up shown in Fig. 9.7 in the discovery of ^{24}Ne. However, instead of a powdered target they directly irradiated neon gas and

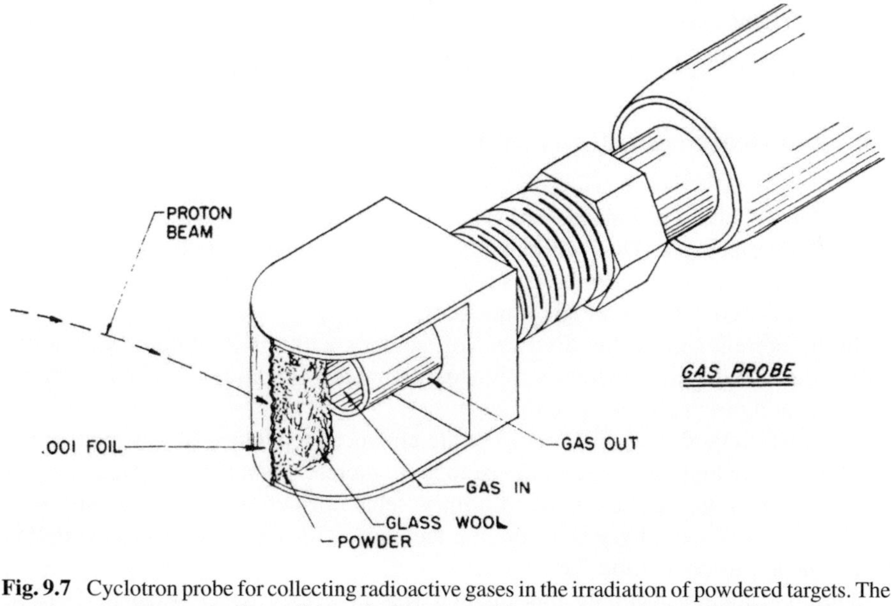

Fig. 9.7 Cyclotron probe for collecting radioactive gases in the irradiation of powdered targets. The gases are swept to a counting cell by a flushing gas. This set-up was designed by Sherr, Muether, and White at Princeton University and used in the discovery of ^{10}C and ^{14}O [106] (Reprinted figure with permission from R. Sherr, H.R. Muether, and M.G. White, *Radioactivity of* C^{10} *and* O^{14}, Phys. Rev. 75 (1949) 282. Copyright 1949 by the American Physical Society.)

they were able to separate ^{22}Ne from sodium and fluorine activities with a cold charcoal trap and measure its 3.38 ± 0.02 min half-life with β- and γ-scintillation spectrometers [140].

With the exception of ^{106}In [153] and the two germanium isotopes ^{64}Ge and ^{65}Ge [167] which were first observed at the University of California at Los Angeles and Michigan State University, respectively, all isotopes in the U.S. were discovered at National Laboratories. In addition to Los Alamos and Oak Ridge mentioned above, ^{195}Tl [137], ^{145}Gd [147], and ^{131}Ce [160] were identified at Brookhaven National Laboratory. Hofstetter and Daly from Purdue University performed experiments at the Argonne 60-in. as well as the Oak Ridge 88-in. cyclotron to identify ^{181}Os [162].

Four additional isotopes were discovered in Canada and Argentina. ^{104}Cd and ^{104}Ag were reported by Johnson from McGill University in Montreal [139] while ^{126}Sb and ^{61}Fe were produced by Fränz et al. [141] and Ricci et al. [142] with the synchrocyclotron of the Comisión Nacional de la Energía Atómica at Buenos Aires, respectively.

In Europe, in addition to Harwell and Amsterdam, isotopes were discovered at the cyclotrons in Russia (^{113}Sb, ^{115}Sb [144], ^{112}Sb, ^{114}Sb, [146], and ^{115}Te [151] in Moscow and ^{109}Sn [158] in Dubna), France (^{142}Pm and ^{142}Sm [148] in Saclay and ^{105}In [168] in Grenoble), and Germany (^{142}Eu [159] and ^{144}Gd [164] in Karlsruhe). The discovery of ^{105}In was submitted in June 1973, but the paper was only published two years later in 1975 [168].

By that time most of the long-lived isotopes close to the valley of stability which could be reached with low-energy charged-particles had been discovered and other production and identification techniques were necessary to expand the chart of nuclei.

9.4 Physical Separation and Identification Techniques

Chemical separation techniques can isolate a specific element but they cannot separate different isotopes. Thus without additional physical separation techniques, the identification of a new isotope always relied on other analytical methods, for example γ-ray spectroscopy of the daughter nuclides.

In the present section the discovery of isotopes with three different techniques will be discussed: mass separation following chemical separation, β-delayed particle emission and missing mass spectroscopy.

The first method is applicable when the chemical separation can be performed quickly relative to the half-life of the isotope of interest. Then the separated element can be placed into the ion source of a mass separator before it decays, thus combining chemical with physical separation techniques. Isotopes discovered with this technique are listed in Table 9.6.

The first attempt to apply this method was described by Ghoshal in 1948 at Berkeley for the identification of ^{109}In. However, it is not clear if mass spectroscopy was successfully used. Although the introduction states "All of these activities have been chemically attributed to indium and have been assigned by mass-spectrograph separation to In^{110}, In^{109}, and In^{111}, respectively" the article describes the mass iden-

Table 9.6 Isotopes discovered by light-charged-particle induced reactions followed by chemical and mass separation: The submission date, first author and reference of the publication, as well as the beam energy (in MeV), target and reaction are listed

Date	First author	Ref.	Energy	Target	Reaction	Isotope
01/05/1948	Ghoshal	[173]	30	^{107}Ag	(α,2n)	^{109}In
12/02/1948	Reynolds	[174]	20–100	^{79}Br, ^{81}Br	(α,xn)	^{81}Rb, ^{82}Rb
08/01/1949	Temmer	[175]	37	^{115}In	(α,3n)	^{116}Sb
10/31/1949	Fink	[176]	60	^{127}I	(α,4n), (α,2n)	^{127}Cs, ^{129}Cs
12/12/1949	Naumann	[177]	50,<30	^{133}Cs	(α,4n), (α,n)	^{133}La, ^{136}La
07/10/1950	Karraker	[178]	60	^{81}Br	(α,2n)	^{83}Rb
03/19/1951	Gransden	[179]	20–90	Ba	(p,xn)	^{131}La, ^{132}La
08/23/1954	Huizenga	[180]	21	^{205}Tl	(d,2n)	^{205}Pb
05/23/1955	Boyd	[181]	22	^{98}Mo	(p,n)	^{98}Tc
08/12/1958	Dahlstrom	[182]	40	^{140}Ce	(p,4n)	^{137}Pr
11/29/1972	Ward	[183]	92	^{186}W	(p,3p)	^{184}Hf
10/28/1975	Wigmans	[184]	40	^{112}Sn	(^{3}He,3n)	^{112}Te

tification from the measured excitation functions. No further details about the mass separator are given and at the end it reads "These conclusions were also corroborated by mass-spectrograph work, though the evidence is not yet conclusive" [173].

Later in the year Reynolds, Karraker, and Templeton used a 60° deflection spectrograph to separate ^{81}Rb and ^{82}Rb produced by irradiating ammonium bromide with α-particles from the Berkeley 60-in. and 184-in. cyclotrons. In addition to the mass spectroscopy they used the photographic transfer technique to demonstrate the radioactivity of the isotopes. In this method a second photographic plate is placed on top of the original plate for a period of time to be exposed to any possible radiation. Figure 9.8a shows the original and the "transfer" plate clearly showing the presence of radioactivity at the location of ^{81}Rb and ^{82}Rb [174].

In 1949, Temmer discovered ^{116}Sb using a calutron to separate it from heavier antimony isotopes [175]. The calutron was originally developed by Lawrence in order to separate uranium ^{235}U from ^{238}U as part of the Manhattan project. The name represents a contraction of California University cyclotron as Lawrence dismantled the 37-in. cyclotron to reuse its magnet for the first calutron prototype [185].

Five additional isotopes were discovered with the 60° mass spectrograph at Berkeley in 1950. Figure 9.8b shows another example of the photographic transfer technique. It shows microphotometer tracings of the original mass spectrograph plate as well as three transfer plates taken at three subsequent later times. The intensity of the 5.5 h ^{127}Cs activity decreases significantly faster compared to the 31 h activity of ^{129}Cs [176].

At McGill University in Montreal, Canada, Gransden and Boyle used a 180° spectrograph to discover ^{131}La and ^{132}La [179] in 1951 while in 1958, Dahlstrom, Foster, and Thompson identified ^{137}Pr with a 90° device [182].

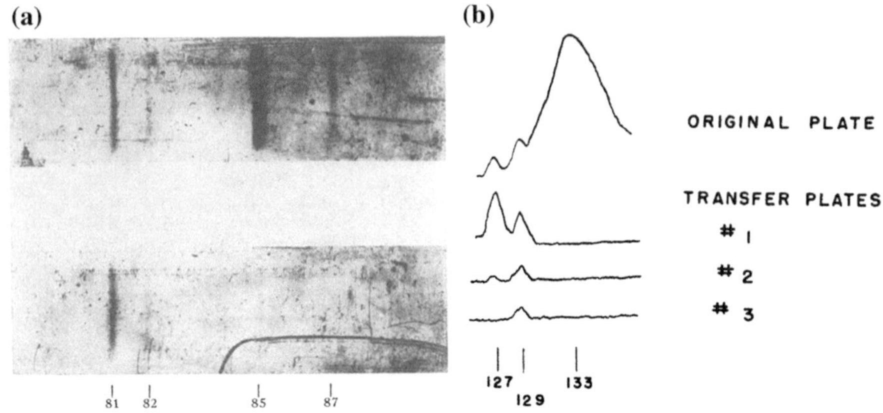

Fig. 9.8 Two examples of the photographic transfer technique. **a** The original mass spectrograph plate on the *top* shows natural occurring ^{85}Rb and ^{87}Rb in addition to the radioactive isotopes ^{81}Rb and ^{82}Rb. The *bottom* plate was placed on top of the original plate for several hours before development in order to locate radioactivity [174] (Reprinted figure with permission from F.L. Reynolds, D.G. Karraker, and D.H. Templeton, Mass spectrographic assignment of rubidium isotopes, Phys. Rev. 75 (1949) 313. Copyright 1949 by the American Physical Society.). **b** Microphotometer tracings of the original and three transfer plates of mass separated cesium. Stable ^{133}Cs was used as a carrier. The transfer plates were exposed following the separation for subsequent intervals of 24 h, 48 h, and 72 h for plate #1, #2, and #3, respectively [176] (Reprinted figure with permission from R.W. Fink, F.L. Reynolds, D.H. Templeton, Neutron-deficient cesium isotopes, Phys. Rev. 77 (1950) 614. Copyright 1950 by the American Physical Society.)

Mass spectrometers were also instrumental in the discovery of very long-lived isotopes where the half-life cannot be determined from radioactivity directly. Two such isotopes were ^{205}Pb and ^{98}Tc which were both separated using 60° mass spectrometers. Huizenga and Stevens discovered ^{205}Pb in 1954 at Argonne National Laboratory [180] while in 1955 Boyd et al. identified ^{98}Tc at Oak Ridge National Laboratory [181].

The final two isotopes discovered in light-charged-particle induced reactions with traditional mass spectroscopy were observed in the 1970s. Ward, Chu, and Cumming identified ^{184}Hf with the Brookhaven isotope separator [183] and Wigmans et al. produced ^{112}Te with the AVF cyclotron of the Vrije Universiteit Amsterdam [184].

Beta-delayed particle emission was discovered in 1949 by Alvarez and isotopes discovered using this special decay mechanism are listed in Table 9.7. Alvarez observed delayed neutrons emitted from ^{17}O following the β-decay of ^{17}N. The proposed decay scheme is shown in Fig. 9.9 [186]. Delayed neutrons had been already observed in neutron-induced fission of uranium, and Bohr and Wheeler had explained the effect as neutrons emitted from excited fragments populated by a β-emitting precursor. However, although this hypothesis was well accepted, the experimental proof of neutrons in coincidence with β-particles was still lacking. Alvarez produced ^{17}N in the reaction ^{19}F(d,α) with 190 MeV deuterons from the 184-in. Berke-

Table 9.7 Isotopes discovered by light-charged-particle induced reactions observed by β-delayed particle emission: The submission date, first author and reference of the publication, as well as the beam energy (in MeV), target and reaction are listed

Date	First author	Ref.	Energy	Target	Reaction	Isotope
03/01/1949	Alvarez	[186]	190	^{19}F	(d,α)	^{17}N
07/24/1950	Alvarez	[187]	32	^{10}B, ^{20}Ne	(p,t), (p,n)	^{8}B, ^{20}Na
07/09/1951	Gardner	[188]	–	^{9}Be, ^{11}B	(d,2p), (p,3p)	^{9}Li
08/14/1963	Barton	[189]	97	^{27}Al, ^{28}Si	(p,p3n)	^{25}Si
06/08/1964	McPherson	[190]	39–85	^{19}F	(p,3n)	^{17}Ne
08/10/1964	McPherson	[191]	39–85	^{23}Na, ^{24}Mg	(p,3n), (p,p3n)	^{21}Mg
10/15/1964	Hardy	[192]	–	^{32}S	(p,p3n)	^{29}S
11/13/1964	Reeder	[193]	31.8	^{32}S, ^{36}Ar	(^{3}He,2n)	^{33}Ar, ^{37}Ca[a]
				^{40}Ca	(^{3}He,2n)	^{41}Ti
11/13/1964	Hardy	[194]	85	^{39}K	(p,3n)	^{37}Ca[a]
08/03/1965	McPherson	[195]	51	^{14}N	(p,2n)	^{13}O
02/01/1979	Äystö	[196]	70	^{24}Mg	(^{3}He,3n)	^{24}Si
03/09/1981	Honkanen	[197]	25	^{58}Ni	(^{3}He,2n)	^{59}Zn
05/17/1982	Cable	[198]	110	^{24}Mg	(^{3}He,p4n)	^{22}Al
01/04/1983	Cable	[199]	110–130	^{28}Si	(^{3}He,p4n)	^{26}P
06/04/1985	Äystö	[200]	135	^{40}Ca	(^{3}He,α4n)	^{35}Ca

[a]Discovered simultaneously

ley cyclotron. He observed a 4.2 s half-life which was determined to be a nitrogen isotope from chemical separations. Subsequently the neutron energy spectrum was determined from the measurement of the ^{16}O recoils.

Two years later, Gardner, Knable, and Moyer, also at Berkeley, reported with ^{9}Li another β-delayed neutron emitter [188].

Beta-delayed α emission in light nuclides had already been observed by Crane et al. in 1935 with the observation that ^{8}Li β-decays to ^{8}Be which then breaks up into two α-particles [201]. In the search for β-delayed proton emission in 1950, Alvarez observed the β-delayed α-emission of ^{8}B and ^{20}Na [187].

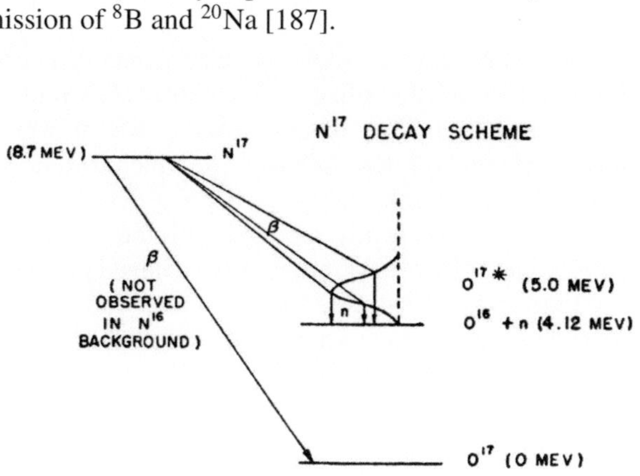

Fig. 9.9 Proposed decay scheme of the first β-delayed neutron emitter ^{17}N discovered by Alvarez at the University of California at Berkeley [186] (Reprinted figure with permission from L.W. Alvarez, *N*^{17}, *a delayed neutron emitter*, Phys. Rev. 75 (1949) 1127. Copyright 1949 by the American Physical Society.)

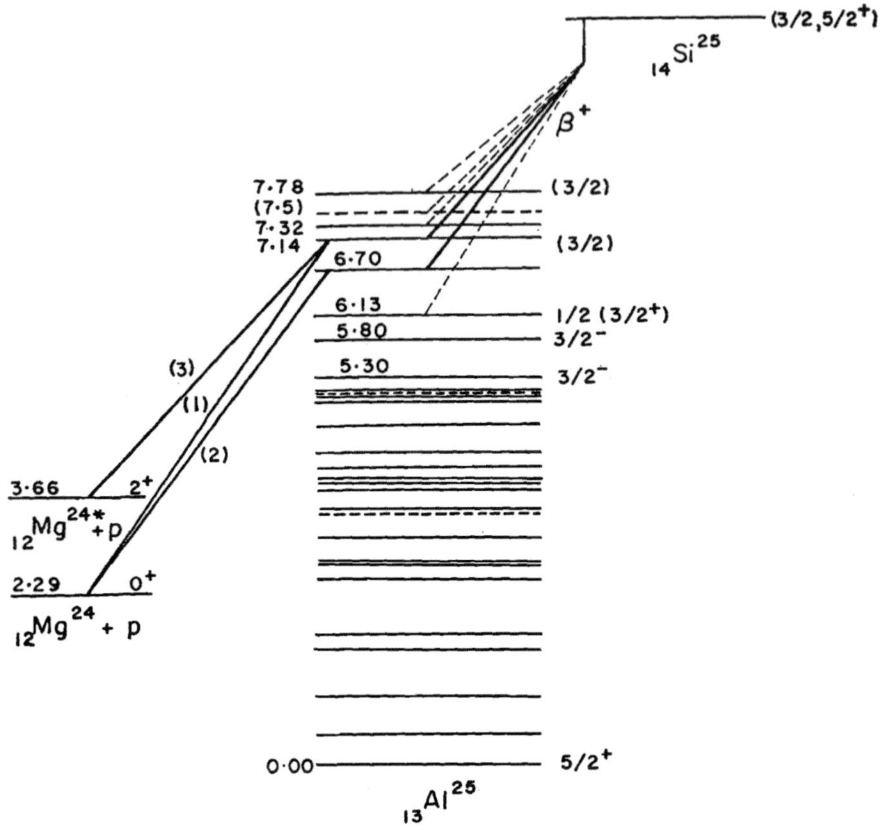

Fig. 9.10 Decay scheme of the new β-delayed proton emitter ^{25}Si proposed by Barton et al. from McGill University [189] (R. Barton et al., *Observation of delayed proton radioactivity*, Can. J. Phys. 41 (1963) 2007, Copyright 2008 Canadian Science Publishing or its licensors, reproduced with permission.)

The first β-delayed proton emitter was not discovered until 13 years later, when Barton et al. irradiated silica (SiO_2) with 97 MeV protons from the McGill University synchrocyclotron. They measured a half-life of 300 ms for the β-decay of ^{25}Si to proton-unbound excited states of ^{25}Al which then decay to ^{24}Mg by the emission of proton. The decay scheme is shown in Fig. 9.10 [189]. Within the next year the group of McPherson and Hardy reported with the discovery of ^{17}Ne [190], ^{21}Mg [191], ^{29}S [192] and ^{37}Ca [194] four more β-delayed proton emitters at McGill University.

At the end of 1964, Reeder, Poskanzer, and Esterlund reported the discovery of the β-delayed proton emitters ^{33}Ar, ^{37}Ca, and ^{41}Ti produced in (^{3}He,2n) reactions at the 60-in. cyclotron at Brookhaven National Laboratory [193]. The discovery of ^{37}Ca was submitted on the same date (11/13/1964) as the paper by Hardy and Verrall [194], and the two papers were published adjacent to each other in the same issue of Physical Review Letters. McPherson, now at Brookhaven National Laboratory,

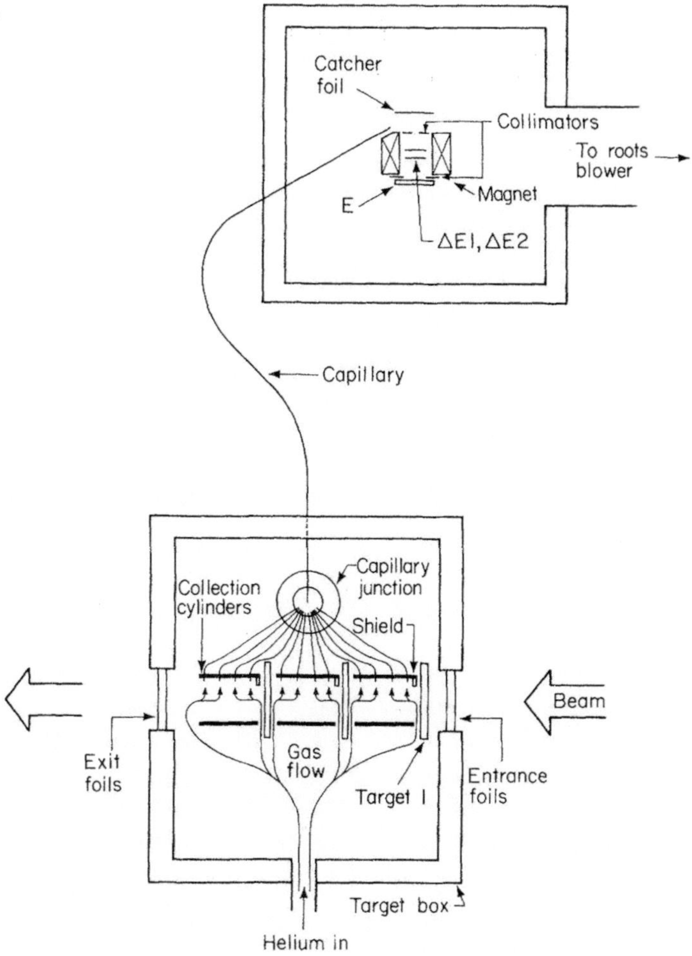

Fig. 9.11 Schematic view of the target, He-jet transport system, and counting chambers used in the discovery of the β-delayed proton emitter ^{22}Al at the University of California at Berkeley [198] (Reprinted figure with permission from M. D. Cable et al., *Beta-delayed proton decay of an odd-odd* $T_z = -2$ *isotope,* ^{22}*Al*, Phys. Rev. C 26 (1982) 1778. Copyright 1982 by the American Physical Society.)

discovered with his collaborators one more β-delayed proton emitter a year later when he identified ^{13}O in the reaction ^{14}N(p,2n) at the Brookhaven linac [195].

It took another 14 years before the next β-delayed proton emitter was discovered. Between 1979 and 1985 five new β-delayed proton emitters were observed at Berkeley (^{24}Si [196], ^{22}Al [198], ^{26}P [199], and ^{35}Ca [200]) and the University of Yväskylä, Finland (^{59}Zn [197]). All five isotopes used a helium gas transport system to move the recoils via a capillary to various detection systems to measure

Table 9.8 Isotopes discovered by light-charged-particle induced reactions missing mass: The submission date, first author and reference of the publication, as well as the beam energy (in MeV), target and reaction are listed

Date	First author	Ref.	Energy	Target	Reaction	Isotope
10/16/1952	Harvey	[202]	14	^{209}Bi	(d,t)	^{208}Bi
05/31/1955	Jarmie	[203]	1.8	^{19}F	(t,p)	^{21}F
06/10/1959	Jarmie	[204]	2.6	^{18}O	(t,p)	^{20}O
12/30/1960	Hinds	[205]	6	^{14}C	(t,p)	^{16}C
07/31/1961	Ajzenberg-Selove	[206]	3.4	^{20}Ne	(^{3}He,n)	^{22}Mg
11/08/1963	Bryant	[207]	25	^{40}Ca	(^{3}He,n)	^{42}Ti
11/19/1964	Cerny	[208]	65	^{12}C	(^{3}He,^{6}He)	^{9}C
12/09/1964	Marinov	[209]	12	^{48}Ca	(d,α)	^{46}K
03/29/1965	Bassani	[210]	40	^{112}Sn	(p,t)	^{110}Sn
10/08/1965	Erskine	[211]	12	^{48}Ca	(α,p)	^{51}Sc
06/13/1966	Williams	[212]	7.5	^{50}Ti	(t,p)	^{52}Ti
10/24/1966	Hardy	[213]	40	^{40}Ca	(p,t)	^{38}Ca
02/03/1969	Cerny	[214]	55	^{28}Si	(p,^{6}He)	^{23}Al
04/26/1972	Proctor	[215]	65–75	^{50}Cr, ^{54}Fe, ^{58}Ni	(^{3}He,^{6}He)	^{47}Cr, ^{51}Fe, ^{55}Ni
12/07/1973	Jelley	[216]	78	^{48}Ca	(α,^{7}Be)	^{45}Ar
			80	^{48}Ca	(^{6}Li,^{8}Be)	^{46}Ar
04/08/1974	Robertson	[217]	156	^{24}Mg	(α,^{8}He)	^{20}Mg
01/08/1976	Benenson	[218]	74	^{40}Ca	(^{3}He,^{8}Li)	^{35}K
05/19/1976	Kashy	[219]	74	^{48}Ca, ^{64}Ni	(^{3}He,^{8}B)	^{43}Cl, ^{59}Mn
10/14/1976	Tribble	[220]	121	^{40}Ca	(α,^{8}He)	^{36}Ca
12/06/1976	Tribble	[221]	110	^{54}Fe, ^{58}Ni	(α,^{8}He)	^{50}Fe, ^{54}Ni
12/17/1976	Benenson	[222]	70, 74	^{32}S, ^{36}Ar	(^{3}He,^{8}Li)	^{27}P, ^{31}Cl
02/21/1978	Kouzes	[223]	80	^{70}Zn	(^{3}He,^{8}B)	^{65}Co
05/24/1978	Kouzes	[224]	56	^{70}Zn	(α,^{7}Be)	^{67}Ni
06/04/1979	Pardo	[225]	70	^{92}Mo	(^{3}He,^{6}He)	^{89}Mo

the delayed protons. Figure 9.11 shows as an example the set-up used by Cable et al. for the discovery of ^{22}Al [198].

A different way to identify a new isotope produced in a nuclear reaction is to deduce its mass from the kinematics instead of its half-life. In a two-body reaction, the mass and energies of excited states of a produced nucleus can be reconstructed from the beam energy of the projectile and the energy and angle of the ejectile. Isotopes identified with this missing mass method are listed in Table 9.8.

In 1952 Harvey irradiated a ^{209}Bi target with 14 MeV deuterons from the MIT cyclotron and measured the energy of outgoing tritons in a triple coincidence proportional counter at angles varying from 30° to 150°. From the measured spectrum shown in Fig. 9.12 he deduced three excited states in ^{208}Bi [202].

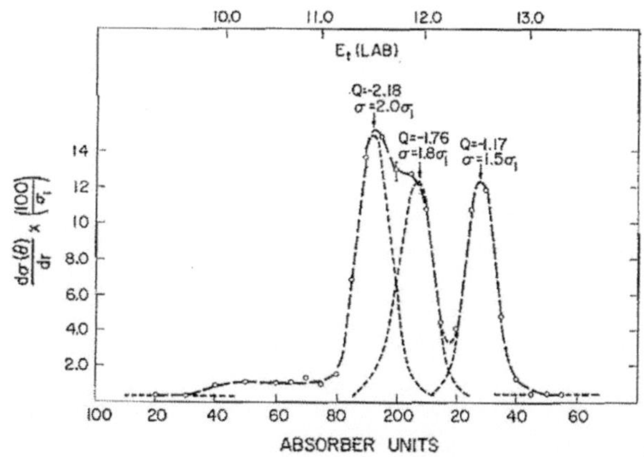

Fig. 9.12 Triton spectrum at 90° from the reaction ^{209}Bi(d,t)^{208}Bi. The Q-values shown in the figure were calculated relative to the −1.1 MeV Q-value of ^{208}Pb for the (d,t) reaction [202] (J.A. Harvey, *Energy levels in lead and bismuth and nuclear shell structure*, Can. J. Phys. 31 (1953) 278, Copyright 2008 Canadian Science Publishing or its licensors, reproduced with permission.)

A few years later, ^{21}F and ^{20}O were discovered by Jarmie [203] and Jarmie and Silbert [204], respectively, with a double focusing magnetic spectrograph following (t,p) reactions at the Los Alamos National Laboratory (LANL) Van de Graaff accelerator. ^{16}C was identified in 1960 with a broad range magnetic spectrograph by Hinds et al. at the Atomic Weapons Research Establishment in Aldermaston, England. In addition to the Q-value and the corresponding mass excess of ^{16}C, Hinds et al. also observed β-delayed neutrons [205]. Three additional isotopes were discovered with relatively simple transfer reactions using spectrographs. ^{46}K and ^{51}Sc were measured by Marinov and Erskine [209] and Erskine et al. [211], respectively, with the Argonne National Laboratory broad range spectrograph and ^{110}Sn was identified by Bassani et al. with the 40-in. 180° magnetic spectrometer at the University of Minnesota linear accelerator [210].

Ajzenberg-Selove, Cranberg, and Dietrich applied the missing mass technique in 1961 to neutron ejectiles in the discovery of ^{22}Mg. They irradiated a ^{20}Ne target with 3.4 MeV ^{3}He ions from the LANL large vertical Van de Graaff generator and determined the energy of the ejected neutrons by measuring their time of flight from the target to a proton recoil plastic scintillator detector [206]. Two years later, Bryant et al. also used the (^{3}He,n) reactions at LANL to identify ^{42}Ti [207].

The first more complex transfer reaction was used by Cerny et al. in 1964 to discover ^{9}C. 65 MeV ^{3}He ions from the Berkeley 88-in. cyclotron irradiated a ^{12}C target and the ^{6}He ejectiles were identified and measured with a ΔE-E semiconductor counter telescope. The ^{6}He spectrum displaying the ground-state peak of ^{9}C is shown in Fig. 9.13 [208].

Similar telescopes were subsequently used to identify ^{52}Ti at LANL [212], ^{38}Ca at the University of Oxford, England [213], and ^{23}Al [214] again at Berkeley. For the discovery of ^{45}Ar and ^{46}Ar in 1973 at Berkeley Jelley et al. used a more sophisticated

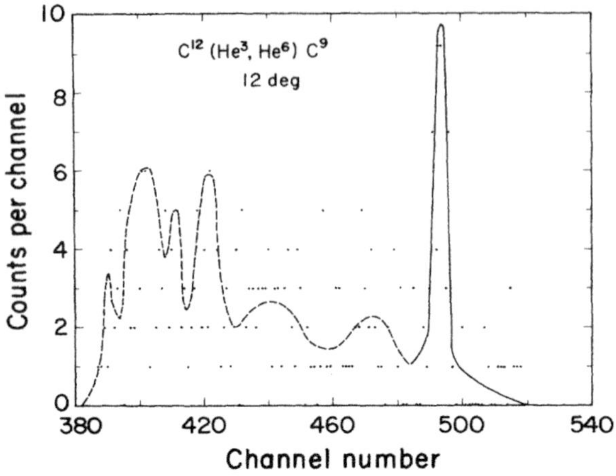

Fig. 9.13 Energy spectrum from the reaction $^{12}C(^{3}He,^{6}He)^{9}C$ at 12°. The *dashed line* at lower channels than the ^{9}C ground-state peak merely represents an average of the scattered counts in this region [208] (Reprinted figure with permission from J. Cerny et al., *Completion of the mass-9 isobaric quartet via the three-neutron pickup reaction* $C^{12}(He^{3}, He^{6})C^{9}$, Phys. Rev. Lett. 13 (1964) 726. Copyright 1964 by the American Physical Society.)

four element telescope consisting of two transmission ΔE detectors and E detector backed up by a reject detector [216].

In the 1970s all remaining isotopes were discovered with complicated transfer reactions using magnetic spectrographs. Robertson used the double focusing magnetic analyzer at the Institut für Kernphysik of the Kernforschungsanlage in Jülich, Germany, to identify ^{20}Mg in 1974 [217]. Between 1972 and 1979 nine isotopes were discovered by the group of Kashy and Benenson with the Enge split-pole magnetic spectrograph at Michigan State University using ^{3}He beams [215, 218, 219, 222, 225]. Tribble and his collaborators also used an Enge split-pole spectrograph in the discovery of ^{36}Ca [220] and ^{50}Fe and ^{54}Ni [221] at the Cyclotron Institute of Texas A&M University in 1976. Two years later, ^{65}Co and ^{67}Ni were identified by Kouzes and Mueller [223] and Kouzes, Mueller, and Yu [224], respectively, with the quadrupole-dipole-dipole-dipole (QDDD) spectrograph at Princeton University.

References

1. R.S. Krishnan, E.A. Nahum, Proc. Camb. Phil. Soc. **36**, 490 (1940)
2. J.J. O'Connor, M.L. Pool, J.D. Kurbatov, Phys. Rev. **62**, 413 (1942)
3. C.T. Hibdon, M.L. Pool, Phys. Rev. **67**, 313 (1945)
4. M.L. Wiedenbeck, Phys. Rev. **70**, 435 (1946)
5. A.D. Schelberg, M.B. Sampson, A.C.G. Mitchell, Rev. Sci. Instrum. **19**, 458 (1948)
6. E.L. Kelly, E. Segrè, Phys. Rev. **75**, 999 (1949)
7. L.W. Alvarez, Phys. Rev. **75**, 1815 (1949)

8. E.L. Hudspeth, C.P. Swann, N.P. Heydenburg, Phys. Rev. **77**, 736 (1950)
9. W.M. Martin, S.W. Breckon, Can. J. Phys. **30**, 643 (1952)
10. N.W. Glass, L.K. Jensen, Phys. Rev. **90**, 320 (1953)
11. J.L.W. Churchill, W.M. Jones, S.E. Hunt, Nature **172**, 460 (1953)
12. J.D. Gow, L.W. Alvarez, Phys. Rev. **94**, 365 (1954)
13. M.M. Winn, Proc. Phys. Soc. A **67**, 949 (1954)
14. N.W. Glass, J.R. Richardson, Phys. Rev. **98**, 1251 (1955)
15. V. Kistiakowsky Fischer, Phys. Rev. **99**, 764 (1955)
16. H. Morinaga, Phys. Rev. **100**, 431 (1955)
17. C.R. Sun, B.T. Wright, Phys. Rev. **109**, 109 (1958)
18. R.G. Wilson, M.L. Pool, Phys. Rev. **118**, 227 (1960)
19. O. Ames et al., Phys. Rev. **118**, 1599 (1960)
20. E.L. Robinson, O.E. Johnson, Phys. Rev. **120**, 1321 (1960)
21. E.L. Robinson, J.I. Rhode, O.E. Johnson, Phys. Rev. **122**, 879 (1961)
22. W.J. Tomlinson III, H.H. Stroke, Phys. Rev. Lett. **8**, 436 (1962)
23. Y. Shida et al., Phys. Lett. **13**, 59 (1964)
24. R. Messlinger, H. Morinaga, C. Signorini, Phys. Lett. **19**, 133 (1965)
25. R.G. Miller, R.W. Kavanagh, Phys. Lett. **22**, 461 (1966)
26. R.E. Berg, J.L. Snelgrove, E. Kashy, Phys. Rev. **153**, 1165 (1967)
27. T. Yamazaki, G.T. Ewan, S.G. Prussin, Phys. Rev. Lett. **20**, 1376 (1968)
28. T.A. Doron, W.A. Lanford, J. Inorg. Nucl. Chem. **34**, 443 (1972)
29. T. Paradellis, A.J. Houdayer, S.K. Mark, Nucl. Phys. A **174**, 617 (1971)
30. K. Miyano, H. Nakahara, C. Gil, J. Phys. Soc. Japan **33**, 1505 (1972)
31. M. Singh et al., Nucl. Phys. A **193**, 449 (1972)
32. R. Iafigliola et al., Can. J. Phys. **52**, 96 (1974)
33. A. Charvet, R. Chery, and R. Duffait, J. Phys. (Paris) **35**, L (1974)
34. K.L. Kosanke et al., Nucl. Instrum. Meth. **124**, 365 (1975)
35. V.N. Burminskii et al., JETP Lett. **22**, 54 (1975)
36. H.C. Hseuh, E.S. Macias, Phys. Rev. C **14**, 345 (1976)
37. K. Oxorn, A.J. Houdayer, S.K. Mark, Z. Phys. A **279**, 289 (1976)
38. J. Deslauriers, S.C. Gujrathi, S.K. Mark, Z. Phys. A **283**, 33 (1977)
39. R. Chiba et al., Phys. Rev. C **17**, 2219 (1978)
40. R.A. Zarifov, O.D. Kovrigin, V.P. Burminskii, Bull. Acad. Sci. USSR Phys. Ser. **43**(1), 76 (1979)
41. N.K. Aras, P.W. Gallagher, W.B. Walters, J. Phys. G **6**, L195 (1980)
42. J. Brunner et al., Helv. Phys. Acta **28**, 475 (1955)
43. E. Merz, Z. Naturforsch. **16a**, 1246 (1961)
44. C.J. Orth et al., Phys. Rev. C **8**, 718 (1973)
45. http://www.nndc.bnl.gov/ensdf/ENSDF, Evaluated Nuclear Structure Data File, maintained by the National Nuclear Data Center at Brookhaven National Laboratory, published in Nuclear Data Sheets (Academic Press, Elsevier Science)
46. E.C. Mallary, M.L. Pool, Phys. Rev. **76**, 1454 (1949)
47. B.G. Kiselev, V.R. Burmistrov, Sov. J. Nucl. Phys. **11**, 137 (1970)
48. J.W. Gofman, G.T. Seaborg, *The Transuranium Elements: Research Papers, Paper 19.14*, National Nuclear Energy Series IV, 14b (McGraw-Hill, New York, 1949), p. 1427
49. L. Jacobson, R. Overstreet, *Radiochemical Studies: The Fission Products, Paper 91*, National Nuclear Energy Series IV, 9 (McGraw-Hill, New York, 1951), p. 735
50. W.H. Sullivan, N.R. Sleight, E.M. Gladrow, Phys. Rev. **70**, 778 (1946)
51. J.J. Howland, D.H. Templeton, I. Perlman, Phys. Rev. **71**, 552 (1947)
52. M.H. Studier, E.K. Hyde, Phys. Rev. **74**, 591 (1948)
53. E.K. Hyde et al., *The Transuranium Elements: Research Papers, Paper 19.17*, National Nuclear Energy Series IV, 14b (McGraw-Hill, New York, 1949), p. 1439
54. D.W. Osborne, R.C. Thompson, Q. Van Winkle, *The Transuranium Elements: Research Papers, Paper 19.11*, National Nuclear Energy Series IV, 14b (McGraw-Hill, New York, 1949), p. 1397

55. W.C. Barber, Phys. Rev. **72**, 1156 (1947)
56. C.E. Leith, A. Bratenahl, B.J. Moyer, Phys. Rev. **72**, 732 (1947)
57. G. Wilkinson, Nature **160**, 864 (1947)
58. G. Wilkinson, Phys. Rev. **73**, 252 (1948)
59. M. Lindner, I. Perlman, Phys. Rev. **73**, 1124 (1948)
60. J.B. Chubbuck, I. Perlman, Phys. Rev. **74**, 982 (1948)
61. R. Overstreet, L. Jacobson, P.R. Stout, Phys. Rev. **75**, 231 (1949)
62. G. Wilkinson, H.G. Hicks, Phys. Rev. **74**, 1733 (1948)
63. G. Wilkinson, H.G. Hicks, Phys. Rev. **75**, 696 (1949)
64. G. Wilkinson, H.G. Hicks, Phys. Rev. **75**, 1370 (1949)
65. D.A. Orth et al., Phys. Rev. **75**, 1100 (1949)
66. G. Wilkinson, H.G. Hicks, Phys. Rev. **75**, 1687 (1949)
67. G. Wilkinson, H.G. Hicks, Phys. Rev. **77**, 314 (1950)
68. L. Marquez, I. Perlman, Phys. Rev. **78**, 189 (1950)
69. W.W. Meinke, G.T. Seaborg, Phys. Rev. **78**, 475 (1950)
70. T.C. Chu, Phys. Rev. **79**, 582 (1950)
71. G. Wilkinson, H.G. Hicks, Phys. Rev. **79**, 815 (1950)
72. G. Wilkinson, Phys. Rev. **80**, 495 (1950)
73. L.S. Germain, Phys. Rev. **80**, 937 (1950)
74. B.J. Stover, Phys. Rev. **80**, 99 (1950)
75. B.J. Stover, Phys. Rev. **81**, 8 (1951)
76. G. Wilkinson, H.G. Hicks, Phys. Rev. **81**, 540 (1951)
77. R.W. Hoff, J.O. Rasmussen, S.G. Thompson, Phys. Rev. **83**, 1068 (1951)
78. V. Kistiakowsky Fischer, Phys. Rev. **87**, 859 (1952)
79. S.V. Castner, D.H. Templeton, Phys. Rev. **88**, 1126 (1952)
80. J.O. Rasmussen, S.G. Thompson, A. Ghiorso, Phys. Rev. **89**, 33 (1953)
81. D.C. Dunlavey, G.T. Seaborg, Phys. Rev. **92**, 206 (1953)
82. W.W.T. Crane, G.M. Iddings, Phys. Rev. **95**, 1702 (1954)
83. J.R. Wilkinson, R.K. Sheline, Phys. Rev. **99**, 752 (1955)
84. V.S. Shirley, W.G. Smith, J.O. Rasmussen, Nucl. Phys. **4**, 395 (1957)
85. C.J. Gallagher Jr., M. Sweeney, J.O. Rasmussen, Phys. Rev. **108**, 108 (1957)
86. K.S. Toth, J.O. Rasmussen, Phys. Rev. **109**, 121 (1958)
87. D.R. Nethaway, A.A. Caretto Jr., Phys. Rev. **109**, 504 (1958)
88. K.S. Toth, J. Inorg. Nucl. Chem. **7**, 1 (1958)
89. R.M. Diamond, J.M. Hollander, Nucl. Phys. **8**, 143 (1958)
90. K.S. Toth, K.T. Faler, J.O. Rasmussen, Phys. Rev. **115**, 158 (1959)
91. R.D. MacFarlane, J. Inorg. Nucl. Chem. **19**, 9 (1961)
92. L.J. Goodman, M.L. Pool, Phys. Rev. **71**, 288 (1947)
93. L.E. Glendenin, R.R. Edwards, Phys. Rev. **71**, 742 (1947)
94. A.C.G. Mitchell, E.T. Jurney, M. Ramsey, Phys. Rev. **71**, 825 (1947)
95. J.E. Edwards, M.L. Pool, Phys. Rev. **72**, 384 (1947)
96. K.D. Coleman, M.L. Pool, Phys. Rev. **72**, 1070 (1947)
97. D.J. Tendam, H.L. Bradt, Phys. Rev. **72**, 1118 (1947)
98. M.L. Pool, N.L. Krisberg, Phys. Rev. **73**, 1035 (1948)
99. W.S. Cowart et al., Phys. Rev. **73**, 1454 (1948)
100. D.T. Eggen, M.L. Pool, Phys. Rev. **74**, 57 (1948)
101. E.E. Motta, G.E. Boyd, Phys. Rev. **74**, 220 (1948)
102. L.L. Woodward, D.A. McCown, M.L. Pool, Phys. Rev. **74**, 761 (1948)
103. L.L. Woodward, D.A. McCown, M.L. Pool, Phys. Rev. **74**, 870 (1948)
104. D.A. McCown, L.L. Woodward, M.L. Pool, Phys. Rev. **74**, 1315 (1948)
105. D.N. Kundu, M.L. Pool, Phys. Rev. **74**, 1775 (1948)
106. R. Sherr, H.R. Muether, M.G. White, Phys. Rev. **75**, 282 (1949)
107. D.N. Kundu, M.L. Pool, Phys. Rev. **75**, 1690 (1949)
108. R.A. Hinshaw, M.L. Pool, Phys. Rev. **76**, 358 (1949)

109. A.C.G. Mitchell et al., Phys. Rev. **76**, 1450 (1949)
110. B.E. Robertson, W.E. Scott, M.L. Pool, Phys. Rev. **76**, 1649 (1949)
111. D.L. Anderson, M.L. Pool, Phys. Rev. **77**, 142 (1950)
112. R.E. Hein, A.F. Voigt, Phys. Rev. **79**, 783 (1950)
113. J.Y. Mei, C.M. Huddleston, A.C.G. Mitchell, Phys. Rev. **79**, 429 (1950)
114. F.D.S. Butement, Proc. Phys. Soc. A **63**, 775 (1950)
115. J.R. Gum, M.L. Pool, Phys. Rev. **80**, 315 (1950)
116. J.K. Long, M.L. Pool, Phys. Rev. **85**, 137 (1952)
117. S.C. Fultz, M.L. Pool, Phys. Rev. **86**, 347 (1952)
118. C.L. Scoville, S.C. Fultz, M.L. Pool, Phys. Rev. **85**, 1046 (1952)
119. A.B. Smith, A.C.G. Mitchell, R.S. Caird, Phys. Rev. **87**, 454 (1952)
120. A. van der Wiel, A.H.W. Aten Jr., Physica **18**, 356 (1952)
121. A.H. Wapstra, N.F. Verster, M. Boelhouwer, Physica **19**, 138 (1953)
122. R.M. Diamond, Phys. Rev. **89**, 1149 (1953)
123. T.H. Handley, E.L. Olson, Phys. Rev. **90**, 500 (1953)
124. B.L. Cohen, Phys. Rev. **91**, 74 (1953)
125. T.H. Handley, E.M. Olson, Phys. Rev. **92**, 1260 (1953)
126. T.H. Handley, E.M. Olson, Phys. Rev. **93**, 524 (1954)
127. R.A. Sharp, R.M. Diamond, Phys. Rev. **93**, 358 (1954)
128. D. Maeder, A.H. Wapstra, Phys. Rev. **93**, 1433 (1954)
129. T.H. Handley, E.M. Olson, Phys. Rev. **94**, 968 (1954)
130. R.M. Diamond, Phys. Rev. **95**, 410 (1954)
131. T.H. Handley, W.S. Lyon, E.M. Olson, Phys. Rev. **98**, 688 (1955)
132. R.A. Naumann, Phys. Rev. **96**, 90 (1954)
133. T.H. Handley, E.L. Olson, Phys. Rev. **96**, 1003 (1954)
134. F.D.S. Butement, E.G. Prout, Phil. Mag. **46**, 357 (1955)
135. A.H.W. Aten Jr., T. De Vries-Hamerling, Physica **21**, 597 (1955)
136. A.H.W. Aten Jr., G.D. de Feyfer, Physica **21**, 543 (1955)
137. J.D. Knight, E.W. Baker, Phys. Rev. **100**, 1334 (1955)
138. L. Lindner, G.A. Brinkman, Physica **21**, 747 (1955)
139. F.A. Johnson, Can. J. Phys. **33**, 841 (1955)
140. B.J. Dropesky, A.W. Schardt, Phys. Rev. **102**, 426 (1956)
141. I. Fränz, R. Radicella, J. Rodriguez, Z. Naturforsch. **11a**, 1038 (1956)
142. E. Ricci, J. Pahissa Campá, N. Nussis, Publs. Com. Nacl. Energia Atomica, Ser. quim **1**, 129 (1957)
143. J.W. Mihelich, B. Harmatz, T.H. Handley, Phys. Rev. **108**, 989 (1957)
144. I.P. Selinov et al., Sov. At. Energy **5**, 1605 (1958)
145. B. Harmatz, T.H. Handley, J.W. Mihelich, Phys. Rev. **114**, 1082 (1959)
146. I.P. Selinov et al., Sov. At. Energy **7**, 1011 (1959)
147. J.R. Grover, Phys. Rev. **116**, 406 (1959)
148. I. Gratot et al., Nucl. Phys. **13**, 302 (1959)
149. B.J. Dropesky, A.W. Schardt, T.T. Shull, Nucl. Phys. **16**, 357 (1960)
150. B. Harmatz, T.H. Handley, J.W. Mihelich, Phys. Rev. **119**, 1345 (1960)
151. I.P. Selinov et al., Sov. Phys. JETP **11**, 1191 (1960)
152. E.G. Funk Jr., J.W. Mihelich, C.F. Schwerdtfeger, Phys. Rev. **120**, 1781 (1960)
153. R.C. Catura, J.R. Richardson, Phys. Rev. **126**, 646 (1962)
154. G. Wolzak, H. Morinaga, Radiochim. Acta **1**, 225 (1963)
155. R. van Lieshout et al., Phys. Lett. **9**, 164 (1964)
156. H. Morinaga, G. Wolzak, Phys. Lett. **11**, 148 (1964)
157. K. Kotajima, K.W. Brockman Jr., G. Wolzak, Nucl. Phys. **65**, 109 (1965)
158. D.E. Khulelidze, V.L. Chikhladze, V.G. Onufriev, Bull. Acad. Sci. USSR Phys. Ser. **29**, 732 (1965)
159. H.P. Malan, H. Münzel, G. Pfennig, Radiochim. Acta **5**, 24 (1966)
160. A.E. Norris, G. Friedlander, E.M. Franze, Nucl. Phys. **86**, 102 (1966)

161. A.H.W. Aten Jr., J.C. Kapteijn, Physica **32**, 1159 (1966)
162. K.J. Hofstetter, P.J. Daly, Phys. Rev. **152**, 1050 (1966)
163. A.H.W. Aten Jr., J.C. Kapteijn, Physica **33**, 705 (1967)
164. K.A. Keller, H. Münzel, Radiochim. Acta **9**, 176 (1968)
165. C.J. Toeset, A.H.W. Aten Jr., Radiochim. Acta **9**, 55 (1968)
166. A.H.W. Aten Jr., J.C. Kapteijn, Radiochim. Acta **12**, 218 (1969)
167. R.G.H. Robertson, S.M. Austin, Phys. Rev. Lett. **29**, 130 (1972)
168. J. Rivier, R. Moret, Radiochim. Acta **22**, 27 (1975)
169. W.H. Sullivan, N.R. Sleight, E.M. Gladrow, *Radiochemical Studies: The Fission Products, Paper 333*, National Nuclear Energy Series IV, 9 (McGraw-Hill, New York, 1951), p. 1958
170. W.H. Sullivan, N.R. Sleight, E.M. Gladrow, *Radiochemical Studies: The Fission Products, Paper 107*, National Nuclear Energy Series IV, 9 (McGraw-Hill, New York, 1951), p. 817
171. W.H. Sullivan, N.R. Sleight, E.M. Gladrow, *Radiochemical Studies: The Fission Products, Paper 335*, National Nuclear Energy Series IV, 9 (McGraw-Hill, New York, 1951), p. 1967
172. R.J. Walen, J. Phys. Radium **10**, 95 (1949)
173. S.N. Ghoshal, Phys. Rev. **73**, 417 (1948)
174. F.L. Reynolds, D.G. Karraker, D.H. Templeton, Phys. Rev. **75**, 313 (1949)
175. G.M. Temmer, Phys. Rev. **76**, 424 (1949)
176. R.W. Fink, F.L. Reynolds, D.H. Templeton, Phys. Rev. **77**, 614 (1950)
177. R.A. Naumann, F.L. Reynolds, I. Perlman, Phys. Rev. **77**, 398 (1950)
178. D.G. Karraker, D.H. Templeton, Phys. Rev. **80**, 646 (1950)
179. M.M. Gransden, W.S. Boyle, Phys. Rev. **82**, 447 (1951)
180. J.R. Huizenga, C.M. Stevens, Phys. Rev. **96**, 548 (1954)
181. G.E. Boyd et al., Phys. Rev. **99**, 1030 (1955)
182. C. Dahlstrom, J.S. Foster, A.L. Thompson, Can. J. Phys. **36**, 1483 (1958)
183. T.E. Ward, Y.Y. Chu, J.B. Cumming, Phys. Rev. C **8**, 340 (1973)
184. M.E.J. Wigmans et al., Phys. Rev. C **14**, 243 (1976)
185. H.D. Smyth, Rev. Mod. Phys. **17**, 351 (1945)
186. L.W. Alvarez, Phys. Rev. **75**, 1127 (1949)
187. L.W. Alvarez, Phys. Rev. **80**, 519 (1950)
188. W.L. Gardner, N. Knable, B.J. Moyer, Phys. Rev. **83**, 1054 (1951)
189. R. Barton et al., Can. J. Phys. **41**, 2007 (1963)
190. R. McPherson, J.C. Hardy, R.E. Bell, Phys. Lett. **11**, 65 (1964)
191. R. McPherson, J.C. Hardy, Can. J. Phys. **43**, 1 (1964)
192. J.C. Hardy, R.I. Verrall, Phys. Lett. **13**, 148 (1964)
193. P.L. Reeder, A.M. Poskanzer, R.A. Esterlund, Phys. Rev. Lett. **13**, 767 (1964)
194. J.C. Hardy, R.I. Verrall, Phys. Rev. Lett. **13**, 764 (1964)
195. R. McPherson et al., Phys. Rev. **140**, B1513 (1965)
196. J. Äystö et al., Phys. Lett. B **82**, 43 (1979)
197. J. Honkanen et al., Nucl. Phys. A **366**, 109 (1981)
198. M.D. Cable et al., Phys. Rev. C **26**, 1778 (1982)
199. M.D. Cable et al., Phys. Lett. B **123**, 25 (1983)
200. J. Äystö et al., Phys. Rev. Lett. **55**, 1384 (1985)
201. H.R. Crane et al., Phys. Rev. **47**, 971 (1935)
202. J.A. Harvey, Can. J. Phys. **31**, 278 (1953)
203. N. Jarmie, Phys. Rev. **99**, 1043 (1955)
204. N. Jarmie, M.G. Silbert, Phys. Rev. Lett. **3**, 50 (1959)
205. S. Hinds et al., Phys. Rev. Lett. **6**, 113 (1961)
206. F. Ajzenberg-Selove, L. Cranberg, F.S. Dietrich, Phys. Rev. **124**, 1548 (1961)
207. H.C. Bryant et al., Nucl. Phys. **53**, 97 (1964)
208. J. Cerny et al., Phys. Rev. Lett. **13**, 726 (1964)
209. A. Marinov, J.R. Erskine, Phys. Lett. **14**, 46 (1965)
210. G. Bassani et al., Phys. Rev. **139**, B830 (1965)
211. J.R. Erskine, A. Marinov, J.P. Schiffer, Phys. Rev. **142**, 633 (1966)

212. D.C. Williams, J.D. Knight, W.T. Leland, Phys. Lett. **22**, 162 (1966)
213. J.C. Hardy, D.J. Skyrme, I.S. Towner, Phys. Lett. **23**, 487 (1966)
214. J. Cerny et al., Phys. Rev. Lett. **22**, 612 (1969)
215. I.D. Proctor et al., Phys. Rev. Lett. **29**, 434 (1972)
216. N.A. Jelley et al., Phys. Rev. C **9**, 2067 (1974)
217. R.G.H. Robertson et al., Phys. Rev. Lett. **32**, 1207 (1974)
218. W. Benenson et al., Phys. Rev. C **13**, 1479 (1976)
219. E. Kashy et al., Phys. Rev. C **14**, 1773 (1976)
220. R.E. Tribble, J.D. Cossairt, R.A. Kenefick, Phys. Rev. C **15**, 2028 (1977)
221. R.E. Tribble et al., Phys. Rev. C **16**, 917 (1977)
222. W. Benenson et al., Phys. Rev. C **15**, 1187 (1977)
223. R.T. Kouzes, D. Mueller, Nucl. Phys. A **307**, 71 (1978)
224. R.T. Kouzes, D. Mueller, C. Yu, Phys. Rev. C **18**, 1587 (1978)
225. R.C. Pardo et al., Phys. Rev. C **21**, 462 (1980)

Chapter 10
Spallation and Charged-Particle Induced Fission

10.1 High-Energy Nuclear Collisions

The first accelerators developed during the 1930s were only able to accelerate particles to about 10 MeV (for protons). This energy corresponds approximately to the Coulomb barrier for heavy elements so reactions induced by charged-particles concentrated on the light and medium mass region. For this reason, most of the radioactive isotopes of heavy elements were discovered using neutrons. Also, fission could only be induced by neutrons either from sources or produced in reactions on lithium or beryllium.

With the rapid development of more powerful accelerators, primarily the construction of larger cyclotrons, fission could also be directly induced with charged-particles (Sect. 10.2). With the completion of the construction of the 184-in. cyclotron at Berkeley in 1947, beam energies of more than 100 MeV became available. It was quickly realized that charged-particles accelerated to these energies could penetrate and excite the nucleus so that it would emit a large number of nucleons in the process. In this "spallation" of the nucleus many new radioactive isotopes could be populated that were not reachable in low-energy reactions (Sect. 10.3). The first report of a spallation reaction was presented in an abstract of the proceedings of the meeting of the American Physical Society in July 1947 by Cunningham et al. [1]. The term spallation for this process was first suggested by Sullivan and Seaborg a month later [2]. Already in October 1947, Serber published the theoretical description of spallation [3].

A major technical breakthrough for the discovery of isotopes using spallation (and charged-particle fission) was the development of the technique later known as ISOL (Isotope Separation On-Line). Kofoed-Hansen and Nielsen developed this method in 1950 at the Copenhagen cyclotron where they quickly extracted fission fragments following the irradiation of uranium and injected them into an ion source of an isotope separator (see Sect. 6.3) [4]. Isotopes discovered in spallation and charged-particle induced fission using the ISOL method for separation and identification are described in Sects. 10.4 and 10.5, respectively.

M. Thoennessen, *The Discovery of Isotopes*,
DOI 10.1007/978-3-319-31763-2_10

10.2 Charged-Particle Fission

For the discovery of isotopes fission induced by charged-particles did not offer a significant advantage over fast neutron fission at reactors (see Sects. 6.2 and 6.4). The fission fragment distribution from high-energy irradiation with deuterons peaks closer to the valley of stability and thus does not have a large excess of neutrons [5]. Also at lower beam energies, the mass distribution for a given element peaks at lower masses in charged-particle induced reactions relative to fast neutron irradiations [6].

At the time the 184-in. cyclotron came online in 1947, almost fifty neutron-rich fission fragments had been discovered at reactors as part of the Manhattan project. The eleven isotopes discovered by charged-particle induced fission and identified following chemical separation are listed in Table 10.1.

Goeckermann and Perlman discovered ^{66}Ni and ^{67}Cu in the fission fragments of bismuth following irradiation with 200 MeV deuterons from the Berkeley 184-in. cyclotron in 1948. The measured half-lives for these isotopes were only mentioned in footnotes [7]. The only other fission fragment discovered using the 184-in. cyclotron was ^{113}Pd reported by Hicks and Gilbert in 1953 [8].

Fission at relatively low incident energies was studied by Alexander, Schindewolf, and Coryell with the cyclotron of the Massachusetts Institute of Technology (MIT). Short-lived palladium and silver isotopes were isolated from the fission products by fast chemical separations, and their decay properties were studied by β- and γ-ray counting and spectroscopy or by periodic extractions of daughter activities [10]. Similarly, a few years later Gleit and Coryell discovered ^{118}Cd [11] and ^{119}Cd [12] also at MIT.

Hagebo, Kjelberg, and Pappas used again high-energy beams (protons) to induce fission of uranium and identified ^{129}Sn in 1961. They linked an 8.8 min half-life to the known 4.6 h antimony activity of ^{129}Sb by successive milking of chemically separated tin [9]. The paper does not specify whether the experiment was performed at the synchrocyclotron of the Institute for Nuclear Research in Amsterdam, The Netherlands, or at the synchrocyclotron of the Gustaf Werner Institute for Nuclear Chemistry in Uppsala, Sweden.

Table 10.1 Isotopes discovered in charged-particle induced fission: The submission date, first author and reference of the publication, as well as the target, beam, and beam energy (MeV) are listed

Date	First author	Ref.	Target	Beam	Energy	Isotope
03/22/1948	Goeckermann	[7]	Bi	d	200	^{66}Ni, ^{67}Cu
10/09/1953	Hicks	[8]	U	d	190	^{113}Pd
07/24/1961	Hagebo	[9]	U	p	170	^{129}Sn
01/31/1958	Alexander	[10]	U	d	14	^{114}Pd, ^{115}Pd, ^{114}Ag, ^{116}Ag, ^{117}Ag
11/02/1960	Gleit	[11]	U	d	14	^{118}Cd
07/20/1961	Gleit	[12]	U	d	14	^{119}Cd

10.3 Spallation

In contrast to charged-particle induced fission, spallation reactions provided a great opportunity to produce a large number of new isotopes. Essentially all isotopes with less protons and/or neutrons than the target could be produced in these reactions. When the final nucleus consists of only a small fraction of the target nucleus the process is also referred to as target fragmentation.

In the previous chapter, the new isotopes produced in light-charged-particle reactions at low beam energies were populated after the emission of only a few nucleons. Spallation reactions occur typically at higher energies and emit more nucleons in the process. There is no clear separation between the two processes. As an approximate guideline spallation was classified as reactions with energies above about 40 MeV per nucleon where more than about five nucleons were emitted.

Isotopes discovered in spallation until 1961 are listed in Table 10.2. Heavy isotopes produced by spallation from uranium and thorium are displayed separately in Table 10.3. Starting in 1957 a large number of isotopes were discovered at the Joint Institute for Nuclear Research at Dubna (Table 10.4). New observations of isotopes at other laboratories after 1961 are listed in Table 10.5. The most violent collisions where only a fraction of the initial target nucleus survives have been used to discover several light neutron-rich isotopes which are shown in Table 10.6.

The advantage of spallation reactions to populate a wide range of nuclides also represents a disadvantage in identifying the isotopes. The number of possible isotopes of a given element which could be isolated by various methods of chemical separations was even larger than in the light-charged-particle induced reactions at lower beam energies. In addition, cross bombardments with different beams and targets could not be used to narrow down the different possibilities. Thus it was even more difficult to determine the exact mass of a measured activity.

The first new isotopes discovered in a spallation reaction were ^{100}Pd, ^{101}Pd, ^{100}Rh and ^{101}Rh at the Berkeley 184-in. cyclotron by Lindner and Perlman. They irradiated antimony targets with 200 MeV deuterons and observed two new chemically separated activities of palladium decaying into radioactive rhodium daughters [13]. Assuming the reactions resulted primarily from the most stable antimony isotope ^{121}Sb, 5 protons and 16 neutrons were removed in the process to populate ^{100}Pd. It should be mentioned that the identification was based on unpublished results from the plutonium project where it had been suggested that similar half-lives originated from ^{100}Rh and ^{101}Rh [54]. However, the results were never published in the refereed literature and the report was also not included in Volume 9 of the National Nuclear Energy Series (NNES) on the radiochemical studies of fission products [55].

Less than two weeks later, Hopkins and Cunningham reported the half-lives of 28 different isotopes after the irradiation of arsenic with 200 MeV deuterons from the Berkeley 184-in. cyclotron. The newly observed isotopes ^{68}Ge and ^{72}Se wereproduced in (d,2p7n) and (d,5n) reactions, respectively [14]. Another eight

Table 10.2 Isotopes discovered in spallation until 1961: The submission date, first author and reference of the publication, as well as the target, beam, and beam energy (MeV) are listed

Date	First author	Ref.	Target	Beam	Energy	Isotope
04/05/1948	Lindner	[13]	Sb	d	200	^{100}Pd, ^{100}Rh^{a}, ^{101}Pd, ^{101}Rh^{a}
04/16/1948	Hopkins	[14]	As	d	200	^{68}Ge, ^{72}Se
06/17/1948	Miller	[15]	Cu	d	190	^{52}Fe, ^{62}Zn
12/29/1949	Neumann	[16]	Pb	d	100	^{198}Bi, ^{199}Bi, ^{200}Bi, ^{201}Bi, ^{203}Bi, ^{199}Pb^{a}, ^{200}Pb^{a}, ^{201}Pb^{a}, ^{201}Tl^{a}
01/05/1950	Hopkins	[17]	As	d	190	^{66}Ge, ^{67}Ge, ^{70}As, ^{70}Se
02/11/1950	Lindner	[18]	Sb	d, α	190, 380	^{109}Cd
04/13/1950	Thomas	[19]	Cs	p	250	^{129}Ba
05/06/1950	Fink	[20]	Cs	p	85	^{128}Ba
08/18/1950	Stover	[21]	La	p	80	^{133}Ce, ^{134}Ce, ^{134}La^{a}
09/07/1950	Rasmussen	[22]	Gd	p	150	^{149}Tb
10/27/1950	Karraker	[23]	Bi, Pb	p, α	–	^{205}Po, ^{204}Po, ^{203}Po, ^{202}Po, ^{205}Bi, ^{202}Bi^{a}
12/18/1950	Barton	[24]	Bi	α	380	^{203}At, ^{205}At, ^{207}At, ^{209}At
01/24/1951	Fink	[25]	Cs	p	96	^{128}Cs^{a}
02/07/1951	Hyde	[26]	Nb	p	100	^{86}Y, ^{86}Zr, ^{88}Zr
03/02/1951	Karraker	[27]	Bi	p	150	^{200}Po, ^{201}Po
07/03/1951	Batzel	[28]	Cu	p	340	^{47}Ca
08/13/1951	Fink	[29]	Au, Hg	p	55–96	^{192}Hg, ^{193}Hg, ^{195}Hg
03/10/1952	Worthington Jr.	[30]	Zn	p	340	^{56}Ni
04/01/1952	Rudstam	[31]	Fe	p	340	^{48}Cr
09/02/1952	Caretto Jr.	[32]	Y	p	130, 240	^{85}Y
09/17/1952	Lindner	[33]	Cs	p	190	^{127}Ba
09/17/1952	Dropesky	[34]	I	p	240	^{121}Xe, ^{122}Xe, ^{123}Xe
04/02/1953	Lindner	[35]	Cl	p	340	^{32}Si
09/17/1953	Gillon	[36]	Au	p	65	^{191}Hg, ^{191}Au^{a}
09/21/1953	Caretto Jr.	[37]	Y	p	240	^{76}Kr
01/19/1954	Michel	[38]	I, Ta	α	100, 350	^{125}Cs, ^{160}Er, ^{166}Yb
02/01/1954	Haldar	[39]	Ag	p	80–170	^{103}Ag
02/04/1954	Mathur	[40]	I	α	130	^{123}Cs
09/20/1954	Andersson	[41]	Tl	p	45–115	^{198}Pb, ^{197}Pb, ^{197}Tl^{a}, ^{196}Tl^{a}
09/27/1954	Nervik	[42]	Ta	p	340	^{169}Lu
02/23/1955	Smith	[43]	Pt	p	120	^{187}Au, ^{188}Au, ^{189}Au
			Ir	p	32	^{189}Pt, ^{189}Ir^{a}
08/02/1955	Roy	[44]	Cu	p	400	^{60}Fe
09/02/1955	Haldar	[45]	Re	p	120–240	^{177}Re, ^{178}Re

(continued)

Table 10.2 (continued)

Date	First author	Ref.	Target	Beam	Energy	Isotope
06/06/1956	Andersson	[46]	Tl	p	45–115	^{195}Pb, ^{196}Pb
04/01/1959	Ware	[47]	Ce	p	240	^{132}Ce
05/15/1959	Toth	[48]	Gd	p	60	^{150}Tb
09/11/1959	Jung	[49]	Hg, Tl	p	80–90	^{194}Pb, ^{194}Tl
11/26/1959	Butement	[50]	Tm	p	230	^{164}Yb
05/24/1960	Butement	[51]	Br	p	280	^{74}Kr, ^{75}Kr
06/28/1960	Andersson	[52]	Hg	p	75–105	^{192}Tl
01/23/1961	Merz	[53]	Ta, W	p	400	^{168}Hf, ^{170}Hf

[a]Populated by subsequent electron capture/β^+-decay

weeks later, Miller, Thompson, and Cunningham measured 21 isotopes in the spallation of copper. The paper lists the various identification methods used after chemical separation: "Identification of radioactive product isotopes was made by these half-life determinations supplemented in most cases by at least one of the following methods: (a) observation of parent daughter relationships when possible; (b) observation of the sign of the radiation by counting through a magnetic field; (c) rough energy determinations by counting through aluminum, beryllium, or lead absorbers; and (d) differential counting through different absorbers" [15]. It also demonstrates the broad range of isotopes produced. The two new isotopes discovered in this reaction, ^{62}Zn and ^{52}Fe were produced in a (d,5n) and a (d,4p11n) reaction, respectively. The lightest isotope identified in this reaction was the previously discovered phosphorus isotope ^{32}P which required a (d,15p34n) reaction.

The broad range of isotopes produced and identified within a single spallation experiment is also demonstrated in Fig. 10.1 [18]. Lindner and Perlman display the chart of nuclides in another interesting way. They plot the elements simultaneously as a function of mass number A and as a function of the difference between the numbers of neutrons and protons N−Z on diagonals. The plot shows that spallation reactions populate nuclides on both sides of the valley of stability. It also includes ^{109}Cd located between two stable cadmium isotopes which had previously not been uniquely identified.

^{109}Cd is another example of the difficulties in assigning credit for the discovery of long-lived isotopes. In 1940 Krishnan had reported a half-life of approximately one year but could not determine whether it was ^{109}Cd or ^{107}Cd [56]. At the 1941 meeting of the American Physical Society Helmholz reported a somewhat shorter half-life of 158 ± 7 d but he was also not able to make a mass assignment [57]. In 1945 Bradt et al. confirmed Helmholz's results, again assigning the activity to either ^{109}Cd or ^{107}Cd [58]. Bradt et al. corrected the half-life value to 330 d at the Meeting of the Swiss Physical Society on May 4, 1946: "The two silver isomers aredecay products

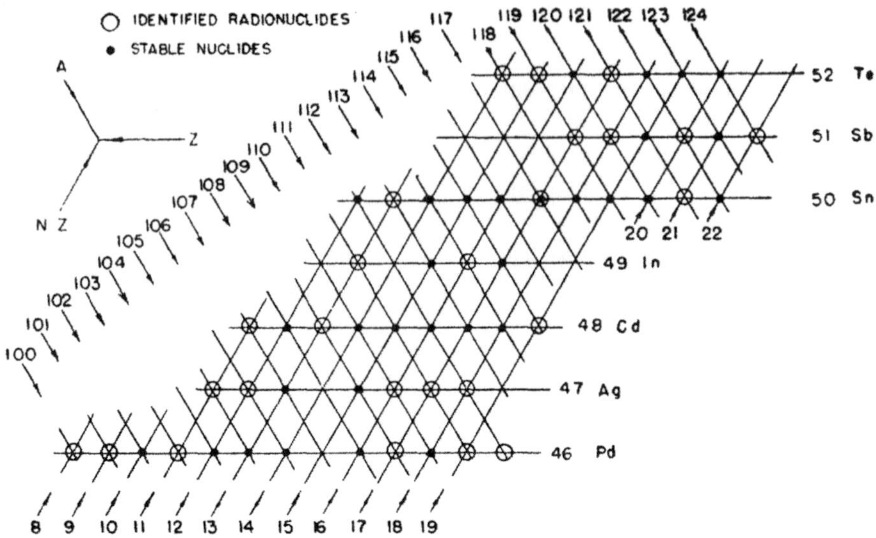

Fig. 10.1 Section of the chart of nuclides as presented by Lindner and Perlman. Nuclides of equal mass numbers are plotted on *top-left* to *bottom-right diagonals*. *Bottom-left* to *top-right diagonals* correspond to nuclides with constant N−Z [18]. The chart includes ^{109}Cd which was correctly identified for the first time (Reprinted figure with permission from M. Lindner and I. Perlman, *The Spallation Products of Antimony Irradiated with High Energy Particles*, Phys. Rev. 78 (1950) 499. Copyright 1950 by the American Physical Society.)

of the 6.7 h $Cd^{107,109}$ and the long-lived $Cd^{107,109}$. We determined the half-life of the latter—correcting Helmholz's value of 158 d—to be 330 d" [59]. The quote shows that a mass assignment was still not made. When less than three months later, on July 15, 1946, Helmholz submitted a paper where he assigned the 158 d activity to ^{109}Cd, he apparently was not aware of the longer half-life reported by Bradt et al. Thus, Helmholz is not credited with the discovery of ^{109}Cd because his measured half-life is significantly shorter than the presently accepted value of 461.4 ± 1.2 d [60].

One could argue that Bradt et al. and Helmholz should jointly be credited with the discovery of ^{109}Cd, as the former measured a half-life close to the correct value while the latter made the correct identification. By the time, Lindner and Perlman observed and identified ^{109}Cd in their spallation experiment they accepted the combined results as the identification of ^{109}Cd: "This [The 330-day Cd^{109}] activity was apparent in the decay curve taken through beryllium soon after the 6.7-hr. Cd^{107} had disappeared. Ultimately it could be seen in the curve taken without absorber after the 42-day Cd^{115} had decayed. Identification was made certain by removal of its daughter, 40-s. Ag^{109m}" [18]. Thus, they were the first to correlate the 330 d cadmium half-life with the isotope with mass 109.

The work by Lindner and Perlman was submitted in February 1950 and it should be mentioned that in May of 1950 Cork et al. reported a half-life of 250 d for ^{109}Cd [61] and in June of 1950 Gum and Pool measured a half-life of 470 days [62].

Until 1951, almost all spallation experiments were performed with the Berkeley 184-in. cyclotron. For example, Neumann and Perlman discovered five neutron-deficient bismuth isotopes and some of their β-decay lead and thallium daughter nuclides [16] and Fink and Templeton assigned an observed barium activity in the reaction ^{133}Cs(p,xn) to ^{128}Ba by mass spectroscopy [20]. Figure 10.2 shows the yields of the large range of isotopes produced in the irradiation of copper with 340 MeV protons. Of the observed isotopes only ^{47}Ca had not previously been identified [28]. The last isotopes discovered at Berkeley in the 1950s were several neutron-deficient gold, platinum and iridium isotopes reported by Smith and Hollander in 1955 [43].

The only other accelerator where new spallation products were observed during this time was the 130-in. cyclotron at the University of Rochester. In 1950, Thomas and Wiig discovered ^{129}Ba [19] and a year later Fink and Wiig identified ^{128}Cs following the β-decay of ^{128}Ba [25]. The group of Wiig continued their spallation experiments for the remainder of the decade and discovered another twelve isotopes: ^{192}Hg, ^{193}Hg, and ^{195}Hg [29] in collaboration with Fink in 1951, ^{85}Y [32] and ^{76}Kr

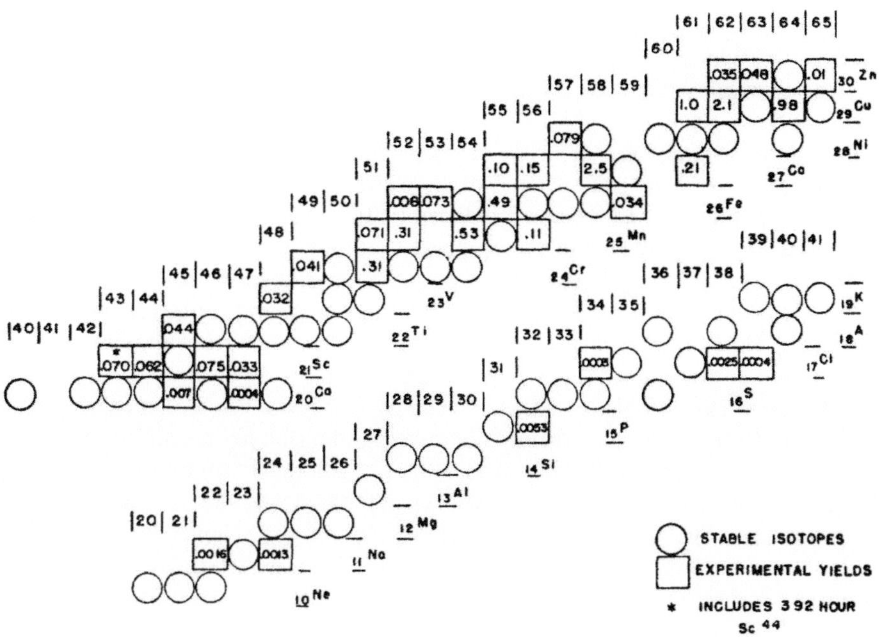

Fig. 10.2 Yields of radioactive isotopes produced in the irradiation of copper with 340 MeV protons. The yields are normalized to the production of ^{61}Cu [28] (Reprinted figure with permission from R.E. Batzel, D.R. Miller, and G.T. Seaborg, *The High Energy Spallation Products of Copper*, Phys. Rev. 84 (1951) 671. Copyright 1951 by the American Physical Society.)

[37] with Caretto Jr. in 1952 and 1953, respectively, ^{121}Xe, ^{122}Xe, and ^{123}Xe [34] with Dropesky also in 1952, ^{103}Ag [39] and ^{177}Re and ^{178}Re [45] with Haldar in 1954 and 1955, respectively, and finally ^{132}Ce [47] with Ware in 1959. Although the results for ^{177}Re and ^{178}Re were submitted on September 2, 1955 to Physical Review, the paper was only published almost one and a half years later on February 15, 1957 [45].

Between 1952 and 1961, three other U.S. institutions contributed to the expansion of the chart of nuclides using spallation reactions. Lindner and Osborne discovered ^{127}Ba by irradiating cesium with 190 MeV protons at Livermore [33], while Gillon et al. irradiated gold foils with 65 MeV protons from the Harvard University cyclotron to identify ^{191}Hg and its daughter ^{191}Au [36]. A few years later, ^{60}Fe and ^{168}Hf and ^{170}Hf were discovered using 400 MeV protons from the synchrocyclotron of the Carnegie Institute of Technology in Pittsburgh by Roy and Kohman [44] and Merz and Caretto [53], respectively.

Outside the U.S., the group of Andersson used the synchrocyclotron from the Gustaf Werner Institute for Nuclear Chemistry at the University of Uppsala, Sweden, to discover several neutron-deficient lead and thallium isotopes produced with (p,xn) reactions on thallium and mercury targets, respectively [41, 46, 49, 52]. In addition, Toth et al. observed ^{150}Tb by irradiating gadolinium oxide targets with 60 MeV protons [48]. Butement and his group discovered ^{164}Yb [50] and ^{74}Kr and ^{75}Kr [51] with over 200 MeV proton beams from the synchrocyclotron of the University of Liverpool in England.

Heavy nuclides produced by the spallation of uranium and thorium primarily decay by the sequential emission of α-particles which represent collateral branches of the naturally occurring radioactive decay chains. They eventually feed into known bismuth, polonium, and lead isotopes of these chains. All isotopes produced first by spallation of uranium and thorium are listed in Table 10.3. The method to produce, detect and identify these nuclides was pioneered at Berkeley in the late 1940s. Studier and Hyde had discovered the $A = 4n + 2$ protactinium side series by irradiating ^{232}Th with 19 MeV deuterons (see Sect. 9.3). Ghiorso, Seaborg, and Hyde then applied the technique to higher beam energies and discovered 33 isotopes from uranium through polonium in a series of spallation experiments until the mid 1950s. As an example, Fig. 10.3 shows the discovery of four isotopes of an $A = 4n + 3$ side series beginning at ^{227}U. In the figure some of the old notations for the isotopes are still used [66].

In 1963, Butement and Robinson identified ^{223}Rn and ^{224}Rn by their β^--decay daughters ^{223}Ra and ^{224}Ra, respectively. They populated the radon isotopes by irradiating thorium with 370 MeV protons from the cyclotron of the University of Liverpool [70]. In the same year Butement, Robinson, and Qaim extended the $A = 4n + 1$ collateral series starting at ^{221}Rn discovered by Momyer and Hyde in 1955 [69] to ^{213}Pb [71].

Hahn, Roche, and Toth irradiated enriched ^{238}U with protons from the Oak Ridge Isochronous Cyclotron (ORIC) to populate ^{229}Np. The discovery of ^{229}Np is included in Sect. 5.4 on transuranium isotopes. The four subsequent newly discovered α-emitters ^{225}Pa, ^{221}Ac, ^{217}Fr, and ^{213}At correspond to a collateral branch of the $A = 4n + 1$ decay chain leading to the known isotope ^{209}Bi [72]. Five years later

Table 10.3 Isotopes discovered in spallation on thorium and uranium: The submission date, first author and reference of the publication, as well as the target, beam, and beam energy (in MeV) are listed

Date	First author	Ref.	Target	Beam	Energy	Isotope
08/09/1948	Ghiorso	[63]	Th	d	80	^{216}At, ^{219}Fr, ^{220}Fr, ^{223}Ac, ^{224}Ac, ^{227}Pa, ^{228}Pa
12/02/1948	Meinke	[64]	Th	d, α	150	^{214}At, ^{216}Rn, ^{217}Rn, ^{218}Fr, ^{220}Ra, ^{221}Ra, ^{222}Ac, ^{224}Th, ^{225}Th, ^{226}Pa, ^{228}U, ^{229}U
11/30/1949	Hyde	[65]	Th	p	350	^{212}Fr, ^{212}Rn^{a}, ^{208}At^{b}
10/15/1951	Meinke	[66]	Th	α	150	^{215}Rn, ^{219}Ra, ^{223}Th, ^{227}U
04/18/1952	Momyer	[67]	Th	p	340	^{209}Rn, ^{210}Rn, ^{211}Rn
04/19/1955	Momyer	[68]	Th	p	340	^{208}Rn, ^{213}Ra
08/29/1955	Momyer	[69]	Th	p	110	^{217}Po, ^{221}Rn
04/03/1963	Butement	[70]	Th	p	370	^{223}Rn, ^{224}Rn
07/23/1963	Butement	[71]	Th	p	370	^{213}Pb
03/01/1968	Hahn	[72]	U	p	32–42	^{213}At, ^{217}Fr, ^{221}Ac, ^{225}Pa
06/07/1973	Viola	[73]	Th	α	140	^{226}U
04/29/1983	Chu	[74]	U	p	28,000	^{233}Ac

[a]Populated by β^{+}-decay
[b]Populated by α-decay

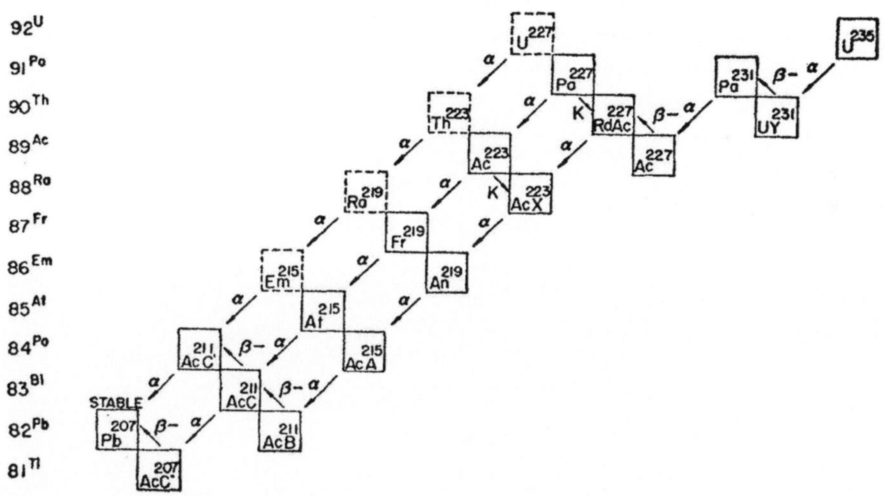

Fig. 10.3 Actinium (A = 4n + 3) family. The artificial collateral radioactive series was produced by ^{232}Th(α,9n)^{227}U and ^{232}Th(d,7n)^{227}Pa. The newly identified isotopes ^{227}U, ^{223}Th, ^{219}Ra, and ^{215}Rn are shown in *dashed boxes* [66] (Reprinted figure with permission from W.W. Meinke, A. Ghiorso, and G.T. Seaborg, *Further Work on Heavy Collateral Radioactive Chains*, Phys. Rev. 85 (1952) 429. Copyright 1952 by the American Physical Society.)

Viola, Minor, and Roche extended a collateral branch of the A = 4n + 2 chain with the discovery of ^{226}U produced in the ^{232}Th(α,10n) reaction at the University of Maryland cyclotron [73]. In 1983, Chu and Zhou irradiated uranium targets with 28 GeV protons from the Alternating Gradient Synchrotron (AGS) at Brookhaven National Laboratory and populated ^{233}Ac in the ^{238}U(p,4p2n) reaction. ^{233}Ac was identified by measuring characteristic γ-rays of the β^--decay daughter ^{233}Th after chemical separation [74].

Figure 10.4 shows a summary of the isotopes discovered in spallation reactions on uranium and thorium.

Two years after the 184-in. cyclotron was built at Berkeley, the construction of a synchrocyclotron was completed at the Joint Institute for Nuclear Research (JINR) at Dubna in the USSR. This "phasotron" shown in Fig. 10.5 was capable of accelerating protons to energies of 660 MeV and the first new isotopes were discovered with this

Table 10.4 Isotopes discovered in spallation at the Joint Institute for Nuclear Research at Dubna: The submission date, first author and reference of the publication, as well as the target are listed. All isotopes were produced with 660 MeV protons from the phasotron

Date	First author	Ref.	Target	Isotope
12/1957[a]	Gorodinskii	[77]	Ta	^{146}Gd, ^{146}Eu[b]
12/11/1957	Kuznetsova	[78]	Sb	^{116}Te, ^{117}Te
08/11/1958[c]	Aron	[79]	Ta	^{167}Lu
12/10/1958	Kalyamin	[80]	Ta	^{155}Ho
02/1959[a]	Antoneva	[81]	Ta	^{149}Eu
08/13/1959	Dneprovskii	[82]	Ta	^{158}Er, ^{158}Ho[b]
03/14/1960	Abdurazakov	[83]	Ta	^{164}Tm
09/23/1960	Lavrukhina	[84]	Ir	^{183}Ir
1086/1960[c]	Baranov	[85]	Au	^{184}Ir, ^{186}Pt, ^{187}Pt
06/20/1961	Abdurazakov	[86]	Ta	^{159}Er
06/11/1963	Abdumalikov	[87]	Ta	^{162}Tm, ^{162}Yb
04/30/1964	Gromov	[88]	Er	^{138}Nd
06/04/1965	Zhelev	[89]	Ta	^{157}Er, ^{157}Ho[b]
04/02/1966	Herrmann	[90]	Er	^{140}Sm, ^{141}Sm
1497/1968[c]	Zhelev	[91]	Gd	^{136}Nd, ^{136}Pr[b]
1133/1969[c]	Arlt	[92]	Ta	^{166}Lu[b], ^{167}Hf, ^{169}Hf
1144/1969[c]	Arlt	[93]	Hg/Re	^{168}Ta, ^{169}Ta, ^{170}Ta, ^{171}Ta
619/1970[c]	Arlt	[94]	Au	^{176}Os, ^{177}Os
706/1970[c]	Abdurazakov	[95]	Gd	^{133}Pr, ^{134}Nd, ^{135}Nd
1472/1970[c]	Vandlik	[96]	Pb	^{188}Tl, ^{190}Tl
04/08/1971	Arlt	[97]	Ag	^{92}Ru
1/1974[a]	Votsilka	[98]	Ag	^{86}Nb
4/1974[a]	Vandlik	[99]	Pb	^{191}Tl

[a]Issue/year of journal publication
[b]Populated by β^+-decay
[c]Page number/year of journal publication

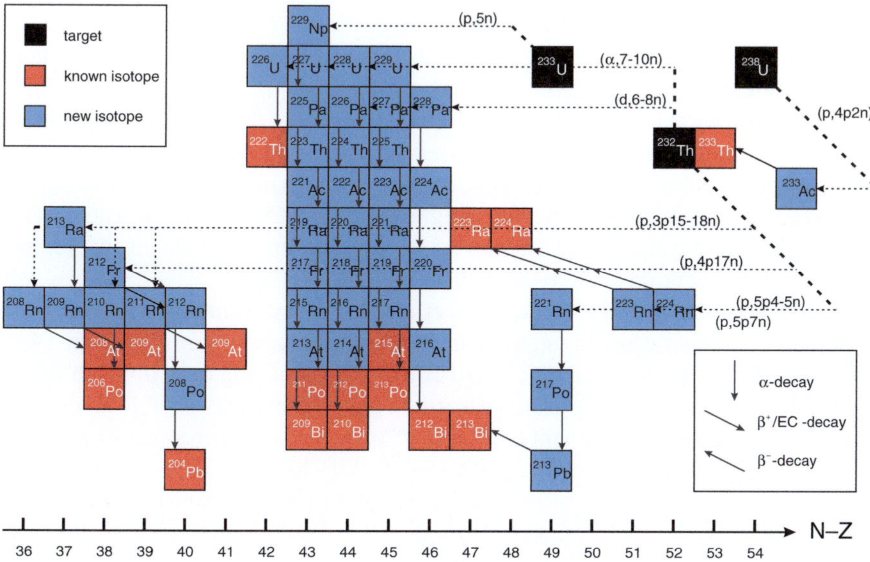

Fig. 10.4 Radioactive decay chains produced in spallation reactions on uranium and thorium. The isotopes of the different elements on the vertical axis are plotted versus N–Z so that the decay α chains follow vertical columns. In this representation, α and deuteron induced reactions are indicated by *vertical dashed lines* and proton induced reactions follow the *dashed diagonal lines*

Fig. 10.5 The phasotron at the Joint Institute for Nuclear Research in Dubna [75] (Reprinted from I.N. Meshkov, A.N. Sissakian, and G.V. Trubnikov, Proceedings of IPAC 2010, slides of TUZRA02, published under Creative Commons license CC-BY 3.0 [76].)

accelerator in 1957. The more than forty isotopes discovered with the phasotron and separated with chemical methods are listed in Table 10.4. All isotopes were produced by irradiating various targets with 660 MeV protons.

The identification of ^{146}Gd and ^{146}Eu represents the first discovery of an isotope in the USSR. Gorodinskii et al. measured a ~60 d activity in the chromatographically separated gadolinium fraction and an increasing ~5 d activity attributed to a europium daughter. In the paper the activities are assigned to mass 145, however, in a note added in proof the authors state: "We are now of the opinion that the Gd and Eu isotopes described in this section actually have mass number 146" [77].

The observation of ^{186}Pt by Baranov et al. [85] is an example of how difficult it can be in certain cases to assign credit for the discovery of an isotope. In 1955, Smith and Hollander assigned a 2.5 h half-life to ^{187}Pt decaying to ^{187}Ir which subsequently decayed to ^{187}Os with a half-life of 14 h [43]. Three years later, Diamond and Hollander revised the assignment of ^{187}Ir to ^{186}Ir by measuring the decay properties of several neutron-deficient iridium isotopes [100]. They actually credit Scharff-Goldhaber et al. quoted as a private communication for the definite assignment of ^{186}Ir. This reassignment of ^{186}Ir directly implied that the 2.5 h platinum half-life originated from ^{186}Pt instead of ^{187}Pt. Diamond and Hollander, however, do not mention platinum isotopes at all, so it is difficult to credit them with the discovery of ^{186}Pt. Baranov et al. were aware of the reassignment of ^{186}Ir and thus considered ^{186}Pt to be a known isotope. Although the assignment of the discovery of ^{186}Pt is debatable, Baranov et al. were the first to measure the 2.5 h half-life and assigning it to the correct isotope.

The last two isotopes identified after chemical separation (^{86}Nb [98] and ^{191}Tl [99]) were reported after an ISOL system had been established at the JINR. Isotopes discovered at Dubna with the online isotope separation facility YASNAPP are discussed in the next section.

While in the first thirteen years of spallation about 130 isotopes were discovered, only an additional 60 isotopes were first identified in spallation reactions using primarily chemical separation techniques in the subsequent twenty-five years. This decline is primarily due to the development of online separation techniques described in the next section. The remaining thirty isotopes of elements lighter than $Z = 85$ (astatine) discovered after 1961 at laboratories other than Dubna are listed in Table 10.5.

The two European institutions which had identified isotopes using spallation reactions in the late 1950s, the University of Uppsala and the University of Liverpool, subsequently discovered a few additional isotopes. Butement and collaborators discovered ^{85}Zr [101] and ^{101}Ag [105] using proton beams of 230 and 340 MeV from the synchrocyclotron at the University of Liverpool and ^{182}Pt, ^{183}Pt, ^{184}Pt [102] and ^{159}Tm [112] were produced with the synchrocyclotron of the Gustaf Werner Institute at the University of Uppsala by Graeffe, and Ekström, Olsmats, and Wannberg, respectively. Also, only one additional isotope was discovered at Berkeley in the mid 1960s. Korteling and Heyde used protons and α-particles up to energies of 880 MeV from the 184-in. cyclotron to produce ^{88}Nb [103].

Even higher beam energies were used to produce new isotopes at Brookhaven National Laboratory in 1965. Poskanzer, Esterlund, and McPherson irradiated cotton

Table 10.5 Isotopes discovered in spallation after 1961: The submission date, first author and reference of the publication, as well as the target, beam, and beam energy (MeV) are listed

Date	First author	Ref.	Target	Beam	Energy	Isotope
06/20/1962	Butement	[101]	Y	p	230	^{85}Zr
03/08/1963	Graeffe	[102]	Ir	p	90	^{182}Pt, ^{183}Pt, ^{184}Pt
06/11/1964	Korteling	[103]	Nb	p, α	720, 880	^{88}Nb
04/27/1965	Brun	[104]	Bi	p	80−155	^{197}Po, ^{198}Po, ^{199}Po
09/15/1965	Butement	[105]	Ag	p	340	^{101}Ag
11/24/1965	Poskanzer	[106]	Cotton	p	2,200	^{8}He
07/05/1967	Bleyl	[107]	Nd	d	50	^{139}Pm
01/23/1969	Toth	[108]	Dy	α	60−73	^{155}Er
06/23/1969	Chu	[109]	Ta	p	28,000	^{147}Tb
09/05/1969	de Boer	[110]	Er	p	54	^{158}Tm, ^{160}Tm
05/15/1970	Toth	[111]	Dy, Er	^{3}He	68−97	^{151}Er, ^{156}Yb, ^{157}Yb
03/18/1971	Ekström	[112]	Er	p	90	^{159}Tm
06/30/1971	van Klinken	[113]	Nd	α	104	^{139}Sm
12/1972[a]	Berlovich	[114]	Hg	p	1,000	^{172}Re, ^{174}Re, ^{175}Os
05/23/1973	Meijer	[115]	Tm	^{3}He	70−80	^{165}Lu
8/1974[a]	Berlovich	[116]	Tl	p	1,000	^{170}Re
01/06/1975	van Klinken	[117]	Pr, Sm	α	80−140	^{135}Pm, ^{139}Eu
03/26/1975	Nowicki	[118]	Pr	α	104	^{137}Pm
08/03/1981	Deslauriers	[119]	Nd	p	60	^{138}Pm
02/19/1986	Lunardi	[120]	Sm	α	65−107	^{142}Gd

[a]Issue/year of journal publication

fibers with 2.2 GeV protons from the Cosmotron to discover ^{8}He. They used the diffusion-loss method where the produced activities diffuse out of the target into a carrier gas which is then transported by expansion into counting chambers [106]. In 1969, Chu, Franz, and Friedlander identified ^{147}Tb by irradiating a tantalum target with 28 GeV protons from the Brookhaven Alternating Gradient Synchrotron (AGS) [109]. In two experiments in 1972 and 1974, Berlovich et al. discovered ^{172}Re, ^{174}Re, ^{175}Os [114] and ^{170}Re [116] with 1 GeV protons from the synchrocyclotron of the Leningrad Institute of Nuclear Physics of the Academy of Sciences of the USSR.

A few laboratories continued their studies at lower energies. Brun, Le Beyec, and Lefort discovered ^{197}Po, ^{198}Po, and ^{199}Po with 80−155 MeV protons from the Orsay synchrocyclotron in 1965 [104] and 50 and 104 MeV proton beams from the Karlsruhe Isochronous Cyclotron were used to identify ^{139}Pm, ^{139}Sm, and ^{137}Pm. While Bleyl, Münzel, and Pfennig still used chemical separation techniques to identify ^{139}Pm in 1967 [107], van Klincken et al. in 1971 identified ^{139}Sm without any separation by measuring γ-ray spectra with several Ge(Li) detectors in a low-background area where the target was moved to within 30s after the irradiation [113]. Four years later Nowicki et al. discovered ^{137}Pm by selecting mass 137 nuclides in an electromagnetic mass separator from chemically separated samples [118].

Gamma-ray spectroscopy without any chemical separation was also used to discover five isotopes in two different laboratories in The Netherlands. In 1969, ^{158}Tm and ^{160}Tm were identified by de Boer et al. following the irradiation of isotopically enriched ^{162}Er targets with 54 MeV protons from the synchrocyclotron at the Instituut voor Kernfysisch Onderzoek in Amsterdam [110]. Four years later, Meijer, de Boer, and Goudsmit observed ^{165}Lu in the irradiation of thulium foils with 70–80 MeV ^{3}He ions also at Amsterdam [115]. In 1975, van Klinken and Feenstra observed ^{135}Pm and ^{139}Eu with 80–140 MeV ^{4}He beams from the cyclotron of the Kernfysich Versneller Instituut at the Rijksuniversiteit Groningen [117].

Toth et al. built several different recoil catcher devices at Oak Ridge National Laboratory to transport radioactive recoil products in front of Si(Au) detectors to measure the subsequent α-emission. In 1969 they discovered ^{155}Er by irradiating dysprosium targets with 60–73 MeV α-particles from the isochronous cyclotron [108] and a year later ^{151}Er, ^{156}Yb, and ^{157}Yb were identified in the irradiation of dysprosium and erbium targets, respectively, with 68–97 MeV ^{3}He ions [111].

In 1981, Deslauriers, Gujrathi, and Mark discovered ^{138}Pm at the Foster Radiation Laboratory of McGill University in Montreal, Canada, by irradiating an enriched ^{142}Nd target with 60 MeV protons [119].

Finally, ^{142}Gd was first observed by Lunardi et al. at the Kernforschungsanlage in Jülich, Germany, by measuring γ-ray spectra following the ^{144}Sm(α,6n) reaction [120]. They considered the production mechanism a fusion-evaporation reaction rather than spallation demonstrating that there is no sharp distinction between these two reaction mechanisms. In the same paper the assignment of ^{152}Gd was confirmed in the fusion-evaporation reaction ^{110}Cd(^{35}Cl, p2n) performed with the Tandem XTU accelerator of the Laboratori Nazionali di Legnaro, Italy.

A major breakthrough for the detection and identification of isotopes far from stability was the 1966 experiment by Poskanzer et al. Instead of analyzing heavy fragments produced in spallation reactions, the experiment was designed to search for new particle-stable isotopes in light fragments produced in collisions of high-energy protons on uranium. All isotopes produced in these "target fragmentation" reactions are listed in Table 10.6.

In the original experiment by Poskanzer et al. 5.3 GeV protons from the Berkeley Bevatron irradiated a uranium foil and the fragments were measured directly in a silicon counter telescope located at 40° consisting of two ΔE detectors, an E detector and a rejection detector. A newly developed particle identifier system based on the signals from the telescope was able to achieve isotope separation of the fragments. Figure 10.6 shows the particle-identifier spectra from the experiment [121]. Poskanzer et al. were able to identify the neutron drip-line nuclides ^{11}Li, ^{12}Be, ^{14}B, and ^{15}B for the first time. In addition, from the non-observation of ^{10}He, ^{10}Li, and ^{13}Be it could be deduced that these isotopes were unbound with respect to neutron emission.

In a follow-up experiment two years later, Poskanzer et al. also observed ^{17}C for the first time [122]. Thomas et al. reported the discovery of ^{19}N and ^{21}O only one month later, utilizing the same method. They irradiated a gold target with 3 GeV protons from the Princeton-Pennsylvania Accelerator (a proton synchrotron) and

Table 10.6 Isotopes discovered in target fragmentation induced by high-energy protons: The submission date, first author and reference of the publication, as well as the target and the proton energy (GeV) are listed

Date	First author	Ref.	Target	Energy	Isotope
11/14/1966	Poskanzer	[121]	U	5.3	^{11}Li, ^{12}Be, ^{14}B, ^{15}B
07/16/1968	Poskanzer	[122]	U	5.5	^{17}C
08/20/1968	Thomas	[123]	Au	3.0	^{19}N, ^{21}O
06/25/1973	Bowman	[124]	U	4.8	^{14}Be, ^{17}B
10/10/1973	Bowman	[125]	U	4.8	^{19}C
04/08/1977	Butler	[126]	U	0.8	^{27}Ne, ^{31}Mg, ^{32}Mg, ^{34}Al, ^{39}P
04/04/1990	Tu	[127]	Th	0.8	^{54}Sc, ^{55}Sc

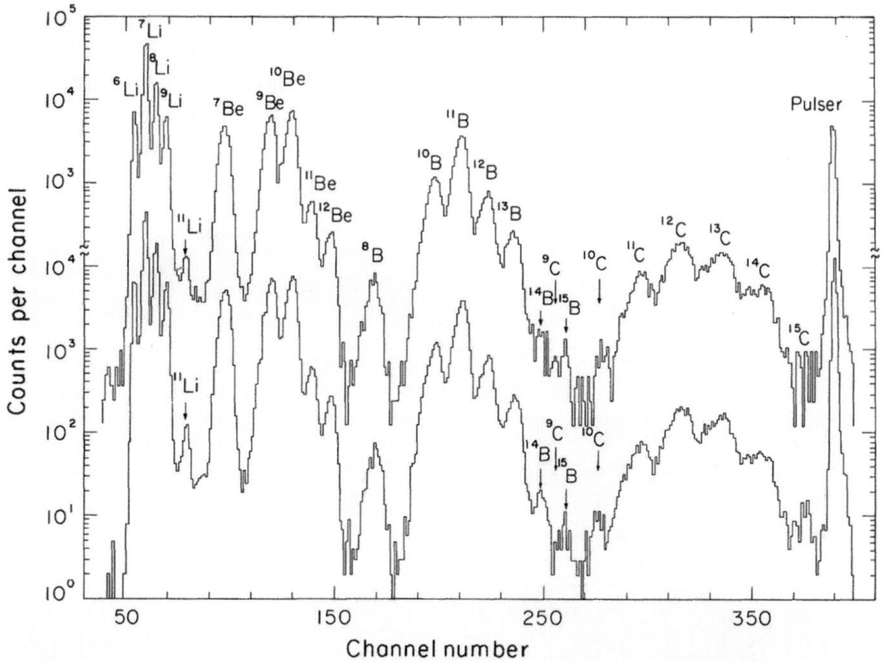

Fig. 10.6 Particle identification spectrum following the target fragmentation reaction of 5.3 GeV protons on uranium. The observed and identified isotopes are indicated in the figure. The *top* and *bottom* spectra were recorded with different solid angle coverage of the detector telescope [121] (Reprinted figure with permission from A.M. Poskanzer et al., *New Isotopes:* ^{11}*Li,* ^{14}*B, and* ^{15}*B*, Phys. Rev. Lett. 17 (1966) 1271. Copyright 1966 by the American Physical Society.)

used a counter telescope of four surface barrier detectors located at 45° to identify the fragments [123].

In 1973, Bowman et al. discovered ^{14}Be at Berkeley. They separated the ΔE and E detector by 25 cm and measured the time of flight between the detectors to determine the velocity of the fragments. The addition of the velocity as another observable significantly improved the isotopic resolution [124].

With the discovery of ^{14}Be—which had been predicted to be unbound—all bound isotopes of the first four elements were known. Figure 10.7 shows the chart of nuclides for elements up to titanium after Bowman et al. also discovered ^{19}C [125]. While the theoretical calculations correctly predicted the neutron drip-line for boron and carbon, they predicted too many nitrogen and oxygen and too few fluorine and neon isotopes to be particle bound.

^{27}Ne, ^{31}Mg, ^{32}Mg, ^{34}Al, and ^{39}P were discovered by Butler et al. in 1977. They irradiated a uranium target with 800 MeV protons from the Clinton P. Anderson Meson Physics Facility (LAMPF) at Los Alamos National Laboratory. The special time-of-flight technique between the fragment telescope and the accelerator radio-frequency structure allowed isotope separation up to mass 40. The paper displayed for the first time an event-by-event two-dimensional particle identification plot shown in Fig. 10.8 where the newly discovered isotopes can clearly be identified [126].

In the discovery of ^{54}Sc and ^{55}Sc in 1990 by Tu et al. the reaction products were measured and identified with a spectrometer instead of a simple detector telescope used in previous experiments. 800 MeV protons from LAMPF irradiated a thorium

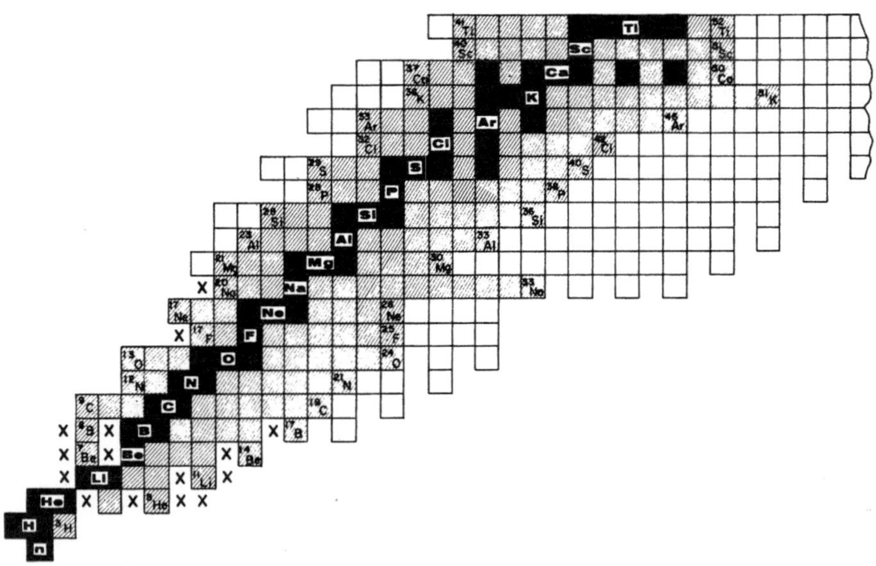

Fig. 10.7 Chart of nuclides displaying isotopes of elements up to titanium known in 1974. Stable isotopes are shown as *black squares* and the most neutron-deficient and most neutron-rich isotope for each element are labeled in the figure. *White squares* correspond to isotopes predicted to be bound but which had not yet been observed. Isotopes that had been shown to be unbound are marked by an X [125] (Reprinted figure with permission from J.D. Bowman et al., *Detection of neutron-excess isotopes of low-Z elements produced in high-energy nuclear reactions*, Phys. Rev. C 9 (1974) 836. Copyright 1974 by the American Physical Society.)

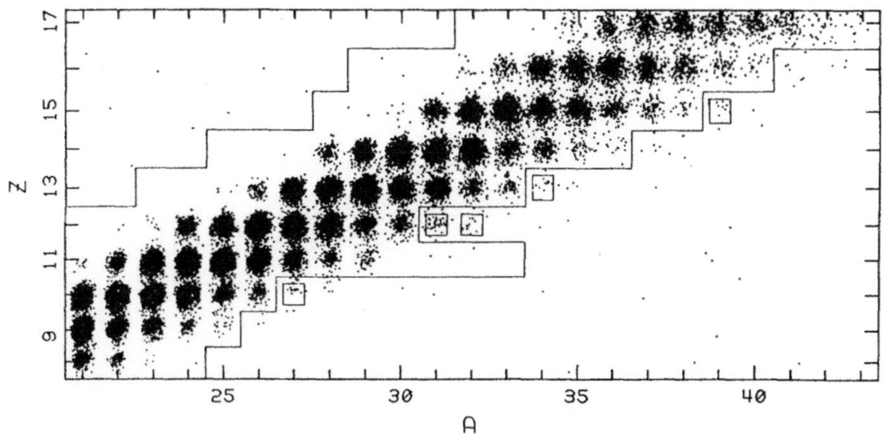

Fig. 10.8 Two-dimensional particle identification plot from the reaction of 800 MeV protons on uranium. Isotopes of different elements are plotted versus the mass number. The *solid lines* denote the limits of previously known isotopes. The individual *squares* indicate the discovery of the new isotopes ^{27}Ne, ^{31}Mg, ^{32}Mg, ^{34}Al, and ^{39}P [126] (Reprinted figure with permission from G.W. Butler et al., *Observation of the new nuclides* ^{27}*Ne*, ^{31}*Mg*, ^{32}*Mg*, ^{34}*Al, and* ^{39}*P*, Phys. Rev. Lett. 38 (1977) 1380. Copyright 1977 by the American Physical Society.)

target and fragments emitted at an angle of 87.5° were analyzed with the Time Of Flight Isochronous (TOFI) spectrometer. The mass excesses for 29 neutron-rich isotopes from chlorine to iron (including ^{54}Sc and ^{55}Sc) were measured for the first time [127].

10.4 Spallation with ISOL

By the mid 1960s essentially all isotopes close to the valley of stability had been discovered as shown in Fig. 10.9. In order to measure the shorter and shorter half-lives of isotopes increasingly further removed from stability it was necessary to develop a method to extract, separate and measure the produced activities as fast as possible. Such a system was first built at CERN in Geneva, Switzerland, in 1964 [128, 129]. This "Isotope Separation On-Line" technique was based on the pioneering work by Kofoed-Hansen and Nielsen in 1950 [4] (see Sect. 6.2).

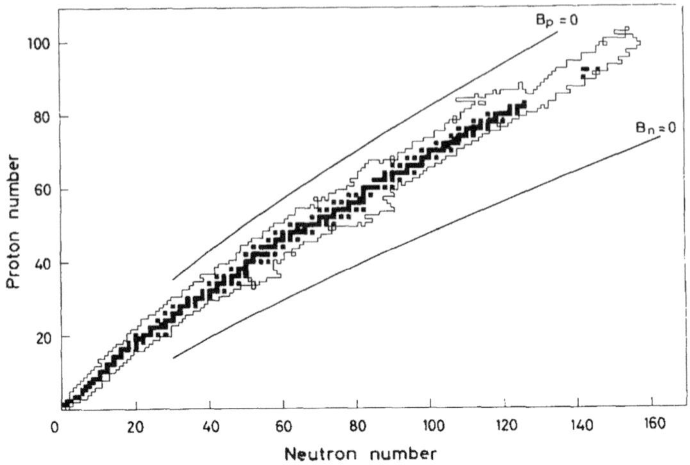

Fig. 10.9 Chart of nuclides in 1965. Stable isotopes are shown as *black squares*. Isotopes within the boundary were known at the time. The neutron and proton *drip-lines* are shown as *straight lines* labeled $B_n = 0$ and $B_p = 0$, respectively [128] (Reprinted from G. Rudstam, *Project for an isotope separator on-line with the CERN synchro-cyclotron*, Nucl. Instrum. Meth. 38 (1965) 282. Copyright 1965, with permission from Elsevier.)

Figure 10.10 shows the schematic of the set-up at CERN. Proton beams of 600 MeV accelerated by the synchrocyclotron irradiated a thick fission or spallation target (several cm long with a diameter of about 1 cm) which is held at a high temperature so that the produced radioactive isotopes diffuse out of the target. They are then pumped through a chemical separation stage directly into an ion source from which they are injected into the mass separator. At the focal plane of the separator, the mass-separated isotopes are collected on a catcher located in front of the detectors which measure the activities [128].

This facility, later called ISOLDE, was very productive and more than 100 new isotopes were discovered in fission and spallation reactions. Table 10.7 lists the isotopes discovered in spallation since 1964. Most of the experiments were performed using 600 MeV protons from the synchrocyclotron.

The addition of a mass separator to the selection of the radioactive isotopes significantly reduced the number of incorrect assignments of new isotopes. It was not necessary any more to deduce the specific isotope from cross-section systematics, excitation functions, or other indirect methods.

The first new isotopes identified at CERN were ^{118}Xe, ^{119}Xe, and ^{120}Xe reported by Andersson, Rudstam, and Sörensen in 1964. The 600 MeV protons irradiated a lanthanum target and xenon diffused directly into the ion source of the separator. Half-lives of 6 ± 1 min, 6 ± 1 min and 40 ± 1 min for ^{118}Xe, ^{119}Xe, and ^{120}Xe,

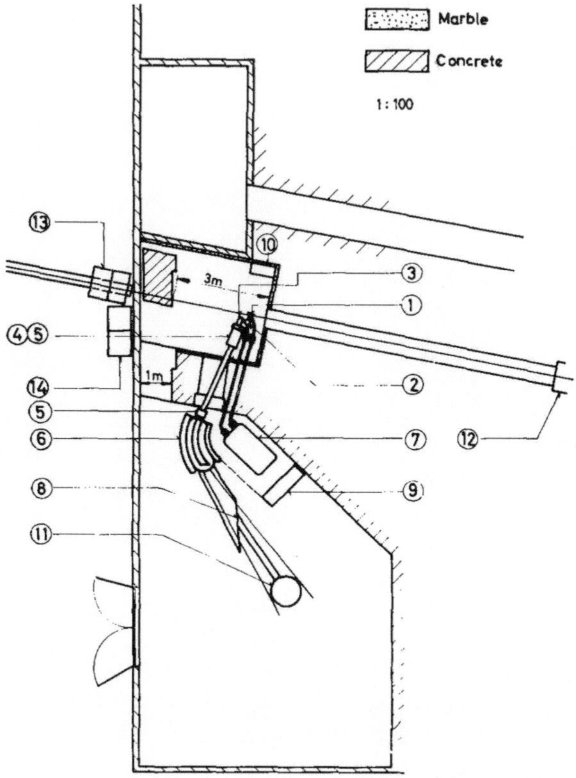

Fig. 10.10 General outline of the separator-on-line equipment (ISOLDE) at CERN in 1964: *1*—target on H.V. and high temperature; *2*—chemical separation apparatus; *3*—ion source; *4*—acceleration chamber; *5*—electrostatic lenses; *6*—separator magnet; *7*—high voltage set with power supplies for ion source and chemical separation apparatus; *8*—focal plane with sniffer stabilization; *9*—control desk; *10*—storage for active targets; *11*—measuring equipment (beta spectrometer); *12*—beam dump; *13*—pair of quadrupole lenses; *14*—shielding blocks. [128] (Reprinted from G. Rudstam, *Project for an isotope separator on-line with the CERN synchro-cyclotron*, Nucl. Instrum. Meth. 38 (1965) 282. Copyright 1965, with permission from Elsevier.)

respectively, were extracted from γ-ray spectra measured with a 3"×3" NaI(Tl) detector. The authors note that the identification would not have been possible with conventional methods, including milking techniques, because the iodine daughter isotopes have similar half-lives [130].

In 1968, Hansen et al. published a broad summary of all isotopes discovered during the first spallation experiments at ISOLDE. A total of seventeen isotopes from argon to radon were identified for the first time using a variety of different targets [131]. Subsequently 52 isotopes were discovered between 1969 and 1991 continuously developing new separation, detection and identification techniques.

Table 10.7 Isotopes discovered in spallation reactions at CERN-ISOLDE: The submission date, first author and reference of the publication, as well as the target, are listed. The incident beams were protons at 600 MeV unless otherwise noted

Date	First author	Ref.	Target	Isotope
06/03/1964	Andersson	[130]	La	^{118}Xe, ^{119}Xe, ^{120}Xe
11/22/1968	Hansen	[131]	Sn, Ce	^{101}Cd, ^{102}Cd, ^{115}Xe, ^{115}I^{a}, ^{116}Xe, ^{117}Xe, ^{117}I^{a}
			Ti, Pb	^{43}Ar, ^{181}Hg, ^{184}Au^{a}, ^{183}Hg, ^{184}Hg
			Th	^{224}Fr^{b}, ^{225}Rn, ^{225}Fr^{b}, ^{226}Rn, ^{226}Fr^{b}
07/04/1969	Chaumont	[132]	Ta, Th, U^{c}	^{76}Rb, ^{118}Cs, ^{119}Cs, ^{120}Cs, ^{121}Cs, ^{122}Cs, ^{124}Cs
10/28/1969	Hnatowich	[133]	Sn	^{100}Ag, ^{100}Cd
03/04/1970	Hansen	[134]	Pb	^{182}Au^{a}, ^{179}Hg, ^{180}Hg
09/03/1970	Hansen	[135]	Pb	^{178}Hg
10/16/1970	Hornshoj	[136]	Th	^{200}Rn
01/24/1972	Hornshoj	[137]	Zr	^{73}Kr
03/02/1972	Ravn	[138]	La	^{117}Cs
05/31/1972	Erdal	[139]	Ge	^{73}Zn, ^{74}Zn
08/03/1972	Klapisch	[140]	U^{c}	^{227}Fr, ^{228}Fr
12/05/1972	Scheidemann	[141]	Sn	^{122}Cd
03/22/1973	Hagberg	[142]	Ce	^{113}Xe
03/23/1974	Ravn	[143]	Th	^{75}Rb, ^{229}Fr, ^{229}Ra
07/09/1975	Westgaard	[144]	Th	^{222}Fr
11/23/1976	D'Auria	[145]	Nb	^{74}Rb
04/05/1977	Husson	[146]	Pb	^{180}Au
07/06/1977	Hagberg	[147]	V	^{32}Ar
09/05/1977	Hagberg	[148]	Ta	^{153}Yb, ^{157}Lu
11/17/1977	Elmroth	[149]	Sn	^{97}Cd, ^{98}Cd, ^{99}Cd
01/24/1978	D'Auria	[150]	La	^{114}Cs, ^{115}Cs
05/20/1980	Ewan	[151]	U	^{201}Fr, ^{202}Fr
08/05/1980	Ewan	[152]	Nb	^{71}Kr
09/23/1983	Ahmad	[153]	U	^{232}Ra
07/09/1984	Calaprice	[154]	Th	^{198}Rn
12/20/1984	Hill	[155]	U	^{231}Fr, ^{231}Ra^{b}
07/29/1986	Borge	[156]	U	^{227}Rn
09/12/1986	Kurcewicz	[157]	U	^{230}Fr
12/23/1988	Burke	[158]	Th	^{216}Bi^{d}, ^{221}At, ^{222}At, ^{223}At
03/17/1989	Borge	[159]	Th	^{228}Rn
02/13/1991	Bouldjedri	[160]	Th	^{182}Tl
08/23/1991	Grawe	[161]	Nb	^{79}Y
11/04/2008	Neidherr	[162]	U^{e}	^{229}Rn

[a]Populated by β^+-decay
[b]Populated by β^--decay
[c]24 GeV beam energy
[d]Populated by α-decay
[e]1.4 GeV beam energy

For example, in 1983 ^{232}Ra was identified by measuring the isotope shift using on-line collinear fast beam laser spectroscopy by Ahmad et al. [153]. The authors did not consider the observation a new discovery assuming that other properties such as the half-life must have been already measured. An isotope shift measurement had never been the first property measured for an isotope. The half-life of ^{232}Ra was only reported three years later [188]. A major upgrade of the facility to ISOLDE II in 1974 [189] was only mentioned in a discovery paper over ten years later by Hill et al. reporting the first observation of ^{231}Fr and ^{231}Ra [155]. The first new isotope discovered with a Penning-trap was ^{229}Rn by Neidherr et al. in 2008. A uranium-carbide target was irradiated with 1.4 GeV protons from the proton synchrotron booster accelerator and radon isotopes were transported into the double Penning-trap mass spectrometer ISOLTRAP. The identification of ^{229}Rn was confirmed with a half-life measurement at the ISOLDE spectroscopy station [162].

In addition to ISOLDE at CERN, several other ISOL facilities contributed to the discovery of new isotopes produced in spallation reactions as listed in Table 10.8.

Prior to the establishment of ISOLDE at CERN some of the major technical developments of isotope separators for analyzing radioactive isotopes were made at the Gustaf Werner Institute for Nuclear Chemistry in Uppsala, Sweden. In 1957, Aagaard et al. discovered ^{118}I and ^{120}I by irradiating cesium chloride with 170 MeV protons from the Uppsala synchrocyclotron. They separated and purified iodine with traditional chemical methods before placing it in the ion source of a mass separator [163].

In 1959 Albouy and Poffé developed a precursor of the ISOL method. They irradiated gold targets with 155 MeV protons from the Orsay synchrocyclotron and new mercury isotopes were produced in (p,8n–13n) reactions. Mercury was separated by fast vacuum distillation and then analyzed with a double magnetic deflection isotope separator. The gold isotopes were identified as daughter products from the mercury activities [164–166]. The method was subsequently used by several Orsay researchers to discover another 8 isotopes between 1963 and 1968 [167–173].

In 1974 the full ISOL system ISOCELE was established at Orsay. Putaux, Obert, and Aguer used the facility to identify ^{157}Tm for the first time [177]. The system was upgraded in 1976 to ISOCELE-2 as shown in Fig. 10.11. Between 1980 and 1988 Liang and collaborators discovered another six isotopes with this facility [180, 184–186].

Two ISOL facilities were constructed in the USSR in the 1970s. ^{77}Rb [174], ^{189}Tl [175], and ^{132}Pr, ^{148}Dy, and ^{161}Yb [176] were discovered with the online separation facility YASNAPP (Nuclear Spectroscopy on the Proton Beam) at the JINR in Dubna between 1972 and 1973. The IRIS (Investigation of Radioactive Isotopes at the Synchrocyclotron) mass separator located at the Leningrad Nuclear Physics Institute in Gatchina, USSR, utilized the 1 GeV proton beam from the synchrocyclotron. In a series of experiments between 1979 and 1982 Alkhazov and collaborators discovered eight isotopes of the rare earth elements lutetium, terbium, dysprosium, and promethium [178, 179, 181–183].

Table 10.8 Isotopes discovered in spallation using the ISOL technique: The submission date, first author and reference of the publication, as well as the target and the proton beam energy (MeV) are listed

Date	First author	Ref.	Target	Energy	Isotope
04/16/1957	Aagaard	[163]	Cs	170	^{118}I, ^{120}I
07/15/1959[a]	Albouy	[164]	Au	155	^{190}Hg, ^{190}Au[b]
05/1960[c]	Poffé	[165]	Au	155	^{189}Hg, ^{188}Hg
07/20/1960	Albouy	[166]	Au	155	^{185}Hg, ^{186}Hg, ^{187}Hg, ^{185}Au[b], ^{186}Au[b], ^{185}Pt[b]
04/02/1963	Santoni	[167]	Ta	155	^{173}W, ^{175}W
06/08/1964[a]	Paris	[168]	Tm	–	^{165}Yb
10/28/1964[a]	Abou-Leila	[169]	Hf	72	^{172}Ta
03/22/1965	Gerschel	[170]	La	150	^{130}Ce
10/08/1965	Lagarde	[171]	Ho	–	^{154}Ho
07/24/1967[a]	Paris	[172]	Tm	–	^{163}Yb
01/11/1968	Rahmouni	[173]	Sb	155	^{114}Te
02/15/1972	Arlt	[174]	Zr	660	^{77}Rb
02/23/1972	Vandlik	[175]	Pb	660	^{189}Tl
10/16/1973	Latuszynski	[176]	Ta	660	^{132}Pr, ^{148}Dy, ^{161}Yb
06/19/1974	Putaux	[177]	Er	157	^{157}Tm
05/24/1979	Alkhazov	[178]	W, Ta	1,000	^{158}Lu, ^{160}Lu, ^{163}Lu
04/01/1980	Alkhazov	[179]	W	1,000	^{159}Lu
06/06/1980	Liang	[180]	Er	200	^{152}Tm
10/10/1980	Alkhazov	[181]	W	1,000	^{145}Tb
03/30/1981	Alkhazov	[182]	W	1,000	^{146}Dy
02/11/1982	Alkhazov	[183]	W, Ta	1,000	^{136}Pm, ^{145}Dy
10/29/1982	Liang	[184]	Mo, Ta	280[d]	^{82}Zr, ^{167}Ta
06/06/1985	Liang	[185]	Lu	280[d]	^{162}Ta, ^{163}Ta
10/11/1988	Liang	[186]	Th	200	^{220}At
01/31/1990	Mezlev	[187]	U	1,000	^{232}Fr, ^{233}Ra, ^{234}Ra

[a] Date of presentation
[b] Populated in subsequent β^+-decay
[c] Month/year of journal publication
[d] ^{3}He beam

The discovery of ^{145}Dy [183] was simultaneously reported by Nolte et al. who produced the isotope in a fusion-evaporation reaction [191] (see Sect. 11.2). Finally, Mezlev et al. discovered ^{232}Fr, ^{233}Ra, and ^{234}Ra by irradiating the 1 GeV proton beam on a uranium carbide target [187].

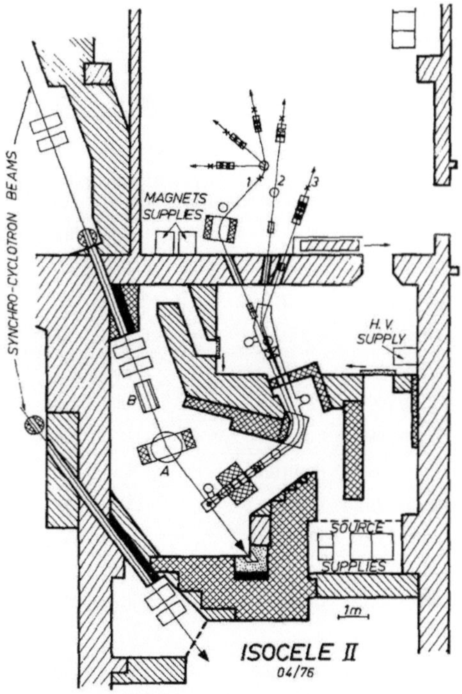

Fig. 10.11 ISOCELE-2 set-up at the Orsay synchrocyclotron [190] (Reprinted from P. Paris et al., *Development of the ORSAY high current on-line separator ISOCELE*, Nucl. Instrum. Meth. 139 (1976) 251. Copyright 1976, with permission from Elsevier.)

10.5 Charged-Particle Fission with ISOL

Isotopes discovered with the ISOL method using charged-particle induced fission are listed in Table 10.9. Since most of the isotopes were discovered at CERN-ISOLDE they are listed separately in Table 10.10. Fission induced by high energy charged-particles produce fragments over a wide range of masses. While the heavy fission fragments are clearly separated from spallation products, the mass distribution of the light fission fragments merge seamlessly with the mass distribution of target fragmentation products. Thus, Table 10.10 also includes the light isotopes discovered at ISOLDE by target fragmentation.

The ISOL method was applied for the first time to identify fragments produced in charged-particle induced fission at Orsay in 1967. Amarel et al. irradiated a uranium target with 150 MeV protons from the synchrocyclotron. The uranium was coated on 100 slabs of heated graphite. Fission fragments recoiled into the graphite and then diffused out. Rubidium and cesium were selectively ionized by surface ionization and guided into a Nier-type mass spectrometer. ^{95}Rb, ^{96}Rb and ^{144}Cs were identified for the first time by measuring β- and γ-rays with a plastic scintillator telescope. In addition, ^{144}Ba and ^{144}La were discovered from the decay of ^{144}Cs [192]. In two subsequent experiments, Amarel et al. and Macias-Marques et al. discovered ^{97}Rb in 1968 [193] and ^{96}Sr in 1970 [194], respectively.

Table 10.9 Isotopes discovered in proton-induced fission of uranium at ISOL facilities: The submission date, first author and reference of the publication, as well as the proton energy (MeV) are listed

Date	First author	Ref.	Energy	Isotope
03/13/1967	Amarel	[192]	150	^{95}Rb, ^{96}Rb, ^{144}Cs, ^{144}Ba[a], ^{144}La[a]
09/03/1968	Amarel	[193]	150	^{97}Rb
11/18/1970	Macias-Marques	[194]	150	^{96}Sr[a]
01/18/1971	Tracy	[195]	50	^{98}Rb, ^{99}Rb, ^{98}Sr, ^{145}Cs, ^{146}Cs
09/15/1987	Äystö	[196]	20	^{115}Rh
10/26/1987	Äystö	[197]	20	^{114}Rh
04/04/1988	Penttilä	[198]	20	^{111}Tc, ^{113}Ru
03/28/1990	Äystö	[199]	20	^{112}Tc
08/20/1990	Penttilä	[200]	20	^{119}Pd
03/19/1991	Leino	[201]	20	^{114}Ru
07/01/1991	Penttilä	[202]	23	^{117}Rh
04/17/1992	Äystö	[203]	20	^{105}Zr, ^{107}Nb, ^{109}Mo, ^{110}Mo, ^{113}Tc, ^{115}Ru
08/06/1992	Janas	[204]	20	^{120}Pd
10/03/1995	Ichikawa	[205]	15	^{166}Tb
03/17/1998	Ichikawa	[206]	20	^{161}Sm, ^{165}Gd
01/19/1999	Asai	[207]	20	^{167}Tb, ^{168}Tb
11/17/2004	Ichikawa	[208]	20	^{159}Pm, ^{162}Sm, ^{166}Gd
12/20/2006	Hayashi	[209]	24	^{163}Eu, ^{164}Eu, ^{165}Eu

[a]Populated by β^--decay

Furthermore, rubidium, strontium, and cesium isotopes were discovered with essentially the same technique using 50 MeV proton beams from the cyclotron of the Institut des Sciences Nucléaires in Grenoble, France, in 1971 by Tracy et al. [195].

A different ISOL facility was built in 1987 at the University of Jyväskylä, Finland. In the Ion Guide Isotope Separator On-Line (IGISOL) the fission fragments recoil into helium gas where they thermalize and are then transferred by helium flow through a differential pumping system into the acceleration stage of a mass separator [210]. The schematic of IGISOL is shown in Fig. 10.12. The separation times with this approach are very fast so that isotopes with millisecond half-lives can be measured. Within the five years between 1987 and 1992, the group of Äystö discovered fifteen new isotopes of elements between zirconium and palladium with IGISOL by fissioning uranium with 20 MeV protons from the MC-20 cyclotron.

In addition, the Jyväskylä group collaborated with researchers from the Katholieke Universiteit in Leuven, Belgium. In 1991, Penttilä et al. discovered ^{117}Rh at the LISOL (Leuven Isotope Separator On Line) facility located at the CYCLONE heavy-ion cyclotron laboratory in Louvain-la-Neuve.

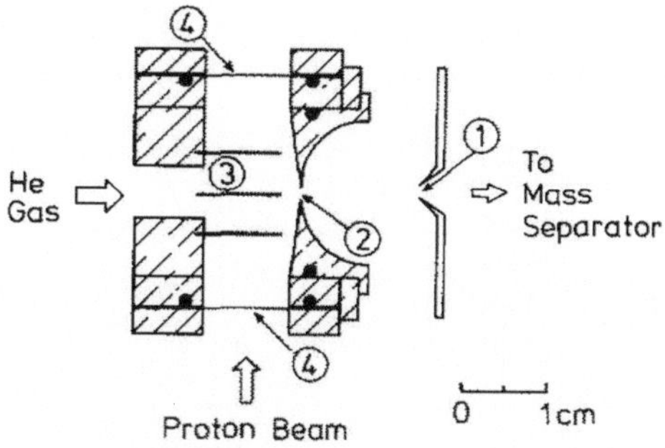

Fig. 10.12 Schematic of the Ion Guide Isotope Separator On-Line (IGISOL) at Jyväskylä: *1* Skimmer, *2* Exit hole, *3* Uranium target, and (4) Havar window [210] (Reprinted from J. Äystö et al., *Separation of fission products by the ion guide fed isotope separator, IGISOL*, Nucl. Instrum. Meth. B 26 (1987) 394. Copyright 1987, with permission from Elsevier.)

In 1996, a gas-jet coupled thermal ion source was installed at the isotope separator on line of the Japan Atomic Energy Research Institute (JAERI) in Tokai, Japan. Between 1995 and 2006 a total of eleven isotopes of the lanthanide elements from promethium through terbium were discovered with the JAERI-ISOL system using protons between 15 and 24 MeV from the tandem accelerator to fission uranium.

Shortly after the installation of ISOLDE was completed at CERN it was realized that online mass spectrometers can also be used to analyze light fragments emitted in high-energy reactions on heavy targets. Since alkali metals diffuse very quickly out of heated graphite and can be ionized by surface ionization, sodium and potassium isotopes could be extracted efficiently.

Isotopes discovered in fission and target fragmentation reactions at ISOLDE are listed in Table 10.10. In three separate papers between 1968 and 1972, Klapisch et al. extended the knowledge of neutron-rich sodium isotopes from mass 27 to 33 [140, 211, 212]. Eleven years later Langevin et al. added two more isotopes (^{34}Na and ^{35}Na) [215]. Klapisch et al. and Langevin et al. also discovered seven neutron-rich potassium isotopes, ^{48}K—^{50}K [140] and ^{51}K—^{54}K [216], respectively.

The separation of potassium isotopes with ISOLDE also led to the discovery of the three calcium neutron-rich isotopes ^{51}Ca [214], ^{52}Ca [217], and ^{53}Ca [216]. They were identified as decay products from the corresponding potassium isotopes.

Heavier alkali elements were produced in fission of uranium. The neutron-rich cesium isotopes from ^{149}Cs through ^{152}Cs were reported by Ravn in a review article for the first time. It is not obvious if these isotopes were actually measured. In the figure caption of a plot showing the cesium isotope yields produced in 600 MeV proton induced fission of ^{238}U Ravn states: "It should be noted that the ISOLDE yields from ^{146}Cs and outwards are the neutron saturation count rates corrected for the estimated neutron branching ratio and normalized to the yield of ^{146}Cs, measured by β^- counting" [213].

The first application of collinear laser spectroscopy to xenon isotopes far from stability identified ^{144}Xe and ^{146}Xe for the first time by Borchers et al. in 1988 [220]. The method had previously been used in the discovery of ^{232}Ra produced in

Table 10.10 Isotopes discovered in fission and target fragmentation reactions at ISOLDE: The submission date, first author and reference of the publication, as well as the target and beam energy (GeV) are listed

Date	First author	Ref.	Target	Energy	Isotope
01/29/1968	Klapisch	[211]	Ta, Ir, U	10.5	^{27}Na
06/16/1969	Klapisch	[212]	Ir, U	24	^{28}Na, ^{29}Na, ^{30}Na, ^{31}Na
08/03/1972	Klapisch	[140]	U	24	^{32}Na, ^{33}Na, ^{48}K, ^{49}K, ^{50}K
03/30/1979	Ravn	[213]	U	0.6	^{149}Cs, ^{150}Cs, ^{151}Cs, ^{152}Cs
03/03/1980	Huck	[214]	U	0.6	^{51}Ca[a]
03/04/1983	Langevin	[215]	Ir	10	^{34}Na, ^{35}Na
07/14/1983	Langevin	[216]	Ir	10	^{51}K, ^{52}K, ^{53}K, ^{54}K, ^{53}Ca[a]
09/04/1984	Huck	[217]	U	0.6	^{52}Ca[a]
10/10/1986	Kratz	[218]	U	0.6	^{130}Cd
03/21/1988	Kratz	[219]	U	0.6	^{93}Br, ^{94}Br[b]
10/11/1988	Borchers	[220]	U	0.6	^{144}Xe, ^{146}Xe
06/18/1991	Balog	[221]	U	0.6	^{101}Rb
07/03/1991	Kratz	[222]	U	0.6	^{78}Cu, ^{79}Cu, ^{81}Zn, ^{84}Ga
12/03/1994	Lhersonneau	[223]	U	1	^{102}Rb
07/03/1995	Fedoseyev	[224]	U	1	^{127}Ag
04/22/1996	Hoff	[225]	U	1	^{133}In, ^{134}In
05/04/2000	Hannawald	[226]	U	1	^{131}Cd, ^{132}Cd
07/13/2000	Kautzsch	[227]	U	1	^{128}Ag, ^{130}Ag
07/15/2000	Kratz	[228]	U	1	^{129}Ag
01/17/2002	Dillmann	[229]	U	1	^{135}In
05/14/2002	Bergmann	[230]	U	1, 1.4	^{145}Xe

[a]Populated by β^--decay
[b]Data also taken at OSTIS at the Institut Laue-Langevin in Grenoble (see Sect. 6.5)

spallation reactions and a schematic of the method is shown in Fig. 10.13. Similarly to the discovery of ^{232}Ra, Borchers et al. did not consider their measurement of the isotope shift in these nuclides the discovery of a new isotope assuming it had been observed previously. However, β-decay properties of ^{144}Xe and ^{146}Xe were not measured until 2003 [230] and the paper does not quote the earlier isotope shift measurement.

Until 1991 the high-energy proton beams were delivered by the proton synchro-cyclotron. With the completion of the proton synchrotron booster in 1972, ISOLDE used 1 GeV or 1.4 GeV proton beams from the booster and the first new isotope discovered utilizing these beams was ^{102}Rb in 1994 [223].

Another significant technical improvement was the development of a Resonance Ionization Laser Ion Source (RILIS) in 1993. The chemically selective ionization significantly improved the purity of the extracted isotopes and a schematic of the

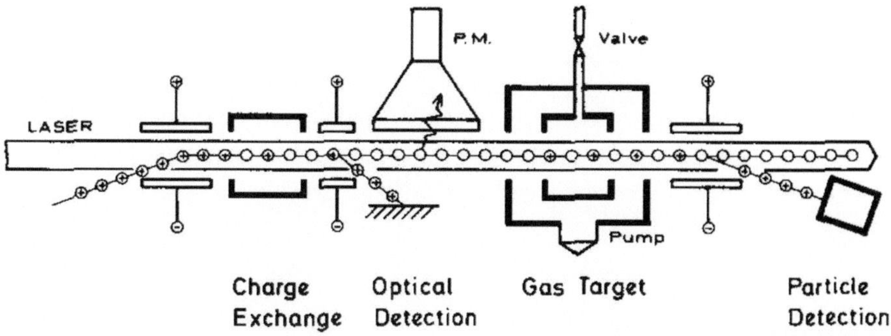

Fig. 10.13 Schematic of the collinear laser spectroscopy set-up used in the discovery of ^{144}Xe and ^{146}Xe at CERN-ISOLDE [220] (Reprinted from W. Borchers et al., *Xenon isotopes far from stability studied by collosional ionization laser spectroscopy*, Phys. Lett. B 216 (1989) 7. Copyright 1989, with permission from Elsevier.)

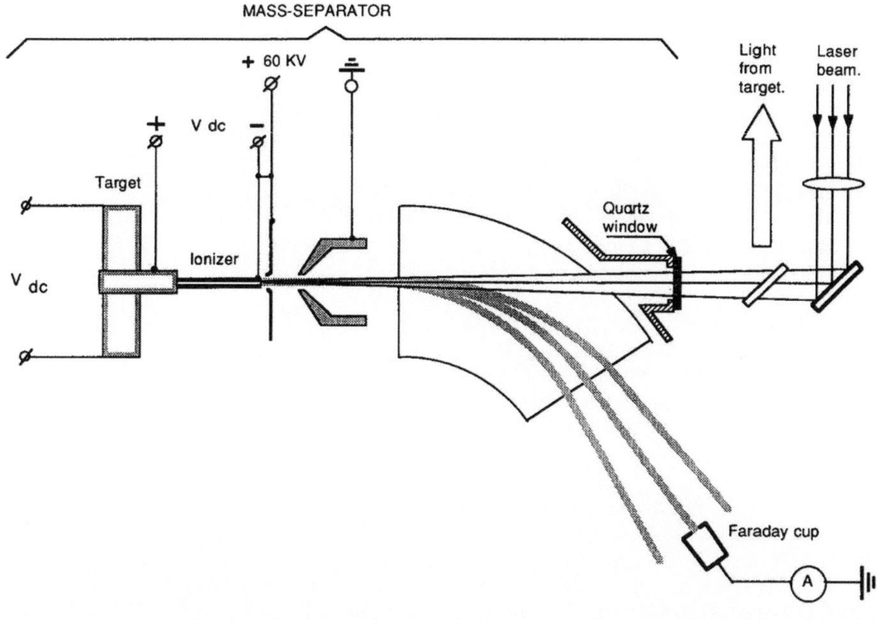

Fig. 10.14 Schematic of the Resonance Ionization Laser Ion Source (RILIS) at CERN-ISOLDE [231]. In 1995, Fedoseyev et al. discovered ^{127}Ag with this set-up [224] (Reprinted from V.I. Mishin et al., *Chemically selective laser ion-source for the CERN-ISOLDE on-line mass separator facility*, Nucl. Instrum. Meth. B 73 (1993) 550. Copyright 1993, with permission from Elsevier.)

set-up is shown in Fig. 10.14 [231]. It was first used in the discovery of ^{127}Ag by Fedoseyev et al. in 1995 [224].

The discovery of ^{135}In by Dillmann et al. in 2002 was not induced directly by proton-fission of uranium. Instead the 1.4 GeV protons from the synchrotron booster irradiated a tantalum converter to produce neutrons. These neutrons then induced fission in an adjacent UC_x/graphite target. ^{135}In was separated using the resonance ionization laser ion source [229].

References

1. B.B. Cunningham et al., Phys. Rev. **72**, 739 (1947)
2. B.G. Harvey, Spallation, in *Chapter 3 in Progress in Nuclear Physics*, ed. by O.R. Frisch, vol. 7 (Pergamon Press, Oxford, 1959), p. 90
3. R. Serber, Phys. Rev. **72**, 1114 (1947)
4. O. Kofoed-Hansen, K.O. Nielsen, Phys. Rev. **82**, 96 (1951)
5. R.H. Goeckermann, I. Perlman, Phys. Rev. **76**, 628 (1949)
6. G. Lhersonneau et al., Eur. Phys. J. A **9**, 385 (2000)
7. R.H. Goeckermann, I. Perlman, Phys. Rev. **73**, 1127 (1948)
8. H.G. Hicks, R.S. Gilbert, Phys. Rev. **94**, 371 (1954)
9. E. Hagebo, A. Kjelberg, A.C. Pappas, J. Inorg. Nucl. Chem. **24**, 117 (1962)
10. J.M. Alexander, U. Schindewolf, C.D. Coryell, Phys. Rev. **111**, 228 (1958)
11. C.E. Gleit, C.D. Coryell, Phys. Rev. **122**, 229 (1961)
12. C.E. Gleit, C.D. Coryell, Phys. Rev. **124**, 1914 (1961)
13. M. Lindner, I. Perlman, Phys. Rev. **73**, 1202 (1948)
14. H.H. Hopkins, B.B. Cunningham, Phys. Rev. **73**, 1406 (1948)
15. D.R. Miller, R.C. Thompson, B.B. Cunningham, Phys. Rev. **74**, 347 (1948)
16. H.M. Neumann, I. Perlman, Phys. Rev. **78**, 191 (1950)
17. H.H. Hopkins, Phys. Rev. **77**, 717 (1950)
18. M. Lindner, I. Perlman, Phys. Rev. **78**, 499 (1950)
19. C.C. Thomas, E.O. Wiig, J. Am. Chem. Soc. **72**, 2818 (1950)
20. R.W. Fink, D.H. Templeton, J. Am. Chem. Soc. **72**, 2818 (1950)
21. B.J. Stover, Phys. Rev. **81**, 8 (1951)
22. J.O. Rasmussen et al., Phys. Rev. **80**, 475 (1950)
23. D.G. Karraker, D.H. Templeton, Phys. Rev. **81**, 510 (1951)
24. G.W. Barton, A. Ghiorso, I. Perlman, Phys. Rev. **82**, 13 (1951)
25. R.W. Fink, E.O. Wiig, J. Am. Chem. Soc. **73**, 2365 (1951)
26. E.K. Hyde, G.D. O'Kelley, Phys. Rev. **82**, 944 (1951)
27. D.G. Karraker, A. Ghiorso, D.H. Templeton, Phys. Rev. **83**, 390 (1951)
28. R.E. Batzel, D.R. Miller, G.T. Seaborg, Phys. Rev. **84**, 671 (1951)
29. R.W. Fink, E.O. Wiig, J. Am. Chem. Soc. **74**, 2457 (1952)
30. W.J. Worthington Jr., Phys. Rev. **87**, 158 (1952)
31. G. Rudstam, P.C. Stevenson, R.L. Folger, Phys. Rev. **87**, 358 (1952)
32. A.A. Caretto Jr., E.O. Wiig, J. Am. Chem. Soc. **74**, 5235 (1952)
33. M. Lindner, R.N. Osborne, Phys. Rev. **88**, 1422 (1952)
34. B. Dropesky, E.O. Wiig, Phys. Rev. **88**, 683 (1952)
35. M. Lindner, Phys. Rev. **91**, 642 (1953)
36. L.P. Gillon, K. Gopalakrishnan, A. de Shalit, Phys. Rev. **93**, 124 (1954)
37. A.A. Caretto Jr., E.O. Wiig, Phys. Rev. **93**, 175 (1954)
38. M.C. Michel, D.H. Templeton, Phys. Rev. **93**, 1422 (1954)
39. B.C. Haldar, E.O. Wiig, Phys. Rev. **94**, 1713 (1954)

40. H.B. Mathur, E.K. Hyde, Phys. Rev. **95**, 708 (1954)
41. G. Andersson et al., Phil. Mag. **46**, 70 (1955)
42. W.E. Nervik, G.T. Seaborg, Phys. Rev. **97**, 1092 (1955)
43. W.G. Smith, J.M. Hollander, Phys. Rev. **98**, 1258 (1955)
44. J.-C. Roy, T.P. Kohman, Can. J. Phys. **35**, 649 (1957)
45. B.C. Haldar, E.O. Wiig, Phys. Rev. **105**, 1285 (1957)
46. G. Andersson, E. Arbman, B. Jung, Ark. Fysik **11**, 297 (1957)
47. W.R. Ware, E.O. Wiig, Phys. Rev. **117**, 191 (1960)
48. K.S. Toth et al., Phys. Rev. **116**, 118 (1959)
49. B. Jung, G. Andersson, Nucl. Phys. **15**, 108 (1960)
50. F.D.S. Butement, P. Glentworth, J. Inorg. Nucl. Chem. **15**, 205 (1960)
51. F.D.S. Butement, G.G.J. Boswell, J. Inorg. Nucl. Chem. **16**, 10 (1960)
52. G. Andersson, I.B. Häller, R. Ringh, J. Inorg. Nucl. Chem. **17**, 15 (1961)
53. E.R. Merz, A.A. Caretto Jr., Phys. Rev. **122**, 1558 (1961)
54. W.H. Sullivan, N.R. Sleight, and E.M. Gladrow, *Rhodium Radioisotopes Induced in Deuteron-Bombarded Ruthenium*, Declassified Plutonium Project Report No. MDDC-918
55. C.D. Coryell, N. Sugarman (eds.), *Radiochemical Studies: The Fission Products*, Nat. Nucl. Ener. Ser. IV, 9, (McGraw-Hill, New York, 1951)
56. R.S. Krishnan, Proc. Camb. Phil. Soc. **36**, 500 (1940)
57. A.C. Helmholz, Phys. Rev. **60**, 160 (1941)
58. H. Bradt et al., Helv. Phys. Acta **18**, 256 (1945)
59. H. Bradt et al., Helv. Phys. Acta **19**, 218 (1946)
60. J. Blachot, Nucl. Data Sheets **107**, 355 (2006)
61. J.M. Cork et al., Phys. Rev. **79**, 938 (1950)
62. J.R. Gum, M.L. Pool, Phys. Rev. **80**, 315 (1950)
63. A. Ghiorso, W.W. Meinke, G.T. Seaborg, Phys. Rev. **74**, 695 (1948)
64. W.W. Meinke, A. Ghiorso, G.T. Seaborg, Phys. Rev. **75**, 314 (1949)
65. E.K. Hyde, A. Ghiorso, G.T. Seaborg, Phys. Rev. **77**, 765 (1950)
66. W.W. Meinke, A. Ghiorso, G.T. Seaborg, Phys. Rev. **85**, 429 (1952)
67. F.F. Momyer et al., Phys. Rev. **86**, 805 (1952)
68. F.F. Momyer, E.K. Hyde, J. Inorg. Nucl. Chem. **1**, 274 (1955)
69. F.F. Momyer, E.K. Hyde, Phys. Rev. **101**, 136 (1956)
70. F.D.S. Butement, V.J. Robinson, J. Inorg. Nucl. Chem. **26**, 1 (1964)
71. F.D.S. Butement, V.J. Robinson, S.M. Qaim, J. Inorg. Nucl. Chem. **26**, 491 (1964)
72. R.L. Hahn, M.F. Roche, K.S. Toth, Nucl. Phys. A **113**, 206 (1968)
73. V.E. Viola, M.M. Minor, C.T. Roche, Nucl. Phys. A **217**, 372 (1973)
74. Y.Y. Chu, M.L. Zhou, Phys. Rev. C **28**, 1379 (1983)
75. I.N. Meshkov, A.N. Sissakian, G.V. Trubnikov, *Proceedings of IPAC 2010, Kyoto*, 23–28 May 2010, TUZRA02, https://accelconf.web.cern.ch/accelconf/IPAC10/talks/tuzra02_talk.pdf
76. http://creativecommons.org/licenses/by/3.0/
77. G.M. Gorodinskii et al., Bull. Acad. Sci. USSR Phys. Ser. **21**, 1611 (1957)
78. M.Ya. Kuznetsova et al., Sov. At. Energy **4**, 766 (1958)
79. P.M. Aron et al., Bull. Acad. Sci. USSR Phys. Ser. **22**, 811 (1958)
80. A.V. Kalyamin, I.Y. Levenberg, B.A. Yakovlev, Sov. At. Energy **6**, 435 (1959)
81. N.A. Antoneva et al., Bull. Acad. Sci. USSR Phys. Ser. **23**, 196 (1959)
82. I.S. Dneprovskii, Sov. At. Energy **8**, 38 (1961)
83. A.A. Abdurazakov et al., Nucl. Phys. **21**, 164 (1960)
84. A.K. Lavrukhina, T.V. Malysheva, B.A. Khotin, Sov. Phys. Dokl. **6**, 236 (1961)
85. V.I. Baranov et al., Bull. Acad. Sci. USSR Phys. Ser. **24**, 1086 (1960)
86. A.A. Abdurazakov et al., Sov. Phys. JETP **14**, 1229 (1962)
87. A. Abdumalikov et al., Phys. Lett. **5**, 359 (1963)
88. K.Ya. Gromov et al. Sov. Phys. JETP **20**, 1104 (1965)
89. Z.T. Zhelev et al., Sov. J. Nucl. Phys. **2**, 682 (1966)
90. E. Herrmann, G. Pfrepper, D. Christov, Radiochim. Acta **7**, 10 (1967)

91. Zh Zhelev et al., Bull. Acad. Sci. USSR Phys. Ser. **32**, 1497 (1968)
92. R. Arlt et al., Bull. Acad. Sci. USSR Phys. Ser. **33**, 1133 (1969)
93. R. Arlt et al., Bull. Acad. Sci. USSR Phys. Ser. **33**, 1144 (1969)
94. R. Arlt et al., Bull. Acad. Sci. USSR Phys. Ser. **34**, 619 (1970)
95. A.A. Abdurazakov et al., Bull. Acad. Sci. USSR Phys. Ser. **34**, 706 (1970)
96. J. Vandlik et al., Bull. Acad. Sci. USSR Phys. Ser. **34**, 1472 (1970)
97. R. Arlt et al., JETP Lett. **13**, 397 (1971)
98. I. Votsilka et al., Bull. Acad. Sci. USSR Phys. Ser. **38**(1), 49 (1974)
99. J. Vandlik et al., Bull. Acad. Sci. USSR Phys. Ser. **38**(4), 26 (1974)
100. R.M. Diamond, J.M. Hollander, Nucl. Phys. **8**, 143 (1958)
101. F.D.S. Butement, G.B. Briscoe, J. Inorg. Nucl. Chem. **25**, 151 (1963)
102. G. Graeffe, Ann. Acad. Sci. Fenn., Ser. A **VI**(128) (1963)
103. R.G. Korteling, E.K. Hyde, Phys. Rev. **136**, B425 (1964)
104. C. Brun, Y. Le Beyec, M. Lefort, Phys. Lett. **16**, 286 (1965)
105. F.D.S. Butement, M.Y. Mirza, J. Inorg. Nucl. Chem. **28**, 303 (1966)
106. A.M. Poskanzer, R.A. Esterlund, R. McPherson, Phys. Rev. Lett. **15**, 1030 (1965)
107. H.-J. Bleyl, H. Münzel, G. Pfennig, Radiochim. Acta **8**, 200 (1967)
108. K.S. Toth et al., Phys. Rev. **183**, 1004 (1969)
109. Y.Y. Chu, E.M. Franz, G. Friedlander, Phys. Rev. **187**, 1529 (1969)
110. F.W.N. de Boer et al., Radiochim. Acta **13**, 118 (1970)
111. K.S. Toth et al., Phys. Rev. C **2**, 1480 (1970)
112. C. Ekström, M. Olsmats, B. Wannberg, Nucl. Phys. A **170**, 649 (1971)
113. J. van Klinken et al., Z. Phys. **246**, 369 (1971)
114. E.E. Berlovich et al., Bull. Acad. Sci. USSR Phys. Ser. **36**, 2165 (1972)
115. B.J. Meijer, F.W.N. de Boer, P.F.A. Goudsmit, Radiochim. Acta **19**, 150 (1973)
116. E.E. Berlovich et al., Bull. Acad. Sci. USSR Phys. Ser. **38**(8), 14 (1974)
117. J. van Klinken, S.J. Feenstra, Phys. Rev. C **12**, 2111 (1975)
118. G.P. Nowicki et al., Nucl. Phys. A **249**, 76 (1975)
119. J. Deslauriers, S.C. Gujrathi, S.K. Mark, Z. Phys, A **303**, 151 (1981)
120. S. Lunardi et al., Z. Phys. A **324**, 433 (1986)
121. A.M. Poskanzer et al., Phys. Rev. Lett. **17**, 1271 (1966)
122. A.M. Poskanzer et al., Phys. Lett. B **27**, 414 (1968)
123. T.D. Thomas et al., Phys. Lett. B **27**, 504 (1968)
124. J.D. Bowman et al., Phys. Rev. Lett. **31**, 614 (1973)
125. J.D. Bowman et al., Phys. Rev. C **9**, 836 (1974)
126. G.W. Butler et al., Phys. Rev. Lett. **38**, 1380 (1977)
127. X.L. Tu et al., Z. Phys. A **337**, 361 (1990)
128. G. Rudstam, Nucl. Instrum. Meth. **38**, 282 (1965)
129. G. Andersson, G. Rudstam, Nucl. Instrum. Meth. **29**, 93 (1964)
130. G. Andersson, G. Rudstam, G. Sorensen, Ark. Fysik **28**, 37 (1965)
131. P.G. Hansen et al., Phys. Lett. B **28**, 415 (1969)
132. J. Chaumont et al., Phys. Lett. B **29**, 652 (1969)
133. D.J. Hnatowich et al., J. Inorg. Nucl. Chem. **32**, 3137 (1970)
134. P.G. Hansen et al., Nucl. Phys. A **148**, 249 (1970)
135. P.G. Hansen et al., Nucl. Phys. A **160**, 445 (1971)
136. P. Hornshoj et al., Nucl. Phys. A **163**, 277 (1971)
137. P. Hornshoj et al., Nucl. Phys. A **187**, 637 (1972)
138. H.L. Ravn, S. Sundell, L. Westgaard, Phys. Lett. B **39**, 337 (1972)
139. B.R. Erdal et al., Nucl. Phys. A **194**, 449 (1972)
140. R. Klapisch et al., Phys. Rev. Lett. **29**, 1254 (1972)
141. O. Scheidemann, E. Hagebo, J. Inorg. Nucl. Chem. **35**, 3055 (1973)
142. E. Hagberg et al., Nucl. Phys. A **208**, 309 (1973)
143. H.L. Ravn et al., J. Inorg. Nucl. Chem. **37**, 383 (1975)
144. L. Westgaard et al., Z. Phys. A **275**, 127 (1975)

145. J.M. D'Auria et al., Phys. Lett. B **66**, 233 (1977)
146. J.P. Husson, C.F. Liang, C. Richard-Serre, J. Phys. (Paris) Lett. **38**, L (1977)
147. E. Hagberg et al., Phys. Rev. Lett. **39**, 792 (1977)
148. E. Hagberg et al., Nucl. Phys. A **293**, 1 (1977)
149. T. Elmroth et al., Nucl. Phys. A **304**, 493 (1978)
150. J.M. D'Auria et al., Nucl. Phys. A **301**, 397 (1978)
151. G.T. Ewan et al., Z. Phys. A **296**, 223 (1980)
152. G.T. Ewan et al., Nucl. Phys. A **352**, 13 (1981)
153. S.A. Ahmad et al., Phys. Lett. B **133**, 47 (1983)
154. F. Calaprice et al., Phys. Rev. C **30**, 1671 (1984)
155. P. Hill et al., Z. Phys. A **320**, 531 (1985)
156. M.J.G. Borge et al., Z. Phys. A **325**, 429 (1985)
157. W. Kurcewicz et al., Nucl. Phys. A **464**, 1 (1987)
158. D.G. Burke et al., Z. Phys. A **333**, 131 (1989)
159. M.J.G. Borge et al., Z. Phys. A **333**, 109 (1989)
160. A. Bouldjedri et al., Z. Phys. A **339**, 311 (1991)
161. H. Grawe et al., Z. Phys. A **341**, 247 (1992)
162. D. Neidherr et al., Phys. Rev. Lett. **102**, 112501 (2009)
163. P. Aagaard et al., J. Inorg. Nucl. Chem. **5**, 105 (1957)
164. G. Albouy et al., Compt. Rend. Acad. Sci. **249**, 407 (1959)
165. N. Poffe et al., J. Phys. Radium **21**, 343 (1960)
166. G. Albouy, M.M. Gusakow, N. Poffe, J. Phys. Radium **21**, 751 (1960)
167. A. Santoni, A. Caruette, J. Valentin, J. Phys. (Paris) **24**, 407 (1963)
168. P. Paris, F. Perrin, Compt. Rend. Acad. Sci. **258**, 5855 (1964)
169. H. Abou-Leila, Compt. Rend. Acad. Sci. **259**, 3003 (1964)
170. G. Gerschel et al., Nuovo Cimento **37**, 1756 (1965)
171. P. Lagarde et al., J. Phys. (Paris) **27**, 116 (1966)
172. P. Paris, F. Perrin, Compt. Rend. Acad. Sci. **265**, 510 (1967)
173. O. Rahmouni, J. Phys. (Paris) **29**, 550 (1968)
174. R. Arlt et al., Nucl. Instrum. Meth. **102**, 253 (1972)
175. T.B. Vandlik et al., JETP Lett. **15**, 271 (1972)
176. A. Latuszynski et al., Nukleonika **20**, 1043 (1974)
177. J.C. Putaux, J. Obert, P. Aguer, Nucl. Instrum. Meth. **121**, 615 (1974)
178. G.D. Alkhazov et al., Z. Phys. A **291**, 397 (1979)
179. G.D. Alkhazov et al., Z. Phys. A **295**, 305 (1980)
180. C.F. Liang et al., Z. Phys. A **297**, 303 (1980)
181. G.D. Alkhazov et al., Acta Phys. Pol. B **12**, 175 (1981)
182. G.D. Alkhazov et al., Acta Phys. Pol. B **12**, 825 (1981)
183. G.D. Alkhazov et al., Z. Phys. A **305**, 185 (1982)
184. C.F. Liang et al., Z. Phys. A **309**, 185 (1982)
185. C.F. Liang et al., Z. Phys. A **321**, 695 (1985)
186. C.F. Liang et al., J. Phys. G **15**, L31 (1989)
187. K.A. Mezlev et al., Z. Phys. A **337**, 109 (1990)
188. K.-L. Gippert et al., Nucl. Phys. A **453**, 1 (1986)
189. H.L. Ravn et al., Nucl. Instrum. Meth. **139**, 267 (1976)
190. P. Paris et al., Nucl. Instrum. Meth. **139**, 251 (1976)
191. E. Nolte et al., Z. Phys. A **306**, 223 (1982)
192. I. Amarel et al., Phys. Lett. B **24**, 402 (1967)
193. I. Amarel, H. Gauvin, A. Johnson, J. Inorg. Nucl. Chem. **31**, 577 (1969)
194. M.-I. Macias-Marques et al., J. Phys. (Paris) **32**, 237 (1971)
195. B.L. Tracy et al., Phys. Lett. B **34**, 277 (1971)
196. J. Äystö et al., Phys. Lett. B **201**, 211 (1988)
197. J. Äystö et al., Nucl. Phys. A **480**, 104 (1988)
198. H. Penttilä et al., Phys. Rev. C **38**, 931 (1988)

199. J. Äystö et al., Nucl. Phys. A **515**, 365 (1990)
200. H. Penttilä et al., Z. Phys. A **338**, 291 (1991)
201. M. Leino et al., Phys. Rev. C **44**, 336 (1991)
202. H. Penttilä et al., Phys. Rev. C **44**, 935 (1991)
203. J. Äystö et al., Phys. Rev. Lett. **69**, 1167 (1992)
204. Z. Janas et al., Nucl. Phys. A **552**, 340 (1993)
205. S. Ichikawa et al., Nucl. Instrum. Meth. A **374**, 330 (1996)
206. S. Ichikawa et al., Phys. Rev. C **58**, 1329 (1998)
207. M. Asai et al., Phys. Rev. C **59**, 3060 (1999)
208. S. Ichikawa et al., Phys. Rev. C **71**, 067302 (2005)
209. H. Hayashi et al., Eur. Phys. J. A **34**, 363 (2007)
210. J. Äystö et al., Nucl. Instrum. Meth. B **26**, 394 (1987)
211. R. Klapisch et al., Phys. Rev. Lett. **20**, 740 (1968)
212. R. Klapisch et al., Phys. Rev. Lett. **23**, 652 (1969)
213. H.L. Ravn, Phys. Rep. **54**, 201 (1979)
214. A. Huck et al., Phys. Rev. C **22**, 2544 (1980)
215. M. Langevin et al., Phys. Lett. B **125**, 116 (1983)
216. M. Langevin et al., Phys. Lett. B **130**, 251 (1983)
217. A. Huck et al., Phys. Rev. C **31**, 2226 (1985)
218. K.-L. Kratz et al., Z. Phys. A **325**, 489 (1986)
219. K.-L. Kratz et al., Z. Phys. A **330**, 229 (1988)
220. W. Borchers et al., Phys. Lett. B **216**, 7 (1989)
221. K. Balog et al., Z. Phys. A **342**, 125 (1992)
222. K.-L. Kratz et al., Z. Phys. A **340**, 419 (1991)
223. G. Lhersonneau et al., Z. Phys. A **351**, 357 (1995)
224. V.N. Fedoseyev et al., Z. Phys. A **353**, 9 (1995)
225. P. Hoff et al., Phys. Rev. Lett. **77**, 1020 (1996)
226. M. Hannawald et al., Phys. Rev. C **62**, 054301 (2000)
227. T. Kautzsch et al., Eur. Phys. J. A **9**, 201 (2000)
228. K.-L. Kratz et al., Hyperfine Interact. **129**, 185 (2000)
229. I. Dillmann et al., Eur. Phys. J. A **13**, 281 (2002)
230. U.C. Bergmann et al., Nucl. Phys. A **714**, 21 (2003)
231. V.I. Mishin et al., Nucl. Instrum. Meth. B **73**, 550 (1993)

Chapter 11
Fusion-Evaporation Reactions

11.1 Heavy Ions

In the quest for superheavy transuranium elements after the Second World War it soon became clear that neutron capture reactions followed by subsequent β-decay would not be able to reach elements beyond fermium (Z = 100). At the time, nuclear reactions were limited to light-particle induced reactions which again were not suitable to produce new elements. By 1950, the heaviest element produced in light-particle reactions was berkelium. Thompson, Ghiorso, and Seaborg had discovered ^{243}Bk by irradiating ^{241}Am with α-particles [1]. If, however, it was possible to accelerate heavier ions, fusion reactions of these heavy ions with heavy targets could lead to new heavier elements. The first acceleration of a heavy-ion beam was achieved by Miller et al. at the Crocker Laboratory of the University of California at Berkeley in 1950. The intensity of the accelerated ^{12}C and ^{16}O nuclei was not sufficient to induce a detectable number of fusion reactions [2]. However, only a couple of months later, Ghiorso et al. succeeded in identifying ^{246}Cf by irradiating ^{238}U with ^{12}C ions. The compound nucleus ^{250}Cf evaporated four neutrons to populate ^{246}Cf which then subsequently was identified by its α-radioactivity [3].

Soon thereafter it was realized that heavy-ion fusion-evaporation reactions were not only an excellent tool to produce new elements but that they also could be used to populate light neutron-deficient nuclides. Since these early attempts fusion-evaporation reactions have become one of the most productive reaction mechanisms to discover new nuclides. In addition to the almost 200 transuranium and superheavy nuclei well over 500 neutron-deficient nuclides were first identified in heavy-ion fusion-evaporation reactions.

The present chapter discusses the discovery of proton-bound neutron-deficient nuclides with Z < 92. The discovery of isotopes of transuranium ($92 \leq Z \leq 100$) and superheavy (Z > 100) elements using fusion-evaporation reactions are described in Chaps. 5 and 12, respectively. Nuclides beyond the proton drip-line are discussed in Chap. 16. In the present chapter the nuclides are arranged according to their

M. Thoennessen, *The Discovery of Isotopes*,
DOI 10.1007/978-3-319-31763-2_11

decay mechanism; β^+-emitters in Sects. 11.2 and 11.3, α-emitters in Sect. 11.4, and β-delayed proton emitters in Sect. 11.5.

Initially, the detection and identification techniques were the same as for light-particle induced and spallation reactions. The element identification was somewhat easier at least for the dominant decay channels because charged-particle evaporation is suppressed relative to neutron evaporation for isotopes located not too far from the line of stability. Thus the most probable element of the residues was identical to the compound nucleus. Nevertheless chemical separation was important to discriminate against nuclei populated by other reaction mechanisms like inelastic scattering or incomplete fusion.

The mass identification was quite uncertain as the incident beam energy was not well known. In addition, the energy could not easily be changed so that excitation functions were measured by inserting different absorber foils into the beam.

A major breakthrough occurred when it was realized that the incoming heavy-ion beam imparted significant recoil velocities to the residues which was sufficient to knock them out of the target. The recoils were caught on catcher foils which were then chemically analyzed or placed directly in front of detectors. Later helium-jets were developed where the fragments recoiled into flowing helium gas which then transported them to catcher foils.

11.2 Beta-Decay

Neutron-deficient isotopes that decay by β^+ emission and that have been discovered in fusion-evaporation reactions are listed in Tables 11.1, 11.2 and 11.3. New isotopes populated by fusion-evaporation typically did not decay back to stable nuclei but to radioactive nuclei that had previously been populated by different methods. Thus the observation of the subsequent known daughter decay could be used to help in the identification of the new isotope.

Table 11.1 lists isotopes that were identified following chemical separation. The first non-transuranium isotope populated in a fusion-evaporation reaction was ^{74}Br in 1953. Hollander irradiated enriched ^{63}Cu and ^{65}Cu targets with ~90 MeV carbon ions from the Berkeley 60-in. cyclotron [4]. He assigned a 36-min activity to ^{74}Br based on the relative intensity observed with the two different copper targets. ^{73}Br was ruled out because there was no evidence for the known 7.1-h activity of the ^{73}Se daughter.

Later in 1953, Rossi et al. discovered ^{119}I also at Berkeley by irradiating palladium targets with a ^{14}N beam. They linked the observed 18-min activity to the known 4.5-d ^{119}Te daughter activity [5]. A year later Kalkstein and Hollander used a mass separator to achieve simultaneous chemical mass separation of ^{126}Cs and thus establish the decay sequence ^{126}Ba$-$^{126}Cs$-$^{126}Xe in the reaction ^{115}In(^{14}N,3n) [6].

The first isotopes discovered in fusion-evaporation reactions in Europe were ^{79}Rb and ^{71}Se in 1956 and 1957, respectively, at the Centre d'Etudes Nucléaires de Saclay, in Gif-sur-Yvette, France. The paper on ^{79}Rb by Beydon et al. [7] was followed

Table 11.1 Isotopes discovered in fusion-evaporation reactions and identified by chemical separation: The submission date, first author and reference of the publication, as well as the target, beam, and the evaporation channel are listed

Date	First author	Ref.	Target	Beam	Evaporation	Isotope
08/11/1953	Hollander	[4]	^{65,63}Cu	^{12}C	3n, 1n	^{74}Br
11/25/1953	Rossi	[5]	^{110,108}Pd	^{14}N	5n, 3n	^{119}I
07/09/1954	Kalkstein	[6]	^{115}In	^{14}N	3n	^{126}Ba, ^{126}Cs[a]
09/28/1956	Beydon	[7]	^{65}Cu	^{16}O	2n	^{79}Rb
01/21/1957[b]	Beydon	[8]	^{nat}Cu	^{14}N	xn	^{71}Se[a]
11/03/1958	Toth	[9]	^{141}Pr	^{12}C	5n	^{148}Tb
04/15/1959	Faler	[10]	^{165}Ho	^{14}N	6n, 5n, 4n	^{173}Ta, ^{174}Ta, ^{175}Ta
05/28/1959	Chackett	[11]	^{184}W	^{14}N	5n	^{193}Tl
05/16/1960	Hoff	[12]	^{71,69}Ga	^{14}N	5n, 3n	^{80}Sr, ^{80}Rb[a]
05/27/1960	Preiss	[13]	^{nat}Mo	^{16}O	2pxn	^{103}Cd
02/23/1961	Diamond	[14]	^{169}Tm	^{16}O	3n	^{182}Ir
11/20/1961	Sheline	[15]	^{121}Sb	^{12}C	7n, 5n, 3n	^{126}La, ^{128}La, ^{130}La
01/10/1962	Preiss	[16]	^{nat}In	^{16}O	pxn	^{123}Ba, ^{125}Ba
03/05/1962	Maxia	[17]	^{75}As	^{12}C	4n, 3n	^{83}Y, ^{84}Y
08/13/1962	Preiss	[18]	^{nat}In	^{16}O	xn	^{127}La, ^{129}La
12/21/1964	Nurmia	[19]	^{58}Ni	^{6}Li	1n	^{63}Ga
07/09/1965	Belyaev	[20]	^{169}Tu	^{15}N	4n	^{180}Os
03/15/1967	Nadjakov	[21]	^{165}Ho	^{16}O	6n, 5n	^{175}Re, ^{176}Re
03/30/1967	Bakhru	[22]	^{nat}Mo	^{11}B	xn	^{99}Ag
1968[c]	Belyaev	[23]	^{169}Tm	^{14}N	4n	^{179}Os
10/17/1969	Murray	[24]	^{59}Co	^{16}O	2n	^{73}Br
1971[c]	Nadjakov	[25]	^{155}Gd	^{20}Ne	5n	^{170}W
04/18/1973	Schmeing	[26]	^{58}Ni	^{16}O	2n	^{72}Kr
09/25/1974	Kaba	[27]	^{54}Fe	^{32}S	2p1n	^{83}Zr
06/19/1981	Bruchertseifer	[28]	^{147}Sm	^{22}Ne	5n, 4n	^{164}Hf, ^{165}Hf
02/10/1982	Bruchertseifer	[29]	^{151}Eu	^{20}Ne	6n	^{165}Ta
07/19/1982	Eichler	[30]	^{151}Eu	^{20}Ne	7n	^{164}Ta

[a]Populated by β^+-decay
[b]Date of session
[c]No date listed

immediately in the same issue of Nuclear Physics by three additional papers by various coauthors from Saclay and Stockholm reporting on the same set of experiments [31–33]. All four papers were submitted on the same day. It is also interesting to note that all authors on the discovery paper of ^{71}Se were female [8].

The husband and wife team of Ken and Alma Chackett discovered ^{193}Tl with the Nuffield cyclotron at the University of Birmingham in England in 1959 [11]. Preiss and collaborators utilized the Yale University heavy-ion linear accelerator to produce ^{103}Cd [13], ^{123}Ba and ^{125}Ba [16], and ^{127}La and ^{129}La [18] between 1960 and 1962.

The other isotopes discovered during this time period were produced at Berkeley with the HILAC (Heavy-Ion Linear Accelerator) where Hoff, Hollander, and Michel used a mass separator in addition to β- and γ-ray measurements to identify ^{80}Sr and its daughter ^{80}Rb [12].

In the subsequent years until 1982 when the last isotope was discovered utilizing chemical separation, the majority of the new isotopes were produced at the Joint Institute for Nuclear Research at Dubna, Russia. While Belyaev and collaborators still utilized the U150 cyclotron to identify ^{180}Os [20] and ^{179}Os [23], the experiments by Nadjakov and collaborators (^{175}Re and ^{176}Re, [21]) and ^{170}W [25] and later by the group of Bruchertseifer (^{164}Hf and ^{165}Hf [28], ^{165}Ta [29], and ^{164}Ta [30]) were performed with the U300 accelerator.

By 1970 the development of high-resolution Ge-detectors allowed the clear identification of isotopes by their γ-ray spectra eliminating the need for chemical separation. Table 11.2 lists the isotopes identified by β- and γ-decay without chemical separation. By that time many institutes had acquired either cyclotrons or various linear accelerators and since the discovery measurements required only a few small detectors, the 52 isotopes published in the 38 papers listed in the table were discovered at 19 different institutions in 9 different countries.

In the first experiments, the produced activities were measured directly at or near the target position of the beam. For example, in 1970, Droste et al. discovered a long-lived isomeric state of ^{137}Nd by placing a Ge(Li) detector and a toroidal electron spectrometer directly next to the target to measure γ-spectra and conversion electrons, respectively. They measured the activities in between the beam pulses from the Dubna U300 accelerator [34]. Later in the same year ^{72}Br was identified by Nolte et al. at the Max Planck Institut für Kernphysik in Heidelberg, Germany. They stopped recoils from the fusion-evaporation reactions in a gold catcher foil and measured off-beam γ-spectra with a Ge(Li) detector [35]. Zioni et al. used a mechanical beam chopper at the EN tandem of the Racah Institute of Physics of the Hebrew University in Jerusalem, Israel, to determine the short half-life (260 ± 6 ms) of ^{46}Cr [37]. For isotopes with longer half-lives longer beam-on and beam-off periods were chosen for the irradiations and measurements, respectively.

Measurements at the target position were limited to the strongest decay channels. In order to populate and detect the weaker channels it was necessary to increase the beam intensities which also increased the background activities. In addition, neutrons which were present in the target area during the irradiation damaged the Ge(LI) detectors. Thus it was necessary to move the irradiated targets to a separate, well shielded detection station. For the identification of the relatively long-lived isotopes ^{87}Nb (3.5 ± 0.2 min) and ^{88}Mo (8.2 ± 0.5 min) with ^{32}S beams from the Rochester University MP tandem Van de Graaff accelerator, Doron and Blann moved the irradiated targets manually in front of a Ge(Li) detector [36].

For shorter-lived activities the helium-jet technique developed in the early 1960s by Macfarlane and Griffioen [72] to measure short-lived α-emitters (see Sect. 11.4), was widely used. It was first applied for the discovery of a β-emitting isotope in 1974 by Newman et al. in the identification of ^{146}Tb with the Oak Ridge Isochronous cyclotron (ORIC) [40]. A schematic diagram of the helium-jet transfer system at

Table 11.2 Isotopes discovered in fusion-evaporation reactions and identified by β- and γ-decay without chemical separation: The submission date, first author and reference of the publication, as well as the target, beam, and the evaporation channel are listed

Date	First author	Ref.	Target	Beam	Evaporation	Isotope
04/15/1970	Droste	[34]	^{119}Sn	^{22}Ne	4n	^{137}Nd
09/01/1970	Nolte	[35]	^{58}Ni	^{16}O	pn	^{72}Br
10/23/1970	Doron	[36]	^{58}Ni	^{32}S	3p, 2p	^{87}Nb, ^{88}Mo
03/11/1971	Zioni	[37]	^{32}S	^{16}O	2n	^{46}Cr
1972[a]	Akhmadzhanov	[38]	^{169}Tm	^{16}O	7n, 5n, 4n	^{178}Ir, ^{180}Ir, ^{181}Ir
04/26/1972	Ladenb.-Bellis	[39]	^{69}Ga	^{14}N	4n	^{79}Sr
08/27/1973	Newman	[40]	^{141}Pr	^{12}C	7n	^{146}Tb
01/10/1974	Nolte	[41]	^{40}Ca	^{32}S	2p1n	^{69}Se
07/09/1974	Newton	[42]	^{181}Ta	^{19}F	7n	^{193}Pb
02/27/1975	Toth	[43]	^{141}Pr	^{14}N	8n	^{147}Dy
05/12/1975	Hamilton	[44]	^{181}Ta	^{16}O	9n	^{186}Tl
01/30/1976	Chojnacki	[45]	^{181}Ta	^{20}Ne	5n	^{196}Bi
05/13/1976	Varley	[46]	^{92}Mo	^{16}O	p3n	^{104}In
08/19/1976	Leber	[47]	^{159}Tb	^{16}O	9n	^{166}Ta
10/08/1976	Parks	[48]	^{48}Ca	^{7}Li	pn	^{53}Ti
01/25/1977	Korschinek	[49]	^{58}Ni	^{32}S	α2p, αpn	^{84}Zr, ^{84}Nb
					2p1n	^{87}Mo
03/25/1977	Hunter	[50]	^{155}Gd	^{14}N	5n	^{164}Lu
06/23/1977	Nathan	[51]	^{48}Ca	^{9}Be	pn	^{55}V
08/17/1977	Burman	[52]	^{151}Eu	^{16}O	5n	^{162}Lu
01/03/1978	Davids	[53]	^{48}Ca	^{11}B	pn	^{57}Cr
03/06/1978	Norman	[54]	^{48}Ca	^{18}O	αpn	^{60}Mn
06/19/1978	Toth	[55]	^{144}Sm	^{10}B	6n, 5n	^{148}Ho, ^{149}Ho
07/05/1978	Alburger	[56]	^{58}Ni	^{10}B, ^{14}N	2n	^{66}As, ^{70}Br
02/18/1980	Deprun	[57]	^{54}Fe	^{32}S	3p1n	^{82}Y
03/27/1980	Murphy	[58]	^{58}Ni	^{14}N	αn	^{67}As
01/28/1981	Lister	[59]	^{58}Ni	^{24,25}Mg	pn	^{80}Y, ^{81}Y
09/15/1981	Sousa	[60]	^{144}Sm	^{10}B	α6n	^{144}Tb
01/22/1982	Schrewe	[61]	^{142}Nd	^{24}Mg	4n, 3n	^{162}Hf, ^{163}Hf
01/25/1982	Gui	[62]	^{90}Zr	^{58}Ni	pn	^{146}Ho
02/11/1982	Nolte	[63]	^{90}Zr	^{58}Ni	2p1n	^{145}Dy
			^{94}Mo	^{58}Ni	2p, pn	^{150}Er, ^{150}Tm
			^{92}Mo	^{58}Ni	3p, 2p, pn	^{147}Ho, ^{148}Er, ^{148}Tm
01/03/1985	Ollivier	[64]	^{112}Sn	^{35}Cl	2p2n	^{143}Tb
08/01/1986	Szymanski	[65]	^{165}Ho	^{16}O	8n	^{173}Re

(continued)

Table 11.2 (continued)

Date	First author	Ref.	Target	Beam	Evaporation	Isotope
07/17/1987	Runte	[66]	^{139}La	^{36}Ar	4n	^{171}Re
07/16/1990	Heiguchi	[67]	^{60}Ni	^{32}S	p2n	^{89}Tc
07/31/1991	Bosch-Wicke	[68]	^{148}Nd	^{36}Ar	p4n	^{179}Ir
08/15/1991	Zhou	[69]	^{58}Ni	^{35}Cl	p2n	^{90}Ru
06/02/1999	Xie	[70]	^{106}Cd	^{36}Ar	p2n	^{139}Tb
11/20/2003	Xu	[71]	^{92}Mo	^{40}Ca	p2n	^{129}Pm

[a]No submission date listed

Brookhaven National Laboratory used in the discovery of ^{80}Y and ^{81}Y in 1981 by Lister et al. is shown in Fig. 11.1.

In a different technique designed for short-lived activities, the target was moved from the irradiation position to the detection position by a pneumatic transfer or "rabbit" system. First developed by Goosman and Alburger in 1972 [73] it was used for the first time to identify a new isotope by Varley, Cunnane, and Gelletly in 1976 at the HILAC of the Schuster Laboratory at the University of Manchester, UK. They actually used three different methods, "namely the study of radiation emitted between accelerator pulses, the use of a mechanical rabbit system and the He-jet recoil transport system" [46] to identify ^{104}In. Later in the year, Parks, Davids, and Pardo from Argonne National Laboratory discovered ^{53}Ti with a multiple rabbit system. The advantage of using several targets is that it reduced the build-up of background radiation from longer-lived activities produced in the irradiation [48].

In yet another approach, Heiguchi et al. from the University of Tsukuba, Japan, identified ^{89}Tc by catching the recoils from the target on tantalum foils positioned on a rotating disk. After the irradiation, the disk was rotated by 180° to move the activities in front of two high-purity Ge (HPGe) detectors before another foil was moved behind the target [67].

The first isotope ever discovered in China was measured with a similar technique. Zhou et al. identified ^{90}Ru at the HI-13 tandem accelerator of the China Institute of Atomic Energy in Beijing in 1991. The residual nuclei recoiled on tantalum foils of a wheel system with 60 positions. The wheel rotated by 174° so that irradiation and detection could be done simultaneously [69].

Online isotope separators had been well established for the discovery of new isotopes in neutron fission (Sect. 6.5), spallation (Sect. 10.4), and charged-particle fission (Sect. 10.5) by the mid-sixties. In the 1970s several ISOL facilities dedicated for fusion evaporations were established all over the world. Isotopes discovered in fusion-evaporation reactions using online isotope and mass separation are listed in Table 11.3. The first new β-emitting isotope discovered in fusion-evaporation reactions using an online-mass separation system was ^{116}I. It was first produced by Gowdy et al. in 1975 with the Oak Ridge Isochronous Cyclotron (ORIC) and

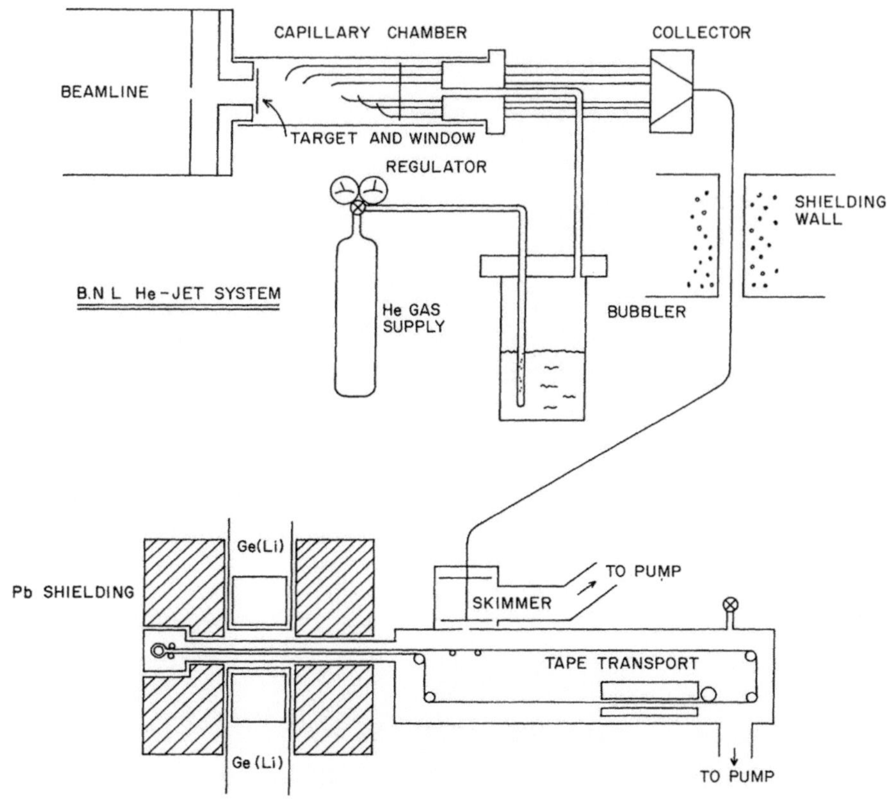

Fig. 11.1 Schematic diagram of the Brookhaven helium-jet transfer system used in the discovery of ^{80}Y and ^{81}Y [59] (Reprinted figure with permission from C.J. Lister et al., *New isotope* ^{80}*Y, and the decays of* ^{79}*Sr,* ^{81}*Y, and* ^{82}*Y*, Phys. Rev. C 24 (1981) 260. Copyright 1981 by the American Physical Society.)

identified using the UNISOR on-line isotope separator [74]. A total of 23 new isotopes were discovered with the BEMS-2 isotope separator at the Dubna U300 heavy-ion cyclotron in three separate experiments by Bogdanov et al. in 1976 [75] and 1978 [76] and by Nowicki et al. in 1980 [79].

^{103}In [77], ^{97}Ag and ^{98}Ag [78], ^{104}Sn [82], and ^{101}In [90] were discovered by the group of Huyse between 1978 and 1988 with the Louvain isotope separator (LISOL) installed at the CYCLONE cyclotron at Louvain-la-Neuve, Belgium. In 1981 Beraud et al. measured ^{102}In with the Grenoble cyclotron [80] and a year later Hagberg et al. discovered ^{71}Br using the Chalk River online-isotope separator in Canada [81].

After 1984, most isotopes were discovered with the online mass separator at the UNILAC at GSI, Darmstadt, by Kirchner, Klepper, Roeckl, and collaborators [83, 87, 97, 98] and the isotope separator OASIS at the SuperHILAC at Berkeley by Nitschke, Wilmarth, and collaborators [85, 86, 88, 92–95]. In 1986, Redon et al. coupled a helium-jet recoil system with the ion source of the on-line mass separator

Table 11.3 Isotopes discovered in fusion-evaporation reactions using online isotope and mass separation: The submission date, first author and reference of the publication, as well as the target, beam, and the evaporation channel are listed

Date	First author	Ref.	Target	Beam	Evaporation	Isotope
09/08/1975	Gowdy	[74]	^{103}Rh	^{16}O	3n	^{116}I
07/23/1976	Bogdanov	[75]	^{102}Pd	^{32}S	3p2n	^{129}Pr
			^{106}Cd	^{32}S	5p3n, 5p2n	^{130}Pr, ^{131}Pr
			^{106}Cd	^{32}S	4pxn	^{130}Nd, ^{131}Nd, ^{132}Nd, ^{133}Nd
			^{106}Cd	^{32}S	3pxn	^{132}Pm, ^{133}Pm, ^{134}Pm
			^{106}Cd	^{32}S	2p2n, 2p1n	^{134}Sm, ^{135}Sm
04/28/1978	Bogdanov	[76]	^{96,98}Ru	^{32}S	2pxn	^{123}La, ^{124}La
					3pxn	^{124}Ce, ^{125}Ce, ^{126}Ce, ^{127}Ce
05/30/1978	Lhersonneau	[77]	^{nat}Mo	^{14}N	xn	^{103}In
08/14/1978	Huyse	[78]	^{92}Mo	^{14}N	2p7n, 2p6n	^{97}Ag, ^{98}Ag
10/18/1980	Nowicki	[79]	^{112}Sn	^{32}S	4p4n, 4p2n	^{136}Sm, ^{138}Sm
					3pxn	^{137}Eu, ^{138}Eu, ^{140}Eu
01/26/1981	Beraud	[80]	^{92}Mo	^{14}N	4n	^{102}In
02/02/1982	Hagberg	[81]	^{40}Ca	^{35}Cl	2p2n	^{71}Br
11/29/1984	Deneffe	[82]	^{92}Mo	^{20}Ne	2p6n	^{104}Sn
11/26/1985	Kleinheinz	[83]	^{96}Ru	^{58}Ni	2p1n	^{151}Yb
06/05/1986	Redon	[84]	^{106}Cd	^{35}Cl	3p1n	^{137}Sm
			^{112}Sn	^{32}S, ^{35}Cl	2p1n, p2n	^{141}Gd, ^{144}Dy
08/04/1986	Toth	[85]	^{96}Ru	^{58}Ni	pn	^{152}Lu
09/23/1986	Wilmarth	[86]	^{92}Mo	^{54}Fe	3p2n	^{141}Tb
12/10/1986	Sekine	[87]	^{12}C	^{40}Ca	p3n	^{48}Mn
03/23/1987	Toth	[88]	^{94}Mo	^{58}Ni	p2n	^{149}Tm
05/11/1987	Kern	[89]	^{92}Mo	^{48}Ti	p3n	^{136}Eu
03/14/1988	Huyse	[90]	^{92}Mo	^{20}Ne	3p8n	^{101}In
07/11/1988	Sekine	[91]	^{92}Mo	^{32}S	p2n	^{121}La
01/09/1989	Vierinen	[92]	^{92}Mo	^{58}Ni	2p3n	^{145}Er
06/26/1989	Nitschke	[93]	^{92}Mo	^{64}Zn	p2n	^{153}Lu
08/31/1990	Firestone	[94]	^{92}Mo	^{54}Fe	3p1n	^{142}Tb
02/23/1993	Toth	[95]	^{92}Mo	^{58}Ni	2p2n	^{146}Er
02/07/1995	Gizon	[96]	^{92}Mo	^{40}Ca	αp	^{127}Pr
02/07/1995	Guglielmetti	[97]	^{58}Ni	^{58}Ni	2n	^{114}Ba
04/14/1997	Janas	[98]	^{60}Ni	^{58}Ni	3n, 2n	^{115}Ba, ^{116}Ba
			^{63}Cu	^{58}Ni	1p2n	^{118}Ba

at the SARA accelerator at Grenoble to identify ^{137}Sm, ^{141}Gd, and ^{144}Dy [84] and in 1988 ^{121}La was discovered by Sekine et al. by mass separation at the Japan Atomic Energy Research Institute (JAERI) in Tokai [91].

11.3 In-Beam γ-Ray Spectroscopy

The discoveries described in the previous section were achieved with techniques developed earlier for isotope discoveries using other probes and reaction mechanisms. They involved the measurements of the β-decay half-lives and were identified based on excitation functions and/or characteristic known radiation of the daughter nuclides. In a different approach, Stephens, Lark, and Diamond from Berkeley realized that heavy-ion fusion-evaporation reactions transfer a large amount of angular momentum to the compound nucleus [99]. If the nuclides populated after particle evaporation are prolate deformed they deexcite by γ-rays of rotational bands. Coupled with excitation function measurements the in-beam detection of these γ-ray cascades can be used to uniquely identify an isotope without measuring the β-decay. Stephens, Lark, and Diamond credit Morinaga and Gugelot [100] for the first observation of γ-rays from a rotational band in (α,4n) reactions. Although this technique was not primarily intended to search for new isotopes but to understand the evolution of nuclear structure in neutron-deficient nuclei, over 40 new nuclides were discovered during these investigations. The isotopes discovered in fusion-evaporation reactions using in-beam γ-ray spectroscopy are listed in Table 11.4.

The first isotopes discovered utilizing this method were ^{166}Hf, ^{172}W, and ^{174}W submitted for publication by Stephens, Lark, and Diamond in 1964 [101]. Until 1973, the Berkeley group was the only one identifying new isotopes with γ-ray spectroscopy. As an example, the ^{124}Ba spectrum and the corresponding rotational band is shown in Fig. 11.2 [103]. In the following 13 years the method was then adopted by nine different laboratories. Anholt, Rasmussen, and Rezanka used γ-ray spectroscopy to identify millisecond isomers. They identified ^{161}Lu using the pulsed beam structure of the Yale heavy ion accelerator to begin the recording of γ-spectra at the end of the 2 ms wide beam pulses [107].

In Germany, isotopes were identified at the tandem accelerators in Heidelberg (^{120}Ba and ^{122}Ba [109]), Munich (^{159}Yb [110] and ^{152}Yb [114]) and the VICKSI (Van-de-Graaff Isochron Cyclotron Kombination für Schwere Ionen) facility at the Hahn-Meitner Institut in Berlin (^{169}W [116]). In addition, ^{151}Tm, ^{78}Sr, and ^{171}W were discovered at Argonne [112], Brookhaven [113], and Louvain-la-Neuve [115], respectively.

The first—of only two—isotopes ever discovered in Australia was ^{167}W in 1985. It was produced by Gerl et al. with the 14UD Pelletron accelerator from the Australian National University in Canberra [117].

Table 11.4 Isotopes discovered in fusion-evaporation reactions using in-beam γ-ray spectroscopy: The submission date, first author and reference of the publication, as well as the target, beam, and the evaporation channel are listed

Date	First author	Ref.	Target	Beam	Evap.	Isotope
08/24/1964	Stephens	[101]	^{159}Tb, ^{165}Ho	^{14}N	7n	^{166}Hf, ^{172}W
			^{169}Tm	^{11}B	6n	^{174}W
09/06/1966	Burde	[102]	^{169}Tm	^{14}N	5n	^{178}Os
09/29/1966	Clarkson	[103]	^{116}Sn, ^{127}I	^{12}C	4n,5n	^{124}Ba, ^{134}Pr
10/09/1967	Ward	[104]	^{120}Sn	^{40}Ar	4n	^{156}Er
			^{122,124}Te	^{40}Ar	4n	^{158}Yb, ^{160}Yb
06/06/1968	Ward	[105]	^{116}Sn	^{16}O	4n	^{128}Ce
01/18/1971	Stephens	[106]	^{144}Sm	^{28}Si	2p2n	^{168}W
04/09/1973	Anholt	[107]	^{148}Sm	^{19}F	6n	^{161}Lu
04/17/1973	Nakai	[108]	^{116}Sn	^{14}N	5n	^{125}La
07/19/1974	Conrad	[109]	^{106,108}Cd	^{16}O	2n	^{120}Ba, ^{122}Ba
09/15/1975	Trautmann	[110]	^{144}Sm	^{18}O	3n	^{159}Yb
05/23/1977	Gizon	[111]	^{116}Sn	^{16}O	3n	^{129}Ce
03/05/1982	Helppi	[112]	^{92}Mo	^{60}Ni	p	^{151}Tm
05/28/1982	Lister	[113]	^{58}Ni	^{24}Mg	2p2n	^{78}Sr
08/04/1982	Nolte	[114]	^{96}Ru	^{58}Ni	2p	^{152}Yb
10/20/1982	Arciszewski	[115]	^{155}Gd	^{20}Ne	4n	^{171}W
12/19/1984	Recht	[116]	^{154}Gd	^{20}Ne	5n	^{169}W
03/23/1985	Gerl	[117]	^{142}Nd	^{28}Si	3n	^{167}W
05/20/1985	Lister	[118]	^{92,92,94}Mo	^{40}Ca, ^{50}Cr	2p2n	^{128}Nd, ^{138}Gd, ^{140}Gd
06/01/1987	Lister	[119]	^{24}Mg	^{58}Ni	2n	^{80}Zr
08/17/1987	Goettig	[120]	^{92}Mo	^{56}Fe	p2n	^{145}Ho
03/19/1988	Kuroyanagi	[121]	^{58,60}Ni	^{28}Si	p2n	^{83}Nb, ^{85}Nb
07/04/1989	Wadsworth	[122]	^{96}Ru	^{40}Ca	2p2n	^{132}Sm
05/17/1990	Lister	[123]	^{12}C, ^{24}Mg	^{58}Ni, ^{54}Fe	2n	^{68}Se, ^{76}Sr
10/22/1990	Gelletly	[124]	^{28}Si	^{58}Ni	2n	^{84}Mo
03/18/1991	Rudolph	[125]	^{58}Ni	^{32}S	p2n, pn	^{87}Tc, ^{88}Tc
07/18/1991	Gross	[126]	^{40}Ca	^{50}Cr	2p2n	^{86}Mo
04/06/1992	de Angelis	[127]	^{92}Mo	^{58}Ni	2pn	^{147}Er
09/14/1993	Seweryniak	[128]	^{54}Fe	^{58}Ni	3p2n	^{107}Sb
12/19/1997	Parry	[129]	^{96}Ru	^{40}Ca	αp	^{131}Pm
02/13/2002	Krolas	[130]	^{92}Mo	^{54}Fe	2p4n	^{140}Dy
03/20/2002	Wilson	[131]	^{64}Zn	^{64}Zn	p2n	^{125}Pr
04/22/2005	Smith	[132]	^{64}Zn	^{64}Zn	α2n	^{122}Ce

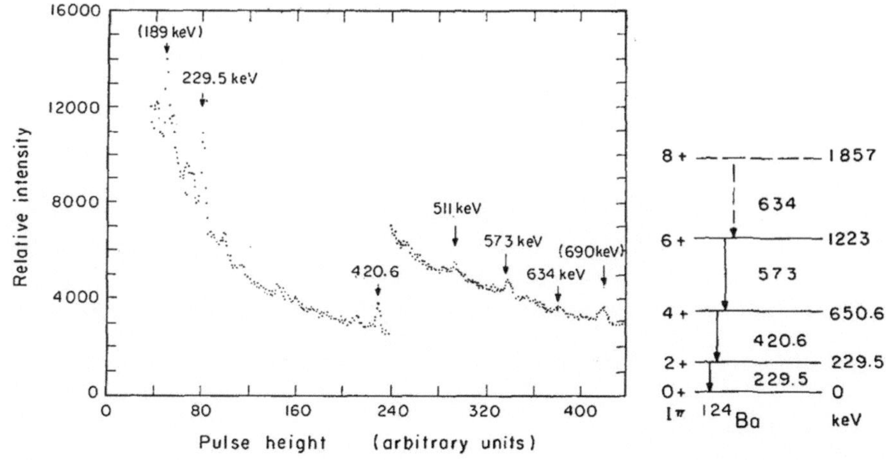

Fig. 11.2 In-beam γ-ray spectrum (*left*) and corresponding level scheme (*right*) of ^{124}Ba [103] (Reprinted from J.E. Clarkson et al., *Collective excitations in neutron-deficient barium, xenon, and cerium isotopes*, Nucl. Phys. A 93 (1967) 272. Copyright 1967, with permission from Elsevier.)

Between 1985 and 1991 the field was again dominated by a single laboratory. With the exception of ^{83}Nb and ^{85}Nb which were discovered by Kuroyanagi et al. [121] at the tandem accelerator of Kyushu University in Fukuoka, Japan, all isotopes during this period were produced with the Daresbury Laboratory Van de Graaff accelerator in the UK. For the identification of ^{128}Nd,^{138}Gd, and ^{140}Gd [118] and ^{145}Ho [120] these weaker residue channels were enhanced by measuring the γ-ray spectra in coincidence with evaporated neutrons and light-charged-particles. In subsequent experiments an increasing number of bismuth germanate (BGO) suppressed germanium detectors were placed at the target position of the Daresbury recoil separator. In the discovery of ^{87}Tc and ^{88}Tc by Rudolph et al. the 20 element array had been named POLYTESSA (TESSA—The Energy Suppression Shield Array) [125].

The combination of high-resolution suppressed γ-ray arrays and light-charged-particle arrays and recoil separators was a sensitive tool to measure γ-ray spectra of residues populated with small cross-sections. In 1992, ^{147}Er was discovered with four Compton suppressed germanium detectors and the Recoil Mass Spectrometer (RMS) of the Laboratori Nazionali di Legnaro (LNL) in Italy by de Angelis et al. [127]. Seweryniak et al. identified ^{107}Sb in 1993 using the NORDBALL array consisting of 15 detectors and a 21 element ΔE silicon detector array at the Tandem Accelerator Laboratory of the Niels Bohr Institute in Roskilde, Denmark [128]. Similarly, ^{131}Pm was identified by Parry et al. at the TASCC (Tandem Accelerator SuperConducting Cyclotron) facility of Chalk River, Canada, in 1997. They used the 8π array, consisting of an outer shell of 20 Compton suppressed HPGe detectors and an inner shell of 71 BGO scintillation detectors. It was used in conjunction with a 4π modular array of cesium iodide detectors for the reaction channel selection [129].

In 2002, Krolas et al. placed the Clover Germanium Detector Array for Recoil Decay Spectroscopy (CARDS) at the focal plane of the Recoil Mass Spectrometer RMS of the Holifield Radioactive Ion Beam Facility (HRIBF) at Oak Ridge in order to measure short-lived isomers and discovered a 7 μs isomer in ^{140}Dy [130].

The sensitivity was improved even further for the discovery of the two most recent isotopes by utilizing 78 75 %-efficient high-purity germanium detectors of Gammasphere coupled to the Microball and the Neutron Shell in addition to measuring the recoils in the Fragment Mass Analyser (FMA). The experiment was performed at the Argonne Tandem Linear Accelerator System (ATLAS) by Wilson et al. [131] and Smith [132] discovering the neutron deficient isotopes ^{125}Pr and ^{122}Ce, respectively.

11.4 Alpha Emitters

As mentioned in the introduction fusion-evaporation reactions were originally intended for the discovery of transuranium elements. These nuclides decay predominantly by the emission of α-particles so that techniques to detect these α-particles were developed in order to identify the new isotopes. The realization that the recoils were leaving the targets after the reaction by Ghiorso et al. was then also used to identify α-emitters in lighter mass regions. Table 11.5 lists α-emitting isotopes discovered in fusion-evaporation and identified following chemical separation or by implantation in a solid catcher.

The second facility to accelerate heavy-ions after the initial success at Berkeley was the 60-in. Nuffield cyclotron of the University of Birmingham, UK. In 1954, Burcham used a ^{14}N beam with a continuous energy distribution up to 120 MeV to irradiate gold foils and measured α-spectra following chemical separation of radon to identify ^{206}Rn and ^{207}Rn. The α-spectrum is shown in Fig. 11.3 [133]. The discovery of ^{149}Dy, and ^{150}Dy, ^{151}Dy by Toth and Rasmussen at the Berkeley 60-in. cyclotron was achieved without chemical separation by measuring the α activities of the irradiated gold foil directly. ^{149}Dy is a β^+-emitter which was identified by the subsequent α-decay of the ^{149}Tb daughter nucleus [134]. Five neutron-deficient astatine isotopes were measured after chemical separation for the first time by Latimer et al. [135] and Hoff et al. [136] using the HILAC at Berkeley. For the discovery of ^{200}At and ^{201}At Hoff, Asaro, and Perlman employed a magnetic spectrograph to measure the energy of the α-particles.

Rotter et al. collected the recoils from "complete amalgamation reactions" of 80 MeV carbon ions irradiating lead and bismuth targets from the Dubna 150 cm cyclotron on an aluminum foil. The collecting foil was tilted by 45° so that it also faced a Au-Si semiconductor detector measuring the α energies from the decays of ^{213}Rn, ^{214}Ra, ^{215}Ra, ^{216}Ac, ^{214}Fr [137]. Also at Dubna—using the U300 cyclotron (U stands for Uskoritel, russian for accelerator)—Sung-Ching-Yang et al. identified ^{216}Pa by slowing the recoils in a gas to collect them on filters which were then pneumatically transported in front of two Si(Au) detectors [138].

Table 11.5 Isotopes discovered in fusion-evaporation and identified by their α-decay following chemical separation or by implantation in a solid catcher: The submission date, first author and reference of the publication, as well as the target, beam, and the evaporation channel are listed

Date	First author	Ref.	Target	Beam	Evap.	Isotope
03/09/1954	Burcham	[133]	^{197}Au	^{14}N	5n, 4n	^{206}Rn, ^{207}Rn
09/13/1957	Toth	[134]	^{141}Pr	^{14}N	6n, 5n, 4n	^{149}Dy^{a}, ^{150}Dy, ^{151}Dy
07/12/1960	Latimer	[135]	^{197}Au	^{12}C	7n, 5n, 3n	^{202}At, ^{204}At, ^{206}At
04/03/1963	Hoff	[136]	^{197}Au	^{12}C	9n, 8n	^{200}At, ^{201}At
12/13/1965	Rotter	[137]	^{208}Pb	^{12}C	6n, 5n, 3n	^{214}Ra, ^{215}Ra, ^{213}Rn^{b}
			^{209}Bi	^{12}C	5n, 3n	^{216}Ac, ^{214}Fr^{b}
07/12/1971	Sung-Ching-Yang	[138]	^{190}Os	^{31}P	5n	^{216}Pa
10/13/1971	Cerny	[139]	^{40}Ca	^{6}Li	2n	^{44}V
06/14/1972	Nomura	[140]	^{208}Pb	^{14}N, ^{12}C	5n, 4n	^{217}Ac, ^{216}Ra
05/22/1973	Hiruta	[141]	^{209}Bi	^{14}N	5n	^{218}Th
06/05/1973	Häusser	[142]	^{207}Pb	^{16}O	4n, 3n	^{219}Th, ^{220}Th
12/22/1995	Batchelder	[143]	^{144}Sm	^{48}Ti	2n	^{190}Po
04/30/1996	Toth	[144]	^{144}Sm	^{40}Ca	4n	^{180}Pb

$^{a}\beta^{+}$-emitter, identified by subsequent α-decay of ^{149}Tb
^{b}Populated by α-decay

In 1971, ^{44}V was discovered by Cerny et al. at the Brookhaven three-stage MP tandem Van de Graaff facility by its β-delayed α-decay [139]. The short-lived ($\sim$100 ns) α-emitters ^{216}Ra, ^{217}Ac [140] and ^{218}Th [141] were identified by Nomura et al. and Hiruta et al. respectively, using the pulsed-beam of the IPCR cyclotron at RIKEN in 1972 and 1973. The activities were measured at the target position with a silicon surface barrier detector and the identification was achieved from excitation function measurements and the observation of α-energies of known daughter nuclides. Essen-

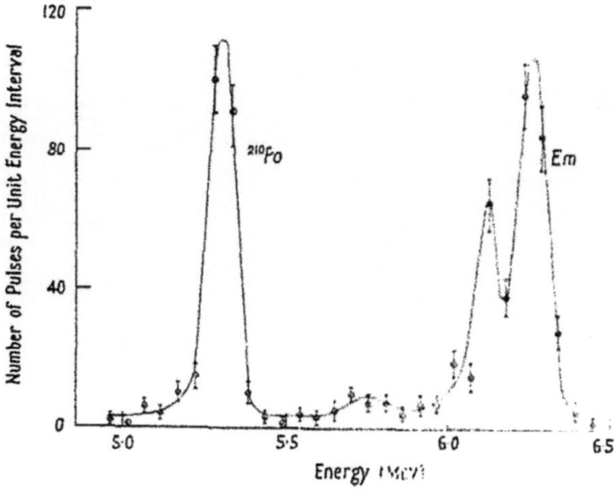

Fig. 11.3 Alpha-spectrum following the 5n, and 4n fusion-evaporation reaction of ^{14}N on ^{197}Au to produce the radon (emanation) isotopes ^{206}Rn and ^{207}Rn, respectively [133] (W.E. Burcham, *The α activity induced in gold by irradiation with nitrogen ions*, Proc. Phys. Soc. A 67 (1954) 555. Copyright IOP Publishing, reproduced with permission, all rights reserved.)

tially at the same time Häusser et al. reported the discovery of ^{219}Th and ^{220}Th using a similar technique at the Chalk River MP tandem by catching the recoils in carbon foil placed behind the target [142].

More than twenty years later, Batchelder et al. and Toth et al. used a rapidly rotating recoil catcher wheel system at the 88-in. Berkeley cyclotron to discover ^{190}Po [143] and ^{180}Pb [144], respectively.

While only about ten isotopes were discovered with solid catchers, well over a hundred isotopes were first identified with helium gas techniques originally developed by Ghiorso et al. for transuranium elements and they are listed in Tables 11.6 and 11.7.

Essentially all isotopes discovered until 1970 listed in Table 11.6 were first identified at Berkeley. Macfarlane and Griffioen first adopted the helium gas technique for α-emitters in the lighter mass region [72, 164] (see Fig. 11.4) and discovered 24 new neutron-deficient nuclides with the HILAC in several papers between 1962 and 1964 [145–149]. Subsequently several other researchers used similar experimental set-ups at the HILAC to identify an additional 65 isotopes until 1970. As a sole author Siivola discovered nine platinum [150], seven iridium [151], four polonium [155], and five gold [159] isotopes. Valli and collaborators added four astatine [152], five radon [153], one francium [154], seven radium [156], seven actinium [158], and five thorium and one protactinium [160] isotopes during this time.

In addition, in 1971, Borggreen, Valli, and Hyde measured the three protactinium isotopes ^{222}Pa, ^{223}Pa, and ^{224}Pa as well as their α-decaying daughters and two of their granddaughters [162]. Borggreen and Hyde then also observed three osmium isotopes (^{172}Os, ^{173}Os, and ^{174}Os) in the same year [163].

The only isotopes not discovered at Berkeley during this period with this method were ^{182}Hg identified by Demin et al. at the U300 Dubna cyclotron [157] and ^{221}Th and ^{222}Th as well as their daughter nuclides ^{217}Ra and ^{218}Ra, and the grand-daughter

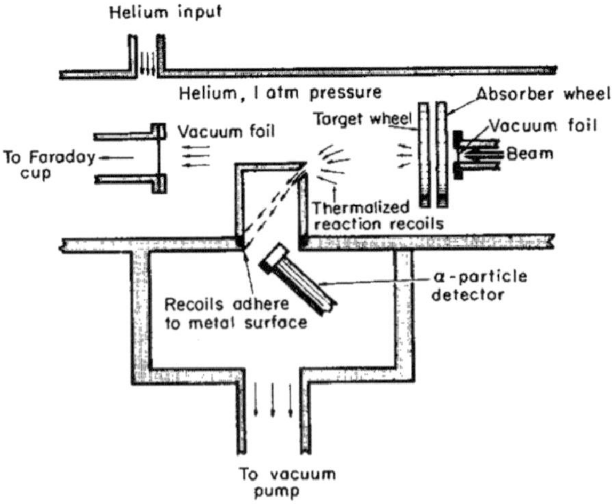

Fig. 11.4 Schematic of the set-up to collect recoils in helium gas by Macfarlane and Griffioen [72] (Reprinted from R.D. Macfarlane and R.D. Griffioen, *System for studying accelerator-produced short-lived alpha emitters*, Nucl. Instrum. Meth. 24 (1963) 461. Copyright 1963, with permission from Elsevier.)

Table 11.6 Isotopes discovered in fusion-evaporation and identified by their α-decay using helium-gas transfer techniques until 1970: The submission date, first author and reference of the publication, as well as the target, beam, and the evaporation channel are listed

Date	First author	Ref.	Target	Beam	Evap.	Isotope
12/10/1962	Macfarlane	[145]	^{141}Pr	^{16}O	7-4n	^{150}Ho, ^{151}Ho, ^{152}Ho, ^{153}Ho
04/22/1963	Macfarlane	[146]	^{142}Nd	^{16}O	6-4n	^{152}Er, ^{153}Er, ^{154}Er
09/20/1963	Griffioen	[147]	^{197}Au	^{16}O	9-6n	^{204}Fr, ^{205}Fr, ^{206}Fr, ^{207}Fr
			^{205}Tl	^{12}C	9-6, 4n	^{208}Fr, ^{209}Fr, ^{210}Fr, ^{211}Fr, ^{213}Fr
11/02/1963	Macfarlane	[148]	^{144}Sm	^{19}F, ^{20}Ne	7n, 6n	^{155}Lu, ^{156}Lu, ^{157}Hf, ^{158}Hf
07/15/1964	Macfarlane	[149]	^{141}Pr	^{20}Ne	8n, 7n	^{153}Tm, ^{154}Tm
			^{144}Sm	^{16}O	6n, 5n	^{154}Yb, ^{155}Yb
01/28/1966	Siivola	[150]	^{nat}Er	^{20}Ne	xn	^{173}Pt, ^{174}Pt, ^{175}Pt, ^{176}Pt, ^{177}Pt
			^{nat}Yb	^{16}O	xn	^{178}Pt, ^{179}Pt, ^{180}Pt, ^{181}Pt
09/09/1966	Siivola	[151]	^{162,164}Er	^{19}F	10n, 9n	^{171}Ir, ^{172}Ir, ^{173}Ir, ^{174}Ir
			^{166}Er	^{19}F	10-8n	^{175}Ir, ^{176}Ir, ^{177}Ir
01/18/1967	Treytl	[152]	^{185}Re	^{20}Ne	9-6n	^{196}At, ^{197}At, ^{198}At, ^{199}At
01/19/1967	Valli	[153]	^{197}Au	^{14}N	10-6n	^{201}Rn, ^{202}Rn, ^{203}Rn, ^{204}Rn, ^{205}Rn
03/19/1967	Valli	[154]	^{197}Au	^{16}O	10n	^{203}Fr
04/17/1967	Siivola	[155]	^{185}Re	^{19}F	11-8n	^{193}Po, ^{194}Po, ^{195}Po, ^{196}Po
04/28/1967	Valli	[156]	^{197}Au	^{19}F	10-7n	^{206}Ra, ^{207}Ra, ^{208}Ra, ^{209}Ra
					6-4n	^{210}Ra, ^{211}Ra, ^{212}Ra
06/20/1967	Demin	[157]	^{170}Yb	^{20}Ne	8n	^{182}Hg
10/06/1967	Valli	[158]	^{197}Au	^{20}Ne	8-5n	^{209}Ac, ^{210}Ac, ^{211}Ac, ^{212}Ac
			^{205}Tl	^{16}O	8-6n	^{213}Ac, ^{214}Ac, ^{215}Ac
10/24/1967	Siivola	[159]	^{168}Yb	^{19}F	10-8n	^{177}Au, ^{178}Au, ^{179}Au
					6n, 4n	^{181}Au, ^{183}Au
06/14/1968	Valli	[160]	^{206}Pb	^{16}O	9-5n	^{213}Th, ^{214}Th, ^{215}Th, ^{216}Th, ^{217}Th
			^{203}Tl	^{20}Ne	6n	^{217}Pa
01/23/1970	Torgerson	[161]	^{208}Pb	^{16}O	3n	^{221}Th, ^{217}Ra^{a}
					2n	^{222}Th, ^{218}Ra^{a}, ^{214}Rn^{a}
05/11/1970	Borggreen	[162]	^{208}Pb	^{19}F	5n	^{222}Pa, ^{218}Ac^{a}
					4n	^{223}Pa, ^{219}Ac^{a}, ^{215}Fr^{a}
					3n	^{224}Pa, ^{220}Ac^{a}, ^{216}Fr^{a}
09/23/1970	Borggreen	[163]	^{164}Er	^{16}O	8-6n	^{172}Os, ^{173}Os, ^{174}Os

[a]Populated by α-decay

Table 11.7 Isotopes discovered in fusion-evaporation reactions and identified by their α-decay using helium-gas transfer techniques after 1970: The submission date, first author and reference of the publication, as well as the target, beam, and the evaporation channel are listed

Date	First author	Ref.	Target	Beam	Evap.	Isotope
08/19/1971	Toth	[165]	^{144,147}Sm	^{14}N	3n, 5n	^{155}Tm, ^{156}Tm
01/02/1972	Toth	[166]	^{144}Sm	^{20}Ne	5-3n	^{159}Hf, ^{160}Hf, ^{161}Hf
03/13/1972	Toth	[167]	^{156}Dy	^{20}Ne	6n, 5n	^{170}Os, ^{171}Os
06/19/1972	Toth	[168]	^{156}Dy	^{20}Ne	7n	^{169}Os
07/10/1972	Gauvin	[171]	^{155}Gd	^{40}Ar	9-5n	^{186}Pb, ^{187}Pb, ^{188}Pb, ^{189}Pb, ^{190}Pb
			^{159}Tb	^{40}Ar	9n, 8n	^{190}Bi, ^{191}Bi
01/09/1973	Bogdanov	[181]	^{96}Ru	^{16}O	4n	^{108}Te
02/08/1973	Eastham	[182]	^{144}Sm	^{24}Mg	7-4n	^{161}W, ^{162}W, ^{163}W, ^{164}W
03/29/1973	Gauvin	[172]	^{159}Tb	^{40}Ar	10n	^{189}Bi
07/09/1973	Le Beyec	[179]	^{181}Ta	^{19}F	9n, 8n	^{191}Pb, ^{192}Pb
12/16/1974	Cabot	[173]	^{141}Pr	^{40}Ca	6n, 5n	^{175}Au, ^{176}Au
			^{150}Sm	^{40}Ca	5n	^{185}Pb
04/28/1975	Toth	[169]	^{156}Dy	^{16}O	7n, 6n	^{165}W, ^{166}W
10/06/1975[a]	Cabot	[174]	^{142}Nd	^{40}Ca	5n	^{177}Hg
06/01/1977	Cabot	[175]	^{106}Cd, ^{110}Cd	^{63}Cu	p2n, 3n	^{166}Os, ^{170}Ir
			^{109}Ag	^{63}Cu	5n, 4n	^{167}Os, ^{168}Os
07/11/1977	Della Negra	[176]	^{182}W	^{20}Ne	10n	^{192}Po
03/21/1978	Cabot	[177]	^{nat}Cd	^{63}Cu	xpxn	^{169}Re, ^{168}Ir, ^{169}Ir, ^{165}Os
05/02/1978	Schrewe	[183]	^{89}Y	^{84}Kr	7n	^{166}Re
12/17/1979	Dufour	[180]	^{148}Sm	^{40}Ca	4n	^{184}Pb
03/26/1981	Della Negra	[178]	^{112}Sn	^{63}Cu	p2n	^{172}Pt
10/24/1988	Toth	[170]	^{94}Mo	^{90}Zr	3n	^{181}Pb
03/16/1992	Meissner	[184]	^{141}Pr	^{32}S	6n, 5n	^{167}Re, ^{168}Re

[a]Populated by α-decay
[b]Date of presentation

^{214}Rn discovered at the Yale University Heavy Ion Accelerator by Torgerson et al. [161].

In the 1970s the technique was most productively used to discover new α-emitting nuclides at Oak Ridge and Orsay (see Table 11.7). Toth and collaborators identified ten isotopes of elements between thulium and osmium using the Oak Ridge Isochronous Cyclotron (ORIC) [165–169] and ^{181}Pb in 1988 [170]. Between 1972 and 1981, the group of Le Beyec discovered twenty-four isotopes of elements between rhenium and bismuth mostly with the ALICE accelerator at Orsay [171–178]. For the discovery of ^{191}Pb and ^{192}Pb they used the Berkeley HILAC [179]. Also at Orsay, Dufour et al. identified ^{184}Pb in 1979 by using an electrostatic deflector to deposit lead recoils stopped in nitrogen gas onto the surface of a solid state detector [180].

Other than in Oak Ridge and Orsay, Bogdanov, Karnaukhov, and Petrov identified ^{108}Te with the 300 cm Dubna cyclotron [181], Eastham and Grant measured several tungsten isotopes at the University of Manchester HILAC [182], and Schrewe et al. reported the observation of ^{166}Re at the GSI UNILAC [183]. In 1992, measurements by Meissner et al. at the VICKSI accelerator of the Berlin Hahn-Meitner Institut resulted in the revision of previous assignments of neutron-deficient rhenium isotopes and they are credited with the discovery of ^{167}Re and ^{168}Re [184].

Although the helium gas jet technique was used in the discovery of many nuclides it had the disadvantage that the identification relied on excitation function measurements and/or the observation of known daughter α-emitters. With the application of the Isotope Separation On-Line (ISOL) method for the fusion-evaporation recoil products it was possible to identify the mass of the fragments and measure the subsequent α-decay. Table 11.8 lists the α-emitting isotopes discovered with the ISOL technique.

Already in 1969, Tarantin, Kabachenko, and Demyanov discovered five bismuth isotopes with the heavy-ion beam mass separator at the U300 Dubna cyclotron [185]. Several years later, Toth et al. used the UNISOR facility at Oak Ridge to identify ^{184}Tl, ^{185}Tl, and ^{187}Tl [186]. All other ISOL measurements of new α-emitters were performed with the GSI online mass separator by the group of Kirchner, Klepper, Roeckl, and collaborators, who observed sixteen new isotopes between 1977 and 1981 as listed in the table.

An even more efficient method to measure and identify the evaporation residues and correlate the subsequent α-emission was to separate the recoils directly, rather than stopping them in an ion source of a mass separator. One of the first such devices was the velocity filter SHIP (Separator for Heavy Ion reaction Products) developed at GSI shown in Fig. 11.5. It was not only instrumental in the discovery of many transuranium and especially superheavy nuclides, SHIP was also used in the discovery of many lighter-mass α-emitting isotopes as listed in Table 11.9. Almost fifty isotopes of the elements between lutetium and uranium were discovered with SHIP since 1978; the most recent one was ^{197}Fr in 2013 [213].

In addition to SHIP at GSI, recoil separators were installed at several other heavy-ion accelerator facilities around the world. Isotopes discovered at these separators are listed in Table 11.10. In 1979, DiRienzo et al. discovered ^{199}Rn at the three-stage tandem accelerator in Brookhaven using an MIT designed velocity selector [214]. The in-flight kinematic separator VASSILISSA located at the U400 Dubna cyclotron was used by Andreyev et al. to discover ^{225}U [215], ^{223}U, ^{224}U [217], ^{218}U [218], and ^{219}U [219] between 1988 and 1993. Seven year later, Malyshev et al. pushed the knowledge of neutron-deficient uranium isotopes even further to ^{217}U [237]. Page et al. and Sellin et al. measured α-decays in coincidence with fragments identified with the Daresbury recoil separator in the discovery of ^{108}I [216] and ^{172}Au [220], respectively.

One of the most productive devices next to SHIP was the gas-filled recoil separator RITU (Recoil Ion Transport Unit) at the K130 cyclotron of the University of Jyväskylä Accelerator Laboratory (JYFL) in Finland. Between 1994 and 2012, the group of Leino and Uusitalo discovered sixteen α-emitting nuclei with RITU.

Table 11.8 Isotopes discovered in fusion-evaporation and identified by their α-decay using Isotope Separation OnLine (ISOL): The submission date, first author and reference of the publication, as well as the target, beam, and the evaporation channel are listed

Date	First author	Ref.	Target	Beam	Evap.	Isotope
10/06/1969	Tarantin	[185]	^{181}Ta	^{20}Ne	9n, 8n, 7n	^{192}Bi, ^{193}Bi, ^{194}Bi
					6n, 4n	^{195}Bi, ^{197}Bi
06/15/1976	Toth	[186]	^{180}W	^{14}N	10n, 9n	^{184}Tl, ^{185}Tl
			^{182}W	^{14}N	9n	^{187}Tl
07/08/1977	Kirchner	[187]	^{63}Cu	^{58}Ni	2p5n	^{114}I
			^{58}Ni	^{58}Ni	4p2n, 2p	^{110}Te, ^{114}Xe
					3p3-0n	^{110}I, ^{111}I, ^{112}I, ^{113}I
07/11/1978	Roeckl	[188]	^{58}Ni	^{58}Ni	2p2n	^{112}Xe
04/17/1979	Schardt	[189]	^{58}Ni	^{58}Ni	2p3n	^{111}Xe, ^{107}T^{a}
01/02/1980	Schrewe	[190]	^{142}Nd	^{48}Ti	p6n, 7n	^{183}Tl, ^{183}Pb
			^{107}Ag	^{84}Kr	3n	^{188}Bi
03/30/1981	Schardt	[191]	^{58}Ni	^{58}Ni	2p4n	^{110}Xe, ^{106}Te^{a}
08/26/1981	Plochocki	[192]	^{58}Ni	^{58}Ni	pn	^{106}Sb^{a}

[a]Populated by α-decay

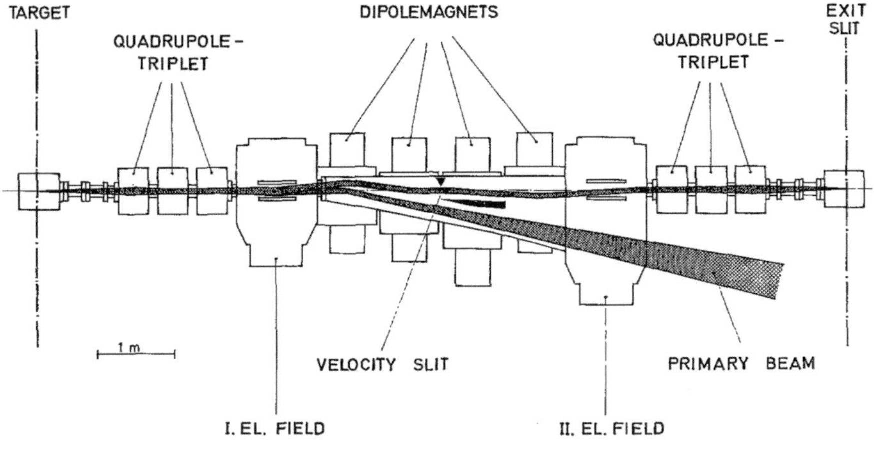

Fig. 11.5 Schematic diagram of the GSI velocity filter [193] (Reprinted from H. Ewald et al., *Report on the heavy ion separator SIS at the GSI*, Nucl. Instrum. Meth. 139 (1976) 223. Copyright 1976, with permission from Elsevier.)

Figure 11.6 shows the measured energies of mother and daughter α-emitters correlated with evaporation residues observed in the discovery of ^{210}Th and ^{211}Th [223]. For the most recent discoveries of ^{161}Os and its α daughter ^{157}W [244] and^{198}Fr [245] the Gamma Recoil Electron Alpha Tagging (GREAT) Spectrometer [249] was installed at the focal plane of RITU.

Table 11.9 Isotopes discovered in fusion-evaporation reactions and identified by their α-decay using the velocity filter SHIP at GSI: The submission date, first author and reference of the publication, as well as the target, beam, and the evaporation channel are listed

Date	First author	Ref.	Target	Beam	Evap.	Isotope
08/25/1978	Schmidt	[194]	^{181}Ta	^{40}Ar	6n, 3n	^{215}Pa, ^{218}Pa
11/27/1978	Hofmann	[195]	^{107}Ag, ^{108}Pd	^{58}Ni	p4n	^{160}W, ^{156}Hf^{a}, ^{161}Ta
			^{109}Ag	^{58}Ni	4n	^{163}Re, ^{159}Ta^{a}
			^{109}Ag	^{58}Ni	3n	^{164}Re, ^{160}Ta^{a}
			^{107}Ag	^{58}Ni^{b}	3n	^{162}Re, ^{158}Ta^{a}
09/05/1979	Vermeulen	[196]	^{176}Hf	^{40}Ar	4n	^{212}Th
02/03/1981	Hofmann	[197]	^{110}Cd	^{58}Ni	5n, 4n	^{163}Os, ^{159}W^{a}, ^{164}Os
			^{nat}Sn	^{58}Ni^{c}	xn	^{168}Pt, ^{169}Pt, ^{170}Pt, ^{171}Pt
			^{nat}Cd	^{58}Ni	xpyn	^{155}Hf^{a}, ^{165}Re
			^{106}Cd	^{58}Ni	2p4n	^{158}W, ^{154}Hf^{a}, ^{154}Lu^{d}
02/24/1983	Schneider	[198]	^{nat}Rb	^{92}Mo	xn	^{173}Au, ^{174}Au
			^{nat}Sr	^{92}Mo	xn	^{175}Hg, ^{176}Hg
			^{89}Y	^{92}Mo	2n	^{179}Tl
05/30/1983	Hingmann	[199]	^{186}W	^{40}Ar	p4n, 4n	^{221}Pa, ^{222}U
09/05/1985	Keller	[200]	^{94}Mo	^{90}Zr	2n	^{182}Pb
11/06/1986	Heßberger	[201]	^{159}Tb	^{51}V	5n	^{205}Ra
02/21/1989	Hofmann	[202]	^{106}Cd	^{58}Ni	2n	^{162}Os
12/22/1992	Quint	[203]	^{100}Mo	^{94}Mo	3n	^{191}Po
12/20/1994	Ninov	[204]	^{170}Er	^{51}V	8n, 7n	^{213}Pa, ^{214}Pa
10/15/1999	Andreyev	[205]	^{142}Nd	^{52}Cr	6n, 5n	^{188}Po, ^{189}Po
03/14/2003	Andreyev	[206]	^{93}Nb	^{94}Mo	3n	^{184}Bi
04/08/2005	Andreyev	[207]	^{144}Sm	^{46}Ti	4n, 3n	^{186}Po, ^{187}Po
10/25/2005	Andreyev	[208]	^{144}Sm	^{51}V	3n	^{192}At
10/02/2006	Andreyev	[209]	^{144}Sm	^{52}Cr	3n, 2n	^{193}Rn, ^{194}Rn
05/04/2009	Andreyev	[210]	^{141}Pr	^{56}Fe	3n	^{194}At
11/13/2009	Andreyev	[211]	^{144}Sm	^{40}Ca	5n	^{179}Pb
07/16/2010	Heredia	[212]	^{147}Sm	^{64}Ni	3n	^{208}Th
01/29/2013	Kalaninova	[213]	^{141}Pr	^{60}Ni	4n	^{197}Fr

[a]Populated by α-decay
[b]These reactions also produced the proton-unbound nuclides ^{161}Re and ^{157}Ta (see Sect. 16.4)
[c]These reactions also produced the proton-unbound nuclides ^{166}Ir and ^{167}Ir (see Sect. 16.4)
[d]Populated by β^{+}-decay

During this time period (1994 through 2006) another ten isotopes were discovered at Argonne National Laboratory with the Fragment Mass Analyzer (FMA) designed and constructed by Davids and Larson [250]. It is interesting to note that out of eight discovery publications only two had first authors from Argonne (^{178}Tl [232] and^{105}Te [242]). The other discoveries were led by researchers affiliated with six different institutions who collaborated with the group of Davids: ^{204}Ra (University of Manchester) [222], ^{181}Tl (Oak Ridge National Laboratory) [225], ^{166}Pt and ^{167}Pt

Table 11.10 Isotopes discovered in fusion-evaporation reactions and identified by their α-decay using recoil separators: The submission date, first author and reference of the publication, as well as the target, beam, and the evaporation channel are listed

Date	First author	Ref.	Target	Beam	Evap.	Isotope
12/28/1979	DiRienzo	[214]	^{169}Tm	^{35}Cl	5n	^{199}Rn
12/16/1988	Andreyev	[215]	^{208}Pb	^{22}Ne	5n	^{225}U
05/14/1990	Page	[216]	^{54}Fe	^{58}Ni	p3n	^{108}I
11/01/1990	Andreyev	[217]	^{208}Pb	^{20}Ne	5n, 4n	^{223}U, ^{224}U
02/04/1992	Andreyev	[218]	^{197}Au	^{27}Al	6n	^{218}U
01/18/1993	Andreyev	[219]	^{197}Au	^{27}Al	5n	^{219}U
07/12/1993	Sellin	[220]	^{106}Cd	^{70}Ge	p3n	^{172}Au
03/10/1994	Leino	[221]	^{175}Lu	^{40}Ar	8n, 7n	^{207}Ac, ^{208}Ac
12/28/1994	Leddy	[222]	^{182}W	^{28}Si	6n	^{204}Ra
03/02/1995	Uusitalo	[223]	^{181}Ta	^{35}Cl	6n, 5n	^{210}Th, ^{211}Th
03/21/1995	Morita	[224]	^{166}Er	^{36}Ar	6n, 5n	^{196}Rn, ^{197}Rn
			^{169}Tm	^{36}Ar	5n	^{200}Fr
12/11/1995	Toth	[225]	^{90}Zr	^{92}Mo	p	^{181}Tl
03/06/1996	Leino	[226]	^{175}Lu	^{35}Cl	7n	^{203}Ra
03/11/1996	Bingham	[227]	^{92}Mo	^{78}Kr	4n, 3n	^{166}Pt, ^{167}Pt
06/13/1996	Ikezoe	[228]	^{182}W	^{32}S	5n	^{209}Th
09/03/1996	Batchelder	[229]	^{97}Mo	^{92}Mo	p2n	^{186}Bi
11/13/1996	Mitsuoka	[230]	^{182}W	^{35}Cl	5n	^{212}Pa
02/17/1997	Uusitalo	[231]	^{144}Sm	^{36}Ar	6n	^{174}Hg
02/27/1997	Carpenter	[232]	^{103}Rh	^{78}Kr	3n	^{178}Tl
09/18/1997	Eskola	[233]	^{175}Lu	^{36}Ar	5n	^{206}Ac
09/10/1998	Batchelder	[234]	^{97}Mo	^{92}Mo	pn	^{187}Bi
03/26/1999	Tagaya	[235]	^{169}Tm	^{36}Ar	α6n, 6n	^{195}At, ^{199}Fr
06/02/1999	Seweryniak	[236]	^{78,80}Kr	^{96}Ru	1n, 3n	^{172}Hg, ^{173}Hg
05/18/2000	Malyshev	[237]	^{182}W	^{40}Ar	5n	^{217}U
12/22/2000	Kettunen	[238]	^{142}Nd	^{56}Fe	3n	^{195}Rn
12/16/2002	Kettunen	[239]	^{141}Pr	^{54,56}Fe	4n	^{191}At, ^{193}At
12/09/2003	Kettunen	[240]	^{96}Ru	^{78}Kr	3n	^{171}Hg
10/27/2004	Uusitalo	[241]	^{141}Pr	^{63}Cu	3n, 2n	^{201}Ra, ^{202}Ra
03/09/2006	Seweryniak	[242]	^{50}Cr	^{58}Ni	3n	^{105}Te
05/16/2006	Liddick	[243]	^{54}Fe	^{58}Ni	3n	^{109}Xe
12/20/2009	Bianco	[244]	^{106}Cd	^{58}Ni	3n	^{161}Os, ^{157}W[a]
11/30/2012	Uusitalo	[245]	^{141}Pr	^{60}Ni	3n	^{198}Fr
12/01/2013	Zhang	[246]	^{169}Tm	^{40}Ca	4n	^{205}Ac
04/09/2015	Ma	[247]	^{180}W	^{40}Ar	4n	^{216}U
07/14/2015	Khuyagbaatar	[248]	^{176}Yb	^{50}Ti	5n	^{221}U

[a]Populated by α-decay

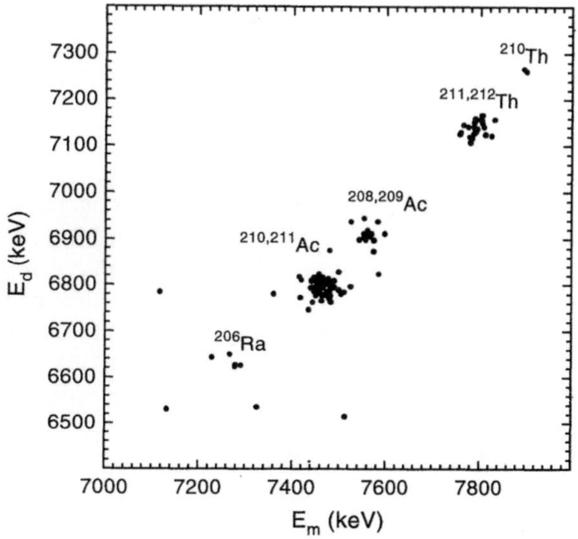

Fig. 11.6 Energies of mother and daughter α-emitters correlated with evaporation residues detected in the gas-filled separator RITU [223] (Reprinted figure with permission from J. Uusitalo et al., *α-decay of the new isotopes ^{210}Th and ^{211}Th*, Phys. Rev. C 52 (1995) 113. Copyright 1995 by the American Physical Society.)

(University of Tennessee) [227], ^{186}Bi (Louisiana State University) [229], ^{187}Bi (Oak Ridge Associated Universities) [234], and ^{172}Hg and ^{173}Hg (University of Maryland) [236].

In Japan, the gas-filled separator GARIS at RIKEN and the recoil mass separator (RMS) at JAERI contributed to the discovery of α-emitting nuclides. At RIKEN, Morita et al. discovered ^{196}Rn, ^{197}Rn, and ^{200}Fr in 1995 [224] and Tagaya et al. identified ^{195}At and ^{199}Fr in 1999 [235]. In 1996, Ikezoe et al. and Mitsuoka et al. discovered ^{209}Th [228] and ^{212}Pa [230], respectively.

In 2006, Liddick et al. employed a digital processing system for the α-particle signals from the decay of ^{109}Xe implanted at the focal plane of the recoil mass separator at Oak Ridge. They were able to separate the first α-decay of ^{109}Xe with a half-life of 13 ± 2 ms from the subsequent α-decay of the daughter nuclide ^{105}Te with a half-life of only 620 ns [243].

Two new α-emitters were discovered at the Heavy Ion Research Facility in Lanzhou (HIRFL), China. Zhang et al. and Ma et al. identified ^{205}Ac in 2013 [246] and ^{216}U in 2015 [247], respectively, with the gas-filled recoil separator SHANS (Spectrometer for Heavy Atoms and Nuclear Structure).

Most recently, Khuyagbaatar et al. used the TransActinide Separator and Chemistry Apparatus (TASCA) at GSI in the discovery of ^{221}U [248].

11.5 Proton Emissions and Delayed Fission

Beta-delayed proton emitters were first observed by Barton et al. in 1963 at McGill University discovering ^{25}Si [251]. Beta-delayed proton emission occurs in neutron-deficient nuclides close to the proton drip-line. In light elements (up to titanium) they can be populated with light-charged-particle beams, however, in order to reach β-delayed proton emitters in heavier elements fusion-evaporation reactions involving heavy ions are necessary. Isotopes discovered by β-delayed proton emission populated in fusion-evaporation reactions are listed in Table 11.11. The initial nuclide is identified by deducing its β^+ half-life from the decay curve measured by the promptly emitted protons from excited unbound states in the daughter nuclide. In most cases other properties of the daughter nuclides were either already known or no information about them could be extracted from the measured proton spectra.

The first β-delayed proton emitters produced in a fusion-evaporation reaction were ^{109}Te and ^{111}Te, discovered by Karnaukhov et al. in 1966 at the U300 Dubna cyclotron [252]. Subsequently, Bogdanov and collaborators identified the β-delayed proton emitters ^{119}Ba [255], ^{116}Cs, ^{121}Ba [256], ^{117}Ba, ^{129}Nd, and ^{133}Sm [75] at Dubna. For the latter two experiments they used the BEMS-2 mass separator to identify the recoils.

In 1970, Cerny et al. discovered ^{49}Fe at the Harwell variable-energy cyclotron [253]. The energy resolution of the semiconductor telescope used for the proton detection was sufficient to measure a discrete resonance in the daughter nuclide ^{49}Mn which had previously not been observed. The decay scheme is shown in Fig. 11.7.

In the same year, Cerny was also involved in two experiments reporting the discovery of proton radioactivity. He was a collaborator in an experiment by Jackson et al. from Oxford University which discovered the decay of an excited isomeric state in ^{53}Co by proton emission. The experiment was performed with the Harwell variable-energy cyclotron to populate the state and the emitted protons were detected with a semiconductor telescope [254]. In addition to this fusion-evaporation reaction, Cerny essentially immediately confirmed the results with the reaction ^{54}Fe(p,2n) at the 88-in. Berkeley cyclotron. The results were submitted on the same day as the paper by Jackson et al. and published adjacent to each other in the same issue of Physics Letters B [281].

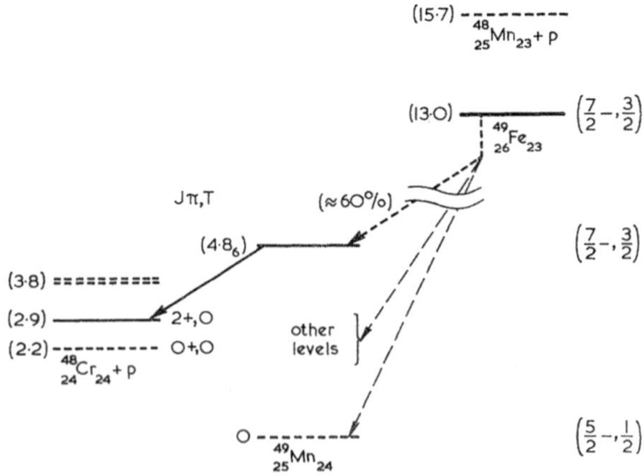

Fig. 11.7 Decay diagram of ^{49}Fe. In addition to the discovery of ^{49}Fe the observation of a discrete level in ^{49}Mn also corresponds to the first observation of this nuclide [253] (Reprinted figure with permission from J. Cerny et al., ^{49}*Fe: A New* $T_Z = 3/2$ *delayed-proton emitter*, Phys. Rev. Lett. 24 (1970) 1128. Copyright 1970 by the American Physical Society.)

Table 11.11 Isotopes discovered in fusion-evaporation reactions and identified by β-delayed proton emission and delayed fission: The submission date, first author and reference of the publication, as well as the target, beam, and the evaporation channel are listed

Date	First author	Ref.	Target	Beam	Evap.	Isotope
01/03/1966	Karnaukhov	[252]	^{92}Mo	^{20}Ne	3n, 1n	^{109}Te, ^{111}Te
04/07/1970	Cerny	[253]	^{40}Ca	^{12}C	3n	^{49}Fe, ^{49}Mn[a]
09/23/1970	Jackson	[254]	^{40}Ca	^{16}O	1p2n	^{53}Co[b]
07/09/1973	Bogdanov	[255]	^{106}Cd	^{16}O	3n	^{119}Ba
02/25/1974	Bogdanov	[256]	^{90}Zr, ^{93}Nb	^{32}S	1p5n, 1p3n	^{116}Cs, ^{121}Ba
03/21/1974	Jackson	[257]	^{32}S	^{16}O	3n	^{45}Cr, ^{45}V[a]
12/02/1975	Vieira	[258]	^{40}Ca	^{16}O, ^{20}Ne	3n	^{53}Ni, ^{57}Zn, ^{57}Cu[a]
05/18/1976	Hardy	[259]	^{40}Ca	^{40}Ca	2p1n	^{77}Sr
07/23/1976	Bogdanov	[75]	^{92}Mo	^{32}S	2p5n	^{117}Ba
			^{102}Pd, ^{106}Cd	^{32}S	2p3n	^{129}Nd, ^{133}Sm
08/11/1980	Nolte	[260]	^{58}Ni	^{40}Ca	2p1n	^{95}Pd
06/25/1981	Tidemand-P.	[261]	^{50}Cr	^{58}Ni	2p3n, 2p1n	^{103}Sn, ^{105}Sn
06/07/1982	Kurcewicz	[262]	^{58,60}Ni, ^{63}Cu	^{40}Ca	2p2n, p3n, 3n	^{94}Pd, ^{96}Ag, ^{100}In
08/23/1982	Hagberg	[263]	^{54}Fe	^{40}Ca	2p1n	^{91}Ru
05/17/1983	Nitschke	[264]	^{90}Zr, ^{92}Mo	^{40}Ca	p3n, αn	^{126}Pr, ^{127}Nd
			^{50}Cr, ^{92}Mo	^{92}Mo, ^{56}Fe	2p1n, αn	^{139}Gd, ^{143}Dy
02/16/1984	Toth	[265]	^{144}Sm	^{12}C	7n	^{149}Er
02/17/1984	Nitschke	[266]	^{58}Ni, ^{92}Mo	^{64}Zn, ^{36}Ar	pn, αpn	^{120}La, ^{122}La
			^{92}Mo	^{36}Ar, ^{54}Fe	αn	^{123}Ce, ^{141}Dy
03/08/1985	Wilmarth	[267]	^{92}Mo	^{40}Ca	3p1n, pn	^{128}Pr, ^{130}Pm
08/25/1986	Hotchkis	[268]	^{40}Ca	^{24}Mg	3n	^{61}Ge, ^{61}Ga[a]
09/23/1986	Wilmarth	[86]	^{92}Mo, ^{96}Ru	^{36}Ar, ^{40}Ca	p3n, 2p3n	^{124}Pr, ^{131}Sm
			^{92}Mo	^{54}Fe	3p3n, 2p2n	^{140}Tb, ^{142}Dy
			^{92}Mo	^{58}Ni	3p2n	^{144}Ho
03/04/1987	Lazarev	[269]	^{144}Sm	^{40}Ca	p3n	^{180}Tl[c]
07/11/1988	Vierinen	[270]	^{92}Mo	^{46}Ti	p3n, p2n	^{134}Eu, ^{135}Eu
01/08/1993	Batchelder	[271]	^{40}Ca	^{28}Si	3n	^{65}Se
07/19/1993	Batchelder	[272]	^{40}Ca	^{36}Ar	3n	^{73}Sr
07/29/1996	Xu	[273]	^{106}Cd	^{32}S	3n	^{135}Gd
02/03/1997	Li	[274]	^{92}Mo	^{32}S	3n	^{121}Ce
09/17/1997	Huang	[275]	^{58}Ni	^{32}S	4p5n	^{81}Zr
04/05/1999	Xu	[276]	^{92}Mo, ^{96}Ru	^{36}Ar	3n	^{125}Nd, ^{129}Sm
			^{96}Ru	^{36}Ar	p3n	^{128}Pm
			^{106}Cd	^{36}Ar	2p3n, 3n	^{137}Gd, ^{139}Dy
05/12/1999	Sonzogni	[277]	^{58}Ni	^{78}Kr	1p4n	^{130}Sm[d]
12/12/2000	Xu	[278]	^{106}Cd	^{40}Ca	p3n	^{142}Ho
03/14/2001	Xu	[279]	^{112}Sn	^{40}Ca	3n	^{149}Yb
07/29/2002	Karny	[280]	^{92}Mo	^{58}Ni	p4n	^{144}Er[d]

[a] β^+-delayed proton emitter
[b] Proton emitter from an isomeric excited state
[c] β^+-delayed fission
[d] Populated by a direct proton emitter

Four years later Jackson et al. observed the next β-delayed proton emitter. In addition to the parent nuclide ^{45}Cr they also identified the proton emitting daughter ^{45}V for the first time with the Chalk River MP tandem accelerator [257]. The intermediate proton-emitting nuclide was also not known in two other cases. At the 88-in. cyclotron at Berkeley, Vieira et al. observed ^{57}Cu in the decay of ^{57}Zn in 1975 [258], and in 1986, Hotchkis et al. discovered ^{61}Ga together with ^{61}Ge [268].

Two more β-delayed proton emitters were discovered with the MP tandem at Chalk River, ^{77}Sr by Hardy et al. in 1976 [259] and ^{91}Ru by Hagberg et al. in 1982 [263]. In the early 1980s six proton emitters were identified in Germany; Nolte and Hick captured the recoils on a rotating wheel behind the target at the Munich MP tandem to discover ^{95}Pd [260] while ^{103}Sn,^{105}Sn [261], ^{94}Pd, ^{96}Ag, and ^{100}In [262] were discovered with the online mass separator at the UNILAC at GSI by Tidemand-Petersson et al. and Kurcewicz et al., respectively.

All identifications of new β-delayed proton emitters in the decade between 1983 and 1993 were achieved at Berkeley. The group of Nitschke and Wilmarth alone discovered fifteen new isotopes using the online isotope separator OASIS at the HILAC [86, 264, 266, 267]. In addition to the discovery of ^{61}Ge mentioned above [268], Toth et al. and Batchelder et al. identified ^{149}Er [265] and ^{65}Se and ^{73}Sr [271, 272], respectively, with the helium-jet technique at the 88-in. cyclotron. ^{73}Sr decayed to an excited state of the proton-unbound nuclide ^{73}Rb which also was identified for the first time in the experiment (see Sect. 16.3) [272].

The helium-jet technique was also employed for the discovery of the ten most recent β-delayed proton emitters, all observed at the sector-focused cyclotron (SFC) of the Institute of Modern Physics in Lanzhou, China [273–276, 278, 279].

Two isotopes were not discovered by subsequent delayed proton emission but by preceding prompt proton emission populating states of the new nuclides. Sonzogni et al. deduced the energy of the first excited 2^+ state in ^{130}Sm from the observation of two peaks in the energy spectrum of the proton emitting parent nuclide ^{131}Eu which was implanted in the focal plane of the FMA at Argonne [277]. Similarly ^{144}Er was identified by observing two proton lines in the decay of ^{145}Tb. Karny et al. used digital data processing to separate the pulses of the short-lived proton emitter ($\sim$3μs) from the implantation signals at the recoil mass separator at Oak Ridge in 2002 [280].

In addition to the isotopes discovered by β-delayed proton emission, one isotope was identified by observing its β-delayed fission decay. Lazarev et al. discovered ^{180}Tl with the U400 Dubna cyclotron in 1987. They irradiated enriched samarium targets deposited on a rotating copper cylinder with ^{40}Ca ions and detected the fission fragments in mica detectors surrounding the cylindrical target. Lazarev et al. argue that the fragments originate from excited states in ^{180}Hg following the β-decay of ^{180}Tl [269].

References

1. S.G. Thompson, A. Ghiorso, G.T. Seaborg, Phys. Rev. **77**, 838 (1950)
2. J.F. Miller et al., Phys. Rev. **80**, 486 (1950)
3. A. Ghiorso et al., Phys. Rev. **81**, 154 (1951)
4. J.M. Hollander, Phys. Rev. **92**, 518 (1953)
5. G.B. Rossi et al., Phys. Rev. **93**, 256 (1954)
6. M.I. Kalkstein, J.M. Hollander, Phys. Rev. **96**, 730 (1954)
7. J. Beydon et al., Nucl. Phys. **2**, 593 (1957)
8. J. Beydon et al., Compt. Rend. Acad. Sci. **244**, 586 (1957)
9. K.S. Toth, J.O. Rasmussen, J. Inorg. Nucl. Chem. **12**, 236 (1960)
10. K.T. Faler, J.O. Rasmussen, Phys. Rev. **118**, 265 (1960)
11. K.F. Chackett, G.A. Chackett, J. Inorg. Nucl. Chem. **13**, 1 (1960)
12. R.W. Hoff, J.M. Hollander, M.C. Michel, J. Inorg. Nucl. Chem. **18**, 1 (1961)
13. I.L. Preiss, P.J. Estrup, R. Wolfgang, Nucl. Phys. **18**, 624 (1960)
14. R.M. Diamond et al., Nucl. Phys. **25**, 248 (1961)
15. R.K. Sheline, T. Sikkeland, R.N. Chanda, Phys. Rev. Lett. **7**, 446 (1961)
16. I.L. Preiss, P.M. Strudler, J. Inorg. Nucl. Chem. **24**, 589 (1962)
17. V. Maxia, W.H. Kelly, D.J. Horen, J. Inorg. Nucl. Chem. **24**, 1175 (1962)
18. I.L. Preiss, P.M. Strudler, R. Wolfgang, Phys. Rev. **129**, 1284 (1963)
19. M.J. Nurmia, R.W. Fink, Phys. Lett. **14**, 136 (1965)
20. B.N. Belyaev et al., Sov. J. Nucl. Phys. **4**, 163 (1967)
21. E. Nadjakov et al., Compt. Rend. Acad. Bulgare Sci. **20**, 533 (1967)
22. H. Bakhru, R.I. Morse, I.L. Preiss, Nucl. Phys. A **100**, 145 (1967)
23. B.N. Belyaev et al., Bull. Acad. Sci. USSR Phys. Ser. **32**, 59 (1968)
24. G. Murray et al., Nucl. Phys. A **142**, 21 (1970)
25. E. Nadjakov et al., Bull. Acad. Sci. USSR Phys. Ser. **35**, 2004 (1971)
26. H. Schmeing et al., Phys. Lett. B **44**, 449 (1973)
27. M. Kaba, K. Miyanon, Radiochim. Acta **21**, 203 (1974)
28. H. Bruchertseifer, B. Eichler, Radiochem. Radioanal. Lett. **48**, 391 (1981)
29. H. Bruchertseifer et al., Radiochem. Radioanal. Lett. **51**, 319 (1982)
30. B. Eichler et al., Radiochem. Radioanal. Lett. **53**, 161 (1982)
31. H. Atterling et al., Nucl. Phys. **2**, 619 (1957)
32. M. Crut et al., Nucl. Phys. **2**, 624 (1957)
33. R. Chaminade et al., Nucl. Phys. **2**, 634 (1957)
34. Ch. Droste et al., Nucl. Phys. A **152**, 579 (1970)
35. E. Nolte et al., Phys. Lett. B **33**, 294 (1970)
36. T.A. Doron, M. Blann, Nucl. Phys. A **161**, 12 (1971)
37. J. Zioni et al., Nucl. Phys. A **181**, 465 (1972)
38. A.I. Akhmadzhanov et al., Bull. Acad. Sci. USSR Phys. Ser. **36**, 1820 (1972)
39. I.M. Ladenbauer-Bellis, H. Bakhru, B. Jones, Can. J. Phys. **50**, 3071 (1972)
40. E. Newman et al., Phys. Rev. C **9**, 674 (1974)
41. E. Nolte et al., Z. Phys. **268**, 267 (1974)
42. J.O. Newton, F.S. Stephens, R.M. Diamond, Nucl. Phys. A **236**, 225 (1974)
43. K.S. Toth et al., Phys. Lett. B **56**, 29 (1975)
44. J.H. Hamilton et al., Phys. Rev. Lett. **35**, 562 (1975)
45. S. Chojnacki et al., Acta Phys. Pol. B **7**, 823 (1976)
46. B.J. Varley, J.C. Cunnane, W. Gelletly, J. Phys. G **3**, 55 (1977)
47. R.E. Leber, P.E. Haustein, I.M. Ladenbauer-Bellis, J. Inorg. Nucl. Chem. **39**, 927 (1977)
48. L.A. Parks, C.N. Davids, R.C. Pardo, Phys. Rev. C **15**, 730 (1977)
49. G. Korschinek et al., Z. Phys. A **281**, 409 (1977)
50. R.C. Hunter et al., Phys. Rev. C **16**, 384 (1977)
51. A.M. Nathan et al., Phys. Rev. C **16**, 1566 (1977)
52. C. Burman, P. Sen, H. Bakhru, Can. J. Phys. **56**, 786 (1978)

53. C.N. Davids et al., Phys. Rev. C **17**, 1815 (1978)
54. E.B. Norman et al., Phys. Rev. C **17**, 2176 (1978)
55. K.S. Toth et al., Phys. Rev. C **19**, 482 (1979)
56. D.E. Alburger, Phys. Rev. C **18**, 1875 (1978)
57. C. Deprun et al., Z. Phys. A **295**, 103 (1980)
58. M.J. Murphy, C.N. Davids, E.B. Norman, Phys. Rev. C **22**, 2204 (1980)
59. C.J. Lister et al., Phys. Rev. C **24**, 260 (1981)
60. D.C. Sousa et al., Phys. Rev. C **25**, 1012 (1982)
61. U.J. Schrewe et al., Phys. Rev. C **25**, 3091 (1982)
62. S.Z. Gui, G. Colombo, E. Nolte, Z. Phys, A **305**, 297 (1982)
63. E. Nolte et al., Z. Phys. A **306**, 223 (1982)
64. T. Ollivier et al., Z. Phys. A **320**, 695 (1985)
65. A. Szymanski et al., Radiochim. Acta **40**, 61 (1986)
66. E. Runte et al., Z. Phys. A **328**, 373 (1987)
67. K. Heiguchi et al., Z. Phys. A **338**, 7 (1991)
68. U. Bosch-Wicke et al., Z. Phys. A **341**, 245 (1992)
69. S. Zhou et al., Chin. J. Nucl. Phys. **13**, 193 (1991)
70. Y. Xie et al., Eur. Phys. J. A **6**, 239 (1999)
71. S.-W. Xu et al., Eur. Phys. J. A **21**, 75 (2004)
72. R.D. Macfarlane, R.D. Griffioen, Nucl. Instrum. Meth. **24**, 461 (1963)
73. D.R. Goosman, D.E. Alburger, Phys. Rev. C **5**, 1252 (1972)
74. G.M. Gowdy et al., Phys. Rev. C **13**, 1601 (1976)
75. D.D. Bogdanov et al., Nucl. Phys. A **275**, 229 (1977)
76. D.D. Bogdanov et al., Nucl. Phys. A **307**, 421 (1978)
77. G. Lhersonneau et al., Phys. Rev. C **18**, 2688 (1978)
78. M. Huyse et al., Z. Phys. A **288**, 107 (1978)
79. M. Nowicki et al., Acta Phys. Pol. B **13**, 879 (1982)
80. B. Beraud et al., Z. Phys. A **299**, 279 (1981)
81. E. Hagberg et al., Nucl. Phys. A **383**, 109 (1982)
82. K. Deneffe et al., J. Phys. G **11**, L59 (1985)
83. P. Kleinheinz et al., Z. Phys. A **322**, 705 (1985)
84. N. Redon et al., Z. Phys. A **325**, 127 (1986)
85. K.S. Toth et al., Phys. Rev. C **35**, 310 (1987)
86. P.A. Wilmarth et al., Z. Phys. A **325**, 485 (1986)
87. T. Sekine et al., Nucl. Phys. A **467**, 93 (1987)
88. K.S. Toth et al., Phys. Rev. C **36**, 826 (1987)
89. B.D. Kern et al., Phys. Rev. C **36**, 1514 (1987)
90. M. Huyse et al., Z. Phys. A **330**, 121 (1988)
91. T. Sekine et al., Z. Phys. A **331**, 105 (1988)
92. K.S. Vierinen et al., Phys. Rev. C **39**, 1972 (1989)
93. J.M. Nitschke et al., Z. Phys. A **334**, 111 (1989)
94. R.B. Firestone et al., Phys. Rev. C **43**, 1066 (1991)
95. K.S. Toth et al., Phys. Rev. C **48**, 445 (1993)
96. A. Gizon et al., Z. Phys. A **351**, 361 (1995)
97. A. Guglielmetti et al., Phys. Rev. C **52**, 740 (1995)
98. Z. Janas et al., Nucl. Phys. A **627**, 119 (1997)
99. F.S. Stephens, N.L. Lark, R.M. Diamond, Phys. Rev. Lett. **12**, 225 (1964)
100. H. Morinaga, P.C. Gugelot, Nucl. Phys. **46**, 210 (1963)
101. F.S. Stephens, N.L. Lark, R.M. Diamond, Nucl. Phys. **63**, 82 (1965)
102. J. Burde, R.M. Diamond, F.S. Stephens, Nucl. Phys. A **92**, 306 (1967)
103. J.E. Clarkson et al., Nucl. Phys. A **93**, 272 (1967)
104. D. Ward, F.S. Stephens, J.O. Newton, Phys. Rev. Lett. **19**, 1247 (1967)
105. D. Ward, R.M. Diamond, F.S. Stephens, Nucl. Phys. A **117**, 309 (1968)
106. F.S. Stephens, J.R. Leigh, R.M. Diamond, Nucl. Phys. A **170**, 321 (1971)

107. R. Anholt, J.O. Rasmussen, I. Rezanka, Nucl. Phys. A **209**, 72 (1973)
108. K. Nakai et al., Phys. Lett. B **44**, 443 (1973)
109. J. Conrad et al., Nucl. Phys. A **234**, 157 (1974)
110. W. Trautmann et al., Phys. Rev. Lett. **35**, 1694 (1975)
111. J. Gizon et al., Nucl. Phys. A **290**, 272 (1977)
112. H. Helppi et al., Phys. Lett. B **115**, 11 (1982)
113. C.J. Lister et al., Phys. Rev. Lett. **49**, 308 (1982)
114. E. Nolte, G. Korschinek, Ch. Setzensack, Z. Phys, A **309**, 33 (1982)
115. H.F.R. Arciszewski et al., Nucl. Phys. A **401**, 531 (1983)
116. J. Recht et al., Nucl. Phys. A **440**, 366 (1985)
117. J. Gerl et al., Nucl. Phys. A **443**, 348 (1985)
118. C.J. Lister et al., Phys. Rev. Lett. **55**, 810 (1985)
119. C.J. Lister et al., Phys. Rev. Lett. **59**, 1270 (1987)
120. L. Goettig et al., Nucl. Phys. A **475**, 569 (1987)
121. T. Kuroyanagi et al., Nucl. Phys. A **484**, 264 (1988)
122. R. Wadsworth et al., Z. Phys. A **333**, 411 (1989)
123. C.J. Lister et al., Phys. Rev. C **42**, R1191 (1990)
124. W. Gelletly et al., Phys. Lett. B **253**, 287 (1991)
125. D. Rudolph et al., J. Phys. G **17**, L113 (1991)
126. C.J. Gross et al., Phys. Rev. C **44**, 2253 (1991)
127. G. de Angelis et al., Z. Phys. A **343**, 121 (1992)
128. D. Seweryniak et al., Phys. Lett. B **321**, 323 (1994)
129. C.M. Parry et al., Phys. Rev. C **57**, 2215 (1998)
130. W. Krolas et al., Phys. Rev. C **65**, 031303 (2002)
131. A.N. Wilson et al., Phys. Rev. C **66**, 021305 (2002)
132. J.F. Smith et al., Phys. Lett. B **625**, 203 (2005)
133. W.E. Burcham, Proc. Phys. Soc. A **67**, 555 (1954)
134. K.S. Toth, J.O. Rasmussen, Phys. Rev. **109**, 121 (1958)
135. R.M. Latimer, G.E. Gordon, T.D. Thomas, J. Inorg. Nucl. Chem. **17**, 1 (1961)
136. R.W. Hoff, F. Asaro, I. Perlman, J. Inorg. Nucl. Chem. **25**, 1303 (1963)
137. H. Rotter et al., Sov. J. Nucl. Phys. **4**, 178 (1967)
138. G. Ya, Sung-Ching-Yang, V.A. Druin, A.S. Trofimov, Sov. J. Nucl. Phys. **14**, 725 (1972)
139. J. Cerny, D.R. Goosman, D.E. Alburger, Phys. Lett. B **37**, 380 (1971)
140. T. Nomura et al., Phys. Lett. B **40**, 543 (1972)
141. K. Hiruta et al., Phys. Lett. B **45**, 244 (1973)
142. O. Häusser et al., Phys. Rev. Lett. **31**, 323 (1973)
143. J.C. Batchelder et al., Phys. Rev. C **54**, 949 (1996)
144. K.S. Toth et al., Z. Phys. A **355**, 225 (1996)
145. R.D. Macfarlane, R.D. Griffioen, Phys. Rev. **130**, 1491 (1963)
146. R.D. Macfarlane, R.D. Griffioen, Phys. Rev. **131**, 2176 (1963)
147. R.D. Griffioen, R.D. MacFarlane, Phys. Rev. **133**, B1373 (1964)
148. R.D. Macfarlane, Phys. Rev. **137**, B1448 (1965)
149. R.D. Macfarlane, Phys. Rev. **136**, B941 (1964)
150. A. Siivola, Nucl. Phys. **84**, 385 (1966)
151. A. Siivola, Nucl. Phys. A **92**, 475 (1967)
152. W. Treytl, K. Valli, Nucl. Phys. A **97**, 405 (1967)
153. K. Valli, M.J. Nurmia, E.K. Hyde, Phys. Rev. **159**, 1013 (1967)
154. K. Valli, E.K. Hyde, W. Treytl, J. Inorg. Nucl. Chem. **29**, 2503 (1967)
155. A. Siivola, Nucl. Phys. A **101**, 129 (1967)
156. K. Valli, W. Treytl, E.K. Hyde, Phys. Rev. **161**, 1284 (1967)
157. A.G. Demin et al., Nucl. Phys. A **106**, 337 (1968)
158. K. Valli, W. Treytl, E.K. Hyde, Phys. Rev. **167**, 1094 (1968)
159. A. Siivola, Nucl. Phys. A **109**, 231 (1968)
160. K. Valli, E.K. Hyde, Phys. Rev. **176**, 1377 (1968)

161. D.F. Torgerson, R.D. MacFarlane, Nucl. Phys. A **149**, 641 (1970)
162. J. Borggreen, K. Valli, E.K. Hyde, Phys. Rev. C **2**, 1841 (1970)
163. J. Borggreen, E.K. Hyde, Nucl. Phys. A **162**, 407 (1971)
164. R.D. Macfarlane, Phys. Rev. **126**, 274 (1962)
165. K.S. Toth, R.L. Hahn, M.A. Ijaz, Phys. Rev. C **4**, 2223 (1971)
166. K.S. Toth et al., Phys. Rev. C **7**, 2010 (1973)
167. K.S. Toth et al., Phys. Rev. C **5**, 2060 (1972)
168. K.S. Toth et al., Phys. Rev. C **6**, 2297 (1972)
169. K.S. Toth et al., Phys. Rev. C **12**, 533 (1975)
170. K.S. Toth, D.M. Moltz, J.D. Robertson, Phys. Rev. C **39**, 1150 (1989)
171. H. Gauvin et al., Phys. Rev. Lett. **29**, 958 (1972)
172. H. Gauvin et al., Nucl. Phys. A **208**, 360 (1973)
173. C. Cabot et al., Nucl. Phys. A **241**, 341 (1975)
174. C. Cabot et al., Compt. Rend. Acad. Sci. **281**, 453 (1975)
175. C. Cabot et al., Z. Phys. A **283**, 221 (1977)
176. S.D. Negra, B. Lagarde, Y. Le Beyec, J. Phys. (Paris) Lett. **38**, L393 (1977)
177. C. Cabot et al., Z. Phys. A **287**, 71 (1978)
178. S.D. Negra et al., Z. Phys. A **300**, 251 (1981)
179. Y. Le Beyec et al., Phys. Rev. C **9**, 1091 (1974)
180. J.P. Dufour et al., Z. Phys. A **294**, 107 (1980)
181. D.D. Bogdanov, V.A. Karnaukhov, L.A. Petrov, Sov. J. Nucl. Phys. **18**, 1 (1974)
182. D.A. Eastham, I.S. Grant, Nucl. Phys. A **208**, 119 (1973)
183. U.J. Schrewe et al., Z. Phys. A **288**, 189 (1978)
184. F. Meissner et al., Z. Phys. A **343**, 283 (1992)
185. N.I. Tarantin, A.P. Kabachenko, A.V. Demyanov, Sov. J. Nucl. Phys. **12**, 248 (1971)
186. K.S. Toth et al., Phys. Lett. B **63**, 150 (1976)
187. R. Kirchner et al., Phys. Lett. B **70**, 150 (1977)
188. E. Roeckl et al., Phys. Lett. B **78**, 393 (1978)
189. D. Schardt et al., Nucl. Phys. A **326**, 65 (1979)
190. U.J. Schrewe et al., Phys. Lett. B **91**, 46 (1980)
191. D. Schardt et al., Nucl. Phys. A **368**, 153 (1981)
192. A. Plochocki et al., Phys. Lett. B **106**, 285 (1981)
193. H. Ewald et al., Nucl. Instrum. Meth. **139**, 223 (1976)
194. K.-H. Schmidt et al., Nucl. Phys. A **318**, 253 (1979)
195. S. Hofmann et al., Z. Phys. A **291**, 53 (1979)
196. D. Vermeulen et al., Z. Phys. A **294**, 149 (1980)
197. S. Hofmann et al., Z. Phys. A **299**, 281 (1981)
198. J.R.H. Schneider et al., Z. Phys. A **312**, 21 (1983)
199. R. Hingmann et al., Z. Phys. A **313**, 141 (1983)
200. J.G. Keller et al., Nucl. Phys. A **452**, 173 (1986)
201. F.P. Heßberger et al., Europhys. Lett. **3**, 895 (1987)
202. S. Hofmann et al., Z. Phys. A **333**, 107 (1989)
203. A.B. Quint et al., Z. Phys. A **346**, 119 (1993)
204. V. Ninov et al., Z. Phys. A **351**, 125 (1995)
205. A.N. Andreyev et al., Eur. Phys. J. A **6**, 381 (1999)
206. A.N. Andreyev et al., Eur. Phys. J. A **18**, 55 (2003)
207. A.N. Andreyev et al., Phys. Rev. C **72**, 014612 (2005)
208. A.N. Andreyev et al., Phys. Rev. C **73**, 024317 (2006)
209. A.N. Andreyev et al., Phys. Rev. C **74**, 064303 (2006)
210. A.N. Andreyev et al., Phys. Rev. C **79**, 064320 (2009)
211. A.N. Andreyev et al., J. Phys. G **37**, 035102 (2010)
212. J.A. Heredia et al., Eur. Phys. J. A **46**, 337 (2010)
213. Z. Kalaninova et al., Phys. Rev. C **87**, 044355 (2013)
214. A.C. DiRienzo et al., Phys. Rev. C **21**, 2101 (1980)

215. A.N. Andreyev et al., Sov. J. Nucl. Phys. **50**, 381 (1989)
216. R.D. Page et al., Z. Phys. A **338**, 295 (1991)
217. A.N. Andreyev et al., Sov. J. Nucl. Phys. **53**, 554 (1991)
218. A.N. Andreyev et al., Z. Phys. A **342**, 123 (1992)
219. A.N. Andreyev et al., Z. Phys. A **345**, 247 (1993)
220. P.J. Sellin et al., Z. Phys. A **346**, 323 (1993)
221. M. Leino et al., Z. Phys. A **348**, 151 (1994)
222. M.J. Leddy et al., Phys. Rev. C **51**, R1047 (1995)
223. J. Uusitalo et al., Phys. Rev. C **52**, 113 (1995)
224. K. Morita et al., Z. Phys. A **352**, 7 (1995)
225. K.S. Toth et al., Phys. Rev. C **53**, 2513 (1996)
226. M. Leino et al., Z. Phys. A **355**, 157 (1996)
227. C.R. Bingham et al., Phys. Rev. C **54**, R20 (1996)
228. H. Ikezoe et al., Phys. Rev. C **54**, 2043 (1996)
229. J.C. Batchelder et al., Z. Phys. A **357**, 121 (1997)
230. S. Mitsuoka et al., Phys. Rev. C **55**, 1555 (1997)
231. J. Uusitalo et al., Z. Phys. A **358**, 375 (1997)
232. M.P. Carpenter et al., Phys. Rev. Lett. **78**, 3650 (1997)
233. K. Eskola et al., Phys. Rev. C **57**, 417 (1998)
234. J.C. Batchelder et al., Eur. Phys. J. A **5**, 49 (1999)
235. Y. Tagaya et al., Eur. Phys. J. A **5**, 123 (1999)
236. D. Seweryniak et al., Phys. Rev. C **60**, 031304 (1999)
237. O.N. Malyshev et al., Eur. Phys. J. A **8**, 295 (2000)
238. H. Kettunen et al., Phys. Rev. C **63**, 044315 (2001)
239. H. Kettunen et al., Eur. Phys. J. A **17**, 537 (2003)
240. H. Kettunen et al., Phys. Rev. C **69**, 054323 (2004)
241. J. Uusitalo et al., Phys. Rev. C **71**, 024306 (2005)
242. D. Seweryniak et al., Phys. Rev. C **73**, 061301 (2006)
243. S.N. Liddick et al., Phys. Rev. Lett. **97**, 802501 (2006)
244. L. Bianco et al., Phys. Lett. B **690**, 15 (2010)
245. J. Uusitalo et al., Phys. Rev. C **87**, 064304 (2013)
246. Z.Y. Zhang et al., Phys. Rev. C **89**, 014308 (2014)
247. L. Ma et al., Phys. Rev. C **91**, 051302 (2015)
248. J. Khuyagbaatar et al., Phys. Rev. Lett. **115**, 242502 (2015)
249. R.D. Page et al., Nucl. Instrum. Meth. B **204**, 634 (2003)
250. C.N. Davids, J.D. Larson, Nucl. Instrum. Meth. B **40/41**, 1224 (1989)
251. R. Barton et al., Can. J. Phys. **41**, 2007 (1963)
252. V.A. Karnaukhov et al., Sov. J. Nucl. Phys. **4**, 327 (1967)
253. J. Cerny et al., Phys. Rev. Lett. **24**, 1128 (1970)
254. K.P. Jackson et al., Phys. Lett. B **33**, 281 (1970)
255. D.D. Bogdanov, V.A. Karnaukhov, L.A. Petrov, Sov. J. Nucl. Phys. **19**, 481 (1974)
256. D.D. Bogdanov et al., Sov. J. Nucl. Phys. **21**, 123 (1975)
257. K.P. Jackson et al., Phys. Lett. B **49**, 341 (1974)
258. D.J. Vieira et al., Phys. Lett. B **60**, 261 (1976)
259. J.C. Hardy et al., Phys. Lett. B **63**, 27 (1976)
260. E. Nolte, H. Hick, Phys. Lett. B **97**, 55 (1980)
261. P. Tidemand-Petersson et al., Z. Phys. A **302**, 343 (1981)
262. W. Kurcewicz et al., Z. Phys. A **308**, 21 (1982)
263. E. Hagberg et al., Nucl. Phys. A **395**, 152 (1983)
264. J.M. Nitschke, M.D. Cable, W.D. Zeitz, Z. Phys, A **312**, 265 (1983)
265. K.S. Toth et al., Phys. Rev. C **30**, 712 (1984)
266. J.M. Nitschke et al., Z. Phys. A **316**, 249 (1984)
267. P.A. Wilmarth et al., Z. Phys. A **321**, 179 (1985)
268. M.A.C. Hotchkis et al., Phys. Rev. C **35**, 315 (1987)

269. YuA Lazarev et al., Europhys. Lett. **4**, 893 (1987)
270. K.S. Vierinen et al., Nucl. Phys. A **499**, 1 (1989)
271. J.C. Batchelder et al., Phys. Rev. C **47**, 2038 (1993)
272. J.C. Batchelder et al., Phys. Rev. C **48**, 2593 (1993)
273. S. Xu et al., Z. Phys. A **356**, 227 (1996)
274. Z. Li et al., Phys. Rev. C **56**, 1157 (1997)
275. W.X. Huang et al., Z. Phys. A **359**, 349 (1997)
276. S.-W. Xu et al., Phys. Rev. C **60**, 061302 (1999)
277. A.A. Sonzogni et al., Phys. Rev. Lett. **83**, 1116 (1999)
278. S.-W. Xu et al., Phys. Rev. C **64**, 017301 (2001)
279. S.-W. Xu et al., Eur. Phys. J. A **12**, 1 (2001)
280. M. Karny et al., Phys. Rev. Lett. **90**, 012502 (2003)
281. J. Cerny et al., Phys. Lett. B **33**, 284 (1970)

Chapter 12
Superheavy Elements

12.1 Overview and Current Status

There is no exact definition for superheavy elements. For the present purpose superheavy elements begin with nobelium (Z = 102). It was the first element which could only be reached with heavy-ion induced fusion-evaporation reactions. All lighter transuranium elements were first observed in proton or α-particle induced reactions or by neutron capture. The first mendelevium (Z = 101) isotope was discovered in the reaction ^{253}Es(α,n)^{256}Md in 1955 by Ghiorso et al. [1] (see Sect. 5.4).

The discovery of isotopes of the superheavy elements can be divided in three phases. Early on (1963 until about 1980) the experimental techniques were more or less the same as the one's developed for the discovery of the lighter transuranium isotopes in heavy-ion reactions. The elements from nobelium through seaborgium were discovered during this time.

A major breakthrough was the development of mass separators and the capability to detect correlated subsequent α-decay emissions. The use of "cold" fusion reactions which emitted only one or two neutrons from the compound nucleus enabled the discovery of six new elements (bohrium through element 113). The acceptance of the element 113 discovery by the fourth Joint Working Party (JWP) was most recently announced by IUPAC in a press release [2] (see Sects. 12.2 and 12.3). Although the recommendations of the JWP still have to be approved by the IUPAC Council and Bureau, they have been adopted in this chapter.

The use of "hot" fusion reactions, where the excitation energy is so high that the compound nucleus emits three to five neutrons, led to the discovery of elements 114–118. The beam and target combinations for these reactions populated fairly neutron-rich compound nuclei and led in some cases to very long α-decay chains (up to 6 sequential α emissions). Although these chains are not connected to any previously known isotopes, the measurement of excitation functions and cross-reactions together with systematics of spontaneous fission half-lives result in a safe assignment of these nuclides.

M. Thoennessen, *The Discovery of Isotopes*,
DOI 10.1007/978-3-319-31763-2_12

12.2 Recoil Collection with Helium Gas

The main difficulty in detecting and identifying fusion-evaporation residues is the separation from the incident beam. As mentioned in Sect. 5.5, in 1958 Ghiorso et al. developed a method to slow down the recoils in helium gas and remove them away from the beam with an electric field. They were then deposited on a moving belt or catcher shown in Fig. 12.1 which moved the recoils into a location where their decay could be detected [3]. Over the years this method was refined to improve the collection and detection efficiency as well as the collection speed.

The isotopes are not uniquely identified in this method. The element is determined under the assumption that the decay of the compound nucleus following the fusion reaction is dominated by neutron evaporation and that proton and α-emission cross-sections are negligible. The mass of the residue is estimated from cross-section systematics, excitation function measurements, and/or cross irradiations with other beam and target combinations.

Five new elements, from nobelium through seaborgium, were discovered with many different variations of this method between 1963 and 1974. Overall about 40 new isotopes were discovered. They are listed in Table 12.1 with the beam and target combination, the evaporation channel and the observed decay mode.

Until 1980, essentially all new isotopes of superheavy elements were discovered at Berkeley and Dubna. The single exception was the discovery of ^{259}No in 1973 where the initial experiments were performed at Oak Ridge National Laboratory [4].

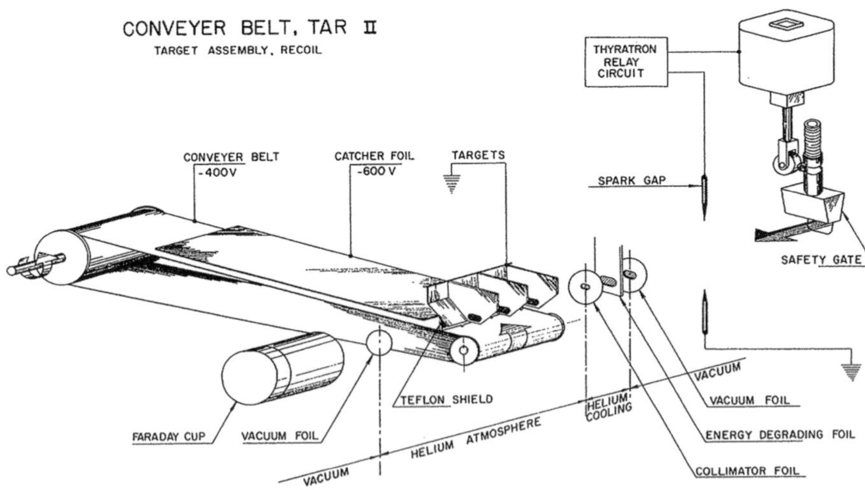

Fig. 12.1 Schematics of the first recoil experiment for superheavy elements produced in fusion-evaporation reactions developed by Ghiorso et al. at Berkeley [3] (Reprinted figure with permission from A. Ghiorso et al., *Element No. 102*, Phys. Rev. Lett. 1 (1958) 18. Copyright 1958 by the American Physical Society.)

Table 12.1 Isotopes of superheavy elements discovered using recoil collection in helium gas: The submission date, first author and reference of the publication as well as the beam, target, evaporation channel, and dominant decay mode (sf = spontaneous fission) are listed

Date	First author	Ref.	Beam	Target	Evap.	Decay	Isotope
05/15/1963	Donets	[5]	^{22}Ne	^{238}U	4n	α	^{256}No
04/20/1965	Donets	[6]	^{18}O	^{243}Am	5n	α	^{256}Lr
12/15/1965	Donets	[7]	^{22}Ne	^{238}U	6n	α	^{254}No
12/15/1965	Zager	[8]	^{15}N	^{243}Am	4n	α	^{254}No
08/20/1966	Mikheev	[9]	^{18}O, ^{16}O	^{239}Pu, ^{242}Pu	5n	α	^{252}No, ^{253}No
10/01/1966	Druin	[10]	^{22}Ne	^{238}U	5n	α	^{255}No
02/09/1967	Ghiorso	[11]	^{12}C, ^{13}C	^{244}Cm, ^{248}Cm	5n, 4n	α	^{251}No, ^{257}No
05/05/1969	Ghiorso	[12]	^{12}C, ^{13}C	^{249}Cf	4n, 3n	α	^{257}Rf, ^{259}Rf
			^{13}C, ^{12}C	^{249}Cf	4n, 3n	sf	^{258}Rf
11/16/1969	Druin	[13]	^{16}O	^{243}Am	4n	α	^{255}Lr
04/17/1970	Ghiorso	[14]	^{15}N	^{249}Cf	4n	α	^{260}Db
04/28/1970	Ghiorso	[15]	^{18}O	^{248}Cm	5n	α	^{261}Rf
06/30/1970	Flerov	[16]	^{22}Ne	^{243}Am	4n	sf	^{261}Db
03/17/1971	Eskola	[17]	^{15}N	^{246}Cm	4n, 3n	α	^{257}Lr, ^{258}Lr
				^{248}Cm	4n, 3n	α	^{259}Lr, ^{260}Lr
05/26/1971	Ghiorso	[18]	^{18}O	^{249}Bk	5n	α	^{262}Db
08/17/1973	Silva	[4]	^{18}O	^{248}Cm	α3n	α	^{259}No
09/02/1974	Oganessian	[19]	^{50}Ti	^{208}Pb, ^{207}Pb	2n	sf	^{256}Rf, ^{255}Rf
09/09/1974	Ghiorso	[20]	^{18}O	^{249}Cf	4n	α	^{263}Sg
09/12/1983	Demin	[21]	^{54}Cr	^{208}Pb	2n	sf	^{260}Sg
09/13/1984	Somerville	[22]	^{16}O, ^{18}O	^{248}Cm	4n	sf	^{260}No[a], ^{260}Rf, ^{262}Rf
12/28/1988	Hulet	[23]	^{13}C	^{248}Cm	3n	sf	^{258}No
04/17/1991	Kratz	[24]	^{18}O	^{249}Bk	4n	sf	^{263}Db
02/23/2000	Wilk	[25]	^{22}Ne	^{249}Bk	4, 5n	α	^{267}Bh, ^{266}Bh
12/07/2000	Gan	[26]	^{22}Ne	^{241}Am	4n	α	^{259}Db
06/25/2002	Kratz	[27]	^{18}O	^{249}Bk	4n	sf	^{263}Rf[b]
03/04/2004	Gan	[28]	^{26}Mg	^{243}Am	4n	α	^{265}Bh
04/19/2006	Dvorak	[29]	^{26}Mg	^{248}Cm	4n	α, sf	^{270}Hs, ^{266}Sg[c]
10/31/2007	Dvorak	[30]	^{26}Mg	^{248}Cm	3n	α, sf	^{271}Hs, ^{267}Sg[c]

[a] Produced by multi-nucleon transfer in the reaction ^{18}O + ^{254}Es
[b] Populated by electron capture from ^{263}Db
[c] Populated by α-decay

However, the discovery paper also included results from Berkeley experiments complementing the Oak Ridge results.

The assignment for the discovery of the first nobelium isotopes was rather difficult. In 1957, Fields et al. [31] assigned the decay of a nobelium isotope to either ^{251}No or ^{253}No, however, a year later, these results could not be replicated by neither Flerov et al. [32] nor Ghiorso et al. [3]. The assignment of a 3 s half-life to ^{254}No proposed by Ghiorso et al. in the same paper was also not correct.

In 1963, Donets et al. irradiated a uranium target with ^{22}Ne beams from the U300 cyclotron at Dubna [5]. They identified a half-life of $\sim$8 s which they assigned to ^{256}No by measuring the α-decay of the ^{252}Fm daughter. This measurement was not credited with the discovery of the element nobelium by the IUPAC-IUPAP Transfermium Working Group, although they stated: "In retrospect, the Dubna results [from the Donets et al. paper] must be considered to be quite probably correct, even though the reported half-life value for ^{256}No is somewhat high" [33, 34]. Thus Donets et al. are credited with the discovery of the isotope ^{256}No.

The discovery of the element nobelium was assigned by the IUPAC-IUPAP Transfermium Working Group [33, 34] to the simultaneous observation of ^{254}No in 1965 by Donets et al. [7] and Zager et al. [8]. Fields et al. had suggested in 1957 the name nobelium [31], and although this original report was not correct the name continued to be used and was officially accepted in 1997 [35–38].

The discovery of the element lawrencium was equally complicated. Ghiorso et al. reported the discovery of element 103 at Berkeley in 1961 with the measurement of an 8$\pm$2 s half-life which they assigned to ^{257}Lr. They suggested the name lawrencium with the symbol Lw [39]. Although the element was correct the mass assignment was incorrect [17]. Four years later, Donets et al. discovered ^{256}Lr at Dubna [6]. The IUPAC-IUPAP Transfermium Working Group gave credit to both groups: "In the complicated situation presented by element 103, with several papers of varying degrees of completeness and conviction, none conclusive, and referring to several isotopes, it is impossible to say other than that full confidence was built up over a decade with credit attaching to work in both Berkeley and Dubna" [33, 34]. The original suggestion of the name lawrencium was adopted but the symbol was later changed to Lr; name and symbol were officially accepted by IUPAC in 1997 [35–38].

Berkeley and Dubna also share the credit for the discovery of element 104: "The chemical experiments in Dubna ([40] with [41]) and the Berkeley experiments [12] were essentially contemporaneous and each show that element 104 had been produced" [33, 34]. Zvara et al. submitted their results on the chemical properties of element 104 on October 14, 1968 and Ghiorso et al. reported the observation of the isotopes ^{257}Rf, ^{258}Rf, and ^{259}Rf on May 5, 1969. Later, the Dubna group suggested the name kurchatovium (Ku) while the Berkeley group suggested rutherfordium (Rf). In 1994, the Commission on Nomenclature of Inorganic Chemistry of IUPAC did not accept either suggestion and recommended dubnium instead [42]. However, this decision was changed to rutherfordium in 1997 [35].

Element 105 was yet again discovered simultaneously at Berkeley and Dubna. While Ghiorso et al. described the discovery of ^{260}Db on April 17, 1970 suggesting the name hahnium with the symbol Ha [14], Flerov et al. reported the observation of ^{261}Db on June 30, 1970 recommending the name nielsbohrium [16]. These names were in use until the controversy was resolved by IUPAC in 1997. In 1994, the Commission on Nomenclature of Inorganic Chemistry of IUPAC had not accepted either of the suggestions and recommended joliotium instead [42]. However, this decision was changed to dubnium in 1997 [35] and officially accepted later in the year [35–38].

The last element discovered with the helium gas technique was seaborgium. In 1974, Ghiorso et al. irradiated a ^{249}Cf target with a ^{18}O beam accelerated by the SuperHILAC at Berkeley and observed α-emission from ^{263}Sg. They submitted their results on September 9, 1974 [20]. Only three days later, Oganessian et al. tentatively assigned spontaneous fission events with a half-life of several milliseconds to ^{259}Sg [43]. This assignment was later shown to be incorrect [21, 44]. The IUPAC-IUPAP Transfermium Working Group gave the credit for the discovery of seaborgium in 1993 to Ghiorso et al.: "Independent work reported in 1974 from Berkeley–Livermore [20] and from Dubna [43] was essentially contemporaneous. The Dubna work is highly important for later developments but does not demonstrate the formation of a new element with adequate conviction, whereas that from Berkeley–Livermore does" [33, 34]. The suggestion for the name seaborgium was not accepted by the 1994 Commission on Nomenclature of Inorganic Chemistry of IUPAC recommending the name rutherfordium instead [42]. This decision was changed in 1997 when the name seaborgium was officially adopted [35–38].

In 2002, Kratz et al. measured a spontaneous fission half-life for ^{263}Rf of about 20 min. ^{263}Rf was formed by electron capture of ^{263}Db which was produced at the Paul Scherrer Institut in Switzerland [27]. However, more recently Dvorak et al. [30] reported a half-life of 8^{+40}_{-4} s without referencing the Kratz et al. results. This apparent discrepancy has not yet been resolved.

The isotopes ^{259}Db and ^{265}Bh were discovered by Gan and collaborators with beams from the Sector Focus Cyclotron at Heavy Ion Research Facility (HIRFL) in Lanzhou, China, in 2000 [26] and 2004 [28], respectively.

12.3 Recoil Separators and Cold Fusion

The discovery of the next seven heavier elements from bohrium to element 113 were made possible by the development of recoil separators for heavy fusion-evaporation residues. The effective suppression of the beam allowed the direct implantation of the residues into position sensitive detectors which were then also used to measure the time and position correlated subsequent decays. With this method it was possible to follow multiple α-emissions thus linking the decay of newly discovered isotopes with known α-emitters populated within the decay chain. In some cases several new α-emitters were discovered before the chain reached previously known isotopes.

Figure 12.2 shows the superheavy isotopes in a different representation of the nuclear chart in order to visualize the various α-decay chains. The isotopes of the different elements are plotted not as a function of neutron numbers, but as a function of the difference between the proton and neutron numbers (N−Z). In this way all α-decay chains appear in a column. The top and bottom panels display the odd-Z and even N−Z elements, respectively. Isotopes which were identified with recoil separators in the region of the chart where the α-decays link to previously known

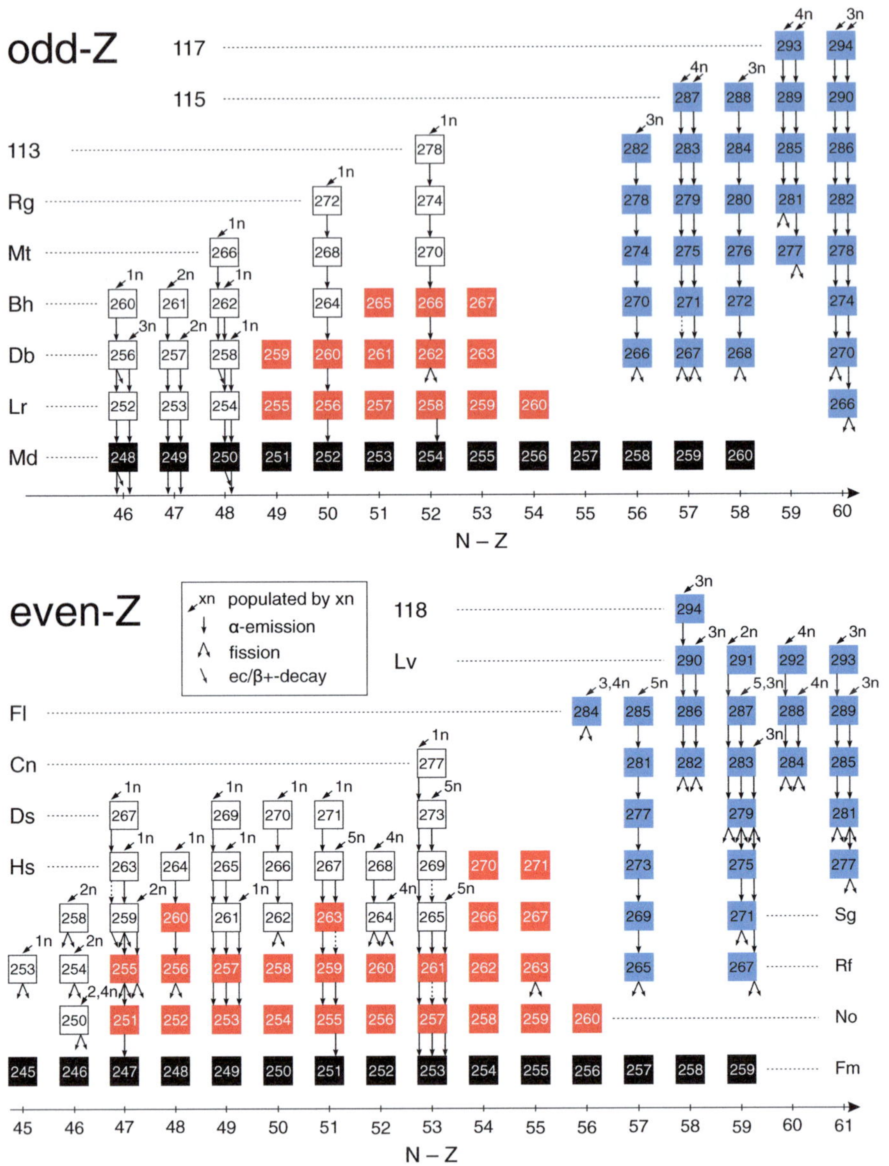

Fig. 12.2 Isotopes of the superheavy elements. Each element corresponds to one row and they are lined up according to the difference of their neutron and proton numbers (N−Z). The *top* and *bottom* panels show the odd-Z and even-Z elements, respectively. *Black squares* represent transuranium isotopes described in Chap. 5. Isotopes shown as *red*, *white*, and *light blue squares* are covered in Sects. 12.2–12.4, respectively. The decay paths shown only correspond to the observed decays of isotopes discovered by α-correlations

isotopes are shown as white boxes and are discussed in the present section. Light blue boxes indicate isotopes where the α-decay chains end in fission of unknown isotopes so that the chains cannot be linked to known isotopes. The discovery of these isotopes will be discussed in the next section. Red and black boxes represent isotopes which were discovered with other methods and have been discussed in the previous section and Chap. 5, respectively.

In addition to the significant technical advances different types of fusion reactions were attempted. Up to then reaction energies well above the Coulomb barrier were chosen in order to maximize the fusion cross-sections. However, these high beam energies then resulted in fairly high excitation energies where the compound nucleus emitted several neutrons which at each step had to compete with fission thus decreasing the cross-section for producing a final heavy nuclide. In the new approach, lower beam energies were selected which reduced the initial fusion cross-section, but increased the survival probability of the fused system as the excitation energy was so low that only one or two neutrons could be emitted. About 30 new isotopes were discovered with this "cold fusion" approach.

All isotopes with $N-Z < 56$ discovered with recoil separators are listed in Table 12.2. Most of these isotopes could be linked to known isotopes by their α-decay, however, some of them fissioned before reaching known isotopes. A total of 38 new isotopes and seven new elements have been discovered since 1981.

The majority of the isotopes were discovered with the velocity filter SHIP (Separator for Heavy Ion reaction Products) at GSI. In 1981, Münzenberg et al. identified for the first time an isotope of the odd-Z element 107. They observed five α-decay chain events which ended in known transitions of ^{250}Fm or ^{250}Md [45]. Also, two of the intermediate α-emitters were not known and represented the discovery of ^{258}Db and ^{254}Lr. The α-decay chain is shown in Fig. 12.3. ^{258}Db was also formed directly in the fusion-evaporation reaction ^{209}Bi(^{50}Ti,n). It is interesting to note that four years later the same group published a paper entitled "The New Isotopes ^{258}105, ^{257}105, ^{254}Lr and ^{253}Lr" [47]. Although they reference the earlier work in the introduction, the previous observation of ^{258}Db is not explicitely mentioned.

In 1982, Münzenberg et al. extended the chain by an additional α-decay to ^{266}Mt corresponding to the new element 109 [46].

The observations of ^{262}Bh and ^{266}Mt were officially accepted by the IUPAC-IUPAP Transfermium Working Group in 1993 as the discovery of these elements [33, 34]. The names bohrium and meitnerium were officially accepted in 1997 [35–38], although the name nielsbohrium (Ns) had been suggested for element 107 [42, 67]. It should be mentioned that the name nielsbohrium had previously also been suggested for element 105 [34]. An earlier observation of spontaneous fission of ^{261}Bh [68] could not be replicated.

A few years later Heßberger et al. and Münzenberg et al. discovered the three isotopes ^{257}Db, ^{253}Lr [47] and ^{261}Bh [48] of the neighboring $N-Z = 47$ chain, respectively. A previous assignment of a 5 s spontaneous fission half-life to ^{257}Db [68] was later reassigned to ^{258}Rf populated by electron capture from ^{258}Db [69].

The discovery of the next odd-Z element was reported in 1994 by Hofmann et al. by detecting three events of a chain containing the three new isotopes ^{272}Rg, ^{268}Mt,

Table 12.2 Isotopes of superheavy elements with N−Z < 56 discovered with recoil separators: The submission date, first author and reference of the publication as well as the target, evaporation channel, and difference between number of neutrons and protons (N−Z) of the chain are listed

Date	First author	Ref.	Beam	Target	Evap.	N−Z	Isotope
Odd-Z							
04/03/1981	Münzenberg	[45]	^{54}Cr, ^{50}Ti	^{209}Bi	1n	48	^{262}Bh, ^{258}Db[a], ^{254}Lr[b]
10/08/1982	Münzenberg	[46]	^{58}Fe	^{209}Bi	1n	48	^{266}Mt
08/13/1985	Heßberger	[47]	^{50}Ti	^{209}Bi	2n	47	^{257}Db, ^{253}Lr[b]
12/12/1988	Münzenberg	[48]	^{54}Cr	^{209}Bi	2n	47	^{261}Bh
12/21/1994	Hofmann	[49]	^{64}Ni	^{209}Bi	1n	50	^{272}Rg, ^{268}Mt[b], ^{264}Bh[b]
01/10/2001	Heßberger	[50]	^{50}Ti	^{209}Bi	3n	46	^{256}Db, ^{252}Lr[b]
07/30/2004	Morita	[51]	^{70}Zn	^{209}Bi	1n	52	^{278}113, ^{274}Rg[b], ^{270}Mt[b]
05/07/2007	Nelson	[52]	^{52}Cr	^{209}Bi	1n	46	^{260}Bh
Even N−Z							
04/19/1984	Münzenberg	[53]	^{58}Fe, ^{54}Cr	^{208}Pb	1n	49	^{265}Hs, ^{261}Sg[a]
06/12/1985	Münzenberg	[44]	^{54}Cr	^{207}Pb	2n	47	^{259}Sg
07/04/1986	Münzenberg	[54]	^{58}Fe	^{207}Pb	1n	48	^{264}Hs
03/31/1994	Lazarev	[55]	^{22}Ne	^{248}Cm	5n	53	^{265}Sg
11/14/1994	Hofmann	[56]	^{62}Ni	^{208}Pb	1n	49	^{269}Ds
11/22/1994	Ghiorso	[57]	^{59}Co	^{209}Bi	1n	47	^{267}Ds
12/21/1994	Hofmann	[49]	^{64}Ni	^{208}Pb	1n	51	^{271}Ds
03/14/1995	Lazarev	[58]	^{34}S	^{238}U	5n	51	^{267}Hs
01/19/1996	Lazarev	[59]	^{34}S	^{244}Pu	5n	53	^{273}Ds
02/21/1996	Hofmann	[60]	^{70}Zn	^{208}Pb	1n	53	^{277}Cn, ^{269}Hs[b]
06/20/1997	Heßberger	[61]	^{51}V	^{209}Bi	2n	46	^{258}Sg
			^{50}Ti	^{206}Pb	2n	46	^{254}Rf
			^{50}Ti	^{204}Pb	1n	45	^{253}Rf
11/27/2000	Hofmann	[62]	^{64}Ni	^{207}Pb	1n	50	^{270}Ds, ^{266}Hs[b], ^{262}Sg[b]
04/27/2001	Oganessian	[63]	^{48}Ca	^{204}Pb, ^{206}Pb	2n, 4n	46	^{250}No
06/20/2006	Gregorich	[64]	^{30}Si	^{238}U	4n	52	^{264}Sg
10/23/2008	Dragojevic	[65]	^{56}Fe	^{208}Pb	1n	47	^{263}Hs
07/02/2010	Nishio	[66]	^{34}S	^{238}U	4n	52	^{268}Hs

[a] Directly populated and populated by α-decay
[b] Populated by α-decay

and ^{264}Bh, and continuing with two other known α-decays [49]. This discovery was officially accepted by the IUPAC-IUPAP Joint Working Party (JWP) in 2003: "In concordance with the criteria established for validating claims, the JWP has agreed

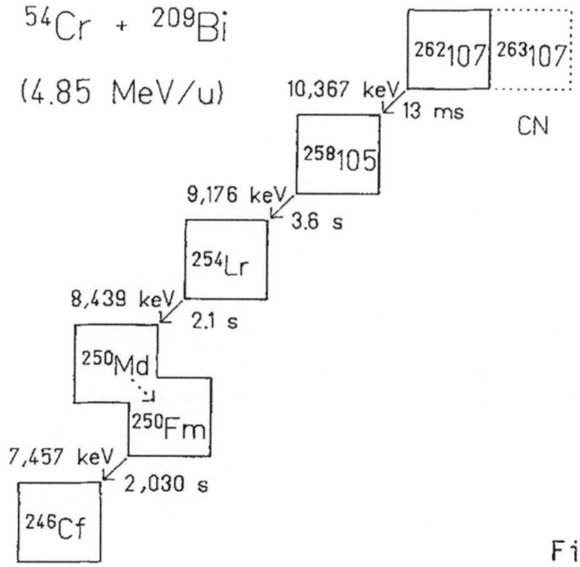

Fig. 12.3 Discovery of bohrium. The decay chain following the population of ^{262}Bh consists of four α-decays and one electron capture leading to ^{246}Cf [45] (With kind permission from Springer Science and Business Media: G. Münzenberg et al., *Identification of element 107 by α correlation chains*, Z. Phys. A 300 (1981) 107.)

that the priority of the Hofmann et al. collaboration's discovery of element 111 at GSI is acknowledged" [70]. The name roentgenium was officially accepted in 2003 [71].

About nine years later Morita et al. reported a single decay chain originating from Z = 113. A ^{70}Zn beam from the RIKEN linear accelerator facility RILAC irradiated bismuth targets. Evaporation residues were identified with the gas-filled separator GARIS and detected with micro-channel plates and a silicon strip detector. Spontaneous fission and α-decay events were recorded with a silicon semiconductor array consisting of a central detector plus four additional silicon strip detectors forming a box [51]. Subsequently, Morita et al. reported two additional events of this decay chain, one in 2007 [72] and one in 2012 [73]. The first two events were not deemed sufficient to claim discovery of element 113 by the 2011 IUPAC-IUPAP JWP: "The work of the collaboration of Morita et al. is very promising but has not met the criteria for discovery owing to the paucity of events, the absence of firm connection(s) to known nuclides, and the inconsistencies noted above" [74]. However, at the end of 2015 IUPAC announced that the JWP had credited the RIKEN group with the discovery in a press release: "The RIKEN collaboration team in Japan have fulfilled the criteria for element Z = 113 and will be invited to propose a permanent name and symbol" [2].

The most recent odd-Z isotope discovery in 2007 extended the N−Z = 46 chain by Nelson et al. [52]. They observed eight decay chains originating from ^{260}Bh with the Berkeley Gas-filled Separator BGS.

The first even N−Z element discovered with a recoil separator was element 108. Münzenberg et al. submitted their measurement of three correlated α-decays originating at ^{265}Hs at GSI on April 19, 1984 [53]. Only two months later, on June 14, 1984, Oganessian et al. reported the observation of $^{263-265}$Hs at Dubna [75]. The Dubna group did not observe the α-decay of the hassium isotopes directly and inferred their formation from known decays of the granddaughters or the great-great-great-granddaughter (^{253}Es of ^{265}Hs). The IUPAC-IUPAP Transfermium Working Group in 1993 gave credit to both groups for the discovery of the element hassium: "The formation of element 108 was established by simultaneous and independent work in Darmstadt [53] and Dubna [75]" [33, 34]. Although initially the name hahnium had been recommended in 1994 [42], ultimately the name hassium was officially accepted in 1997 [35–38].

The discovery of element 110 was equally close. On November 14, 1994, Hofmann et al. reported the extension of the $N-Z=49$ chain to element 110 and observed ^{269}Ds at GSI [56]. Only eight days later, on November 22, 1994, Ghiorso et al. submitted a paper describing the observation of a single event of ^{267}Ds at Berkeley [57]. The IUPAC-IUPAP JWP credited the GSI researchers with the discovery in 2001: "In accordance with the criteria for the discovery of elements, previously established by the 1992 IUPAC-IUPAP Transfermium Working Group, it was determined that the claim by the Hofmann et al. research collaboration for the discovery of element 110 at GSI has fulfilled those criteria" [76]. The name darmstadtium was officially accepted in 2003 [77]. It should be mentioned that in a note added in proof of the GSI paper it was mentioned that three additional chains had been observed. The observation of one of the decay chains was later retracted [78].

The discovery of ^{271}Ds is difficult to assign. It was first mentioned in the refereed literature by Hofmann et al. in the introduction of the article announcing the discovery of element 111: "In a recent experiment we produced the nucleus ^{269}110 by the reaction ^{62}Ni + ^{208}Pb [56]. In a succeeding experiment we investigated the reaction ^{64}Ni + ^{208}Pb and observed the heavier isotope ^{271}110 [79]" [49]. However, no further details about the experiments nor the results were mentioned referring to an internal report. The decay sequence of the single observed event was published in a scientific journal only three years later in a review article by S. Hofmann [80]. In order to acknowledge the contributions of all co-authors Ref. [49] is credited with the discovery of ^{271}Ds as the author list is identical with the list given in the internal report.

The heaviest even $N-Z$ element discovered in a "cold" fusion reaction was element 112 at GSI. Hofmann et al. observed two chains starting at ^{277}Cn, where one of them consisted of six α-decays ending in ^{253}Fm as shown in Fig. 12.4 [60]. The first chain had to be retracted in 2002 [78]. In 2009, the JWP accepted this observation as the discovery of element 112 [81]. Claims for the observation of element 112 in tungsten targets irradiated with 24 GeV protons at CERN [82] in 1984 were deemed not credible [76]. The name copernicium was officially accepted in 2010 [83].

A month earlier, Lazarev et al. had discovered a chain starting with the ^{277}Cn daughter ^{273}Ds at Dubna. Although in a table they identify ^{269}Hs in one of the five chains, they dismiss this assignment because the energy of the α-decay is too high by 0.5 MeV as compared to experimental systematics and theoretical predictions [59]. Thus the discovery of ^{269}Hs is also credited to Hofmann et al. [60].

The first superheavy isotope discovered in a "cold" fusion reaction at Berkeley was ^{267}Ds. Due to a malfunctioning of the electronics the decay of the also unknown

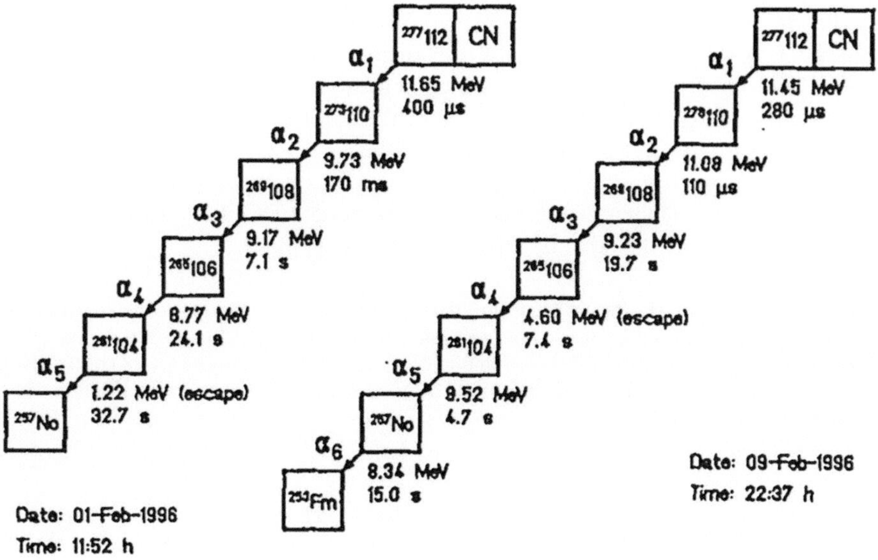

Fig. 12.4 Discovery of copernicium. The two recorded decay chains originating from ^{277}Cn consist of five and six α-decays leading to ^{257}No and ^{253}Fm, respectively [60] (With kind permission from Springer Science and Business Media: S. Hofmann et al., *The new element 112*, Z. Phys. A 354 (1996) 229.)

daughter nuclide ^{263}Hs was not observed [57]. It took another 14 years before Dragojevic et al.—also at Berkeley—observed six events of an α-decay chain originating at ^{263}Hs [65].

The most neutron deficient isotopes with N$-$Z of 45 and 46 decayed directly by spontaneous fission, so the assignment could not be linked to subsequent α-emission. The spontaneous fission half-lives of ^{258}Sg, ^{254}Rf, and ^{253}Rf [61] and ^{250}No [63] were measured at GSI and Dubna, respectively.

12.4 Hot Fusion and Not-Connected Decay Chains

The discovery and assignment of isotopes of the heaviest elements was complicated because the observed decay chains ended in fission events of isotopes which were unknown. Thus, it was impossible to independently confirm the assignments of the evaporation channels, and the specific masses of the observed isotopes are still somewhat uncertain. The difficulty in the mass determination is demonstrated by the initial incorrect identification of flerovium and livermorium isotopes, observed in the fusion-evaporation reactions ^{48}Ca + ^{244}Pu and ^{48}Ca + ^{248}Cm, respectively. Presently, ten separate decay chains have generally been accepted. There are five chains each originating from even N$-$Z and odd-Z elements and one isotope (^{284}Fl) which was identified by its immediate fission decay.

The 56 isotopes that were discovered in hot fusion reactions are listed in Table 12.3. All of them were produced in fusion-evaporation reactions with ^{48}Ca beams. The majority of the isotopes in each of the ten α-decay chains were reported in six major publications submitted in 2003, 2007, and 2010. Nine of the ten chains were discovered by Oganessian et al. at the Joint Institute for Nuclear Research at Dubna with beams from the U400 cyclotron. The fusion-evaporation residues were analyzed with the Dubna Gas-Filled Recoil Separator DGFRS and implanted in position-sensitive semiconductor detectors in the focal plane.

Table 12.3 Isotopes of superheavy elements discovered by α-decay chains not connected to known isotopes: The submission date, first author and reference of the publication as well as the target, evaporation channel, and difference between number of neutrons and protons (N−Z) of the chain are listed. The isotope listed first was populated directly in the fusion-evaporation reaction while the others were populated by subsequent α-decay unless otherwise indicated

Date	First author	Ref.	Target	Evap.	N−Z	Isotope
Odd-Z						
09/08/2003	Oganessian	[84]	^{243}Am	4n	57	^{287}115, ^{283}113, ^{279}Rg, ^{275}Mt, ^{267}Db
				3n	58	^{288}115, ^{284}113, ^{280}Rg, ^{276}Mt, ^{272}Bh, ^{268}Db
04/26/2007	Oganessian	[85]	^{237}Np	3n	56	^{282}113, ^{278}111, ^{274}Mt, ^{270}Bh, ^{266}Db
03/15/2010	Oganessian	[86]	^{249}Bk	4n	59	^{293}117, ^{289}115, ^{285}113, ^{281}Rg
				3n	60	^{294}117, ^{290}115, ^{286}113, ^{282}Rg, ^{278}Mt, ^{274}Bh, ^{270}Db
09/01/2012	Oganessian	[87]	^{243}Am	4n	57	^{271}Bh^{a}
04/06/2013	Oganessian	[88]	^{249}Bk	3n	59	^{277}Mt^{a}
02/22/2014	Khuyagbaatar	[89]	^{249}Bk	4n	60	^{266}Lr^{a}
Even N−Z						
12/01/2003	Oganessian	[90]	^{245}Cm	3n	58	^{290}Lv, ^{286}Fl, ^{282}Cn
			^{245}Cm, ^{244}Pu	2n, 5n	59	^{291}Lv, ^{287}Fl^{b}, ^{283}Cn, ^{279}Ds
			^{244}Pu	4n	60	^{288}Fl, ^{284}Cn
			^{248}Cm, ^{244}Pu	3n	61	^{293}Lv^{c}, ^{289}Fl^{b,c}, ^{285}Cn^{c}, ^{281}Ds^{c}
08/09/2004	Oganessian	[91]	^{242}Pu, ^{238}U	3n	59	^{275}Hs^{a}, ^{271}Sg, ^{267}Rf^{d}
			^{248}Cm	4n	60	^{292}Lv
01/31/2006	Oganessian	[92]	^{249}Cf	3n	58	^{294}118
12/22/2009	Düllmann	[93]	^{244}Pu	3n	61	^{277}Hs^{a}
07/23/2010	Ellison	[94]	^{242}Pu	5n	57	^{285}Fl, ^{281}Cn, ^{277}Ds, ^{273}108, ^{269}106, ^{265}104
08/31/2015	Utyonkov	[95]	^{239}Pu, ^{240}Pu	3n, 4n	56	^{284}Fl

[a]Populated by α-decay
[b]Directly populated and populated by α-decay
[c]Re-assigned from an earlier experiment
[d]Only observed with ^{238}U target

The first two odd-Z chains were reported in 2003 by forming the compound nucleus ^{291}115 populating ^{287}115 and ^{288}115 in 4n and 3n reactions, respectively [84]. One decay chain for ^{287}115 and three chains for ^{288}115 were observed. The chains consisted of five consecutive α-particle emissions and ended with fission events of dubnium. In the case of the odd–odd nucleus ^{268}Db it seems likely that this isotope undergoes electron capture and the spontaneous fission originates from the even–even nucleus ^{268}Rf, similar to the α-decay chain of ^{282}113 ending in ^{266}Db/^{266}Rf [85]. Figure 12.5 shows the four decay chains [84]. The decay of ^{271}Bh was not observed, however, in a later experiment it was identified in the same decay sequence [87].

While the discovery of the two Z = 115 isotopes was not accepted as the discovery of the element 115 by the 2011 IUPAC-IUPAP JWP because they are not connected to any known nuclei and the chemical analysis cannot "distinguish the properties of Groups 4 and 5 elements in this region with confidence" [74], IUPAC announced at the end of 2015 in a press release that "The collaboration between the Joint Institute for Nuclear Research in Dubna, Russia; Lawrence Livermore National Laboratory, California, USA; and Oak Ridge National Laboratory, Oak Ridge, Tennessee, USA have fulfilled the criteria for element Z = 115, 117 and will be invited to propose permanent names and symbols" [2].

The population of ^{283}113 and ^{284}113 in the α-decay from the corresponding Z = 115 isotopes in the 2003 paper by Oganessian et al. also represents the first observation of Z = 113 isotopes submitted to a refereed journal (9/8/2003) [84]. The discovery of ^{278}113 in the cold fusion reaction by Morita et al. described in the previous section was only submitted almost a year later on 7/30/2004. However the 2011 JWP did not credit the paper with the discovery of element 113: "The results are encouraging but do not meet the criteria for discovery because of the paucity of events, the lack of connections to known nuclides, and the absence of cross-bombardments" [74].

Three years later Oganessian et al. discovered five isotopes of the odd-Z N−Z = 56 chain [85]. They observed two chains starting at ^{282}113. The decay mode of the last isotope of one chain was not clear: "The first decay chain was terminated by SF decay with an apparent lifetime of 31.7 min. The origin of this decay can be spontaneous fission of ^{266}Db, or its electron capture with a lifetime of 31.7 min followed by the relatively short-lived spontaneous fission of the even–even isotope ^{266}Rf" [85]. Nevertheless, the discovery of ^{266}Db is credited to Oganessian et al. because of the half-life measurement. The second chain terminated earlier with the α-decay of ^{270}Bh.

Two isotopes of the new element 117 were discovered in 2010 in the fusion-evaporation reactions ^{249}Bk(^{48}Ca,3n)^{294}117 and ^{249}Bk(^{48}Ca,4n)^{293}117 with five events and one event, respectively [86]. The chains originating in ^{294}117 consisted of six α-decays and terminated with the spontaneous fission of ^{270}Db. The one event for ^{293}117 was significantly shorter and ended with the fission of ^{281}Rg. Oganessian et al. repeated the experiment a few years later and recorded six additional events for the ^{293}117 chain. In two of these chains ^{281}Rg emitted an α-particle and the fission of the new isotope ^{277}Mt was reported [88].

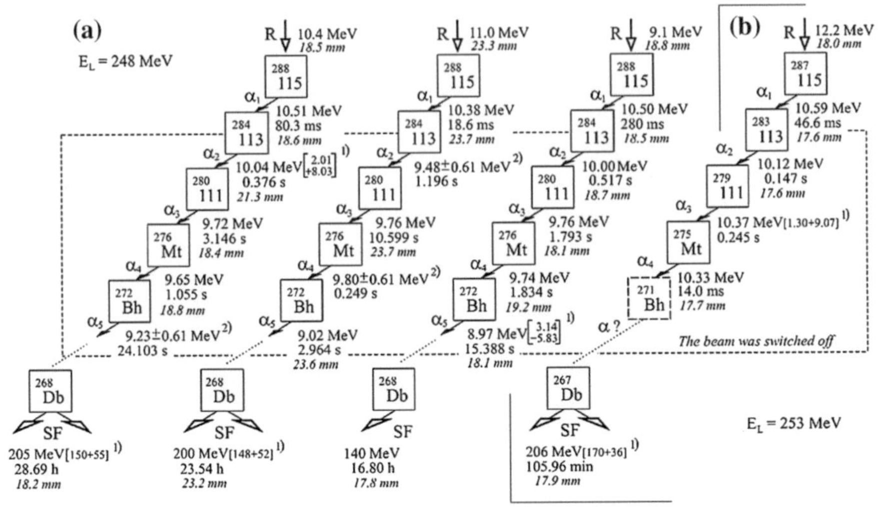

Fig. 12.5 Discovery of Z = 115 isotopes. Three decay chains for ^{288}115 and one chain for ^{287}115 was identified. The decay of ^{271}Bh in the ^{287}115 chain was not observed [84] (Reprinted figure with permission from Yu.Ts. Oganessian et al., *Experiments on the synthesis of element 115 in the reaction* ^{243}*Am(*^{48}Ca,xn)$^{291-x}$115, Phys. Rev. C 69 (2004) 021601. Copyright 2004 by the American Physical Society.)

The chain originating ^{294}117 was recently also extended by an additional α-decay using the same reaction. Khuyagbaatar et al. observed two events of the ^{294}117 decay chain with the gas-filled recoil separator TASCA at GSI and detected α-emission of ^{270}Db and the subsequent fission of ^{266}Lr [89]. This represents the only discovery of an isotope of an odd-Z element within the unconnected island of superheavy elements that was not measured at Dubna.

As mentioned above, the discovery of element 117 was recently credited to the work by Oganessian et al. [2].

The discovery of the even N−Z chains in principle precedes the observation of the odd-Z isotopes, however, as mentioned earlier the initial isotopic assignment of flerovium and livermorium was incorrect. In 2000, Oganessian et al. assigned two decay chains to the 4n channel in the reaction ^{48}Ca + ^{244}Pu originating from ^{288}Fl [96, 97]. A year later they claimed the discovery of livermorium in the reaction ^{48}Ca + ^{248}Cm [98, 99]. The decay chain matched the flerovium chain suggesting also the population of ^{292}Lv in the 4n channel.

Essentially simultaneously with the first observation of the first odd-Z chains, Oganessian et al. reported excitation function measurements of ^{48}Ca + ^{244}Pu and results from the new reaction ^{48}Ca + ^{245}Cm [90] in 2003. They obtained a consistent description of all observed decay chains when they reassigned the previously reported 4n channels to 3n channels. Thus the decay chain ^{292}Lv $\rightarrow^{288}$Fl $\rightarrow^{284}$Cn $\rightarrow^{280}$Ds was reassigned to ^{293}Lv $\rightarrow^{289}$Fl $\rightarrow^{285}$Cn $\rightarrow^{281}$Ds. This reassignment also affected two additional decay chains in Ref. [100]. These two chains were also mentioned in a note added in proof in Ref. [99].

In the same paper, Oganessian et al. also reported the observation of two new decay chains from livermorium (^{290}Lv and ^{291}Lv) and one new chain from flerovium (^{288}Fl). These chains ended with fission events in ^{282}Cn, ^{279}Ds, and ^{284}Cn, respectively.

Earlier reports of ^{287}Fl [101] and ^{289}Fl [101, 102] could not be replicated; neither were previous observations of the copernicium isotopes ^{283}Cn [101, 103, 104] and ^{285}Cn [101, 102]. Two years after Oganessian et al. reported the discovery of ^{283}Cn Gregorich et al. was not able to replicate the results. However, in 2007 Hofmann et al. confirmed the Oganessian et al. data [105]. A comprehensive review of the discovery of these isotopes is presented in Ref. [106].

Later in 2004, Oganessian et al. repeated the ^{48}Ca + ^{248}Cm reaction at a higher energy and reported the observation of six decay chains in the 4n channel. The chains originated in ^{292}Lv and ended after two α-decays with the spontaneous fission of ^{284}Cn [91]. In the same paper, Oganessian et al. also explored the $N-Z = 59$ chain further with the reactions ^{48}Ca + ^{242}Pu and ^{48}Ca + ^{238}U. In addition to several chains terminating in the fission of ^{283}Cn, they observed one chain containing four α-decays in each of these two reactions. In the first one the decay started with ^{287}Fl and ended with the fission of ^{271}Sg, while the second one started at ^{283}Cn and terminated with the fission of ^{267}Rf.

The discoveries of element 114 and 116 have been accepted by the IUPAC-IUPAP Joint Working Party in 2011 [74] and recently been named flerovium and livermorium, respectively [107].

Two years later, Oganessian et al. discovered element 118 in the fusion reaction ^{249}Cf(^{48}Ca,3n)^{294}118 [92]. They observed three decay events where the decay of the daughter nuclides agreed with the decay chain originating from ^{290}Lv. One of the three measured decay chains had been mentioned in a previous publication by the same group [91] referring to internal reports [108]. In another publication it was speculated that two events could have resulted from either ^{294}118 or ^{295}118 [109].

While the discovery of the element 118 was not accepted by the 2011 JWP: "The three events reported for the $Z = 118$ isotope have very good internal redundancy but with no anchor to known nuclei do not satisfy the criteria for discovery" [74], in the press release at the end of 2015 IUPAC announced that "The collaboration between the Joint Institute for Nuclear Research in Dubna, Russia, and Lawrence Livermore National Laboratory, California, USA, have fulfilled the criteria for element $Z = 118$ and will be invited to propose a permanent name and symbol" [2].

In 2009, the $N-Z = 61$ chain was extended by one α-decay by Düllmann et al. [93]. ^{289}Fl was formed in the ^{48}Ca + ^{244}Pu 3n fusion-evaporation reaction from the UNILAC accelerator at GSI. Reaction products as well as α-emission and spontaneous fission decays were measured with the detection system of the gas-filled recoil separator TASCA. One chain consisting of three α-particle emissions and terminating with spontaneous fission from ^{277}Hs was detected. An earlier reported observation of the spontaneous fission of ^{277}Hs [102] could not be replicated.

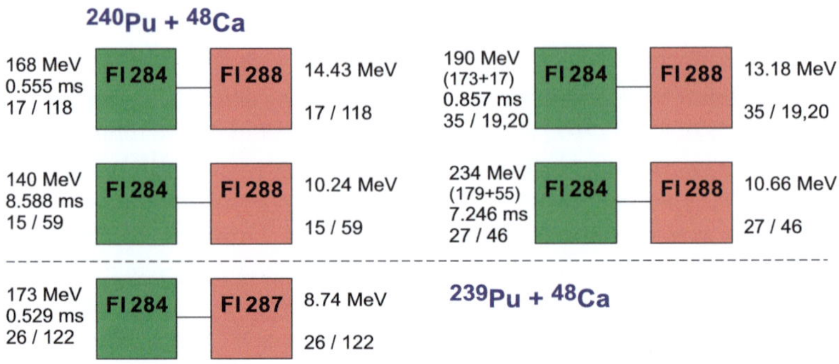

Fig. 12.6 Decay properties of nuclei assigned to ^{284}Fl observed in the ^{240}Pu + ^{48}Ca reaction (*top*) and ^{239}Pu + ^{48}Ca reaction (*bottom*). The *right rows* for each chain show the evaporation residue (in *pink*) energies and strip numbers (front and back). The *left rows* provide the energies, time intervals between events, and strip numbers of the spontaneous fission fragments (in *green*). The decay chain shown in the *top-right* corner was registered in two back strips [95] (Reprinted figure with permission from V.K. Utyonkov et al., *Experiments on the synthesis of superheavy nuclei* ^{284}*Fl and* ^{285}*Fl in the* ^{239,240}*Pu* +^{48}*Ca reactions*, Phys. Rev. C 92 (2015) 034609. Copyright 2015 by the American Physical Society.)

The last chain which is still unconnected to any known isotope was discovered in 2010 by Ellison et al. with ^{48}Ca beams from the Berkeley 88-in. cyclotron irradiating a ^{242}Pu target [94]. Evaporation residues were selected with the Berkeley Gas-filled Separator (BGS) and the fragments were detected in multiwire proportional counters and silicon strip detectors. Subsequent radioactive decay events were recorded in the strip detectors and additional silicon chips forming a five-sided box. They observed a single chain assigned to the 5n evaporation channel of ^{285}Fl. Five subsequent α-particles were detected ending with a fission event from ^{265}Rf.

This chain corresponds to the same chain which in 1999 had been fraudulently claimed to have discovered elements 116 and 118 [110]. The three events following the fusion of ^{86}Kr +^{208}Pb forming ^{293}118 in the 1n channel had apparently been fabricated and the results had to be retracted [111].

Finally, the most recent observation of a superheavy isotope was the discovery of ^{284}Fl produced in the 3n and 4n evaporation reaction on ^{239}Pu and ^{240}Pu targets, respectively. A total of five events were observed which immediately fissioned with the properties shown in Fig. 12.6 [95].

References

1. A. Ghiorso et al., Phys. Rev. **98**, 1518 (1955)
2. International Union of Pure and Applied Chemistry. Discovery and Assignment of Elements with Atomic Numbers 113, 115, 117 and 118, press release, 30 Dec 2015
3. A. Ghiorso et al., Phys. Rev. Lett. **1**, 18 (1958)
4. R.J. Silva et al., Nucl. Phys. A **216**, 97 (1973)

5. E.D. Donets, V.A. Shchegolev, V.A. Ermakov, Sov. At. Energy **14**, 540 (1963)
6. E.D. Donets, V.A. Shchegolev, V.A. Ermakov, Sov. At. Energy **19**, 995 (1965)
7. E.D. Donets, V.A. Shchegolev, V.A. Ermakov, Sov. At. Energy **20**, 257 (1966)
8. B.A. Zager et al., Sov. At. Energy **20**, 264 (1966)
9. V.L. Mikheev et al., Sov. At. Energy **22**, 93 (1967)
10. V.A. Druin et al., Sov. At. Energy **22**, 135 (1967)
11. A. Ghiorso, T. Sikkeland, M.J. Nurmia, Phys. Rev. Lett. **18**, 401 (1967)
12. A. Ghiorso et al., Phys. Rev. Lett. **22**, 1317 (1969)
13. V.A. Druin, Sov. J. Nucl. Phys. **12**, 146 (1971)
14. A. Ghiorso et al., Phys. Rev. Lett. **24**, 1498 (1970)
15. A. Ghiorso et al., Phys. Lett. B **32**, 95 (1970)
16. G.N. Flerov et al., Sov. At. Energy **29**, 967 (1970)
17. K. Eskola et al., Phys. Rev. C **4**, 632 (1971)
18. A. Ghiorso et al., Phys. Rev. C **4**, 1850 (1971)
19. Yu. Ts, Oganessian et al., Sov. At. Energy **38**, 492 (1975)
20. A. Ghiorso et al., Phys. Rev. Lett. **33**, 1490 (1974)
21. A.G. Demin et al., Z. Phys. A **315**, 197 (1984)
22. L.P. Somerville et al., Phys. Rev. C **31**, 1801 (1985)
23. E.K. Hulet et al., Phys. Rev. C **40**, 770 (1989)
24. J.V. Kratz et al., Phys. Rev. C **45**, 1064 (1992)
25. P.A. Wilk et al., Phys. Rev. Lett. **85**, 2697 (2000)
26. Z.G. Gan et al., Eur. Phys. J. A **10**, 21 (2001)
27. J.V. Kratz et al., Radiochim. Acta **91**, 59 (2003)
28. Z.G. Gan et al., Eur. Phys. J. A **20**, 385 (2004)
29. J. Dvorak et al., Phys. Rev. Lett. **97**, 242501 (2006)
30. J. Dvorak et al., Phys. Rev. Lett. **100**, 132503 (2008)
31. P.R. Fields et al., Phys. Rev. **107**, 1460 (1957)
32. G.N. Flerov et al., Sov. Phys. Dokl. **3**, 546 (1958)
33. R.C. Barber et al., Prog. Part. Nucl. Phys. **29**, 453 (1992)
34. Transfermium Working Group, Pure Appl. Chem. **65**, 1757 (1993)
35. Commission on Nomenclature of Inorganic Chemistry, Pure Appl. Chem. **69**, 2471 (1997)
36. Summary of Actions at the 39th IUPAC Council Meeting. Geneva, Switzerland, http://old.iupac.org/news/archives/1997/councils.html. Accessed 29–30 Aug 1997
37. IUPAC Adopts Final Recommendations for Names of Transfermium Elements, http://old.iupac.org/news/archives/1997/august97.html
38. Element Names, news release. J. Chem. Educ. **74**, 1258 (1997)
39. A. Ghiorso et al., Phys. Rev. Lett. **6**, 473 (1961)
40. I. Zvara et al., Radiokhimiya **11**, 163 (1969)
41. I. Zvara et al., J. Inorg. Nucl. Chem. **32**, 1885 (1970)
42. Commission on Nomenclature of Inorganic Chemistry, Pure Appl. Chem. **66**, 2419 (1994)
43. Y.T. Oganessian et al., JETP Lett. **20**, 265 (1974)
44. G. Münzenberg et al., Z. Phys. A **322**, 227 (1985)
45. G. Münzenberg et al., Z. Phys. A **300**, 107 (1981)
46. G. Münzenberg et al., Z. Phys. A **309**, 89 (1982)
47. F.P. Heßberger et al., Z. Phys. A **322**, 557 (1985)
48. G. Münzenberg et al., Z. Phys. A **333**, 163 (1989)
49. S. Hofmann et al., Z. Phys. A **350**, 281 (1995)
50. F.P. Heßberger et al., Eur. Phys. J. A **12**, 57 (2001)
51. K. Morita et al., J. Phys. Soc. Jpn. **73**, 2593 (2004)
52. S.L. Nelson et al., Phys. Rev. Lett. **100**, 022501 (2008)
53. G. Münzenberg et al., Z. Phys. A **317**, 235 (1984)
54. G. Münzenberg et al., Z. Phys. A **324**, 489 (1986)
55. Yu.A. Lazarev et al., Phys. Rev. Lett. **73**, 624 (1994)
56. S. Hofmann et al., Z. Phys. A **350**, 277 (1995)

57. A. Ghiorso et al., Phys. Rev. C **51**, R2293 (1995)
58. Yu.A. Lazarev et al., Phys. Rev. Lett. **75**, 1903 (1995)
59. Yu.A. Lazarev et al., Phys. Rev. C **54**, 620 (1996)
60. S. Hofmann et al., Z. Phys. A **354**, 229 (1996)
61. F.P. Heßberger et al., Z. Phys. A **359**, 415 (1997)
62. S. Hofmann et al., Eur. Phys. J. A **10**, 5 (2001)
63. Y.T. Oganessian et al., Phys. Rev. C **64**, 054606 (2001)
64. K.E. Gregorich et al., Phys. Rev. C **74**, 044611 (2006)
65. I. Dragojevic et al., Phys. Rev. C **79**, 011602 (2009)
66. K. Nishio et al., Phys. Rev. C **82**, 024611 (2010)
67. P. Armbruster et al., Responses on the report "Discovery of the Transfermium Elements". Pure Appl. Chem. **65**, 1815 (1993)
68. Y.T. Oganessian et al., Nucl. Phys. A **273**, 505 (1976)
69. F.P. Heßberger et al., J. Less-Common Met. **122**, 445 (1986)
70. P.J. Karol et al., Pure Appl. Chem. **75**, 1601 (2003)
71. J. Corish, G.M. Rosenblatt, Pure Appl. Chem. **76**, 2101 (2004)
72. K. Morita et al., J. Phys. Soc. Jpn. **76**, 045001 (2007)
73. K. Morita et al., J. Phys. Soc. Jpn. **81**, 103201 (2012)
74. R.C. Barber et al., Pure Appl. Chem. **83**, 1485 (2011)
75. Y.T. Oganessian et al., Z. Phys. A **319**, 215 (1984)
76. P.J. Karol et al., Pure Appl. Chem. **73**, 959 (2001)
77. J. Corish, G.M. Rosenblatt, Pure Appl. Chem. **75**, 1613 (2003)
78. S. Hofmann et al., Eur. Phys. J. A **14**, 147 (2002)
79. S. Hofmann et al., GSI Nachrichten 11–94 Nov. 1994 and to be published
80. S. Hofmann, Rep. Prog. Phys. **61**, 639 (1998)
81. R.C. Barber et al., Pure Appl. Chem. **81**, 1331 (2009)
82. A. Marinov et al., Phys. Rev. Lett. **52**, 2209 (1984)
83. K. Tatsumi, J. Corish, Pure Appl. Chem. **82**, 753 (2010)
84. Y.T. Oganessian et al., Phys. Rev. C **69**, 021601 (2004)
85. Y.T. Oganessian et al., Phys. Rev. C **76**, 011601 (2007)
86. Y.T. Oganessian et al., Phys. Rev. Lett. **104**, 142502 (2010)
87. Y.T. Oganessian et al., Phys. Rev. C **87**, 014302 (2013)
88. Y.T. Oganessian et al., Phys. Rev. C **87**, 054621 (2013)
89. J. Khuyagbaatar et al., Phys. Rev. Lett. **112**, 172501 (2014)
90. Y.T. Oganessian et al., Phys. Rev. C **69**, 054607 (2004)
91. Y.T. Oganessian et al., Phys. Rev. C **70**, 064609 (2004)
92. Y.T. Oganessian et al., Phys. Rev. C **74**, 044602 (2006)
93. Ch.E. Düllmann et al., Phys. Rev. Lett. **104**, 252701 (2010)
94. P.A. Ellison et al., Phys. Rev. Lett. **105**, 182701 (2010)
95. V.K. Utyonkov et al., Phys. Rev. C **92**, 034609 (2015)
96. Y.T. Oganessian et al., Phys. Rev. C **62**, 041604 (2000)
97. Y.T. Oganessian et al., Phys. At. Nuclei **63**, 1679 (2000)
98. Y.T. Oganessian et al., Phys. Rev. C **63**, 011301 (2001)
99. Y.T. Oganessian, V.K. Utyonkov, K.J. Moody, Phys. At. Nuclei **64**, 1349 (2001)
100. Y.T. Oganessian et al., Eur. Phys. J. A **15**, 201 (2002)
101. Y.T. Oganessian et al., Nature **400**, 242 (1999)
102. Y.T. Oganessian et al., Phys. Rev. Lett. **83**, 3154 (1999)
103. Y.T. Oganessian et al., Eur. Phys. J. A **5**, 63 (1999)
104. Y.T. Oganessian et al., Eur. Phys. J. A **19**, 3 (2004)
105. S. Hofmann et al., Eur. Phys. J. A **32**, 251 (2007)
106. Y.T. Oganessian et al., J. Phys. G **34**, R165 (2007)
107. R.D. Loss, J. Corish, Pure Appl. Chem. **84**, 1669 (2012)
108. Y.T. Oganessian et al., JINR Communications D7-2002-287 (2002); Lawrence Livermore National Laboratory Report UCRL-ID-151619 (2003)
109. Y.T. Oganessian et al., Nucl. Phys. A **734**, 109 (2004)
110. V. Ninov et al., Phys. Rev. Lett. **83**, 1104 (1999)
111. V. Ninov et al., Phys. Rev. Lett. **89**, 039901 (2002)

Chapter 13
Spontaneous Fission

13.1 Spontaneous Fission Sources

Thermal neutron fission contributed to a significant number of isotope discoveries between 1940 and 1970. Fission was the only mechanism available to populate neutron-rich isotopes far from the valley of stability. Most of the isotopes were discovered by thermal neutron-induced fission of ^{235}U in reactors. Spontaneous fission was discovered by G.N. Flerov and K.A. Petrzhak reported in a very short article in 1940: "With 15 plates ionization chambers adjusted for detection of uranium fission products we observed 6 pulses per hour which we ascribe to spontaneous fission of uranium. A series of control experiments seem to exclude other possible explanations. Energy of pulses and absorption properties coincide with fission products of uranium bombarded by neutrons. No pulses were found with UX and Th. Mean lifetime of uranium follows ten to sixteen or seventeen years" [1]. This was the whole article which was even shorter than the paper about the discovery of ^{233}U [2] mentioned in Sect. 7.2. Techniques to utilize spontaneous fission to produce and study neutron-rich nuclides were only developed about thirty years later.

Isotopes which spontaneously fission with reasonably long half-lives to be practical as sources still predominantly decay by α-emission and the spontaneous fission branch typically accounts for only a few percent. The most common spontaneous fission sources ^{252}Cf ($T_{1/2}$ = 2.65 y) and ^{248}Cm ($T_{1/2} = 3.48 \times 10^5$ y) have spontaneous fission branches of 3.1 % and 8.4 %, respectively. However, because of the shorter half-life of ^{252}Cf, the specific activity is about 15,000 times higher than that of ^{248}Cm.

The interest in spontaneous fission sources as compared to neutron-induced fission of uranium is primarily due to the different mass distribution of the fission fragments. Because the masses of ^{252}Cf and ^{248}Cm are larger than the mass of ^{235}U, the mass distributions of their fission fragments is shifted towards higher masses. Figure 13.1 shows a comparison of the fragment mass distributions of spontaneous fission of ^{252}Cf and thermal neutron-induced fission of ^{235}U. The fractional yield for the spontaneous

M. Thoennessen, *The Discovery of Isotopes*,
DOI 10.1007/978-3-319-31763-2_13

Fig. 13.1 Fractional fission fragment yields for thermal neutron fission of ^{235}U and spontaneous fission of ^{252}Cf [3]. The *light blue* area indicates the mass range of isotopes discovered in spontaneous fission of ^{252}Cf and ^{248}Cm

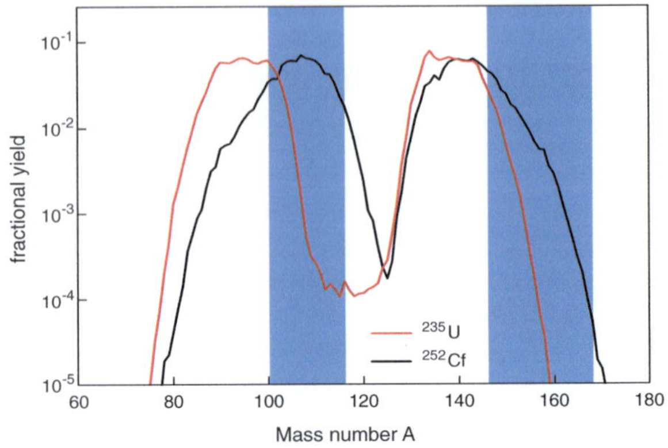

fission of ^{252}Cf in the mass regions around 110 and 160 is more than two orders of magnitude larger than the fission yield from thermal fission of ^{235}U.

^{252}Cf is also more neutron-rich than ^{235}U so that the mass distribution of the fission fragments for a given element peaks at more neutron-rich isotopes. Thus there was an opportunity to discover new isotopes in the spontaneous fission. The light blue shaded area in Fig. 13.1 represents the mass range of isotopes discovered in fission of ^{252}Cf and ^{248}Cm.

The shift of the populated nuclides towards more neutron-rich isotopes also posed a problem in the identification. The half-lives of these isotopes are short (typically less than 10 s) so that fast chemical separation or physical identification techniques were essential.

13.2 Isotopes Discovered in Fission of ^{252}Cf and ^{248}Cm

Isotopes first identified in spontaneous fission are listed in Table 13.1. Similar to thermal neutron fission of uranium it is very difficult to identify the fission fragments. By the late 1960s solid state detectors and the associated electronics had advanced sufficiently to allow coincidence experiments between the detected fission fragments and their emitted radiation.

At Berkeley in 1969, Watson et al. placed a ^{252}Cf source between two phosphorus diffused silicon fragment detectors which were used for mass identification. A lithium drifted silicon detector recorded K X-rays for element determination. In addition, conversion electrons were measured with an electron spectrometer. Several γ-ray transitions with a resolution of ± 1 keV were assigned to unique elements (exact) and mass (± 1 amu) [4]. In addition, half-lives on the order of a few nanoseconds were measured by varying the source position relative to the electron detector. The identification of the new isotopes ^{101}Zr, ^{102}Zr, and ^{110}Ru was later confirmed by independent measurements. At the present time no other experiment has been performed to verify the results for ^{116}Rh.

Table 13.1 Isotopes discovered in spontaneous fission: The submission date, first author and reference of the publication, as well as source and the source strength are listed

Date	First Author	Ref.	Source	Strength	Isotope
11/11/1969	Watson	[4]	^{252}Cf	10 μg	^{101}Zr, ^{102}Zr, ^{110}Ru, ^{116}Rh
05/12/1970	Cheifetz	[5]	^{252}Cf	10^5/min	^{100}Zr, ^{112}Ru, ^{116}Pd
06/05/1970	John	[6]	^{252}Cf	46 μCi	^{108}Tc, ^{111}Ru
07/21/1970	Wilhelmy	[7]	^{252}Cf	10^5/min	^{146}Ba, ^{150}Ce, ^{154}Nd, ^{158}Sm
06/02/1971	Hopkins	[8]	^{252}Cf	0.08 μg	^{103}Nb, ^{104}Nb, ^{112}Rh, ^{113}Rh, ^{149}Ce
07/06/1971	Cheifetz	[9]	^{252}Cf	10^5/min	^{120}Cd
11/01/1971	Hopkins	[10]	^{252}Cf	0.08 μg	^{109}Rh
10/13/1981	Gehrke	[11]	^{252}Cf	600 μg	^{163}Gd
11/17/1981	Greenwood	[12]	^{252}Cf	600 μg	^{155}Pm
03/01/1982	Gehrke	[13]	^{252}Cf	600 μg	^{168}Dy
08/09/1982	Greenwood	[14]	^{252}Cf	600 μg	^{165}Tb
02/17/1987	Greenwood	[15]	^{252}Cf	400 μg	^{153}Pr, ^{153}Nd, ^{156}Nd, ^{157}Pm, ^{158}Pm, ^{162}Eu
07/20/1987	Greenwood	[16]	^{252}Cf	400 μg	^{164}Gd
02/05/1990	Hotchkis	[17]	^{248}Cm	5 mg	^{104}Zr
02/10/1997	Hoellinger	[18]	^{248}Cm	5 mg	^{151}Ce
07/01/2011	Van Schelt	[19]	^{252}Cf	150 μCi	^{155}Pr, ^{157}Nd

The Berkeley group refined the method during the next year and correctly identified another eight new isotopes in three separate publications by Cheifetz et al. [5, 9] and Wilhelmy et al. [7]. The main improvement was the direct electrodeposition of the ^{252}Cf onto one of the fission fragment detectors (F1) as shown in Fig. 13.2. While γ- or X-rays from the fragment recorded in F1 were emitted at rest, γ- or X-rays from the fragment moving towards the other detector (F2) were emitted in-flight and thus Doppler shifted.

In a similar set-up, John, Guy, and Wesolowski from the Lawrence Radiation Laboratory at Livermore measured in addition to the γ-ray energy also the time between the fission fragment implantation and the γ-ray signal. They were sensitive to half-lives between 3 ns and 2 μs. The elements were identified by matching γ-ray transitions with previously observed transitions by Watson et al. [4]. John, Guy, and Wesolowski, also changed previous mass assignments of ^{107}Tc and ^{111}Ru by Watson et al. to ^{108}Tc and ^{111}Ru, respectively [6].

The same technique with improved resolutions was used in 1971 to identify six additional neutron-rich fragments in the spontaneous fission of ^{252}Cf. Hopkins et al. from the Center of Nuclear Studies at the University of Texas at Austin discovered ^{103}Nb, ^{104}Nb, ^{112}Rh, ^{113}Rh, and ^{149}Ce [8] and ^{109}Rh [10].

Ten years later, the group of Gehrke and Greenwood from the Idaho National Engineering Laboratory (INEL) in Idaho Falls developed a fast radiochemical separation technique to identify four more ^{252}Cf fission fragments. The Elemental Separation On-Line (ESOL) facility consisted of a helium gas-jet transport system which was

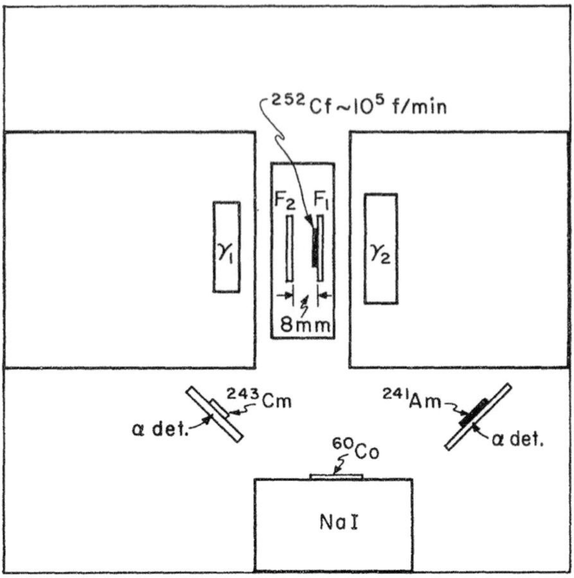

Fig. 13.2 Schematic representation of the experiment set-up of Cheifetz et al. measuring γ- and X-rays in coincidence with both fission fragments from a ^{252}Cf source [5] (Reprinted figure with permission from E. Cheifetz et al., *Experimental Information Concerning Deformation of Neutron Rich Nuclei in the A* $\sim$ 100 *Region*, Phys. Rev. Lett. 25 (1970) 38. Copyright 1970 by the American Physical Society.)

coupled to a microprocessor-controlled high-performance liquid chromatography (HPLC) system. They discovered relatively long-lived isotopes at the upper tail of the mass distribution of the heavy fragments: ^{163}Gd (68 $\pm$ 3 s, [11]) ^{155}Pm (48 $\pm$ 4 s, [12]) ^{168}Dy (8.5 $\pm$ 5 min, [13]), and ^{165}Tb (2.11 $\pm$ 0.10 min, [14]).

In 1987, Greenwood et al. installed an Isotope Separation On-Line (ISOL) system at the INEL which utilizes a mass separator instead of chemical separation. While the ESOL system was limited to half-lives larger than about 30 s, the new ISOL set-up could measure half-lives of a few seconds. Seven new isotopes from the heavy mass peak of ^{252}Cf between ^{153}Pr and ^{164}Gd were subsequently discovered [15, 16].

The increasing availability of large high-resolution Ge-detector arrays in the 1990 s offered a new opportunity to identify fission fragments by measuring their prompt radiation. Hotchkis et al. placed a ^{248}Cm spontaneous fission source at the center of the Argonne-Notre Dame γ-ray facility consisting of ten bismuth-germanate (BGO) Compton-suppressed Ge detectors. The direct correlation between the mass of the heavy and light fragments shown in Fig. 13.3 allowed the identification of γ-ray transitions in ^{104}Zr by gating on known transitions in the complementary ^{142}Ba fragment [17].

Similarly, Hoellinger et al. discovered ^{151}Ce in 1997 with the EUROGM II array at the Centre de Recherches Nucleáires in Strasbourg. The array consisted of 52 escape-suppressed Ge detectors. They also used a ^{248}Cm source and measured seven transitions of a rotational band in ^{151}Ce [18].

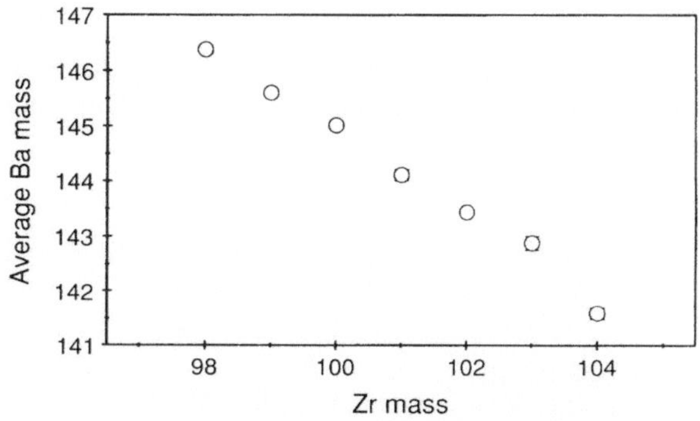

Fig. 13.3 Mass relationship between the light and heavy fragments from ^{248}Cm spontaneous fission. The average mass of complementary barium fragments are plotted versus the corresponding zirconium isotopes detected in coincidence [17] (Reprinted figure with permission from M.A.C. Hotchkis et al., *New neutron-rich nuclei* ^{103,104}*Zr and the A* $\sim$ *100 region of deformation*, Phys. Rev. Lett. 64 (1990) 3123. Copyright 1990 by the American Physical Society.)

Fig. 13.4 Schematic layout of the Canadian Penning Trap for high-precision mass measurements of ^{252}Cf fission fragments [19] (Reprinted figure with permission from J. Van Schelt et al., *Mass measurements near the r-process path using the Canadian Penning Trap mass spectrometer*, Phys. Rev. C 85 (2012) 045805. Copyright 2012 by the American Physical Society.)

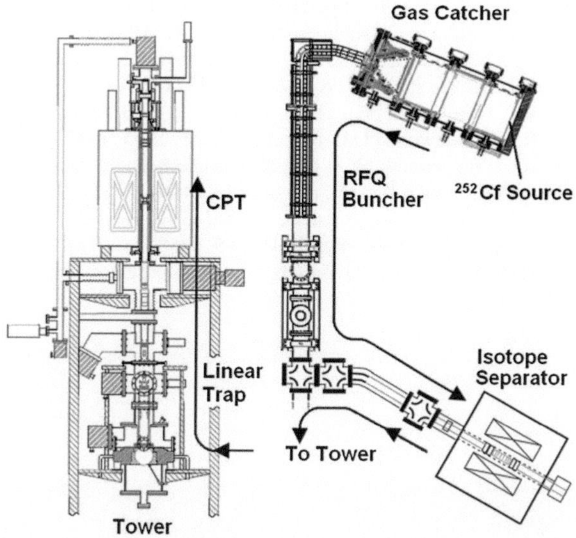

The two most recent spontaneous fission fragments discovered were measured with the Canadian Penning Trap (CPT) located at Argonne National Laboratory. Van Schelt et al. placed a 150 μCi ^{252}Cf source inside a helium gas catcher. The fission fragments were extracted through a nozzle into a radio-frequency quadrupole (RFQ) cooler and buncher before they were injected into an isotope separator Penning trap. The schematics of the set-up is shown in Fig. 13.4. The mass of the newly identified fragments ^{155}Pr and ^{157}Nd were then measured in the CPT [19].

It should be mentioned that in 1992, Knitter, Hambsch, and Budtz-Jorgensen had reported the discovery of 59 new neutron-rich isotopes in the spontaneous fission of ^{252}Cf [20]. However, although they calculated the isotope yields by measuring fission-fragment mass- and nuclear charge distributions as a function of total kinetic energy, they did not identify individual fragments nor did they measure any properties of these fragments.

References

1. G.N. Flerov, K.A. Petrzhak, Phys. Rev. **58**, 89 (1940)
2. G.T. Seaborg, J.W. Gofman, R.W. Stoughton, Phys. Rev. **71**, 378 (1947)
3. A.C. Wahl, At. Data Nucl. Data Tables **39**, 1 (1988)
4. R.L. Watson et al., Nucl. Phys. A **141**, 449 (1970)
5. E. Cheifetz et al., Phys. Rev. Lett. **25**, 38 (1970)
6. W. John, F.W. Guy, J.J. Wesolowski, Phys. Rev. C **2**, 1451 (1970)
7. J.B. Wilhelmy et al., Phys. Rev. Lett. **25**, 1122 (1970)
8. F.F. Hopkins et al., Phys. Rev. C **4**, 1927 (1971)
9. E. Cheifetz et al., Phys. Rev. C **4**, 1913 (1971)
10. F.F. Hopkins et al., Phys. Rev. C **5**, 1015 (1972)
11. R.J. Gehrke et al., Radiochim. Acta **31**, 1 (1982)
12. R.C. Greenwood et al., Radiochim. Acta **30**, 57 (1982)
13. R.J. Gehrke et al., Z. Phys. A **306**, 363 (1982)
14. R.C. Greenwood et al., Phys. Rev. C **27**, 1266 (1983)
15. R.C. Greenwood et al., Phys. Rev. C **35**, 1965 (1987)
16. R.C. Greenwood, M.A. Lee, R.A. Anderl, Radiochim. Acta **43**, 129 (1988)
17. M.A.C. Hotchkis et al., Phys. Rev. Lett. **64**, 3123 (1990)
18. F. Hoellinger et al., Phys. Rev. C **56**, 1296 (1997)
19. J. Van Schelt et al., Phys. Rev. C **85**, 045805 (2012)
20. H.-H. Knitter, F.-J. Hambsch, C. Budtz-Jorgensen, Nucl. Phys. A **536**, 221 (1992)

Chapter 14
Heavy Ion Transfer and Deep Inelastic Reactions

14.1 Dissipative Reactions

Soon after the first heavy ions were accelerated in 1940 they were used to produce new isotopes with fusion-evaporation reactions (see Chap. 11). This reaction mechanism dominated the search for new isotopes for the next thirty years. Most of the proton-rich isotopes during this time period were discovered with this method.

Nuclides on the neutron-rich side of the valley of stability were not accessible with fusion reactions. While isotopes close to stability could be reached with light-charged particle transfer reactions, more neutron-rich nuclides could only be produced in fission which was limited to medium mass elements. It was soon realized that transfer reactions did not have to be limited to light-charged-particles and the missing mass method first used to discover an isotope in 1953 by Harvey [1] (see Sect. 9.4) was also used to measure new isotopes in multi-nucleon transfer reactions with heavy-ion beams.

The particle-identifier system based on ΔE-E measurements with semiconductor detectors developed by Poskanzer et al. in 1966 [2] enabled the exploration of light very neutron-rich systems. In the first experiment high-energy protons irradiated uranium targets producing the neutron-rich isotopes by target fragmentation of the uranium nuclides (see Sect. 10.3). Soon after it was realized that deep inelastic reactions with heavy ions could also be used to populate neutron-rich nuclides. In these dissipative collisions, two heavy ions interact with each other exchanging multiple nucleons. When heavy neutron-rich targets are used on average more neutrons than protons are transferred to the beam.

However, the limited resolution of the particle-identifier system prohibited its application to heavier elements. In 1969, Artukh et al. solved the problem by placing a magnetic spectrometer in front of the particle telescopes [3]. The determination of the magnetic rigidity added another parameter which significantly improved the isotopic resolution.

In the early 1980s deep inelastic reactions induced at significantly higher beam energies were used to produce heavy neutron-rich nuclides. Isotopes which were

M. Thoennessen, *The Discovery of Isotopes*,
DOI 10.1007/978-3-319-31763-2_14

difficult to produce with other reaction mechanisms were identified with the isotope separation online (ISOL) method which had been developed in 1951 by Kofoed-Hansen and Nielsen [4] (see Sect. 6.2).

14.2 Target-Like Fragments

The missing mass method allowed mass measurements of isotopes without actually detecting them directly. In a two-body reaction the mass of the target-like fragment can be deduced from the kinematic properties of the projectile and ejectile. Missing mass experiments require thin targets and high resolution mass spectrometers and were used in only a few cases to discover new isotopes with heavy ions. In a different technique, new target-like fragments have been identified in transfer reactions without measuring the kinematics of the ejectile, but rather by chemical separations of the irradiated targets. Table 14.1 lists target-like new isotopes produced in multi-nucleon transfer reactions.

Already in 1956, Allison, Murphy, and Norbeck determined the mass of ^{13}B by measuring protons at 90° with a CsI(Tl) scintillating crystal at the Van de Graaff accelerator of the Enrico Fermi Institute for Nuclear Studies at the University of Chicago [5]. Three target-like new fragments were identified by reconstructing the missing mass from energy measurements of the projectile-like fragments in magnetic separators. In 1977, Bhatia et al. used the Q3D spectrometer at the MP tandem of the Max Planck Institut für Kernphysik at Heidelberg, Germany, in the discovery of ^{68}Ni [6]. The excitation energy spectrum is shown in Fig. 14.1. Six years later, ^{69}Ni was identified by Dessagne et al. with the double-focusing magnetic spectrometer (BACCHUS) at the MP tandem of the Institut de Physique Nucléaire d'Orsay, France [7].

Table 14.1 Target-like isotopes discovered in multi-nucleon transfer reactions: The submission date, first author and reference of the publication as well as the reaction, beam energy (MeV), and transferred nucleons (# trans. p/n) are listed

Date	First author	Ref.	Reaction	Energy	# trans. p/n	Isotope
03/28/1956	Allison	[5]	^{7}Li(^{7}Li,p)	1.61	+4n + 2p	^{13}B
01/20/1977	Bhatia	[6]	^{70}Zn(^{18}O,^{20}Ne)	81–84	−2p	^{68}Ni
12/01/1983	Dessagne	[7]	^{70}Zn(^{14}C,^{15}O)	72	+1n − 2p	^{69}Ni
10/04/1992	Zhao	[8]	^{198}Pt(^{18}O,^{17}F)	140	+2n − 1p	^{199}Ir
07/01/1993	Zhang	[9]	^{nat}Pb(^{12}C,X)	30 × A[b]	+2n − 2p	^{208}Hg[a]
04/03/1995	Yuan	[10]	^{nat}U(^{18}O,X)	50 × A[b]	+2n − 1p	^{239}Pa[a]
09/05/1997	Zhang	[11]	^{nat}Pb(^{18}O,X)	600	+3n − 2p	^{209}Hg[a]
10/22/1997	Yuan	[12]	^{nat}W(^{18}O,X)	60 × A[b]	+2n − 2p	^{186}Hf[a]
08/12/1998	He	[13]	^{nat}U(^{18}O,X)	60 × A[b]	+2n − 2p	^{238}Th[a]

[a] Assuming transfer from the most abundant target isotope
[b] A denotes the atomic weight of the beam particles

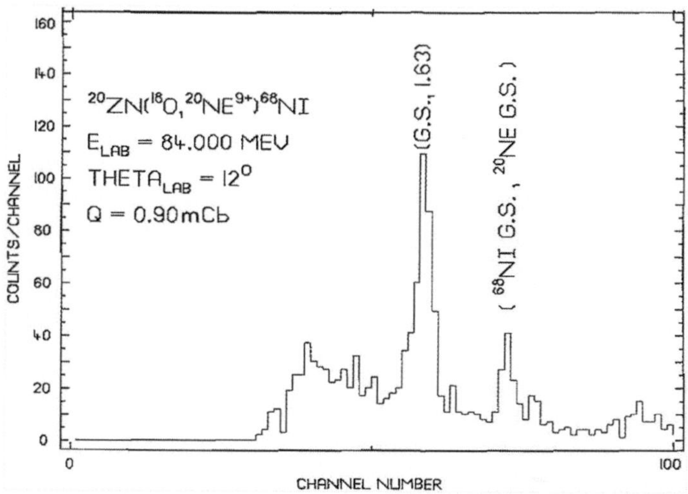

Fig. 14.1 Excitation energy spectrum of ^{68}Ni. The ground-state transition appears twice, once if the ^{20}Ne ejectile remains in its ground-state and if it is in its first excited state [6] (With kind permission from Springer Science and Business Media: T.S. Bhatia et al., *Masses of* ^{62}*Fe and the new isotopes* ^{68}*Ni from* (^{18}*O*, ^{20}*Ne*) *reactions*, Z. Phys. A 281 (1977) 65.)

Zhao et al. discovered ^{199}Ir with the high-resolution QMG/2 magnetic spectrometer at the Nuclear Structure Facility of Daresbury Laboratory in 1992 [8].

Between 1993 and 1998, five additional isotopes of heavy elements were discovered in multi-nucleon transfer reactions at the Heavy Ion Research Facility of the Institute of Modern Physics in Lanzhou, China. Heavy targets were irradiated with high-energy carbon or oxygen beams and the isotopes were identified by β-decay and γ-ray spectra following chemical separation of the targets [9–13].

Recently, Watanabe et al. proposed multi-nucleon transfer reactions as an efficient tool to produce neutron-rich isotopes far from stability around the N = 126 shell closure. They extracted production cross-sections for these isotopes in the reaction ^{136}Xe + ^{198}Pt at 8 MeV/nucleon at GANIL using the VAMOS spectrometer and the EXOGAM array. Among several other new isotopes, Watanabe et al. present evidence of the production of the new N = 126 nuclides ^{200}W and ^{201}Re. However, because they only deduced their existence from measured energies and angles of the projectile-like fragments they did not claim the discovery of these isotopes [14].

14.3 Beam-Like Fragments

In addition to the pioneering work by Artukh et al. at Dubna, researchers at Orsay, Berkeley, and Legnaro contributed to the discovery of isotopes using deep inelastic collisions and direct particle identification between 1969 and 2009. The 47 isotopes detected with this method using magnetic spectrometers are listed in Table 14.2.

Table 14.2 Neutron-rich isotopes discovered with magnetic spectrometer and direct particle identification in transfer reactions and deep inelastic collisions: The submission date, first author and reference of the publication as well as the beam, beam energy (in MeV), target, and angle setting of the spectrometer/detectors are listed

Date	First author	Ref.	Beam	Ener.	Target	Angle	Isotope
08/19/1969	Artukh	[3]	^{18}O	122	^{232}Th	60°	^{18}C, ^{20}N, ^{22}O
12/10/1969	Artukh	[15]	^{22}Ne	174	^{232}Th	40°	^{23}F, ^{24}F, ^{25}Ne, ^{26}Ne
04/14/1970	Artukh	[16]	^{22}Ne	174	^{232}Th	40°	^{21}N, ^{23}O, ^{24}O, ^{25}F
08/19/1971	Artukh	[17]	^{40}Ar	290	^{232}Th	40°	^{29}Mg, ^{30}Mg, ^{31}Al, ^{32}Al, ^{33}Al, ^{33}Si, ^{34}Si, ^{35}Si, ^{36}Si, ^{35}P, ^{36}P, ^{37}P, ^{38}P, ^{39}S, ^{40}S, ^{41}Cl, ^{42}Cl
10/03/1978	Auger	[18]	^{40}Ar	263	^{238}U	60°	^{37}Si, ^{40}P, ^{41}S, ^{42}S
02/26/1980	Guerreau	[19]	^{40}Ar	340	^{238}U	26°	^{54}Ti, ^{56}V, ^{58}Cr, ^{59}Cr, ^{61}Mn, ^{63}Fe, ^{64}Fe
06/02/1980	Breuer	[20]	^{56}Fe	465	^{238}U	40°	^{52}Sc, ^{53}Sc, ^{56}Ti, ^{57}V, ^{58}V, ^{60}Cr, ^{55}Ti
09/25/2009	Söderström	[21]	^{82}Se	460	^{170}Er	52°	^{170}Dy

The question that arose with the event-by-event identification technique of individually detected nuclides was how many events were necessary to claim discovery. There is no general answer to this question, as it depends on several different factors, for example, the resolution of the detectors, the quality of the charge and mass separation, and the number of background events. In deep inelastic reactions, typically about ten events were deemed sufficient for the discovery of a new isotope.

In four separate experiments between 1969 and 1971, Artukh et al. induced deep inelastic collisions with the U300 heavy-ion cyclotron at Dubna and identified reaction products with a ΔE-E semiconductor telescope in the focal plane of a magnetic spectrometer [3, 15–17]. For most of the 28 newly discovered isotopes, Artukh et al. detected more than 10 events. The smallest number observed was for ^{30}Mg with about 6 events [17]. As an example, Fig. 14.2 shows the particle identification plot for the discovery of ^{18}C, ^{20}N, and ^{22}O [3].

About eight years later Auger et al. extended the method to heavier elements using the ALICE facility at Orsay. Instead of a spectrometer Auger et al. measured the time-of-flight of the fragments and used two ΔE measurements for the identification [18]. The observation of about three events each for ^{38}Si and ^{41}P was not considered sufficient for claiming discovery: "There are some hints of the existence of ^{38}Si and ^{41}P but their presence is not completely conclusive (less than 10 events)" [18].

In 1980 Guerreau et al. increased the beam energy and moved the detector telescope to a smaller angle. They reported the discovery of another seven isotopes of even heavier elements (up to iron) [19]. Less than ten events were observed for ^{55}Ti, ^{62}Mn, and ^{65}Fe which were again labeled as "only hints" needing confirmation in further experiments.

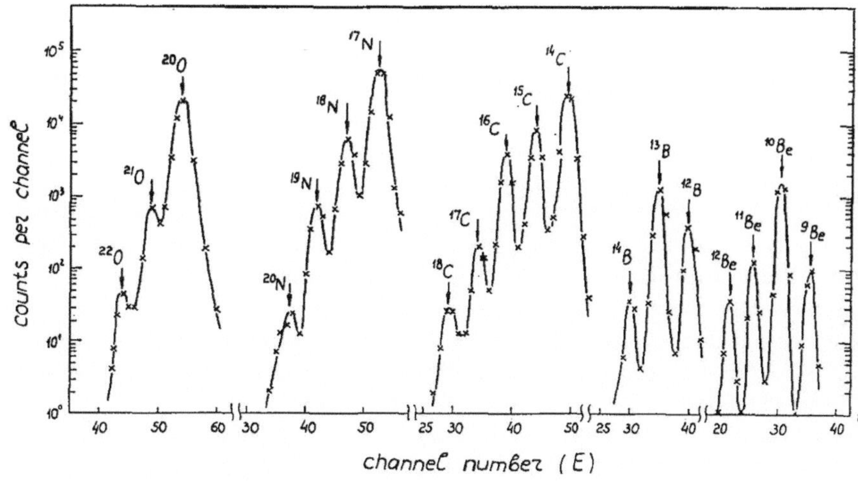

Fig. 14.2 One-dimensional particle identification spectrum showing the discovery of ^{18}C, ^{20}N, and ^{22}O at Dubna [3] (Reprinted from A.G. Artukh et al., *New isotopes ^{22}O, ^{20}N, and ^{18}C transfer reactions with heavy ions*, Nucl. Phys. A 137 (1969) 348. Copyright 1969, with permission from Elsevier.)

Only a few months later Breuer et al. reported their results of a similar experiment performed with the SuperHILAC accelerator at Berkeley [20]. They claimed the discovery of thirteen (six tentative) new isotopes, however, six of them had already been reported by Guerreau et al. in the meantime. Although, Breuer et al. quote the paper by Guerreau et al., they still claimed the discovery of these isotopes quoting an earlier internal report [22].

Recently, Söderström et al. have revitalized the method to separate and identify neutron-rich isotopes with a magnetic spectrometer following deep inelastic collisions in 2009. They irradiated a ^{170}Er target with a 460 MeV ^{82}Se beam from the Legnaro Tandem XTU-ALPI accelerator complex. Beam-like fragments were identified with the magnetic spectrometer PRISMA and the population of target-like fragments was deduced from γ-ray spectra measured with the CLARA γ-ray detector array. A 163 keV γ-ray was tentatively assigned to ^{170}Dy because it was detected in coincidence with the 777 keV γ-ray of the binary partner ^{82}Kr [21].

14.4 ISOL

In 1951 Kofoed-Hansen and Nielsen developed the isotope separation on-line (ISOL) method for separating and identifying fragments from uranium fission [4]. Later on this method was successfully applied at many dedicated facilities to discover new isotopes in spallation reactions (see Sect. 10.4), neutron-induced and spontaneous fission (see Sects. 6.5 and 13.2), and fusion-evaporation reactions (see Sects. 11.2 and 11.4).

Between 1981 and 1993, 21 neutron-rich isotopes were discovered at GSI and Berkeley by applying the ISOL method to separate and identify reaction products from deep inelastic collisions as listed in Table 14.3.

Eighteen of the new isotopes were discovered at GSI using heavy-ion beams from the UNILAC accelerator. Fragments were produced in tungsten targets or sandwich-type targets of tungsten-tantalum layers and extracted from a thermal or a FEBIAD (Forced Electron Beam Induced Arc Discharge) ion source. The GSI online mass separator was used to separate the ions and deposit them in front of X-ray, γ-ray and β-ray detectors measuring the half-life of the isotopes. All isotopes listed in Table 14.3 with the exception of the isotopes discovered by Chasteler et al. were discovered at GSI.

When Runte et al. reported the discovery of ^{62}Mn and the three copper isotopes ^{71}Cu, ^{72}Cu and ^{73}Cu, they also claimed the discovery of ^{63}Fe [25]. They did acknowledge the previous observation of Guerreau et al. [19] using particle-identification techniques. However, at the time, the identification of a new isotope using this technique was not yet accepted as sufficient evidence to claim discovery.

Three isotopes identified with the ISOL method and deep inelastic reactions were discovered at Berkeley. Chasteler et al. produced ^{171}Ho and ^{174}Er [29] and ^{169}Dy [31] with the SuperHILAC and the OASIS mass separation facility.

Table 14.3 Neutron-rich isotopes discovered with ISOL and produced in deep inelastic collisions: The submission date, first author and reference of the publication as well as the beam, beam energy (MeV/nucleon), and target

Date	First author	Ref.	Beam	Energy	Target	Isotope
10/23/1981	Kirchner	[23]	^{136}Xe	9	Ta/W	^{181}Lu, ^{182}Lu, ^{179}Yb
11/08/1982	Rykaczewski	[24]	^{136}Xe	11.7	Ta/W	^{183}Lu
11/29/1982	Runte	[25]	^{76}Ge	9	W	^{62}Mn, ^{71}Cu, ^{72}Cu, ^{73}Cu
09/13/1985	Bosch	[26]	^{76}Ge	11.4	W	^{63}Mn, ^{66}Co, ^{67}Co
12/05/1985	Gippert	[27]	^{238}U	11.4	Ta/W	^{232}Ac, ^{234}Ac
05/22/1987	Runte	[28]	^{186}W	15	Ta/W	^{180}Yb
12/14/1988	Chasteler	[29]	^{176}Yb	8.5	W	^{174}Er, ^{171}Ho
01/13/1989	Rykaczewski	[30]	^{186}W	15[a]	Ta/W	^{184}Lu, ^{177}Tm
06/25/1990	Chasteler	[31]	^{170}Er	8.5	W	^{169}Dy
07/31/1990	Becker	[32]	^{136}Xe	11.6	W[b]	^{172}Ho
06/28/1993	Wennemann	[33]	^{208}Pb	11.4	W	^{205}Au

[a] Also ^{136}Xe and ^{186}W at 11.7 MeV $\times$ A and ^{238}U at 11.4 MeV $\times$ A
[b] Enriched ^{186}W

References

1. J.A. Harvey, Can. J. Phys. **31**, 278 (1953)
2. A.M. Poskanzer et al., Phys. Rev. Lett. **17**, 1271 (1966)
3. A.G. Artukh et al., Nucl. Phys. A **137**, 348 (1969)
4. O. Kofoed-Hansen, K.O. Nielsen, Phys. Rev. **82**, 96 (1951)
5. S.K. Allison, P.G. Murphy, E. Norbeck, Phys. Rev. **102**, 1182 (1956)
6. T.S. Bhatia et al., Z. Phys. A **281**, 65 (1977)
7. P.H. Dessagne et al., Nucl. Phys. A **426**, 399 (1984)
8. K. Zhao et al., Chin. Phys. Lett. **10**, 265 (1993)
9. L. Zhang et al., Phys. Rev. C **49**, R592 (1994)
10. S. Yuan et al., Z. Phys. A **352**, 235 (1995)
11. L. Zhang et al., Phys. Rev. C **58**, 156 (1998)
12. S. Yuan et al., Phys. Rev. C **57**, 1506 (1998)
13. J. He et al., Phys. Rev. C **59**, 520 (1999)
14. Y.X. Watanabe et al., Phys. Rev. Lett. **115**, 172503 (2015)
15. A.G. Artukh et al., Phys. Lett. B **31**, 129 (1970)
16. A.G. Artukh et al., Phys. Lett. B **32**, 43 (1970)
17. A.G. Artukh et al., Nucl. Phys. A **176**, 284 (1971)
18. P. Auger et al., Z. Phys. A **289**, 255 (1979)
19. D. Guerreau et al., Z. Phys. A **295**, 105 (1980)
20. H. Breuer et al., Phys. Rev. C **22**, 2454 (1980)
21. P.-A. Söderström et al., Phys. Rev. C **81**, 034310 (2010)
22. V.E. Viola Jr., University of Maryland Report No. ORO-5172-13 (1979)
23. R. Kirchner et al., Nucl. Phys. A **378**, 549 (1982)
24. K. Rykaczewski et al., Z. Phys. A **309**, 273 (1983)
25. E. Runte et al., Nucl. Phys. A **399**, 163 (1983)
26. U. Bosch et al., Phys. Lett. B **164**, 22 (1985)
27. K.-L. Gippert et al., Nucl. Phys. A **453**, 1 (1986)
28. E. Runte et al., Z. Phys. A **328**, 119 (1987)
29. R.M. Chasteler et al., Z. Phys. A **332**, 239 (1989)
30. K. Rykaczewski et al., Nucl. Phys. A **499**, 529 (1989)
31. R.M. Chasteler et al., Phys. Rev. C **42**, R1171 (1990)
32. K. Becker et al., Nucl. Phys. A **522**, 557 (1991)
33. Ch. Wennemann et al., Z. Phys. A **347**, 185 (1994)

Chapter 15
Projectile Fragmentation and Fission

15.1 A New Paradigm

The development of the projectile fragmentation method to produce new radioactive isotopes initiated a new era not only in the discovery of isotopes but also for the production and utilization of beams of unstable isotopes. The capability to identify individual isotopes by the ΔE-E [1] (and later adding time-of-flight information [2]) method was first applied in the target fragmentation of uranium targets with high energy protons (see Sect. 10.3). The method was subsequently applied to deep-inelastic collisions where the reaction products were separated with a magnet located at (see Sect. 14.3) 40° [3].

The availability of high-energy heavy-ion beams opened up the opportunity to fragment projectiles and detect them at zero degrees. The fragments were separated from the beam with a zero degree spectrometer and identified with the ΔE-E method [4]. An advantage of high-beam energies for an unambiguous identification is that the electrons are completely stripped from the fragments. If the spectrometer is an integral part of the beamline system, the fragments can be utilized as a secondary beam for subsequent reaction studies. This method requires a fairly powerful accelerator and a large spectrometer. Nevertheless, because of the pioneering work at Berkeley several dedicated radioactive beam facilities based on projectile fragmentation were built during the last 30 years.

As was already discussed in the previous chapter, it is not possible to set a fixed number of events necessary to claim the discovery of an isotope. It depends case by case on the quality of the separation and selectivity as well as the number of background events. No common standards or guidelines have been established so it depends on the confidence of the researchers in their data. New isotopes observed by only a few counts (in some cases a single count) are reported as evidence, tentative, or displayed in parentheses. In most cases these isotopes are still not close to the drip-line, so their existence is not questioned. Typically these initial observations are tacitly accepted in later papers so in general the first measurements are credited with the discovery.

M. Thoennessen, *The Discovery of Isotopes*,
DOI 10.1007/978-3-319-31763-2_15

Almost 500 new isotopes have been discovered in projectile fragmentation and projectile fission reactions extending the nuclear chart towards the neutron and the proton drip-line.

15.2 Light Neutron-Rich Isotopes

New bound neutron-rich isotopes from boron through iron were discovered in projectile fragmentation reactions with beams ranging from ^{40}Ar to ^{86}Kr. Table 15.1 lists the discovered isotopes together with the beam, the beam energy (in MeV/nucleon), and the order of magnitude of the beam intensity (in pps).

In four separate papers between 1978 and 1984, the discovery of 19 isotopes were reported from projectile fragmentation experiments with the Berkeley Bevalac. The fragments were analyzed with a zero-degree spectrometer and detected in several detector telescopes. About ten counts were considered sufficient for claiming discovery. Three events for ^{33}Mg were judged to be too small by Symons et al. [4], however, when Westfall et al. observed about 50 events in the subsequent experiment they did not consider it a new isotope anymore [5]. Also, Westfall et al. were apparently not aware of the discovery of ^{37}Si, ^{40}P, and ^{41,42}S in deep-inelastic reactions by Auger et al. earlier in the year [23]. Figure 15.1 shows the first particle identification plot from a projectile fragmentation reaction by Symons et al. [4]. At the time the Berkeley experiments were sensitive to cross-sections of about 100 μb.

Stevenson and Price detected about five counts each for ^{21}C, ^{23}N, and ^{25}O and interpreted this as evidence for possible particle stability of these isotopes: "In addition, at ^{21}C, ^{23}N, and ^{25}O there are bumps consisting of about five counts clearly outside the Gaussian envelopes of the adjacent lighter isotopes" [6]. However, of these three only ^{23}N is particle stable as later determined by Langevin et al. at GANIL [8].

In the following five years (1985–1990) 44 new isotopes were measured with the zero degree achromatic spectrometer LISE at GANIL. The observation of four events for ^{29}F [11] and ^{32}Ne [12] were deemed sufficient to claim discovery while the observation of ^{54}Sc and ^{66}Mn with two and three events, respectively, was reported as tentative [9]. Only three days after the discovery of ^{47}Ar [9], Benenson et al. submitted their mass measurement of this isotope [24].

The new isotopes reported in a 1989 experiment by Lewitowicz et al. were not populated in pure fragmentation reactions [13]. For example, in order to produce ^{48}S, it was not only necessary to remove four protons from the ^{48}Ca beam but also to add four neutrons picked-up from the target. The apparent observation of ^{51}Cl was later questioned and could have been masked by the presence of the hydrogen-like ion ^{48}Cl^{16+} [21].

Another interesting case is the discovery of ^{48}Ar. It most certainly was observed in 1988 by Guillemaud-Mueller et al., however, it was assumed to be known. The magnetic rigidity during the experiment was actually set too high and the two-dimensional particle identification plot presented in the paper does not show any evidence of ^{48}Ar [11]. The observation of ^{48}Ar can be extracted from a figure in an internal report [25], however, the data were never published. Thus, the discovery must be credited

Table 15.1 Light neutron-rich isotopes: The submission date, first author and reference of the publication as well as the beam, beam energy E (MeV/nucleon), and beam intensity I (pps) are listed

Date	First author	Ref.	Beam	E	I	Isotope
11/01/1978	Symons	[4]	^{40}Ar	205	10^6	^{28}Ne, ^{35}Al
10/15/1979	Westfall	[5]	^{48}Ca	213	10^7	^{22}N, ^{26}F,^{33}Mg, ^{34}Mg, ^{36}Al, ^{37}Al, ^{38}Si, ^{39}Si, ^{41}P, ^{42}P, ^{43}S, ^{44}S, ^{44}Cl, ^{45}Cl
03/05/1981	Stevenson	[6]	^{48}Ca	213	10^7	^{20}C, ^{27}F
08/02/1984	Musser	[7]	^{56}Fe	670	10^5	^{19}B
10/15/1984	Langevin	[8]	^{40}Ar	44	10^{11}	^{23}N,^{29}Ne, ^{30}Ne
06/25/1985	Guillemaud-Mueller	[9]	^{86}Kr	33	10^{10}	^{47}Ar, ^{57}Ti, ^{59}V, ^{60}V, ^{61}Cr, ^{62}Cr, ^{64}Mn, ^{65}Mn, ^{65}Fe, ^{66}Fe, ^{67}Fe, ^{68}Fe, ^{68}Co, ^{69}Co, ^{70}Co
04/01/1986	Pougheon	[10]	^{40}Ar	44	10^{11}	^{22}C
09/12/1988	Guillemaud-Mueller	[11]	^{48}Ca	55	10^{10}	^{29}F, ^{35}Mg, ^{36}Mg, ^{38}Al, ^{39}Al, ^{40}Si, ^{41}Si, ^{43}P, ^{44}P, ^{45}S, ^{46}S, ^{47}S, ^{46}Cl, ^{47}Cl, ^{48}Cl, ^{49}Cl, ^{49}Ar, ^{50}Ar, ^{51}Ar
10/31/1989	Guillemaud-Mueller	[12]	^{48}Ca	44	10^{10}	^{32}Ne
11/24/1989	Lewitowicz	[13]	^{48}Ca	44	10^{10}	^{42}Si, ^{45}P, ^{46}P, ^{48}S
10/09/1991	Weber	[14]	^{86}Kr	500	10^7	^{58}Ti, ^{61}V, ^{63}Cr, ^{64}Cr, ^{66}Mn, ^{69}Fe, ^{71}Co, ^{72}Co, ^{75}Ni
07/22/1996	Sakurai	[15]	^{50}Ti	80	10^{10}	^{31}Ne, ^{37}Mg
08/18/1998	Sakurai	[16]	^{40}Ar	94	10^{11}	^{31}F
05/28/2002	Notani	[17]	^{48}Ca	64	10^{11}	^{34}Ne, ^{37}Na,^{38}Mg, ^{40}Al, ^{41}Al, ^{43}Si
01/22/2004	Grévy	[18]	^{48}Ca	60	10^{12}	^{48}Ar
03/30/2007	Tarasov	[19]	^{48}Ca	142	10^{11}	^{44}Si
06/06/2007	Baumann	[20]	^{48}Ca	141	10^{11}	^{40}Mg, ^{42}Al, ^{43}Al
01/15/2009	Tarasov	[21]	^{76}Ge	132	10^{11}	^{52}Ar, ^{53}Ar,^{57}Ca, ^{58}Ca, ^{59}Sc, ^{60}Sc, ^{61}Sc, ^{62}Ti, ^{63}Ti, ^{65}V, ^{66}V, ^{68}Cr, ^{70}Mn, ^{50}Cl, ^{51}Cl, ^{55}K, ^{56}K
02/11/2013	Tarasov	[22]	^{82}Se	139	10^{11}	^{64}Ti,^{67}V, ^{69}Cr, ^{70}Cr, ^{72}Mn, ^{75}Fe

to the first experiment which specifically measured ^{48}Ar which was not until 2004 by Grévy et al. Again the reaction probably is better characterized as a nucleon exchange reaction (removing 2 protons, adding 2 neutrons from a ^{48}Ca beam) rather than a projectile fragmentation reaction. Grévy et al. separated the fragments with the upgraded LISE3 spectrometer and measured the half-life of ^{48}Ar by implanting fragments in a double-sided silicon strip detector and detecting correlated β-particles in plastic scintillators [18].

Nine new isotopes were discovered in 1991 at GSI with the heavy-ion synchrotron SIS and the FRS fragment separator [14]. The observation of ^{63}Cr, ^{72}Co, and ^{75}Ni

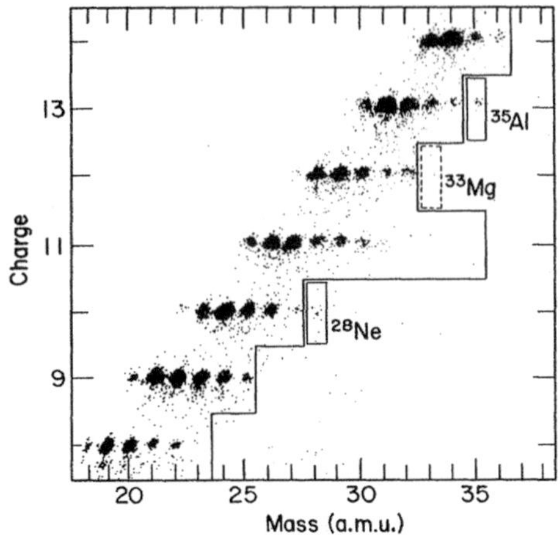

Fig. 15.1 Particle identification plot following the reaction of 205 MeV/nucleon ^{40}Ar on a carbon target . The new isotopes ^{28}Ne and ^{35}Al are marked [4] (Reprinted figure with permission from T. J. M. Symons et al., *Observation of new neutron-rich isotopes by fragmentation of 205-MeV/nucleon Ar ions*, Phys. Rev. Lett. 42 (1979) 40. Copyright 1979 by the American Physical Society.)

with three, three, and two events, respectively, was quoted as indicative evidence. This judgment was considered conservative because of an extremely small background rate. The experiment measured cross-sections as low as 10^{-6} mb, about a factor of 10,000 smaller than the first Berkeley measurements.

The next major advance in delineating the neutron drip-line towards heavier elements was achieved with the RIKEN Ring Cyclotron and the RIPS spectrometer between 1996 and 2002. Nine new isotopes from fluorine through magnesium were observed. Only a few events—^{34}Ne (2), ^{37}Na (3), ^{43}Si (4) [17] and ^{37}Mg (3) [15]—were considered sufficient for clear identification. In the 2002 paper, Notani et al. [17] did not explicitly mention the observation of ^{40}Al and ^{41}Al because its discovery was attributed to a previous publication of a conference proceeding by Sakurai et al. [26]. A year earlier Sakurai et al. had also mentioned the observation of a single count which "could possibly be associated with ^{40}Al" [15]. The RIKEN experiments lowered the observation limit by another factor of 10,000 to about 0.1 pb (10^{-10} mb).

The smallest cross-section observed so far was reached in 2013 with the detection of a single event of ^{75}Fe. The corresponding cross-section was calculated to be $3.4^{+5.5}_{-2.6}\times10^{-12}$ mb [22]. This extreme sensitivity was achieved with a two-step separation system. Figure 15.2 shows the set-up at the National Superconducting Cyclotron Laboratory (NSCL) at Michigan State University where the A1900 fragment separator was coupled to the S800 analysis system as a second separation stage. The probability that the single event of ^{43}Al observed by Baumann et al. in 2007 was caused by a possible contribution from the neighboring ^{42}Al was determined to be 2.4×10^{-3} [20]. Overall, 27 new isotopes were discovered between 2007 and 2013 at the NSCL.

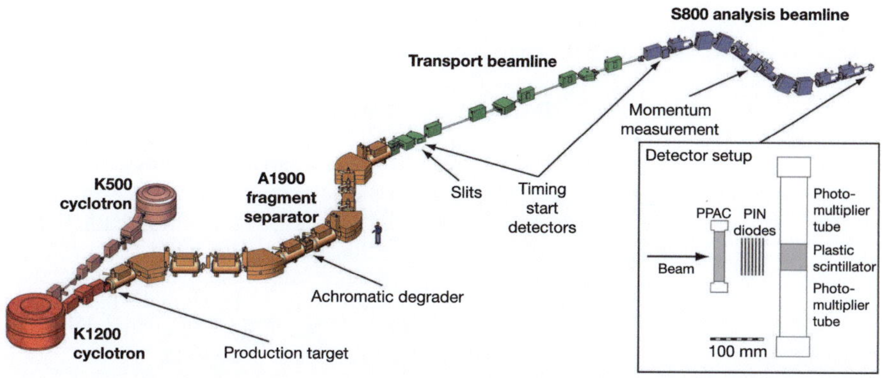

Fig. 15.2 Schematics of the 2-stage separation set-up at the NSCL at Michigan State University [20] (Reprinted by permission from Macmillan Publishers Ltd: T. Baumann et al., *Discovery of* ^{40}*Mg and* ^{42}*Al suggests neutron drip-line slant towards heavier isotopes*, Nature 449 (2007) 1022, copyright (2007).)

15.3 Projectile Fission

Another major breakthrough was the realization that at even higher energies the reaction products are so forward focused that the acceptance of the magnetic separators was sufficiently large to detect fission fragments from high-energy uranium beams. At these high energies, fission is induced by the relativistically contracted Coulomb field. Only two facilities (GSI and RIKEN) currently have the required beam energies to extend the chart of nuclides even further by identifying in-flight fission products. The 240 isotopes discovered with this process are listed in Table 15.2.

In three papers in 1994, 1995, and 1997, a total of 117 neutron-rich new isotopes were discovered at GSI. ^{238}U ions were accelerated in the UNILAC and the heavy-ion synchrotron SIS to an energy of 750 MeV/nucleon and in-flight fission fragments were separated with the FRS. The particle identification plot from the first experiment where 52 new isotopes were discovered is shown in Fig. 15.3. While the first paper listed isotopes with less than six events in brackets [27], the two later papers claimed discovery with only three (^{78}Ni [28]) and two (^{64}V and ^{72}Fe [29]) events. In the 1995 paper [28] Engelmann et al. did not consider ^{76}Ni a new nucleus quoting an internal report [36]. The results of this report were subsequently published in a conference proceedings [37], however, publication in a refereed journal occurred only three years later [38].

It is curious to note that the discovery of ^{155}Pr, ^{156}Pr, ^{157}Nd, and ^{158}Nd was reported by the same group in contribution to a conference proceedings in 1996 where Czajkowski et al. stated: “In this first experiment, 54 new isotopes were discovered, ranging from $^{86}_{32}$Ge to $^{158}_{60}$Nd” [39]. These nuclides were included as newly discovered in a figure of the chart of nuclides. However, in the original publication only 50 new isotopes were listed and there was no evidence for the observation of any

Table 15.2 Isotopes produced in projectile fission: The submission date, first author and reference of the publication as well as the ^{238}U beam energy E (MeV/nucleon) and beam intensity I (pps) are listed

Date	First author	Ref.	E	I	Isotope
03/03/1994	Bernas	[27]	750	10^5	^{86}Ge, ^{88}As, ^{89}As, ^{90}Se, ^{95}Kr, ^{96}Kr, ^{103}Y, ^{104}Y, ^{105}Y, ^{106}Zr, ^{107}Zr, ^{108}Nb, ^{109}Nb, ^{110}Nb, ^{111}Mo, ^{112}Mo, ^{113}Mo, ^{114}Tc, ^{115}Tc, ^{116}Ru, ^{117}Ru, ^{118}Ru, ^{118}Rh, ^{119}Rh, ^{120}Rh, ^{121}Rh, ^{121}Pd, ^{122}Pd, ^{123}Pd, ^{125}Ag, ^{126}Ag, ^{135}Sn, ^{136}Sn, ^{137}Sn, ^{137}Sb, ^{138}Sb, ^{139}Sb, ^{139}Te, ^{140}Te, ^{141}Te, ^{142}Te, ^{143}I, ^{144}I, ^{147}Xe, ^{150}Ba, ^{151}Ba, ^{151}La, ^{152}La, ^{153}La, ^{153}Ce, ^{154}Ce, ^{155}Ce
06/20/1995	Engelmann	[28]	750	10^7	^{73}Co, ^{74}Co, ^{75}Co, ^{80}Cu, ^{76}Ni, ^{77}Ni, ^{78}Ni
07/25/1997	Bernas	[29]	750	10^7	^{54}Ca, ^{55}Ca, ^{56}Ca, ^{56}Sc, ^{57}Sc, ^{58}Sc, ^{59}Ti, ^{60}Ti, ^{61}Ti, ^{62}V, ^{63}V, ^{64}V, ^{65}Cr, ^{66}Cr, ^{67}Cr, ^{67}Mn, ^{68}Mn, ^{69}Mn, ^{70}Fe, ^{71}Fe, ^{72}Fe, ^{82}Zn, ^{83}Zn, ^{85}Ga, ^{86}Ga, ^{87}Ge, ^{88}Ge, ^{89}Ge, ^{90}As, ^{91}As, ^{92}As, ^{92}Se, ^{93}Se, ^{94}Se, ^{95}Br, ^{96}Br, ^{97}Br, ^{97}Kr, ^{98}Kr, ^{99}Kr, ^{100}Kr, ^{103}Sr, ^{104}Sr, ^{105}Sr, ^{106}Y, ^{107}Y, ^{108}Zr, ^{109}Zr, ^{110}Zr, ^{111}Nb, ^{112}Nb, ^{113}Nb, ^{114}Mo, ^{116}Tc, ^{117}Tc, ^{119}Ru, ^{122}Rh, ^{124}Pd
06/04/2008	Ohnishi	[30]	345	10^7	^{125}Pd, ^{126}Pd
03/19/2010	Ohnishi	[31]	345	10^9	^{71}Mn, ^{73}Fe, ^{74}Fe, ^{76}Co, ^{79}Ni, ^{81}Cu, ^{82}Cu, ^{84}Zn, ^{85}Zn, ^{87}Ga, ^{90}Ge, ^{95}Se, ^{98}Br, ^{101}Kr, ^{103}Rb, ^{106}Sr, ^{107}Sr, ^{108}Y, ^{109}Y, ^{111}Zr, ^{112}Zr, ^{114}Nb, ^{115}Nb, ^{115}Mo, ^{116}Mo, ^{117}Mo, ^{118}Tc, ^{119}Tc, ^{120}Tc, ^{120}Ru, ^{121}Ru, ^{122}Ru, ^{123}Ru, ^{124}Ru, ^{123}Rh, ^{124}Rh, ^{125}Rh, ^{126}Rh, ^{127}Pd, ^{128}Pd, ^{133}Cd, ^{138}Sn, ^{140}Sb, ^{143}Te, ^{145}I, ^{148}Xe, ^{152}Ba
12/02/2011	Kurcewicz	[32]	1000	10^8	^{158}Nd, ^{159}Nd, ^{160}Nd, ^{161}Nd, ^{160}Pm, ^{161}Pm, ^{162}Pm, ^{163}Pm, ^{163}Sm, ^{164}Sm, ^{165}Sm, ^{166}Eu, ^{167}Eu, ^{168}Eu, ^{167}Gd, ^{168}Gd, ^{169}Gd, ^{170}Gd, ^{169}Tb, ^{170}Tb, ^{171}Tb, ^{172}Tb, ^{171}Dy, ^{172}Dy, ^{173}Dy, ^{174}Dy, ^{173}Ho, ^{174}Ho, ^{175}Ho, ^{176}Ho, ^{176}Er, ^{177}Er, ^{178}Er, ^{178}Tm, ^{179}Tm, ^{180}Tm, ^{181}Tm, ^{181}Yb, ^{182}Yb, ^{183}Yb, ^{184}Yb, ^{185}Yb, ^{185}Lu, ^{186}Lu, ^{187}Lu, ^{188}Lu, ^{190}Hf, ^{193}Ta, ^{194}Ta, ^{195}W, ^{196}W, ^{197}W, ^{198}Re, ^{199}Re, ^{202}Os, ^{203}Os, ^{205}Ir, ^{206}Pt, ^{207}Pt, ^{208}Pt
12/11/2012	Wang	[33]	345	10^{10}	^{131}Ag
03/28/2014	Xu	[34]	345	10^{10}	^{77}Co, ^{80}Ni
10/14/2014	Lorusso	[35]	345	10^{10}	^{118}Mo, ^{121}Tc, ^{127}Rh, ^{129}Pd, ^{132}Ag, ^{134}Cd, ^{136}In, ^{137}In, ^{139}Sn, ^{141}Sb, ^{144}Te

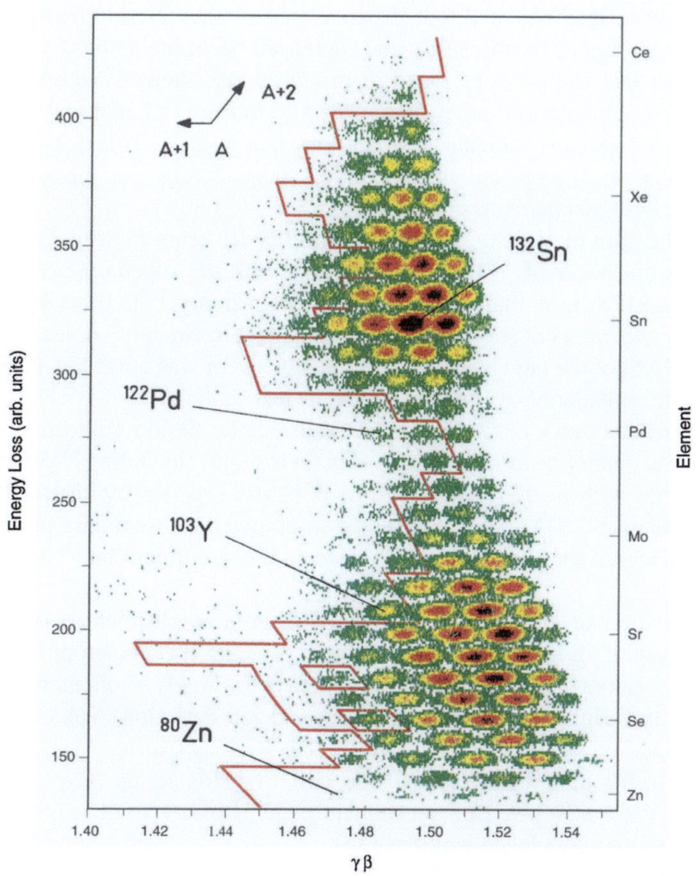

Fig. 15.3 Particle identification plot for the projectile fission of ^{238}U at 750 MeV/nucleon at GSI [27]. The *red line* indicates the limit of known isotopes at the time. For a given element the quantity on the x-axis is proportional to the velocity and thus inverse proportional to the mass (Reprinted from M. Bernas et al., *Projectile fission at relativistic velocities: a novel and powerful source of neutron-rich isotopes well suited for in-flight isotopic separation*, Phys. Lett. B 331 (1994) 19. Copyright 1994, with permission from Elsevier.)

praseodymium or neodymium isotopes [27]. A similar chart with these four isotopes included also was published in another conference proceedings [40] and the 1997 paper [29].

While ^{157}Nd [41] and ^{158}Nd [32] have been discovered in the meantime, ^{155}Pr and ^{156}Pr should still be considered unobserved.

In the proceedings Czajkowski et al. also showed the discovery of ^{122}Rh and ^{124}Pd and did not include the discovery of ^{108}Y, ^{115}Mo, ^{118}Tc, and ^{120}Ru [39] while the final publication by Bernas et al. did not include the former but listed the latter isotopes [29].

Fourteen years later Kurcewicz et al. discovered another 60 isotopes taking advantage of an even higher beam energy and increased beam intensities [32]. It should be mentioned that the isotopes listed in the table also include isotopes produced by projectile fragmentation rather than projectile fission. Calculations of these two production mechanisms indicate that isotopes below $Z = 70$ and above $Z = 72$ were produced by fission and fragmentation, respectively. For the isotopes with $Z = 70-72$, both mechanisms contribute.

During the commissioning of the new RIKEN RI beam factory, Ohnishi et al. reported the discovery of ^{125}Pd and ^{126}Pd [30] with the superconducting in-flight separator BigRIPS. Less than two years later the primary ^{238}U beam intensity had increased by two orders of magnitude facilitating the discovery of another 47 isotopes [31]. For several of the new isotopes only a single event was observed, however, the probability for misidentification was quoted as less than 0.6 %.

The recent discovery of ^{141}Ag combined projectile fission with projectile fragmentation. Secondary beams of neutron-rich tin isotopes (including ^{134}Sn and ^{135}Sn) were produced with in-flight fission from ^{238}U and separated with BigRIPS. The secondary beam (~230 MeV/nucleon) then impinged on a secondary target in front of the Zero Degree spectrometer which selected and identified the ^{131}Ag fragments [33].

The utilization of BigRIPS and the Zero Degree spectrometer as a two stage separation system allowed the latest discovery of neutron-rich isotopes [34]. At the end of the separators the isotopes were implanted in the highly segmented Wide-range Active Silicon Strip Stopper Array for Beta and ion detection (WAS3ABi) which was surrounded by γ-ray detector EUROBALL-RIKEN Cluster Array (EURICA). In addition to the new discovery of ^{77}Co and ^{80}Ni this set-up also measured the half-lives of these isotopes. In a subsequent paper, Lorusso et al. reported the first observation of another eleven isotopes [35].

15.4 Heavy Neutron-Rich Isotopes

The clean identification of heavy nuclides from the fragmentation of (even heavier) beams requires very high beam energies and the discovery of such nuclides has so far only been achieved at the SIS accelerator at GSI. During the period of 1998 through 2011, 83 new isotopes between hafnium and actinium were discovered. Table 15.3 lists the discovered isotopes together with the beam, the beam energy (in MeV/nucleon), and the order of magnitude of the beam intensity (in pps).

The difficulty in producing and identifying neutron-rich nuclides in this heavy mass region around lead is demonstrated by the fact that no new lead isotopes were discovered in over 90 years. Rutherford and Barnes identified ^{214}Pb in 1904 and it was not until 1998 that Pfützner et al. discovered ^{215}Pb [42]. Pfützner et al. did not consider this observation the discovery of a new isotope, giving credit to a paper by van Duppen et al. However, van Duppen et al. did not actually report any results stating: “In conclusion, we have presented a new method that allows detailed decay-

Table 15.3 Heavy neutron-rich isotopes discovered by projectile fragmentation: The submission date, first author and reference of the publication as well as the beam, beam energy E (MeV/nucleon), and beam intensity I (pps) are listed

Date	First author	Ref.	Beam	E	I	Isotope
07/07/1998	Pfützner	[42]	^{238}U	1000	10^6	^{210}Hg, ^{211}Tl, ^{212}Tl, ^{215}Pb, ^{217}Bi, ^{218}Bi, ^{219}Po, ^{220}Po
08/11/1999	Benlliure	[43]	^{197}Au	950	10^6	^{187}Hf, ^{188}Hf, ^{187}Ta, ^{188}Ta, ^{189}Ta, ^{191}W, ^{192}W, ^{193}Re, ^{194}Re,
03/03/2003	Taieb	[44]	^{238}U	1000	–	^{235}Ac
11/02/2007	Kurtukian-Nieto	[45]	^{208}Pb	1000	10^7	^{195}Re
03/31/2008	Steer	[46]	^{208}Pb	1000	10^8	^{196}Re, ^{198}Os, ^{199}Os, ^{200}Ir, ^{201}Ir, ^{202}Ir, ^{203}Pt, ^{204}Pt
09/18/2009	Alkhomashi	[47]	^{208}Pb	1000	10^7	^{189}Hf, ^{190}Ta, ^{191}Ta, ^{192}Ta, ^{193}W, ^{194}W
03/11/2010	Chen	[48]	^{238}U	670	10^6	^{213}Tl, ^{221}Po, ^{222}Po, ^{224}At, ^{236}Ac
07/30/2010	Alvarez-Pol	[49]	^{238}U	1000	10^8	^{205}Pt, ^{207}Au, ^{208}Au, ^{209}Au, ^{210}Au, ^{211}Hg, ^{212}Hg, ^{213}Hg, ^{214}Hg, ^{215}Hg, ^{216}Hg, ^{214}Tl, ^{215}Tl, ^{216}Tl, ^{217}Tl, ^{216}Pb, ^{217}Pb, ^{218}Pb, ^{219}Pb, ^{220}Pb, ^{219}Bi, ^{220}Bi, ^{221}Bi, ^{222}Bi, ^{223}Bi, ^{224}Bi, ^{223}Po, ^{224}Po, ^{225}Po, ^{226}Po, ^{227}Po, ^{225}At, ^{226}At, ^{227}At, ^{228}At, ^{229}At, ^{230}Rn, ^{231}Rn, ^{233}Fr
03/12/2011	Morales	[50]	^{208}Pb	1000	10^8	^{200}Os, ^{201}Os, ^{203}Ir, ^{204}Ir, ^{206}Au
08/02/2011	Steer	[51]	^{208}Pb	1000	10^7	^{197}Re

spectroscopy studies of the neutron-rich 'east of ^{208}Pb' using the pulsed release from the ISOLDE targets... It has been successfully applied in a recent experiment where two new isotopes (^{215}Pb and ^{217}Bi) were identified [10]..." [52]. Reference [10] was listed as K. Rykaczewski et al., to be published and has only appeared in a conference proceedings [53].

Although all isotopes in the present section were produced at GSI, the assignment of the discovery of new isotopes between ytterbium and platinum is rather difficult. In her 2007 Ph.D. thesis Kurtukain-Nieto, described the discovery of 25 new isotopes [54], however, the data were not published in a refereed journal until recently (2014) [55]. Only ^{195}Re was presented as part of the description of a new analysis method to determine β-decay half-lives in 2008 [45]. Instead, the results were presented in multiple conference proceedings [56–62] and preprints [63, 64]. During that time a different group, who was aware of the data because some of the researchers were co-authors on the publications of both groups began to present data for some of the

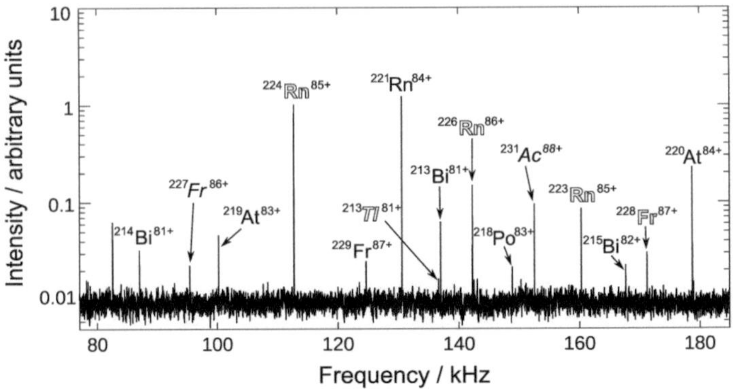

Fig. 15.4 Schottky frequency spectrum measured in the storage cooler ring ESR at GSI. Isotopes with previously unknown masses and the newly discovered isotope ^{213}Tl are indicated by outlined element names. [48] (Reprinted from L. Chen et al., *Discovery and investigation of heavy neutron-rich isotopes with time-resolved Schottky spectrometry in the element range from thallium to actinium*, Phys. Lett. B 691 (2010) 234. Copyright 2010, with permission from Elsevier.)

isotopes without specifically claiming discovery or mentioning them [46, 47, 50, 51]. Some of the isotopes described by Morales et al. were produced in (p,n) reactions rather than by projectile fragmentation [50]. In 2012, Kurcewicz et al. presented the observation of all the isotopes in this mass region [32] as described in the previous section.

The only experiment that utilized a somewhat lower beam energy (670 MeV/nucleon) employed a different identification method following the separation with the FRS. The mass and half-lives of ^{213}Tl, ^{221}Po, ^{222}Po, ^{224}At, and ^{236}Ac were measured with time-resolved Schottky Mass Spectrometry in the storage-cooler ring ESR [48]. The section of the spectrum which includes the first observation of ^{213}Tl is shown in Fig. 15.4.

15.5 Proton-Rich Isotopes

The successful application of projectile fragmentation to produce new neutron-rich isotopes encouraged efforts to also expand the nuclear chart towards more proton-rich nuclides using this method. Since the first measurements in 1985, projectile fragmentation has developed to become the predominant mechanism to produce nuclides at and beyond the proton drip-line for all elements up to antimony. The almost 80 long-lived proton-rich isotopes that have been discovered using projectile fragmentation are listed in Table 15.4. In addition, the three two-proton emitter ^{45}Fe, ^{54}Zn, ^{58}Ni as well five unbound short-lived proton emitters have been identified by projectile fragmentation, as described in Sects. 16.3 and 16.4, respectively.

The first application of the projectile fragmentation process to produce proton-rich isotopes was presented by Langevin et al. in 1986. Fragments from a 77 MeV/nucleon ^{40}Ca beam at GANIL were separated with the LISE spectrometer and ^{23}Si, ^{27}S, and

Table 15.4 Proton-rich isotopes discovered by projectile fragmentation: The submission date, first author and reference of the publication as well as the beam, beam energy E (MeV/nucleon), and beam intensity I (pps) are listed

Date	First author	Ref.	Beam	E	I	Isotope
12/30/1985	Langevin	[65]	^{40}Ca	77	10^{10}	^{23}Si, ^{27}S, ^{31}Ar
12/22/1986	Pougheon	[66]	^{58}Ni	55	10^{10a}	^{43}V, ^{44}Cr, ^{46}Mn, ^{47}Mn, ^{48}Fe, ^{50}Co, ^{51}Co, ^{52}Co, ^{55}Cu, ^{56}Cu, ^{51}Ni, ^{52}Ni
02/09/1987	Saint-Laurent	[67]	^{36}Ar	85	10^{11}	^{22}Si
07/03/1990	Détraz	[68]	^{58}Ni	65	10^{10a}	^{39}Ti
12/27/1990	Mohar	[69]	^{78}Kr	65	–	^{65}As, ^{67}Se, ^{75}Sr, ^{62}Ge, ^{63}Ge
06/01/1992	Borrel	[70]	^{58}Ni	69	10^{10a}	^{43}Cr, ^{46}Fe, ^{47}Fe
08/14/1992	Yennello	[71]	^{92}Mo	70	–	^{78}Y, ^{82}Nb, ^{85}Mo, ^{86}Tc, ^{89}Ru
07/12/1993	Winger	[72]	^{78}Kr	75	–	^{66}Se
04/15/1994	Hencheck	[73]	^{106}Cd	60	10^{9}	^{88}Ru, ^{90}Rh, ^{91}Rh, ^{92}Rh, ^{93}Rh, ^{92}Pd, ^{93}Pd, ^{94}Ag, ^{95}Ag
04/27/1994	Schneider	[74]	^{124}Xe	1095	10^{7}	^{98}In, ^{99}In, ^{100}Sn
05/27/1994	Blank	[75]	^{58}Ni	650	10^{7}	^{50}Ni
06/07/1994	Lewitowicz	[76]	^{112}Sn	63	10^{10}	^{101}Sn, ^{102}Sn
10/25/1994	Rykaczewski	[77]	^{112}Sn	63	10^{10}	^{87}Ru, ^{89}Rh, ^{91}Pd, ^{93}Ag, ^{103}Sb, ^{104}Sb
01/12/1995	Blank	[78]	^{78}Kr	73	10^{11}	^{64}As, ^{69}Kr, ^{70}Kr, ^{60}Ga, ^{74}Sr
07/25/1996	Blank	[79]	^{58}Ni	600	–	^{42}Cr, ^{49}Ni^{b}
04/23/1998	Janas	[80]	^{92}Mo	60	10^{10a}	^{77}Y, ^{79}Zr, ^{83}Mo
10/31/2000	Giovinazzo	[81]	^{58}Ni	75	10^{11}	^{55}Zn, ^{56}Zn
01/01/2001^{c}	Kienle	[82]	^{112}Sn	1000	10^{7}	^{76}Y, ^{78}Zr
09/04/2004	Liu	[83]	^{238}U	1000	10^{7}	^{219}Pa, ^{220}Pa
02/19/2005	Blank	[84]	^{58}Ni	74.5	10^{12}	^{54}Zn
09/22/2005	Stolz	[85]	^{78}Kr	140	10^{10}	^{60}Ge, ^{64}Se
06/27/2008	Bazin	[86]	^{112}Sn	120	10^{10}	^{96}Cd
10/28/2011	Hinke	[87]	^{124}Xe	1000	10^{9}	^{95}Cd, ^{97}In, ^{99}Sn
06/16/2015	Ciemny	[88]	^{78}Kr	150	10^{11}	^{59}Ge

[a] Converted from electric current assuming fully stripped
[b] The two-proton emitter ^{45}Fe was also reported in this publication (see Sect. 16.4)
[c] Publication date

^{31}Ar were identified for the first time [65]. In a subsequent experiment, Pougheon et al. reported the first observation of another twelve proton-rich isotopes [66]. One event corresponding to ^{47}Fe and two events of ^{49}Co were deemed insufficient to prove its existence. While ^{47}Fe was discovered a few years later [70], ^{49}Co was demonstrated to be unbound with a half-life shorter than 300 ns [75].

After the completion of the A1200 fragment separator at the NSCL at Michigan State University, 21 new proton-rich isotopes were discovered with the

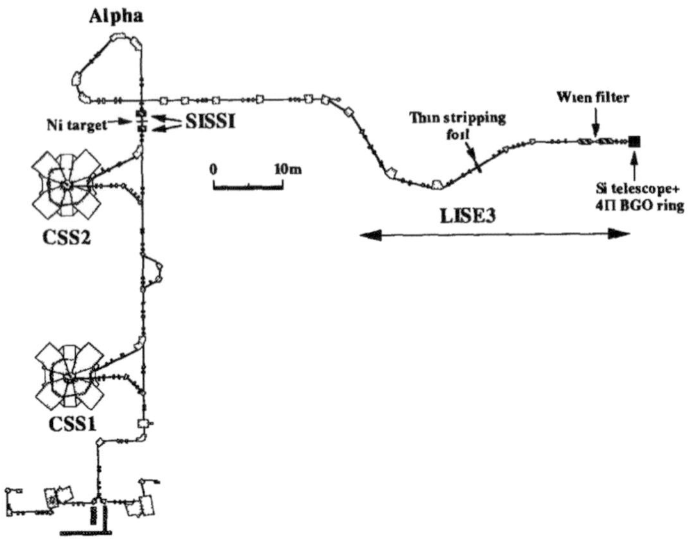

Fig. 15.5 Upgraded projectile fragmentation set-up at GANIL. Fragmentation-like reaction products were separated and identified in-flight with the SISSI (Superconducting Intense Source for Secondary Ions) device and the magnetic spectrometers Alpha and LISE3 [76] (Reprinted from M. Lewitowicz et al., *Identification of the doubly-magic nucleus* ^{100}Sn *in the reaction* ^{112}Sn + ^{nat}Ni *at 63 MeV/nucleon*, Phys. Lett. B 332 (1994) 20. Copyright 1994, with permission from Elsevier.)

device between 1990 and 1994 [69, 71–73]. The initial observation of a few counts for ^{69}Br reported by Mohar et al. [69] could later not be replicated [78]. In addition to cleanly identify nine isotopes listed in Table 15.4, Hencheck et al. observed a few events for ^{77}Y, ^{79}Zr, ^{81}Nb, ^{85}Tc, ^{87}Ru, ^{91}Pd, and ^{93}Ag, however: "Although these events all satisfy the stringent gating requirements mentioned above, we do not wish to state that such few events constitute proof that these nuclei were identified for the first time in this experiment as they could, at that level, be the result of 'contamination' from neighboring peaks in the Z and Q spectra" [73]. Rykaczewski et al. and Janas et al. later confirmed the observation of ^{87}Ru and ^{91}Pd [77] and ^{77}Y and ^{79}Zr [80], respectively, while Janas et al. did not find any evidence for the existence of ^{81}Nb and ^{85}Tc [80]. Both experiments were performed with the LISE3 (Ligne d'Ions Super-Epluchés) spectrometer at GANIL. Rykaczewski et al. also observed eight events of ^{93}Ag and it is not clear why they did not claim discovery as they did for the observation of ^{103}Sb with nine events [77]. Subsequent publications assumed ^{93}Ag to be known [89, 90] so credit for its discovery is given to Rykaczewski et al.

In 1994 GANIL expanded its capabilities for fragmentation beams by implementing the Superconducting Intense Source for Secondary Ions (SISSI), the alpha magnetic spectrometer and a velocity filter to LISE, as shown in Fig. 15.5. With these improvements another eighteen new isotopes were discovered between 1994 and 2005. The observation of ^{101}Sn and ^{102}Sn by Lewitowicz et al. [76] was submitted less than two month after Schneider et al. had identified these isotopes when reporting the discovery of ^{100}Sn [74]. Schneider et al. acknowledged the work by Lewitowicz et al.: "Recently, an alternative production mechanism has been tested by using intermediate-energy projectile-fragmentation of 58 A $\cdot$ MeV ^{112}Sn at the LISE separator at GANIL. In this study, all the light ^{112}Sn isotopes down to ^{101}Sn could be identified", referring to an article in the GANIL newsletter Nouvelles de GANIL [91].

Fifteen new isotopes were discovered at GSI produced with beams from the SIS synchrotron and separated with the FRS between 1994 and 2011. The isotopes ^{76}Y and ^{78}Zr were identified with only two events and one event, respectively [82]. The discovery of the only two isotopes heavier than antimony (^{219}Pa and ^{220}Pa) produced by projectile fragmentation was not acknowledged as such [83]. For ^{219}Pa the authors refer to the 1996 Table of Isotopes [92] which only cites a 1987 conference proceedings [93]. For ^{220}Pa the ENSDF database [94] lists only the 1987 proceedings [93] and an internal report [95]. The particle identification plot in the recent paper by Hinke et al. clearly shows evidence for ^{95}Cd, ^{97}In, and ^{99}Sn, however, their discovery is not explicitly mentioned [87]. The observation of these isotopes in this work is substantiated by two Ph.D. theses [90, 96].

Four additional isotopes were observed for the first time with the upgraded Coupled Cyclotron Facility at Michigan State University. Three and four events were identified for ^{60}Ge and ^{64}Se, respectively, with the A1900 fragment separator [85] while the discovery of ^{96}Cd was achieved by using in addition the Radio Frequency Fragment Separator (RFFS) [86]. Ten years after the discovery of ^{60}Ge an order of magnitude increase in the primary beam intensity of ^{78}Kr enabled the discovery of ^{59}Ge [88].

References

1. A.M. Poskanzer et al., Phys. Rev. Lett. **17**, 1271 (1966)
2. A.M. Poskanzer et al., Phys. Lett. B **27**, 414 (1968)
3. A.G. Artukh et al., Phys. Lett. B **31**, 129 (1970)
4. T.J.M. Symons et al., Phys. Rev. Lett. **42**, 40 (1979)
5. G.D. Westfall et al., Phys. Rev. Lett. **43**, 1859 (1979)
6. J.D. Stevenson, P.B. Price, Phys. Rev. C **24**, 2102 (1981)
7. J.A. Musser, J.D. Stevenson, Phys. Rev. Lett. **53**, 2544 (1984)
8. M. Langevin et al., Phys. Lett. B **150**, 71 (1985)
9. D. Guillemaud-Mueller et al., Z. Phys. A **322**, 415 (1985)
10. F. Pougheon et al., Europhys. Lett. **2**, 505 (1986)
11. D. Guillemaud-Mueller et al., Z. Phys. A **332**, 189 (1989)
12. D. Guillemaud-Mueller et al., Phys. Rev. C **41**, 937 (1990)
13. M. Lewitowicz et al., Z. Phys. A **335**, 117 (1990)
14. M. Weber et al., Z. Phys. A **343**, 67 (1992)
15. H. Sakurai et al., Phys. Rev. C **54**, R2802 (1996)
16. H. Sakurai et al., Phys. Lett. B **448**, 180 (1999)
17. M. Notani et al., Phys. Lett. B **542**, 49 (2002)
18. S. Grévy et al., Phys. Lett. B **594**, 252 (2004)
19. O.B. Tarasov et al., Phys. Rev. C **75**, 064613 (2007)

20. T. Baumann et al., Nature **449**, 1022 (2007)
21. O.B. Tarasov et al., Phys. Rev. Lett. **102**, 142501 (2009)
22. O.B. Tarasov et al., Phys. Rev. C **87**, 054612 (2013)
23. P. Auger et al., Z. Phys. A **289**, 255 (1979)
24. W. Benenson et al., Phys. Lett. B **162**, 87 (1985)
25. W. Zhan et al., Nouvelles du GANIL No. 25, April 1988; and to be published
26. H. Sakurai et al., Nucl. Phys. A **616**, 311c (1997)
27. M. Bernas et al., Phys. Lett. B **331**, 19 (1994)
28. C. Engelmann et al., Z. Phys. A **352**, 351 (1995)
29. M. Bernas et al., Phys. Lett. B **415**, 111 (1997)
30. T. Ohnishi et al., J. Phys. Soc. Jpn. **77**, 083201 (2008)
31. T. Ohnishi et al., J. Phys. Soc. Jpn. **79**, 073201 (2010)
32. J. Kurcewicz et al., Phys. Lett. B **717**, 371 (2012)
33. H. Wang et al., Chin. Phys. Lett. **30**, 042501 (2013)
34. Z.Y. Xu et al., Phys. Rev. Lett. **113**, 032505 (2014)
35. G. Lorusso et al., Phys. Rev. Lett. **114**, 192501 (2015)
36. F. Ameil et al., GSI Sci. Rep. **94–1**, 25 (1995)
37. F. Ameil et al., in *Proceedings of the International Conference on Exotic Nuclei and Atomic Masses*, Arles, France, 19–23 June 1995, p. 537
38. F. Ameil et al., Eur. Phys. J. A **1**, 275 (1998)
39. S. Czajkowski et al., in *Proceedings of the International Conference on Exotic Nuclei, Atomic Masses, ENAM'95*, Arles, France, 1995, Editions Frontieres (1996), p. 553
40. M. Bernas et al., Nucl. Phys. A **616**, 352c (1997)
41. J. Van Schelt et al., Phys. Rev. C **85**, 045805 (2012)
42. M. Pfützner et al., Phys. Lett. B **444**, 32 (1998)
43. J. Benlliure et al., Nucl. Phys. A **660**, 87 (1999)
44. J. Taieb et al., Nucl. Phys. A **724**, 413 (2003)
45. T. Kurtukian-Nieto, J. Benlliure, K.-H. Schmidt, Nucl. Instrum. Meth. A **589**, 472 (2008)
46. S.J. Steer et al., Phys. Rev. C **78**, 061302 (2008)
47. N. Alkhomashi et al., Phys. Rev. C **80**, 064308 (2009)
48. L. Chen et al., Phys. Lett. B **691**, 234 (2010)
49. H. Alvarez-Pol et al., Phys. Rev. C **82**, 041602 (2010)
50. A.I. Morales et al., Phys. Rev. C **84**, 011601 (2011)
51. S.J. Steer et al., Phys. Rev. C **84**, 044313 (2011)
52. P. Van Duppen et al., Nucl. Instrum. Meth. B **134**, 267 (1998)
53. K. Rykaczewski et al., AIP Conf. Proc. **455**, 581 (1998)
54. T. Kurtukain-Nieto, *Production and β decay half-lives of heavy neutron-rich nuclei approaching the stellar nucleosynthesis r-process path around $A = 195$*. Ph.D. Thesis, Universidade de Santiago de Compostela, 2007, http://fpsalmon.usc.es/PhDthesis/PhD_teresa.pdf
55. T. Kurtukian-Nieto et al., Phys. Rev. C **89**, 024616 (2014)
56. T. Kurtukian-Nieto et al., AIP Conf. Proc. **802**, 73 (2005)
57. J. Benlliure et al., Eur. Phys. J. ST **150**, 309 (2007)
58. T. Kurtukian-Nieto et al., Pos(NIC IX) 008 (2007)
59. T. Kurtukian-Nieto et al., in *Proceedings of the International Conference on Nuclear Data for Science and Technology*, 2007, doi:10.1051/ndata:07685
60. T. Kurtukian-Nieto et al., Nucl. Phys. A **827**, 587c (2009)
61. T. Kurtukian-Nieto, J. Phys. Conf. Ser. **202**, 012012 (2010)
62. J. Benlliure et al., Pos(NIC XI) 084 (2010)
63. T. Kurtukian-Nieto et al., arXiv:0711.0101 (2007)
64. J. Benlliure et al., arXiv:1004.0265 (2010)
65. M. Langevin et al., Nucl. Phys. A **455**, 149 (1986)
66. F. Pougheon et al., Z. Phys. A **327**, 17 (1987)
67. M.G. Saint-Laurent et al., Phys. Rev. Lett. **59**, 33 (1987)
68. C. Détraz et al., Nucl. Phys. A **519**, 529 (1990)

69. M.F. Mohar et al., Phys. Rev. Lett. **66**, 1571 (1991)
70. V. Borrel et al., Z. Phys. A **344**, 135 (1992)
71. S.J. Yennello et al., Phys. Rev. C **46**, 2620 (1992)
72. J.A. Winger et al., Phys. Rev. C **48**, 3097 (1993)
73. M. Hencheck et al., Phys. Rev. C **50**, 2219 (1994)
74. R. Schneider et al., Z. Phys. A **348**, 241 (1994)
75. B. Blank et al., Phys. Rev. C **50**, 2398 (1994)
76. M. Lewitowicz et al., Phys. Lett. B **332**, 20 (1994)
77. K. Rykaczewski et al., Phys. Rev. C **52**, 2310 (1995)
78. B. Blank et al., Phys. Rev. Lett. **74**, 4611 (1995)
79. B. Blank et al., Phys. Rev. Lett. **77**, 2893 (1996)
80. Z. Janas et al., Phys. Rev. Lett. **82**, 295 (1999)
81. J. Giovinazzo et al., Eur. Phys. J. A **11**, 247 (2001)
82. P. Kienle et al., Prog. Part. Nucl. Phys. **46**, 73 (2001)
83. Z. Liu et al., Nucl. Instrum. Meth. A **543**, 591 (2005)
84. B. Blank et al., Phys. Rev. Lett. **94**, 232501 (2005)
85. A. Stolz et al., Phys. Lett. B **627**, 32 (2005)
86. D. Bazin et al., Phys. Rev. Lett. **101**, 252501 (2008)
87. C.B. Hinke et al., Nature **486**, 341 (2012)
88. A.A. Ciemny et al., Phys. Rev. C **92**, 014622 (2015)
89. R. Krücken, A.I.P. Conf, Proc. **1072**, 52 (2008)
90. K. Straub, Ph.D. thesis, *Zerfallseigenschaften von Nukliden in der Umbebung von* ^{100}Sn, Technical University Munich, unpublished (2011)
91. M. Lewitowicz et al., Nouvelles de GANIL **48**, 7 (1993)
92. R. B. Firestone et al., *Table of Isotopes*, 8th edn. (Wiley, New Jersey, 1996)
93. T. Faestermann et al., in *Proceedings of the 5th International Conference Nuclei Far from Stability*, Rosseau Lake, Canada, K12 (1987)
94. http://www.nndc.bnl.gov/ensdf/ ENSDF, Evaluated Nuclear Structure Data File, maintained by the National Nuclear Data Center at Brookhaven National Laboratory, published in Nuclear Data Sheets (Academic Press, Elsevier Science)
95. H. Miyatake et al., Inst. Nucl. Study, Univ. Tokyo, Ann. Rept., 1986, p. 37 (1987)
96. C.B. Hinke, Ph.D. Thesis, *Spectroscopy of the Doubly Magic Nucleus* ^{100}Sn *and its Decay*, Technical University Munich, unpublished (2010)

Chapter 16
Unbound Isotopes

16.1 Definition of Unbound Isotopes

In the previous chapters, the discovery of stable and radioactive nuclei has been discussed. The radioactive isotopes decayed by the emission of β^+, β^-, α-particles, or fission with lifetimes between a few milliseconds and millions of years. The further away these isotopes are from the stable isotopes the larger is the imbalance between protons and neutrons leading to shorter and shorter half-lives. At one point this imbalance becomes so large that the nuclide cannot bind the last neutron or proton anymore and it becomes unbound with respect to nucleon emission and decays by the emission of a neutron or a proton.

The boundaries between bound and unbound nuclides are called the drip-lines and are defined by the nucleon separation energies (S_n or S_p) crossing zero. For neutron-rich nuclei this boundary is very sharp. Bound nuclides at the drip-line have half-lives of the order of milliseconds (the shortest observed β^- half-life has been measured for ^{35}Na with 1.5 ms [1]) while the half-life of unbound nuclides is shorter than $\sim 10^{-20}$ s. Only recently, first evidence for a half-life in the range of picoseconds has been reported in the two-neutron decay of ^{26}O [2].

Such short lifetimes raise the question of the existence of a nuclide. In order to define such an existence it is instructive to take the definition of a chemical element as guidance. The International Union of Pure and Applied Chemistry (IUPAC) set a lower limit of 10^{-14} s for the existence of an element because "This lifetime is chosen as a reasonable estimate of the time it takes a nucleus to acquire its outer electrons. It is not considered self-evident that talking about an element makes sense if no outer electrons, bearers of the chemical properties, are present" [3]. Similarly the definition for the existence of a nuclide should be related to the typical timescales of nuclear motion. Nuclear rotation and vibration times are of the order of 10^{-22}s which can be considered a characteristic nuclear timescale [4]. Therefore, the limit for the existence of a nuclide should be set at about 10^{-22} s [5].

M. Thoennessen, *The Discovery of Isotopes*,
DOI 10.1007/978-3-319-31763-2_16

Nuclides which survive for only such short times cannot be measured directly and indirect methods have to be employed. In these methods the existence of a nuclide is demonstrated by a resonance in an excitation energy or decay energy spectrum from which the separation energy can be deduced. The width of the resonance (Γ) is related to the lifetime (τ) by the uncertainty principle: $\tau = \hbar/\Gamma$. Thus lifetimes of more than 10^{-22} s correspond to resonance widths smaller than about 6.6 MeV. Sometimes these systems are only referred to as resonances rather than nuclides.

Long nuclear interaction times can also be determined in scattering experiments where the measured scattering length a can be approximately related to the decay energy: $E \simeq \hbar^2/2\mu a^2$ where μ is the reduced mass of the system. For small values of a—corresponding to large decay energies—the distributions get broad and cannot be distinguished from non-resonant interactions. For unbound states the scattering length is negative and it has been argued that scattering lengths which are less negative than about −5 fm probably should not be considered for the identification and discovery of a new isotope [6].

The proton drip-line is closer to the stable nuclides than the neutron drip-line because of the repulsive Coulomb force among the protons. Together with the attractive nuclear force they form a barrier which can lead to long half-lives for isotopes beyond the proton drip-line. For light nuclides, the barriers are small so that the lifetimes are short, similar to the lifetimes at the neutron drip-line. However, for heavier nuclides the barriers can be large leading to half-lives that can be measured directly and correspond to proton radioactivity. Half-lives longer than about 10^{-12} s are typically considered as radioactivity [4, 7].

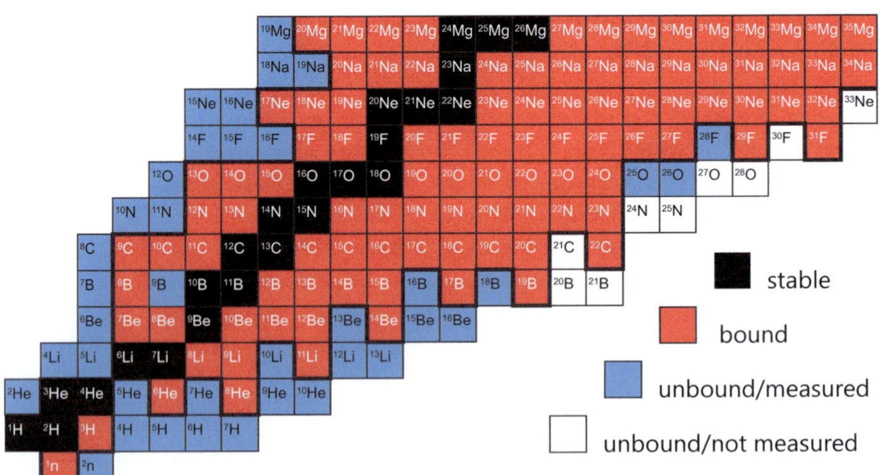

Fig. 16.1 Section of the chart of nuclides (Z ≤ 12) showing the observed proton- and neutron-unbound isotopes (*light blue boxes*) and isotopes which have been shown to be unbound but have not been measured yet

In the following three subsections the discovery of neutron-unbound isotopes, proton-unbound isotopes and proton-radioactive isotopes are discussed. Heavy proton unbound nuclides where the barrier is so large and the proton half-life so long that they decay by β^+ or α-emission rather than proton emission have already been discussed in the previous sections. Figure 16.1 shows a section of the nuclear chart highlighting light neutron- and proton-unbound isotopes ($Z \leq 12$) that have been studied up to now.

16.2 Neutron-Unbound Isotopes

The twenty neutron-unbound isotopes that have been observed up to date are listed in Table 16.1. The neutron drip-line has only been crossed for the lightest elements up to fluorine.

Table 16.1 Neutron-unbound isotopes: The submission date, first author and reference of the publication as well as the reaction, detection method (mm: missing mass, im: invariant mass), and the number of emitted neutrons (#n) are listed

Date	First author	Ref.	Method	Reaction	#n	Isotope
04/20/1937	Williams	[8]	range	^{7}Li(d,^{4}He)	1n	^{5}He
01/25/1965	Haddock	[9]	scat.l.	^{2}H(π^-,γ)	1n	^{2}n
03/13/1967	Stokes	[10]	mm	^{7}Li(t,^{3}He)	1n	^{7}He
09/04/1975	Wilcox	[11]	mm	^{9}Be(^{9}Be,^{8}B)	1n	^{10}Li
03/25/1981	Sennhauser	[12]	mm	^{7}Li(π^-,tπ)	1n	^{4}H
08/27/1982	Aleksandrov	[13]	mm	^{14}C(^{7}Li,^{8}B)	1n	^{13}Be
11/22/1983	Aleksandrov	[14]	mm	^{7}Li(^{7}Li,^{8}B)	3n	^{6}H
12/08/1986	Seth	[15]	mm	^{9}Be(π^-,π^+)	1n	^{9}He
10/15/1993	Korsheninnikov	[16]	im	^{11}Li($-$1p)	2n	^{10}He
12/27/1999	Kalpakchieva	[17]	mm	^{14}C(^{14}C,^{12}N)	1n	^{16}B
03/27/2001	Korsheninnikov	[18]	mm	^{6}He(p,2p)[a]	2n	^{5}H
10/08/2002	Korsheninnikov	[19]	mm	^{8}He(p,2p)[a]	4n	^{7}H
12/22/2007	Hoffman	[20]	im	^{26}F($-$1p)	1n	^{25}O
03/20/2008	Aksyutina	[21]	im	^{14}Be($-$1p)	2n	^{13}Li
08/16/2009	Spyrou	[22]	im	^{19}C($-$1p)	1n	^{18}B
12/04/2009	Hall	[23]	im	^{14}B($-$2p)	1n	^{12}Li
09/23/2011	Christian	[24]	im	^{29}Ne($-$1p)	1n	^{28}F
10/20/2011	Spyrou	[25]	im	^{17}B($-$1p)	2n	^{16}Be
10/23/2011	Lunderberg	[26]	im	^{27}F($-$1p)	2n	^{26}O
04/22/2013	Snyder	[27]	im	^{14}Be(d,p)[a]	1n	^{15}Be

[a]Inverse kinematics

The first neutron-unbound isotope was discovered in 1937. Williams, Shepherd, and Haxby used 275 keV deuterons to form ^{5}He in the reaction ^{7}Li + ^{2}H → ^{5}He + ^{4}He at the University of Minnesota [8]. They determined that ^{5}He was unbound by 0.93 MeV from the range spectrum of the α-particles. In a follow-up paper they extracted a mean life of about 6×10^{-20} s [28].

It took almost another 30 years before other neutron-unbound systems were studied. The first accurate measurement of the most basic neutron-rich system, the interaction of two neutrons, was reported by Haddock et al. in 1965. Negative pions produced by the 184-in. Berkeley cyclotron irradiated a liquid deuterium target and all three outgoing particles—two neutrons and a γ-ray—were detected. The experimental set-up is shown in Fig. 16.2 [9]. From the data a scattering length of -16.4 ± 1.9 fm was deduced. This kinematically complete experiment can be considered to be the first reliable and accurate measurement of the "di-neutron" system [29].

Unbound isotopes which are fairly close to the line of stability can be identified in two-body transfer reactions using stable beams on stable targets with the missing mass method (see Sects. 9.4 and 14.1). In this method it is not necessary to detect the decay products of the unbound isotopes. The decay energy and the width of the isotopes are calculated from the measured kinematic parameters of the incoming beam and the outgoing ejectile. For example, in 1967, Stokes and Young used the missing mass method for the first time to produce ^{7}He in a charge-exchange reaction. Tritons were accelerated to 22 MeV with the Los Alamos three stage tandem accelerator and irradiated a ^{7}Li target. Outgoing ^{3}He ejectiles were identified with a ΔE gas proportional counter and a surface-barrier E detector. This transfer of charge changes the ^{7}Li target nuclides into ^{7}He which then instantaneously decays into ^{6}He and a neutron [10].

Light ions of ^{7}Li, ^{9}Be and ^{14}C were subsequently used to identify four additional isotopes with the missing mass method. The reactions leading to ^{10}Li at Berkeley [11] and ^{13}Be at the Kurchatov Institute of Atomic Energy in Moscow, Russia, [13] did not populate the ground-states but excited states. The excitation energy spectrum measured for ^{10}Li is shown in Fig. 16.3. ^{6}H—also observed in Moscow—was the first isotope not located directly at the drip-line and decayed by the emission of three neutrons [14]. In 1999, Kalpakchieva et al. identified ^{16}B in a two-proton pick-up reaction at the Hahn-Meitner Institute in Berlin, Germany [17].

Two additional isotopes were discovered with the missing mass method using pions as the beam. ^{4}H was produced by pion absorption on ^{7}Li and subsequent breakup into a triton and ^{4}H at the Swiss Institute for Nuclear Research in Villingen, Switzerland [12] and Seth et al. populated ^{9}He in the pion double charge-exchange reaction on ^{9}Be at the EPICS channel of the Los Alamos Meson Physics Facility (LAMPF) [15].

Korsheninnikov et al. employed the missing mass method in inverse kinematics with secondary beams to measure ^{5}H and ^{7}H. ^{6}He and ^{8}He beams irradiated cryogenic hydrogen targets forming ^{5}H and ^{7}H, respectively, by transferring a proton from the beam to the target. The target "di-proton" decays into two protons which were then measured in coincidence to reconstruct the decay energy of ^{5}H [18] and ^{7}H [19]. The ^{5}H experiment was performed in 2001 with the fragment separator ACCULINA

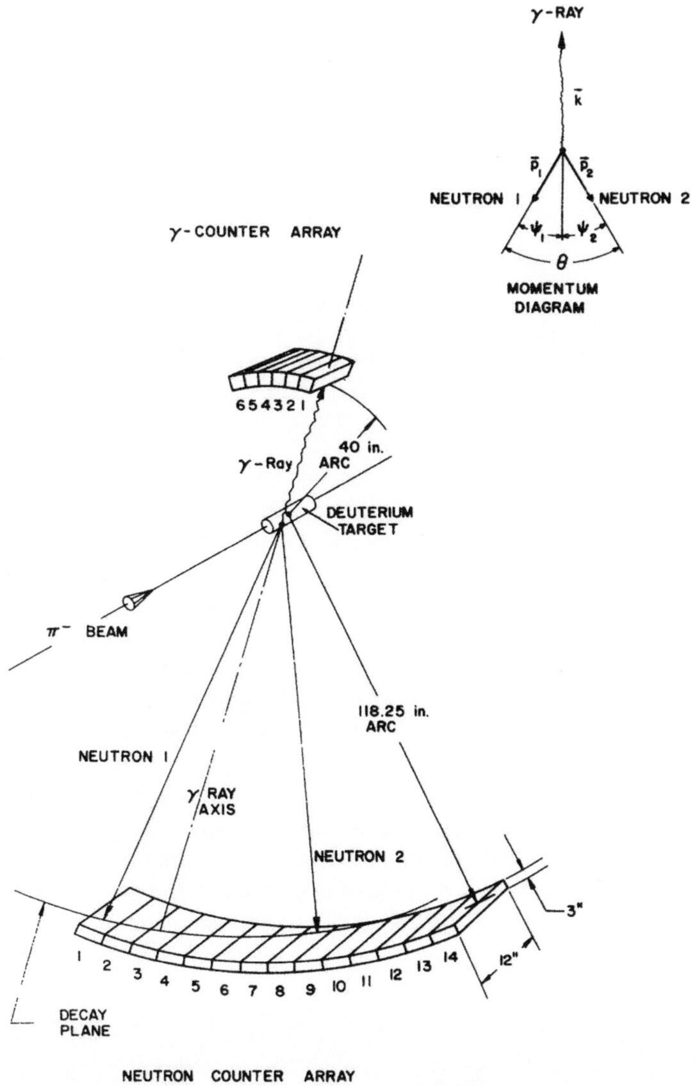

Fig. 16.2 Experimental set-up to measure the scattering length of the "di-neutron" by Haddock et al. [9] (Reprinted figure with permission from R.P. Haddock et al., *Measurement of the neutron-neutron s-wave scattering length from the reaction* $\pi^- + d \rightarrow 2n + \gamma$, Phys. Rev. Lett. 14 (1965) 318. Copyright 1965 by the American Physical Society.)

at the Joint Institute for Nuclear Research in Dubna and ^{7}H was discovered a year later at the fragment separator RIPS at RIKEN in Japan. It should be mentioned that the extracted decay energy of 1.7 ± 0.3 MeV for ^{5}H has been reproduced in a few other secondary beam experiments [30] but has not been replicated in pion induced reactions [31]. Also an earlier 1963 report that ^{5}H was bound with respect to neutron

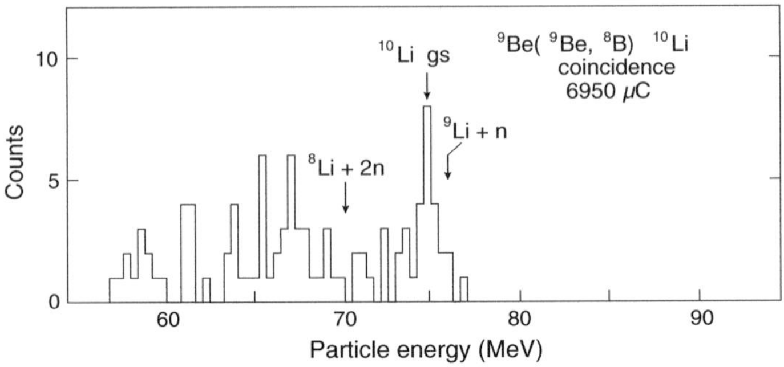

Fig. 16.3 Missing mass spectrum for the reaction ^{9}Be(^{9}Be, ^{8}B)^{10}Li measured by Wilcox et al. [11]. The observed resonance corresponds to an excited state (Adapted from [11])

emission [32] could not be confirmed [33–35]. The ^{7}H experiment reported excess counts close to threshold which were tentatively attributed to a ^{7}H resonance without quoting any values for the decay energy and width [19]. This observation was later confirmed with a measurement of a resonance at 570^{+420}_{-210} keV [36].

The increasing availability of high-intensity secondary beams opened up the possibility to explore nuclides beyond the drip-line which are further and further removed from stability. The high energy beams from fragmentation facilities have the advantage that the produced unbound nuclei decay in-flight so that it is possible to detect all decay fragments at forward angles with high efficiency. The decay energy and width can then be reconstructed just from the energy and angles of the decay products. In this invariant mass method it is not necessary to measure the incoming beam energy.

The first invariant mass measurement of an unbound neutron-rich system was performed in 1993 at the RIPS facility at RIKEN. Korsheninnikov et al. used a one-proton removal reaction from a ^{11}Li beam to populate ^{10}He which then decayed to ^{8}He and two neutrons [16]. Figure 16.4 shows the schematic of the experimental set-up.

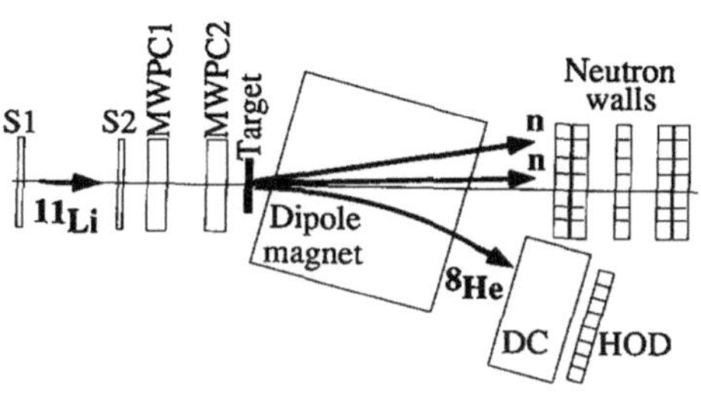

Fig. 16.4 Experimental set-up to measure the decay of ^{10}He [16]. The charged ^{8}He fragments are deflected with a dipole magnet and the momentum is then measured in coincidence with the two neutrons emitted in the decay (Reprinted from A.A. Korsheninnikov et al., *Observation of ^{10}He*, Phys. Lett. B 326 (1994) 31. Copyright 1994, with permission from Elsevier.)

Subsequently three other two-neutron unbound isotopes including their one-neutron unbound sub-systems were measured with the invariant mass methods at fragmentation facilities. In 2008, Aksyutina et al. reported a state at 1.47 ± 0.31 MeV in ^{13}Li and a scattering state with $a_s = 13.7 \pm 1.6$ fm in ^{12}Li [21] using the Large Area Neutron Detector (LAND) at GSI. Due to the limited detection efficiency of LAND for low energy two-neutron emissions they did not identify the ^{13}Li ground-state [37], and the misidentified events contaminated their ^{12}Li result. A 2013 reanalysis of a subsequent experiment on ^{12}Li performed in 2009 with the Modular Neutron Array (MoNA) at the NSCL at Michigan State University by Hall et al. [23] which had used the Aksyutina et al. results yielded a scattering length of < -4 fm for ^{12}Li [37]. At the time, Hall et al. had measured two additional states in ^{10}Li and thus were credited with its discovery.

The first direct two-neutron decay—i.e. the decay through the intermediate sub-system was energetically not possible—was measured in 2011 in the decay of ^{16}Be [25]. The confirmation that the decay could not proceed via ^{15}Be was published two years later, when the decay energy of ^{15}Be was measured with a (d,p) reaction in inverse kinematics [27].

The observation of a very small decay energy for two-neutron unbound ^{26}O by Lunderberg et al. in 2011 [26] was submitted only three days after the ^{16}Be results. It led to the speculation of two-neutron radioactivity in ^{26}O where the data were consistent with a decay lifetime of $4.5^{+1.1}_{-1.5}$(stat) ± 3(syst) ps [2]. The ^{25}O subsystem had been detected five years earlier [20] after an initial report in 1981 that ^{25}O might be bound [38] was demonstrated to be incorrect [39].

The heaviest neutron-unbound system measured so far is ^{28}F which was identified in 2011 by Christian et al. [24] With the exception of ^{13}Li which was first identified with the LAND set-up at GSI, all new unbound isotopes reported during the last ten years were discovered with the Modular Neutron Array (MoNA) at the NSCL at Michigan State University.

16.3 Proton-Unbound Isotopes

The situation for light proton-rich nuclides beyond the proton drip-line is very similar to neutron-rich nuclides beyond the neutron drip-line. They have very short half-lives so that their decay cannot be observed directly. Most of the isotopes were reachable with transfer reactions of stable beams and targets because for light isotopes the proton drip-line is close to stable isotopes. The decay energy could then be determined with the missing mass method. The latest discoveries were obtained with secondary beams where several different detection methods were employed. Table 16.2 lists all twenty-three proton-unbound isotopes with lifetimes less than 10^{-12} s discovered so far.

Table 16.2 Proton-unbound isotopes with lifetimes less than 10^{-12} s: The submission date, first author and reference of the publication as well as the reaction, detection method (ps: phase shift analysis, mm: missing mass, im: invariant mass, β^+p: β-delayed proton emission, if: in-flight decay), and the number of emitted protons (#p) are listed

Date	First author	Ref.	Method	Reaction	#p	Isotope
08/11/1936	Tuve	[40]	ps	p(p,p′)	1p	^{2}He
08/30/1940	Haxby	[41]	mm	^{9}Be(p,n)	1p	^{9}B
04/30/1941	Heydenburg	[42]	scatt.	^{4}He(p,p′)	1p	^{5}Li
06/01/1957	Bogdanov	[43]	mm	^{6}Li(p,n)	2p	^{6}Be
11/08/1963	Bryant	[44]	mm	^{14}N(^{3}He,n)	1p	^{16}F
07/16/1965	Cerny	[45]	mm	^{6}Li(p,t)	1p	^{4}Li
11/10/1967	McGrath	[46]	mm	^{10}B(^{3}He,^{6}He)	3p	^{7}B
02/03/1969	Cerny	[47]	mm	^{24}Mg(p,^{6}He)	1p	^{19}Na
03/11/1974	Benenson	[48]	mm	^{14}N(^{3}He,^{6}He)	1p	^{11}N
04/08/1974	Robertson	[49]	mm	^{12}C(α,^{8}He)	4p	^{8}C
04/26/1977	Holt	[50]	mm	^{16}O(π^+,π^-)	2p	^{16}Ne
10/20/1977	KeKelis	[51]	mm	^{16}O(α,^{8}He)	2p	^{12}O
			mm	^{20}Ne(^{3}He,^{8}Li)	1p	^{15}F
12/14/1987	Woods	[52]	mm	^{40}Ca(^{14}N,^{15}C)	1p	^{39}Sc
07/19/1993	Batchelder	[53]	β^+p	^{40}Ca(^{36}Ar,3n)	1p	^{73}Rb
12/21/2001	Lépine-Szily	[54]	mm	^{10}B(^{14}N,^{14}B)	1p	^{10}N
09/30/2003	Zerguerras	[55]	im	^{20}Mg(−1p1n)	1p	^{18}Na
03/27/2007	Mukha	[56]	if	^{20}Mg(−1n)	2p	^{19}Mg
04/23/2010	Goldberg	[57]	scatt.	^{13}O(p,p′)[a]	1p	^{14}F
12/17/2010	Rogers	[58]	im	^{70}Se(+1p−2n)	1p	^{69}Br
01/17/2014	Wamers	[59]	im	^{17}Ne(−2n)	1p	^{15}Ne
08/06/2015	Mukha	[60]	if	^{31}Ar(−1n)	2p, 1p	^{29}Cl, ^{30}Ar

[a] Inverse kinematics

The determination of an attractive nuclear force between two protons is even more difficult than between two neutrons because of the presence of the repulsive Coulomb force. It can be argued that the first detailed analysis and extraction of such a proton-proton interaction was demonstrated in the scattering experiments of Tuve, Heydenburg, and Hafstad at the Department of Terrestrial Magnetism of the Carnegie Institution of Washington, Washington D.C., in 1936 [40] together with the accompanying phase-shift analysis by Breit et al. [61]. Later analyses extracted a scattering length of about −7.7 fm [62].

The ground-state of ^{5}Li was also determined in a scattering experiment at the Carnegie Institution. In 1941, Heydenburg and Ramsey observed a resonance at about 2 MeV by irradiating a helium gas target with protons [42].

The first missing mass experiment to determine the mass excess or the decay energy of unbound isotopes was performed by Haxby et al. in 1940 at the Westinghouse Research Laboratories in East Pittsburgh, Pennsylvania. They used protons accelerated by the Westinghouse pressure electrostatic generator to irradiate a ^{9}Be target and deduced the separation energy from the onset of neutron emission as a function of beam energy [41]. The first nucleus unbound with respect to two-proton emission was ^{6}Be, again populated in a charge exchange reaction. In 1957, Bogdanov et al. observed with ^{6}Be the first nuclide which was unbound with respect to two-proton emission by irradiating enriched ^{6}Li with 9.6 MeV protons from the 1.5 m cyclotron of the U.S.S.R. Academy of Sciences Nuclear Energy Institute. They deduced the excitation energy of ^{6}Be from the time-of-flight of the emitted neutrons [43].

The only three- and four-proton emitters identified so far are ^{7}B and ^{8}C which were discovered at Berkeley by McGrath, Cerny, and Norbeck in 1967 [46] and at the Kernforschungsanlage Jülich, Germany, by Robertson et al. in 1974 [49], respectively.

Three other one-proton emitters (^{4}Li [45], ^{19}Na [47], and ^{15}F [51]) as well as one additional two-proton emitter (^{12}O [51]) were discovered between 1965 and 1977 at Berkeley. The first resonances measured in ^{16}F at the Los Alamos variable energy cyclotron by Bryant et al. in 1963 [44] and ^{11}N at the cyclotron laboratory at Michigan State University by Benenson et al. in 1974 [48] corresponded to excited states.

In 1977, Holt et al. discovered ^{16}Ne in a double pion charge-exchange reaction on ^{16}O using LAMPF at Los Alamos National Laboratory [50]. The heaviest unbound proton emitter identified with a missing mass measurement is ^{39}Sc discovered by Woods et al. in 1987 with the 14UD pelletron accelerator at the Australian National University in Canberra, Australia [52]. Only about six weeks later, Mohar et al. reported their independent identification of ^{39}Sc [63]. The last unbound isotope measured with the missing mass method was ^{10}N in 2001 by Lépine-Szily et al. at GANIL [54].

The last isotope beyond the proton drip-line populated utilizing a stable beam was the measurement of an excited state of ^{73}Rb by β-delayed proton emission following the fusion-evaporation reaction ^{40}Ca(^{36}Ar, 3n) in 1993 by Batchelder et al. at Berkeley [53].

So far, seven proton-unbound isotopes have been discovered with secondary beams. Three of them were identified with the invariant mass method. In 2003, Zerguerras et al. produced a secondary beam of ^{20}Mg with the SISSI solenoids and the ALPHA spectrometer at GANIL and reconstructed two resonances in ^{18}Na [55]. The decay energy of the ground-state of ^{69}Br was measured in 2010 by Rogers et al. with the S800 spectrometer at the NSCL [58]. ^{69}Br had initially been reported as bound [64], however, a subsequent experiment established an upper limit for the half-life of 100 ns [65]. In 2014 Wamers et al. populated ^{15}Ne in the two-neutron removal reaction at GSI. They measured ^{13}O fragments in coincidence with two protons to extract the ground-state energy of ^{15}Ne [59].

A new innovative technique was developed by Mukha et al. in 2007 to discover the unbound isotope ^{19}Mg [56]. ^{19}Mg was populated by neutron removal from a secondary ^{20}Mg beam and decayed by the emission of two protons in flight. The lifetime was then measured by reconstructing the vertex of the decay to ^{17}Ne and two protons. This method was also recently applied to identify the decay of ^{30}Ar

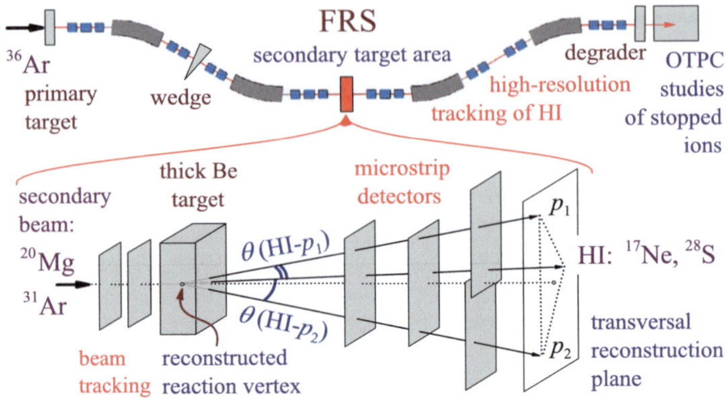

Fig. 16.5 *Upper part* The experiment layout at the FRS fragment separator. *Lower part* Sketch of the detector set-up at the secondary-target area measuring trajectories of the incoming ^{31}Ar (^{20}Mg) ions and their decay products ^{28}S (^{17}Ne) and protons p_1 and p_2 [60] (Reprinted figure with permission from I. Mukha et al., *Observation and spectroscopy of new proton-unbound isotopes* ^{30}*Ar and* ^{29}*Cl: An interplay of prompt two-proton and sequential decay*, Phys. Rev. Lett. 115 (2015) 202501. Copyright 2015 by the American Physical Society.)

Fig. 16.6 Experimental set-up to measure the decay of ^{14}F at the Cyclotron Institute of Texas A&M University [57] (Reprinted from V.Z. Goldberg et al., *First observation of* ^{14}*F*, Phys. Lett. B 692 (2010) 307, published under Creative Commons license CC-BY 3.0 [66].)

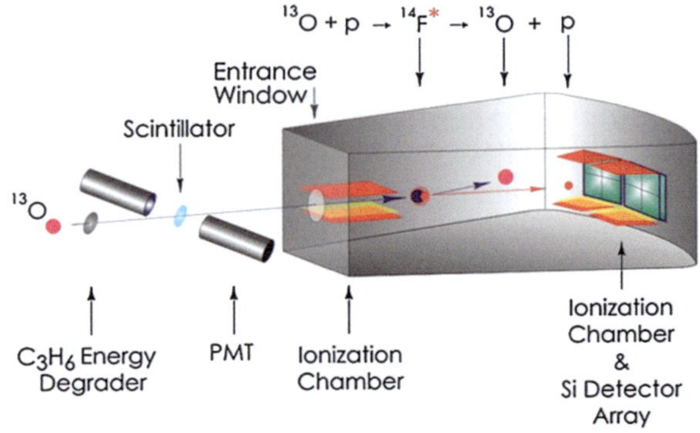

populated in the one-neutron removal reaction from ^{31}Ar. In addition, the decay energy of the intermediate isotope ^{29}Cl was extracted [60]. Figure 16.5 shows the experimental layout at GSI and a schematic of the detector set-up.

In yet another different approach, Goldberg et al. utilized the "thick target inverse kinematics" method to measure proton elastic scattering of ^{13}O [57]. A secondary ^{13}O beam was stopped in a methane gas target and the excitation function in ^{14}F was deduced from the measured energy of the elastically scattered protons as shown in Fig. 16.6.

16.4 Proton Radioactivity

Proton radioactivity was first observed in the decay of an excited isomeric state in ^{63}Co in 1970 (see Sect. 11.5) and the first ground-state proton emitter was only discovered eleven years later in 1981. Although already predicted in 1960 by Goldanskii [4], the even more exotic decay mode of two-proton radioactivity was not discovered until 2002.

All one-proton radioactive isotopes have been discovered in fusion-evaporation reactions. In most cases the residues were selected with a mass separator and then implanted into a catcher or directly into a silicon detector to measure the subsequently emitted protons. The half-life was then determined from the time difference between implantation and decay. In contrast, the three presently known two-proton radioactive isotopes have first been observed in projectile fragmentation reactions. All one- and two-proton radioactive nuclides are listed in Table 16.3.

The table shows the increasing sensitivity of the fusion-evaporation measurements. While the first discoveries utilized 1p2n evaporation reactions with cross-sections on the order of tens of microbarns, the latest discoveries used reactions evaporating up to 6 neutrons with cross-sections which are about a factor of 100,000 smaller. The projectile fragmentation experiments are even more sensitive reaching cross-sections as low as 50 fb.

Proton radioactive isotopes between antimony and bismuth are shown in the section of the chart of nuclides presented in Fig. 16.7. All isotopes to the left of the drip-line indicated by the thick black line are unbound with respect to proton emission. However, due to the Coulomb barrier, not all of them decay by proton emission. The figure shows that proton-radioactivity has only been observed in odd-Z isotopes and that in the mass region between lanthanum and terbium there is still a gap of unknown isotopes between the last β^+ or α-emitting unbound isotope and the first proton emitter. The figure also demonstrates that the proton drip-line has not even been reached for the even-Z isotopes between cerium and hafnium.

The first four isotopes listed in Table 16.3 were not discovered by their proton emission. Hofmann et al. measured the α-decay of ^{157}Ta and ^{161}Re in 1978 [67] and ^{166}Ir and ^{167}Ir in 1981 [68] with the velocity separator SHIP (Separator for Heavy Ion reaction Products) at GSI. It took almost twenty years until the proton decay branches for these nuclides were observed [79, 95].

The first ground-state proton-radioactive isotope discovered was ^{151}Lu in 1981. Hofmann et al. used SHIP to separate the evaporation residues and implanted them into position-sensitive surface barrier detectors [69]. Figure 16.8 shows the spectrum of the particles emitted after the implantation clearly showing the proton line of the decay of ^{151}Lu. The proton character of the activity was ensured by a ΔE-E measurement. Only a month later, Klepper et al. submitted the discovery of ^{147}Tm using the GSI on-line mass separator [70]. The residues were implanted into carbon foils placed in front of surface barrier detector telescopes to measure subsequent proton emissions.

Table 16.3 Proton-radioactive isotopes: The submission date, first author and reference of the publication as well as the reaction and the cross-section (σ) are listed

Date	First author	Ref.	Reaction	σ	Isotope
11/27/1978	Hofmann	[67]	^{107}Ag(^{58}Ni, 4n)	–[a]	^{161}Re, ^{157}Ta[b]
02/03/1981	Hofmann	[68]	Sn(^{58}Ni, 1pxn)	–[a]	^{166}Ir, ^{167}Ir
12/02/1981	Hofmann	[69]	^{96}Ru(^{58}Ni, 1p2n)	70 μb	^{151}Lu
01/08/1982	Klepper	[70]	^{92}Mo(^{58}Ni, 1p2n)	30 μb	^{147}Tm
12/16/1983	Faestermann	[71]	^{58}Ni(^{58}Ni, 1p2n)	30 μb	^{113}Cs
			^{54}Fe(^{58}Ni,1p2n)	40 μb	^{109}I
11/21/1991	Page	[72]	^{106}Cd(^{58}Ni, 1p3n)	1 μb	^{160}Re, ^{156}Ta[b]
01/15/1993	Sellin	[73]	^{96}Ru(^{58}Ni, 1p3n)	–	^{150}Lu
04/01/1993	Livingston	[74]	^{92}Mo(^{58}Ni, 1p3n)	–	^{146}Tm
11/01/1993	Tighe	[75]	^{50}Cr(^{58}Ni, 1p2n)	50 μb	^{105}Sb
11/08/1993	Page	[76]	^{58}Ni(^{58}Ni, 1p3n)	500 nb	^{112}Cs
09/18/1995	Davids	[77]	^{95}Mo(^{92}Mo, 1p1n)	100 nb	^{185}Bi
07/25/1996	Blank	[78]	^{58}Ni(^{9}Be, X)[c]	900 fb[d]	^{45}Fe
01/13/1997	Davids	[79]	^{96}Ru(^{78}Kr, 1p2n)	2 μb	^{171}Au
			^{92}Mo(^{78}Kr, 1p4n)	0.2 μb	^{165}Ir
11/19/1997	Davids	[80]	^{96}Ru(^{40}Ca, 1p4n)	90 nb	^{131}Eu
			^{92}Mo(^{54}Fe, 1p4n)	250 nb	^{141}Ho
12/01/1997	Batchelder	[81]	^{92}Mo(^{58}Ni, 1p4n)	500 nb	^{145}Tm
02/24/1999	Rykaczewski	[82]	^{92}Mo(^{54}Fe, 1p5n)	13 nb	^{140}Ho
03/09/1999	Poli	[83]	^{102}Pd(^{78}Kr, 1p2n)	10 nb	^{177}Tl
10/20/1999	Blank	[84]	^{58}Ni(^{nat}Ni, X)[c]	50 fb[d]	^{48}Ni
10/03/2000	Soramel	[85]	^{64}Zn(^{58}Ni, 1p4n)	200 nb	^{117}La
10/13/2003	Davids	[86]	^{58}Ni(^{78}Kr, 1p5n)	9 nb	^{130}Eu
12/09/2003	Kettunen	[87]	^{96}Ru(^{78}Kr, 1p3n)	90 nb	^{170}Au
			^{102}Pd(^{78}Kr, 1p3n)	3 nb	^{176}Tl
02/20/2004	Woods	[88]	^{92}Mo(^{50}Cr, 1p6n)	3 nb	^{135}Tb
02/19/2005	Blank	[89]	^{58}Ni[c]	100 fb	^{54}Zn
04/12/2005	Robinson	[90]	^{92}Mo(^{36}Ar, 1p6n)	300 pb	^{121}Pr
07/04/2006	Joss	[91]	^{106}Cd(^{58}Ni, 1p4n)	–	^{159}Re
03/14/2007	Page	[92]	^{106}Cd(^{58}Ni, 1p4n)	–	^{155}Ta[b]
12/11/2012	Drummond	[93]	^{92}Mo(^{78}Kr, 1p5n)	4 nb	^{164}Ir

[a] Discovered by α-decay, proton decay branch was discovered later (see text)
[b] Populated by α-decay
[c] Projectile fragmentation reaction
[d] Measured in a later experiment [94]

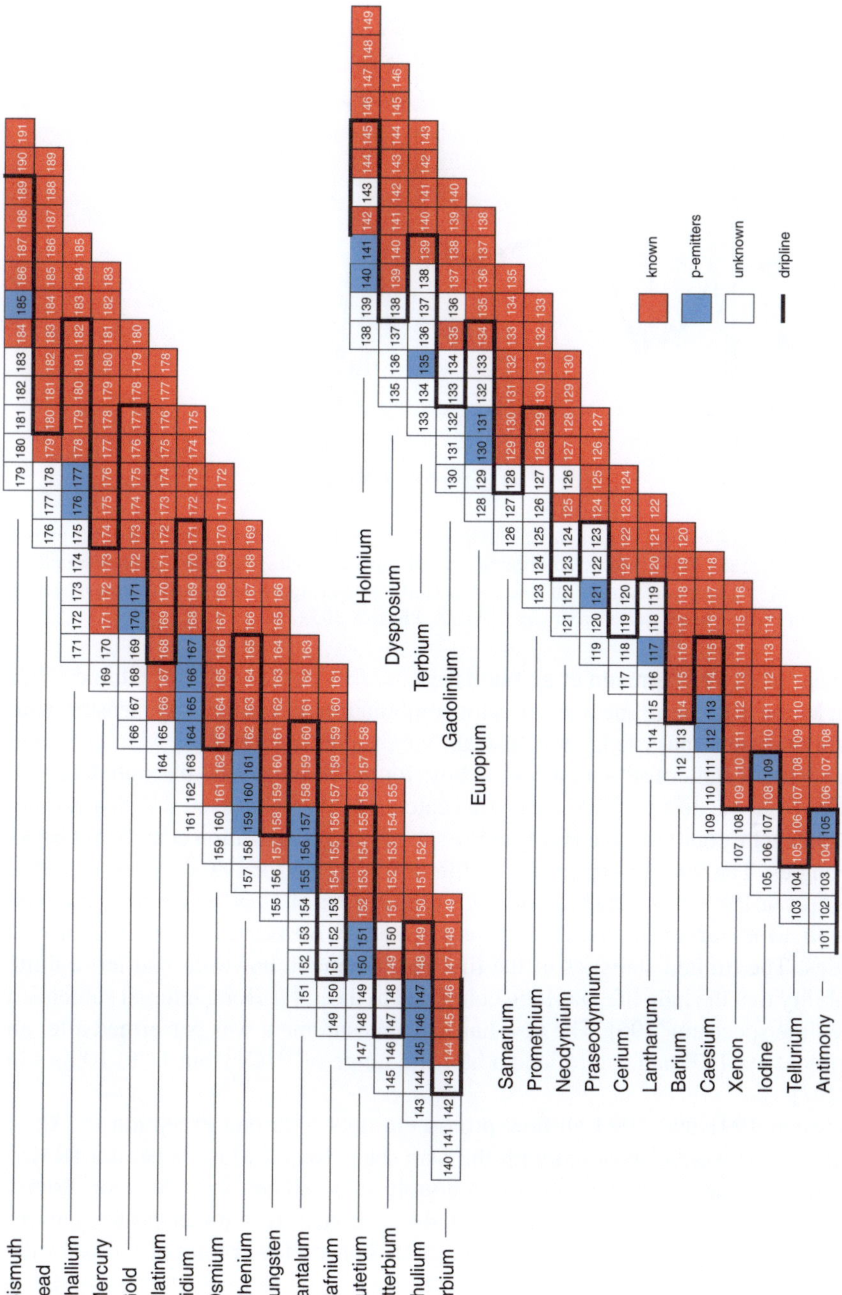

Fig. 16.7 Section of the chart of nuclides which shows the proton-radioactive isotopes

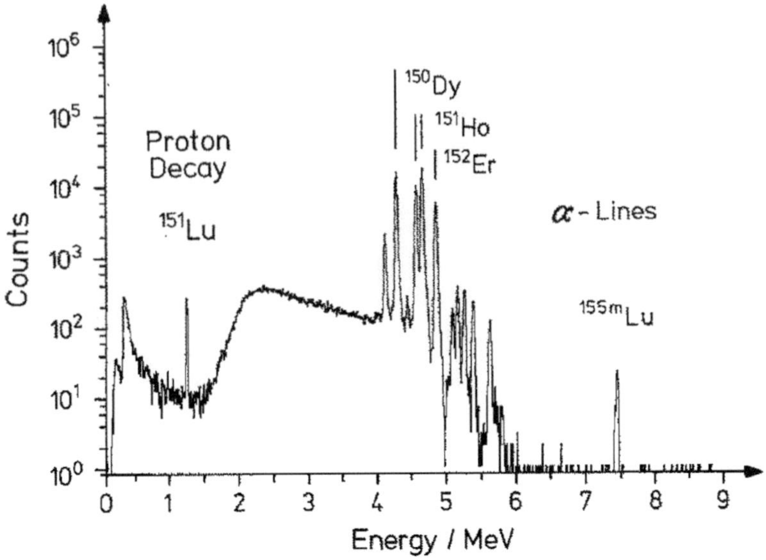

Fig. 16.8 Energy spectrum of particles emitted from evaporation residues implanted at the end of the velocity separator SHIP [69]. In addition to several α-particle lines, a proton line corresponding to the decay of ^{151}Lu is indicated (With kind permission from Springer Science and Business Media: S. Hofmann et al., *Proton radioactivity of* ^{151}*Lu*, Z. Phys. A 305 (1982) 111.)

A year later, Faestermann et al. published the first observation of ^{109}I and ^{113}Cs [71] at Munich Tandem-linear accelerator combination. It was the only measurement that did not use a mass separator. Instead, the evaporation residues were collected on an annular catcher foil with a central hole for the beam to pass through. Charged particles from the decay of particles in the catcher foil were measured with a parallel plate avalanche counter and a Bragg curve spectroscopy ionization chamber at backward angles. The initially reported half-life of $0.9^{+1.3}_{-0.4}$ μs for ^{113}Cs was corrected to 33 ± 7 μs in 1987: "The half-life of ^{113}Cs of the order of 1 μs, which was observed in our first measurement, could not be confirmed by measurements with improved statistics. The limited statistics of the first measurement, however, had left a finite probability that the half-life was long compared to the 1.4 μs long interval subtended by the time spectrum" [96]. The new half-life measurement was performed after an experiment at SHIP had determined a half-life range of ^{113}Cs from 10 to 100 μs in 1984 [97].

Between 1991 and 1993 all new proton emitters with one exception (^{105}Sb at Berkeley [75]) were discovered with the Daresbury Recoil Mass Separator (RMS) where the residues were implanted in a double-sided silicon strip detector. Proton emission of ^{156}Ta was observed in the α-decay branch of another proton emitter (^{160}Re) [72]. In 1989 Hofmann et al. had reported only a lower limit of 10 ms for the β-decay half-life of a high-spin state of ^{156}Ta [98].

The next five proton emitters were discovered between 1995 and 1997 by Davids et al. with the Fragment Mass Separator (FMS) at Argonne National Laboratory.

Also, Poli et al. (^{177}Tl, [83]), Davids et al. (^{130}Eu [86]), Woods et al. (^{135}Tb [88]), and Robinson et al. (^{121}Pr [90]) identified an additional four proton emitters until 2005 at the FMS. The observation of ^{121}Pr [90] was incompatible with a 1973 report of ground-state proton radioactivity from ^{121}Pr with a half-life of 1 s by Bogdanov et al. [99]. If correct, that observation would have represented the discovery of proton radioactivity from the ground-state.

During this time two proton emitters (^{145}Tm [81] and ^{140}Ho [82]) were measured with the recoil mass separator (RMS) at Oak Ridge. The measured half-life of $3.5 \pm 1.0\,\mu$s for ^{145}Tm represents the shortest observed proton-radioactivity so far [81]. The only proton-radioactive nuclide discovered in Italy was ^{117}La which was produced with the Tandem plus LINAC accelerator of the Laboratori Nazionali di Legnaro in Padova and identified using their RMS [85].

In 2003, Kettunen et al. reported the discovery of the first proton emitters in Finland. They identified ^{170}Au and ^{176}Tl with the gas-filled recoil separator RITU at the Jyväskylä accelerator laboratory. For the three most recent discoveries, the isotopes were implanted in the Gamma Recoil Electron Alpha Tagging (GREAT) Spectrometer. After the proton emission of ^{159}Re had been measured in 2006 by Joss et al. [91], an α-decay branch was observed a year later. This decay branch populated ^{155}Ta which then again was observed to decay by proton emission [92]. The measured decay properties contradicted an earlier observation of a proton branch of ^{155}Ta [100] which was speculated to be contaminated with α-decay events from an isomeric state in ^{155}Lu [92].

Finally, Table 16.3 also lists the three two-proton radioactive isotopes ^{45}Fe, ^{48}Ni, and ^{54}Zn. Blank et al. discovered ^{45}Fe [78] and ^{48}Ni [84] at GSI in 1996 and at GANIL in 1999, respectively, in projectile fragmentation without being able to measure their decay properties. The discovery of two-proton radioactivity of ^{45}Fe was almost simultaneously reported from GSI and GANIL in 2002. Pfützner et al. submitted their result from the fragmentation of a 600 MeV/nucleon ^{58}Ni beam at SIS/FRS on May 17, 2002 [101], while Giovinazzo et al. submitted their results of the fragmentation of 75 MeV/nucleon ^{58}Ni at the SISSI-LISE3 facility only four days later [102]. The two-proton radioactivity of ^{48}Ni was discovered by Dossat et al. at GANIL in 2005. In the same year Blank et al. discovered ^{54}Zn also at GANIL and extracted a two-proton decay half-life of $3.2^{+1.8}_{-0.8}$ ms from the space and time correlations between the implantation of ^{54}Zn and subsequently emitted protons [89].

References

1. M. Langevin et al., Phys. Lett. B **125**, 116 (1983)
2. Z. Kohley et al., Phys. Rev. Lett. **110**, 152501 (2013)
3. IUPAC Transfermium Working Group, Pure Appl. Chem. **63**, 879 (1991)
4. V.I. Goldanskii, Annu. Rev. Nucl. Part. Sci. **16**, 1 (1960)
5. M. Thoennessen, Rep. Prog. Phys. **67**, 1187 (2004)
6. T. Baumann, A. Spyrou, M. Thoennessen, Rep. Prog. Phys. **75**, 036301 (2012)
7. J. Cerny, J.C. Hardy, Annu. Rev. Nucl. Part. Sci. **27**, 323 (1977)

8. J.H. Williams, W.G. Shepherd, R.O. Haxby, Phys. Rev. **51**, 888 (1937)
9. R.P. Haddock et al., Phys. Rev. Lett. **14**, 318 (1965)
10. R.H. Stokes, P.G. Young, Phys. Rev. Lett. **18**, 611 (1967)
11. K.H. Wilcox et al., Phys. Lett. B **59**, 142 (1975)
12. U. Sennhauser et al., Phys. Lett. B **103**, 409 (1981)
13. D.V. Aleksandrov et al., Sov. J. Nucl. Phys. **37**, 474 (1983)
14. D.V. Aleksandrov et al., Sov. J. Nucl. Phys. **39**, 323 (1984)
15. K.K. Seth et al., Phys. Rev. Lett. **58**, 1930 (1987)
16. A.A. Korsheninnikov et al., Phys. Lett. B **326**, 31 (1994)
17. R. Kalpakchieva et al., Eur. Phys. J. A **7**, 451 (2000)
18. A.A. Korsheninnikov et al., Phys. Rev. Lett. **87**, 092501 (2001)
19. A.A. Korsheninnikov et al., Phys. Rev. Lett. **90**, 082501 (2003)
20. C.R. Hoffman et al., Phys. Rev. Lett. **100**, 152502 (2008)
21. Y. Aksyutina et al., Phys. Lett. B **666**, 430 (2008)
22. A. Spyrou et al., Phys. Lett. B **683**, 129 (2010)
23. C.C. Hall et al., Phys. Rev. C **81**, 021302 (2010)
24. G. Christian et al., Phys. Rev. Lett. **108**, 032501 (2012)
25. A. Spyrou et al., Phys. Rev. Lett. **108**, 102501 (2012)
26. E. Lunderberg et al., Phys. Rev. Lett. **108**, 142503 (2012)
27. J. Snyder et al., Phys. Rev. C **88**, 031303 (2013)
28. J.H. Williams, W.G. Shepherd, R.O. Haxby, Phys. Rev. **52**, 390 (1937)
29. W. von Witsch et al., Nucl. Phys. A **329**, 141 (1979)
30. M.S. Golovkov et al., Phys. Rev. C **72**, 064612 (2005)
31. Y.B. Gurov et al., Bull. Rus. Acad. Sci. Phys. **73**, 139 (2009)
32. B.M.K. Nefkens, Phys. Rev. Lett. **10**, 55 (1963)
33. R.J. Cence, C.N. Waddell, Phys. Rev. **128**, 1788 (1962)
34. A. Schwarzschild et al., Phys. Rev. **133**, B1 (1964)
35. N.K. Sherman, P. Barreau, Phys. Lett. **9**, 151 (1964)
36. M. Caamano et al., Phys. Rev. Lett. **99**, 062502 (2007)
37. Z. Kohley et al., Phys. Rev. C **87**, 011304 (2013)
38. J.D. Stevenson, P.B. Price, Phys. Rev. C **24**, 2102 (1981)
39. M. Langevin et al., Phys. Lett. B **150**, 71 (1985)
40. M.A. Tuve, N.P. Heydenburg, L.R. Hafstad, Phys. Rev. **50**, 806 (1936)
41. R.O. Haxby et al., Phys. Rev. **58**, 1035 (1940)
42. N.P. Heydenburg, N.F. Ramsey, Phys. Rev. **60**, 42 (1941)
43. G.F. Bogdanov et al., J. Nucl. Energy II **8**, 148 (1958)
44. H.C. Bryant et al., Nucl. Phys. **53**, 97 (1964)
45. J. Cerny, C. Détraz, R.H. Pehl, Phys. Rev. Lett. **15**, 300 (1965)
46. R.L. McGrath, J. Cerny, E. Norbeck, Phys. Rev. Lett. **19**, 1442 (1967)
47. J. Cerny et al., Phys. Rev. Lett. **22**, 612 (1969)
48. W. Benenson et al., Phys. Rev. C **9**, 2130 (1974)
49. R.G.H. Robertson et al., Phys. Rev. Lett. **32**, 1207 (1974)
50. R.J. Holt et al., Phys. Lett. B **69**, 55 (1977)
51. G.J. KeKelis et al., Phys. Rev. C **17**, 1929 (1978)
52. C.L. Woods et al., Nucl. Phys. A **484**, 145 (1988)
53. J.C. Batchelder et al., Phys. Rev. C **48**, 2593 (1993)
54. A. Lépine-Szily et al., Phys. Rev. C **65**, 054318 (2002)
55. T. Zerguerras et al., Eur. Phys. J. A **20**, 389 (2004)
56. I. Mukha et al., Phys. Rev. Lett. **99**, 182501 (2007)
57. V.Z. Goldberg et al., Phys. Lett. B **692**, 307 (2010)
58. A.M. Rogers et al., Phys. Rev. Lett. **106**, 252503 (2011)
59. F. Wamers et al., Phys. Rev. Lett. **112**, 132501 (2014)
60. I. Mukha et al., Phys. Rev. Lett. **115**, 202501 (2015)
61. G. Breit, E.U. Condon, R.D. Present, Phys. Rev. **50**, 825 (1936)

62. J.D. Jackson, J.M. Blatt, Rev. Mod. Phys. **22**, 77 (1950)
63. M.F. Mohar et al., Phys. Rev. C **38**, 737 (1988)
64. M.F. Mohar et al., Phys. Rev. Lett. **66**, 1571 (1991)
65. B. Blank et al., Phys. Rev. Lett. **74**, 4611 (1995)
66. http://creativecommons.org/licenses/by/3.0/
67. S. Hofmann et al., Z. Phys. A **291**, 53 (1979)
68. S. Hofmann et al., Z. Phys. A **299**, 281 (1981)
69. S. Hofmann et al., Z. Phys. A **305**, 111 (1982)
70. O. Klepper et al., Z. Phys. A **305**, 125 (1982)
71. T. Faestermann et al., Phys. Lett. B **137**, 23 (1984)
72. R.D. Page et al., Phys. Rev. Lett. **68**, 1287 (1992)
73. P.J. Sellin et al., Phys. Rev. C **47**, 1933 (1993)
74. K. Livingston et al., Phys. Lett. B **312**, 46 (1993)
75. R.J. Tighe et al., Phys. Rev. C **49**, R2871 (1994)
76. R.D. Page et al., Phys. Rev. Lett. **72**, 1798 (1994)
77. C.N. Davids et al., Phys. Rev. Lett. **76**, 592 (1996)
78. B. Blank et al., Phys. Rev. Lett. **77**, 2893 (1996)
79. C.N. Davids et al., Phys. Rev. C **55**, 2255 (1997)
80. C.N. Davids et al., Phys. Rev. Lett. **80**, 1849 (1998)
81. J.C. Batchelder et al., Phys. Rev. C **57**, R1042 (1998)
82. K. Rykaczewski et al., Phys. Rev. C **60**, 011301 (1999)
83. G. Poli et al., Phys. Rev. C **59**, R2979 (1999)
84. B. Blank et al., Phys. Rev. Lett. **84**, 1116 (2000)
85. F. Soramel et al., Phys. Rev. C **63**, 031304 (2001)
86. C.N. Davids et al., Phys. Rev. C **69**, 011302 (2004)
87. H. Kettunen et al., Phys. Rev. C **69**, 054323 (2004)
88. P.J. Woods et al., Phys. Rev. C **69**, 051302 (2004)
89. B. Blank et al., Phys. Rev. Lett. **94**, 232501 (2005)
90. A.P. Robinson et al., Phys. Rev. Lett. **95**, 032502 (2005)
91. D.T. Joss et al., Phys. Lett. B **641**, 34 (2006)
92. R.D. Page et al., Phys. Rev. C **75**, 061302 (2007)
93. M.C. Drummond et al., Phys. Rev. C **89**, 064309 (2014)
94. C. Dossat et al., Phys. Rev. C **72**, 054315 (2005)
95. R.J. Irvine et al., Phys. Rev. C **55**, R1622 (1997)
96. A. Gillitzer et al., Z. Phys. A **326**, 107 (1987)
97. S. Hofmann et al., in *Proceedings of the 7th International Conference on Atomic Masses and Fundamental Constants, AMCO-7 Darmstadt, 1984*, vol. 26, ed. by O. Klepper (THD, Schriftenreiche Wissenschaft und Technik, Darmstadt, 1984), p. 184
98. S. Hofmann et al., Z. Phys. A **333**, 107 (1989)
99. D.D. Bogdanov et al., Sov. J. Nucl. Phys. **16**, 491 (1973)
100. J. Uusitalo et al., Phys. Rev. C **59**, R2975 (1999)
101. M. Pfützner et al., Eur. Phys. J. A **14**, 279 (2002)
102. J. Giovinazzo et al., Phys. Rev. Lett. **89**, 102501 (2002)

Chapter 17
Summary and Outlook

17.1 Present Status

Until the end of 2015 a total of 3211 different isotopes have been identified. This includes 44 short-lived ($\leq 10^{12}$ s) unbound isotopes. Figure 17.1 shows the evolution of the number of discovered isotopes as a function of time. The top panel shows the integral, cumulative number of known isotopes as a function of years since the discovery of radioactivity in 1896. The total number of isotopes (black, solid line) is broken down into four broad areas of nuclides: stable and near-stable (red, short-dashed lines), neutron-deficient (purple, dot-dashed lines), neutron-rich (green, long-dashed lines) and transuranium (blue, dotted lines) nuclides. Near-stable nuclides are defined as all nuclides between the most neutron-deficient and neutron-rich stable isotopes of a given element. This also includes isotopes of the four radioactive decay chains and isotopes between them. Lighter and heavier radioactive isotopes of these elements are then classified as neutron-deficient and neutron-rich, respectively.

Almost all stable and near-stable nuclides had been discovered by the mid 1950s. Within this classification the last near-stable isotope discovered was ^{222}Fr. It was identified only in 1975 by Westgaard et al. at CERN in the spallation of 600 MeV protons on a thorium target [1] (see Sect. 10.4).

The difficulty of producing heavy neutron-rich nuclides is demonstrated by the fact that it took about 60 years to discover ^{213}Pb after the first observation of ^{212}Pb and ^{214}Pb. The latter isotopes were discovered in the radioactive decay chains by Rutherford in 1899 [3] and 1903 [4], respectively (see Sect. 2.2). Butement, Robinson, and Qaim observed ^{213}Pb only in 1963 by irradiating thorium with 370 MeV protons from the cyclotron of the University of Liverpool [5] (see Sect. 10.3).

The total number of discovered isotopes on the proton-rich side of the valley of stability has been slowly saturating over the last twenty years. Presently there are almost 1300 proton-rich nuclides known. The proton drip-line has been reached—and crossed in many cases—for all elements, so the number of isotopes left to be discovered is more and more limited.

M. Thoennessen, *The Discovery of Isotopes*,
DOI 10.1007/978-3-319-31763-2_17

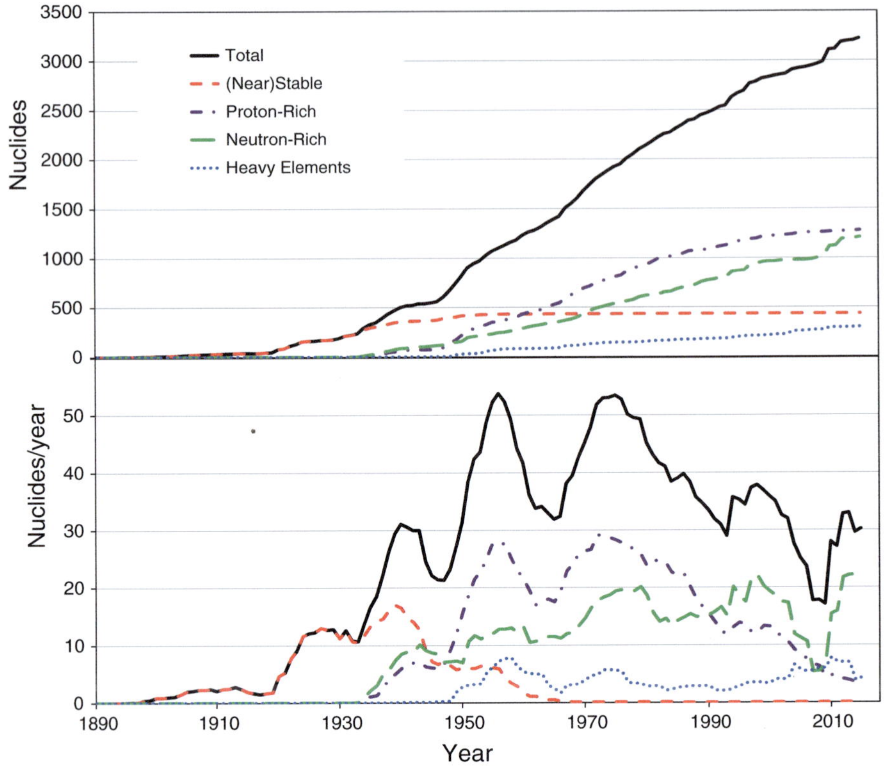

Fig. 17.1 Discovery of nuclides as a function of year. The *top panel* shows the cumulative sum of the discovered nuclides while the *bottom panel* shows the 10-year running average of the number of nuclides discovered per year. The total number of nuclides is shown by the *black, solid lines*. Separately plotted are near-stable (*red, short-dashed lines*), neutron-deficient (*purple, dot-dashed lines*), neutron-rich (*green, long-dashed lines*) and transuranium (*blue, dotted lines*) nuclides. The first version of this figure including isotopes discovered up to the end of 2011 was published in Ref. [2]

On the other hand, the number of observed neutron-rich isotopes continues to increase. With over 1200 discovered isotopes there are now almost as many neutron-rich isotopes known as there are proton-rich isotopes.

As can be seen in the figure, the number of discoveries has not been increasing at a constant rate. This is more obvious by plotting the rate of new nuclides identified per year, shown in the bottom panel of Fig. 17.1. The figure displays the 10-year running average in order to smooth out year-to-year fluctuations. The apparent peaks can directly be correlated with major advances in accelerator technologies or new developments of experimental techniques.

Figure 17.2 shows the cumulative number of discovered isotopes (top panel) and the discovery rate (bottom panel, 10-year running average) for the various production methods.

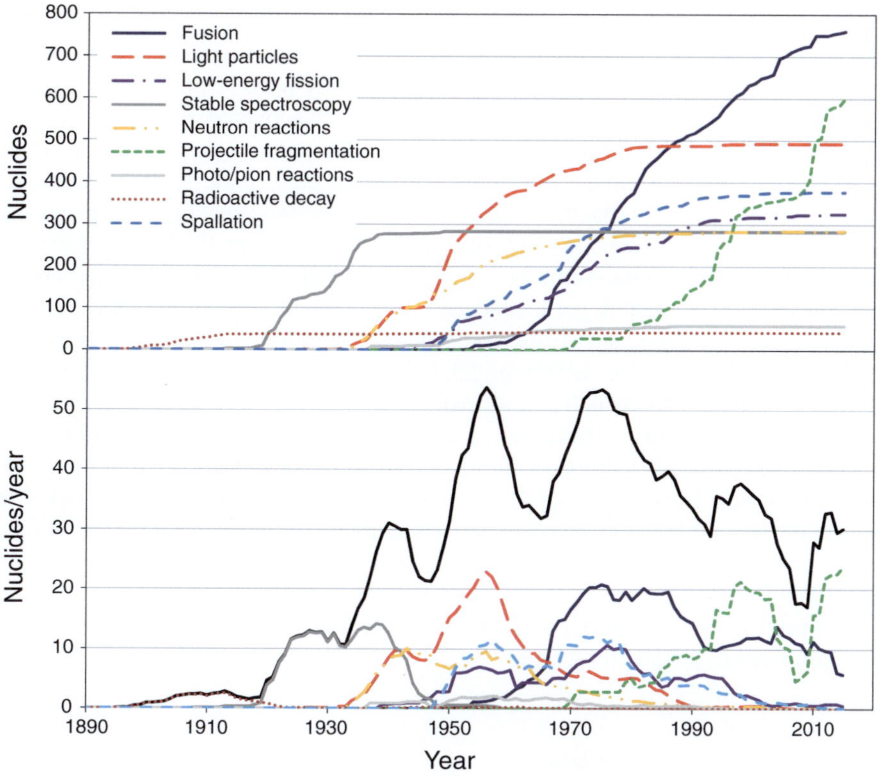

Fig. 17.2 Discovery of nuclides by various production mechanisms as a function of year. The *top panel* shows the cumulative sum of the discovered nuclides while the *bottom panel* shows the 10-year running average of the number of nuclides discovered per year. The line-types and colors for the different mechanisms are explained in the legend. The *bottom panel* also shows the total number of isotopes discovered per year

The last two isotopes discovered in traditional mass spectroscopy of essentially stable isotopes were ^{190}Pt ($T_{1/2} = 6.5 \times 10^{11}$ y) and ^{50}V ($T_{1/2} = 1.4 \times 10^{17}$ y). Duckworth, Black, and Woodcock identified ^{190}Pt at Wesleyan University in Middletown, Connecticut, in March 1949 [6], while ^{50}V was simultaneously discovered by Hess and Inghram at Argonne National Laboratory [7] and by Leland at the University of Minnesota [8] later in the same year (see Sect. 3.4).

^{206}Hg was the last natural radioactive nuclide discovered. It was detected in the α-decay branch of ^{210}Pb by Nurmia et al. at the University of Helsinki, Finland, in 1961 [9] (see Sect. 2.4).

Light-charged-particles and neutrons were utilized in the discovery of new isotopes for the last time in 1998 and 2003, respectively. Laue et al. produced ^{231}Pu in the ^{233}U(^{3}He,5n) reaction at the Berkeley 88-in. cyclotron [10] (see Sect. 5.4) and ^{197}Os was discovered by Xu et al. at the Institute of Modern Physics in Lanzhou, China, in the reaction ^{198}Pt(n,2p) ([11] (see Sect. 7.3).

Essentially, the only production techniques that are currently still being used to discover new isotopes are fusion-evaporation and projectile fragmentation reactions, where the latter also includes projectile fission. They have been, overall, also the most productive methods with 760 and 600 isotopes discovered, respectively.

The correlation of the peaks with the different production methods is demonstrated in the bottom panel of Fig. 17.2. The first significant peak in the 1920s is due to the development of Aston's first mass spectrograph. His second spectrograph generated a second peak for the spectroscopy of stable isotopes in the 1930s. However, the overall peak in the 1930s—and the continuation until the late 1950s—was due to the construction of the first particle accelerators. The significant dip in the 1940s is probably due to the Second World War. The largest production rate over a 10-year period ever was reached in 1956 with 53.8 discoveries per year.

The next major peak which almost matched this rate (53.5 isotopes/year) was due to the broad availability of heavy-ion accelerators in the 1970s. Since then the production rate has steadily dropped, only interrupted by the development of projectile fragmentation at Berkeley, GANIL, Michigan State University, GSI, and RIKEN in the 1990s. By 2009 the rate had dropped to less than 20 isotopes/year which was the lowest rate since 1935. Major upgrades at RIKEN, Michigan State University, and GANIL resulted in the most recent increase in productivity.

Overall the average production rate has been about 30 isotopes per year which also corresponds to the present rate.

17.2 Future Perspectives

The presently known 3211 nuclides still constitute less than 50 % of all nuclides that potentially could exist. Recently, Erler et al. estimated that about 7000 bound nuclides could exist for elements up to 118 [12]. Figure 17.3 shows the chart of nuclides where the stable and primordial isotopes are shown as black squares, presently known isotopes in red and unknown isotopes predicted to exist by Erler et al. in light blue.

The approximately 2000 nuclides in the region of $Z > 82$ and $N > 184$ will most probably never be produced. In addition, another 500 very neutron-rich isotopes above $Z \sim 50$ will most likely stay out of reach. Thus about 1500 mostly neutron-rich nuclides can be expected to be discovered in the future. If one includes proton-unbound nuclides along the proton drip-line with half-lives as short as nano-seconds, another 300 should be accessible.

As mentioned in the previous section only fusion-evaporation and projectile fragmentation/fission are presently used to produce new isotopes. It is probable that these two reaction mechanisms continue to be the only viable options. Figure 17.3 shows that neutron-rich isotopes are expected to be in reach only by projectile fragmentation or projectile fission reactions. Nuclides at and beyond the proton drip-line can be produced with either projectile fragmentation or fusion-evaporation reactions. In the future, fusion-evaporation reactions with radioactive beams could be an alternative to produce nuclides which cannot be populated in fusion-evaporation reactions using stable targets and stable beams.

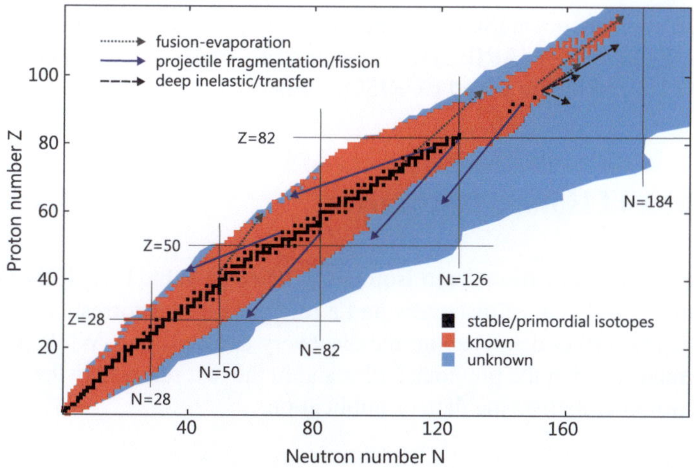

Fig. 17.3 Chart of nuclides. Stable and primordial nuclides are shown in *black*, presently known isotopes in *red* and unknown isotopes predicted to exist by Erler et al. [12] in *light blue*. Possible production mechanisms to discover new nuclides in the different regions are indicated by *arrows*. The arrows for projectile fragmentation (*dark blue*) begin at the primary projectile while the arrows for fusion-evaporation reactions (*dotted dark grey*) and deep inelastic/transfer (*dashed black*) begin at the target (adapted from [13])

While the search for new superheavy elements continues to rely on fusion-evaporation reactions [14–16], more neutron-rich isotopes of the presently known superheavy elements could be produced with other reaction mechanisms. For example, it has been suggested to use deep-inelastic reactions or multi-nucleon transfer reactions [17–19] or fusion-evaporation reactions with radioactive beams [19, 20] to populate these isotopes.

It is obvious that major future progress requires new facilities with advanced accelerators and newly developed new separation and detection techniques. The first of such "next generation" projectile fragmentation radioactive beam facilities (the Radioactive Ion-Beam Factory RIBF at RIKEN [21] is already in operation, while the Facility for Antiproton and Ion Research FAIR at GSI [22], and the Facility for Rare Isotope Beams FRIB at MSU [23, 24]) are currently under construction. For the discovery of new isotopes these facilities also rely on next generation fragment separators: BIG-RIPS [25, 26], the Super FRS [27], and the FRIB fragment separator [28], respectively.

Two additional major facilities are in the planning phase. The High-Intensity Heavy Ion Accelerator Facility (HIAF) is being designed at the Institute of Modern Physics of the Chinese Academy of Sciences in China [29] while in Korea plans for the Korean Rare Isotope Beams Accelerator Facility (KRIA) are progressing [30].

Some of these primarily fragmentation facilities will also have the capability to deliver low-energy radioactive beams which could potentially be used for the fusion-evporation reactions mentioned above. Such beams might also be available with sufficient intensities at ISOL facilities, for example, among others, the Advanced Rare IsotopE Laboratory (ARIEL) which is currently under construction at TRIUMF in Canada [31] and the proposed EURISOL project in Europe [32].

17.3 Table of Isotope Discoveries

A summary of all 3211 discovered isotopes is listed in Table 17.1. It lists the laboratory, country, and year of discovery and a reference to the corresponding table in the book where further details about the discovery can be found. While the years of discovery mentioned in the previous tables and in the text referred to the submission date, the current table lists the date of publication.

Table 17.1 List of all isotopes discovered until the end of 2015. The first author of the paper reporting the discovery, the reference, the laboratory/location and country where the isotopes were discovered, year of discovery, and the table in the book where further details are listed. The year of discovery corresponds to the date of publication

Isotope	First author	Ref.	Laboratory	Country	Year	Table
^{1}n	J. Chadwick	[33]	Cambridge	UK	1932	4.1
^{2}n	R.P. Haddock	[34]	Berkeley	USA	1965	16.1
^{1}H	A.J. Dempster	[35]	Chicago	USA	1918	3.1
^{2}H	H.C. Urey	[36]	Columbia	USA	1932	3.5
^{3}H	M.L. Oliphant	[37]	Cambridge	UK	1934	4.2
^{4}H	U. Sennhauser	[38]	Villigen	Switzerland	1981	16.1
^{5}H	A.A. Korsheninnikov	[39]	Dubna	Russia	2001	16.1
^{6}H	D.V. Aleksandrov	[40]	Moscow	Russia	1984	16.1
^{7}H	A.A. Korsheninnikov	[41]	RIKEN	Japan	2003	16.1
^{2}He	M.A. Tuve	[42]	Carnegie Institute	USA	1936	16.2
^{3}He	M.L. Oliphant	[37]	Cambridge	UK	1934	3.5
^{4}He	E. Rutherford	[43]	Manchester	UK	1908	3.4
^{5}He	J.H. Williams	[44]	Minnesota	USA	1937	16.1
^{6}He	T. Bjerge	[45]	Copenhagen	Denmark	1936	4.1
^{7}He	R.H. Stokes	[46]	Los Alamos	USA	1967	16.1
^{8}He	A.M. Poskanzer	[47]	Brookhaven	USA	1965	10.5
^{9}He	K.K. Seth	[48]	Los Alamos	USA	1987	16.1
^{10}He	A.A. Korsheninnikov	[49]	RIKEN	Japan	1994	16.1
^{4}Li	J. Cerny	[50]	Berkeley	USA	1965	16.2
^{5}Li	N.P. Heydenburg	[51]	Carnegie Institute	USA	1941	16.2
^{6}Li	F.W. Aston	[52]	Cambridge	UK	1921	3.2
^{7}Li	F.W. Aston	[52]	Cambridge	UK	1921	3.2

(continued)

Table 17.1 (continued)

Isotope	First author	Ref.	Laboratory	Country	Year	Table
^{8}Li	H.R. Crane	[53]	Caltech	USA	1935	4.2
^{9}Li	W.L. Gardner	[54]	Berkeley	USA	1951	9.7
^{10}Li	K.H. Wilcox	[55]	Berkeley	USA	1975	16.1
^{11}Li	A.M. Poskanzer	[56]	Berkeley	USA	1966	10.6
^{12}Li	C.C. Hall	[57]	Michigan State	USA	2010	16.1
^{13}Li	Yu. Aksyutina	[58]	Darmstadt	Germany	2008	16.1
^{6}Be	G.F. Bogdanov	[59]	Moscow	Russia	1958	16.2
^{7}Be	R.B. Roberts	[60]	Carnegie Institute	USA	1938	4.2
^{8}Be	J.D. Cockcroft	[61]	Cambridge	UK	1932	4.2
^{9}Be	G.P. Thomson	[62]	Cambridge	UK	1921	3.4
^{10}Be	M.L. Oliphant	[63]	Cambridge	UK	1935	4.2
^{11}Be	M.J. Nurmia	[64]	Arkansas	USA	1958	7.3
^{12}Be	A.M. Poskanzer	[56]	Berkeley	USA	1966	10.6
^{13}Be	D.V. Aleksandrov	[65]	Moscow	Russia	1983	16.1
^{14}Be	J.D. Bowman	[66]	Berkeley	USA	1973	10.6
^{15}Be	J. Snyder	[67]	Michigan State	USA	2013	16.1
^{16}Be	A. Spyrou	[68]	Michigan State	USA	2012	16.1
^{7}B	R.L. McGrath	[69]	Berkeley	USA	1967	16.2
^{8}B	L.W. Alvarez	[70]	Berkeley	USA	1950	9.7
^{9}B	R.O. Haxby	[71]	Westinghouse	USA	1940	16.2
^{10}B	F.W. Aston	[72]	Cambridge	UK	1920	3.2
^{11}B	F.W. Aston	[72]	Cambridge	UK	1920	3.2
^{12}B	H.R. Crane	[73]	Caltech	USA	1935	4.2
^{13}B	S.K. Allison	[74]	Chicago	USA	1956	14.1
^{14}B	A.M. Poskanzer	[56]	Berkeley	USA	1966	10.6
^{15}B	A.M. Poskanzer	[56]	Berkeley	USA	1966	10.6
^{16}B	R. Kalpakchieva	[75]	Berlin	Germany	2000	16.1
^{17}B	J.D. Bowman	[66]	Berkeley	USA	1973	10.6
^{18}B	A. Spyrou	[76]	Michigan State	USA	2010	16.1
^{19}B	J.A. Musser	[77]	Berkeley	USA	1984	15.1
^{8}C	R.G.H. Robertson	[78]	Juelich	Germany	1974	16.2
^{9}C	J. Cerny	[79]	Berkeley	USA	1964	9.8
^{10}C	R. Sherr	[80]	Princeton	USA	1949	9.4
^{11}C	H.R. Crane	[81]	Caltech	USA	1934	4.2
^{12}C	F.W. Aston	[82]	Cambridge	UK	1919	3.2
^{13}C	A.S. King	[83]	Carnegie Institute	USA	1929	3.5
	A.S. King	[84]	Carnegie Institute	USA	1929	3.5
^{14}C	T.W. Bonner	[85]	Caltech	USA	1936	4.4
^{15}C	E.L. Hudspeth	[86]	Carnegie Institute	USA	1950	9.1

(continued)

Table 17.1 (continued)

Isotope	First author	Ref.	Laboratory	Country	Year	Table
^{16}C	S. Hinds	[87]	Aldermaston	UK	1961	9.8
^{17}C	A.M. Poskanzer	[88]	Berkeley	USA	1968	10.6
^{18}C	A.G. Artukh	[89]	Dubna	Russia	1969	14.2
^{19}C	J.D. Bowman	[90]	Berkeley	USA	1974	10.6
^{20}C	J.D. Stevenson	[91]	Berkeley	USA	1981	15.1
^{21}C						
^{22}C	F. Pougheon	[92]	GANIL	France	1986	15.1
^{10}N	A. Lépine-Szily	[93]	GANIL	France	2002	16.2
^{11}N	W. Benenson	[94]	Michigan State	USA	1974	16.2
^{12}N	L.W. Alvarez	[95]	Berkeley	USA	1949	9.1
^{13}N	I. Curie	[96]	Paris	France	1934	4.1
^{14}N	F.W. Aston	[97]	Cambridge	UK	1920	3.2
^{15}N	S.M. Naude	[98]	Chicago	USA	1929	3.5
^{16}N	W.D. Harkins	[99]	Chicago	USA	1933	4.1
^{17}N	L.W. Alvarez	[100]	Berkeley	USA	1949	9.7
^{18}N	L.F. Chase Jr.	[101]	Lockeed Palo Alto	USA	1964	7.3
^{19}N	T.D. Thomas	[102]	Princeton	USA	1968	10.6
^{20}N	A.G. Artukh	[89]	Dubna	Russia	1969	14.2
^{21}N	A.G. Artukh	[103]	Dubna	Russia	1970	14.2
^{22}N	G.D. Westfall	[104]	Berkeley	USA	1979	15.1
^{23}N	M. Langevin	[105]	GANIL	France	1985	15.1
^{12}O	G.J. KeKelis	[106]	Berkeley	USA	1978	16.2
^{13}O	R. McPherson	[107]	Brookhaven	USA	1965	9.7
^{14}O	R. Sherr	[80]	Princeton	USA	1949	9.4
^{15}O	M.S. Livingston	[108]	Berkeley	USA	1934	4.3
^{16}O	A.J. Dempster	[35]	Chicago	USA	1918	3.1
^{17}O	P.M.S. Blackett	[109]	Cambridge	UK	1925	3.5
^{18}O	W.F. Giauque	[110]	Berkeley	USA	1929	3.5
^{19}O	M.E. Nahmias	[111]	Paris	France	1936	4.1
^{20}O	N. Jarmie	[112]	Los Alamos	USA	1959	9.8
^{21}O	T.D. Thomas	[102]	Princeton	USA	1968	10.6
^{22}O	A.G. Artukh	[89]	Dubna	Russia	1969	14.2
^{23}O	A.G. Artukh	[103]	Dubna	Russia	1970	14.2
^{24}O	A.G. Artukh	[103]	Dubna	Russia	1970	14.2
^{25}O	C.R. Hoffman	[113]	Michigan State	USA	2008	16.1
^{26}O	E. Lunderberg	[114]	Michigan State	USA	2012	16.1
^{14}F	V.Z. Goldberg	[115]	Texas A&M	USA	2010	16.2

(continued)

Table 17.1 (continued)

Isotope	First author	Ref.	Laboratory	Country	Year	Table
^{15}F	G.J. KeKelis	[106]	Berkeley	USA	1978	16.2
^{16}F	H.C. Bryant	[116]	Los Alamos	USA	1964	16.2
^{17}F	L. Wertenstein	[117]	Warsaw	Poland	1934	4.1
^{18}F	L.A. DuBridge	[118]	Rochester	USA	1937	4.2
^{19}F	F.W. Aston	[72]	Cambridge	UK	1920	3.2
^{20}F	H.R. Crane	[53]	Caltech	USA	1935	4.2
^{21}F	N. Jarmie	[119]	Los Alamos	USA	1955	9.8
^{22}F	F.J. Vaughn	[120]	Lockeed Palo Alto	USA	1965	7.3
^{23}F	A.G. Artukh	[121]	Dubna	Russia	1970	14.2
^{24}F	A.G. Artukh	[121]	Dubna	Russia	1970	14.2
^{25}F	A.G. Artukh	[103]	Dubna	Russia	1970	14.2
^{26}F	G.D. Westfall	[104]	Berkeley	USA	1979	15.1
^{27}F	J.D. Stevenson	[91]	Berkeley	USA	1981	15.1
^{28}F	G. Christian	[122]	Michigan State	USA	2012	16.1
^{29}F	D. Guillemaud-Mueller	[123]	GANIL	France	1989	15.1
^{30}F						
^{31}F	H. Sakurai	[124]	RIKEN	Japan	1999	15.1
^{15}Ne	F. Wamers	[125]	Darmstadt	Germany	2014	16.2
^{16}Ne	R.J. Holt	[126]	Los Alamos	USA	1977	16.2
^{17}Ne	R. McPherson	[127]	McGill	Canada	1964	9.7
^{18}Ne	J.D. Gow	[128]	Berkeley	USA	1954	9.1
^{19}Ne	J.G. Fox	[129]	Princeton	USA	1939	4.2
^{20}Ne	F.W. Aston	[130]	Cambridge	UK	1913	3.2
^{21}Ne	T.R. Hogness	[131]	Berkeley	USA	1928	3.4
^{22}Ne	F.W. Aston	[130]	Cambridge	UK	1913	3.2
^{23}Ne	M.E. Nahmias	[111]	Paris	France	1936	4.1
^{24}Ne	B.J. Dropesky	[132]	Los Alamos	USA	1956	9.5
^{25}Ne	A.G. Artukh	[121]	Dubna	Russia	1970	14.2
^{26}Ne	A.G. Artukh	[121]	Dubna	Russia	1970	14.2
^{27}Ne	G.W. Butler	[133]	Los Alamos	USA	1977	10.6
^{28}Ne	T.J.M. Symons	[134]	Berkeley	USA	1979	15.1
^{29}Ne	M. Langevin	[105]	GANIL	France	1985	15.1
^{30}Ne	M. Langevin	[105]	GANIL	France	1985	15.1
^{31}Ne	H. Sakurai	[135]	RIKEN	Japan	1996	15.1
^{32}Ne	D. Guillemaud-Mueller	[136]	GANIL	France	1990	15.1
^{33}Ne						
^{34}Ne	M. Notani	[137]	RIKEN	Japan	2002	15.1
^{18}Na	T. Zerguerras	[138]	GANIL	France	2004	16.2

(continued)

Table 17.1 (continued)

Isotope	First author	Ref.	Laboratory	Country	Year	Table
^{19}Na	J. Cerny	[139]	Berkeley	USA	1969	16.2
^{20}Na	L.W. Alvarez	[70]	Berkeley	USA	1950	9.7
^{21}Na	E. Pollard	[140]	Yale	USA	1940	4.2
^{22}Na	O.R. Frisch	[141]	Copenhagen	Denmark	1935	4.1
^{23}Na	A.J. Dempster	[35]	Chicago	USA	1918	3.1
^{24}Na	E. Fermi	[142]	Rome	Italy	1934	4.1
^{25}Na	O. Huber	[143]	Zurich	Switzerland	1943	4.6
^{26}Na	M.J. Nurmia	[144]	Arkansas	USA	1958	7.3
^{27}Na	R. Klapisch	[145]	CERN	Switzerland	1968	10.10
^{28}Na	R. Klapisch	[146]	CERN	Switzerland	1969	10.10
^{29}Na	R. Klapisch	[146]	CERN	Switzerland	1969	10.10
^{30}Na	R. Klapisch	[146]	CERN	Switzerland	1969	10.10
^{31}Na	R. Klapisch	[146]	CERN	Switzerland	1969	10.10
^{32}Na	R. Klapisch	[147]	CERN	Switzerland	1972	10.10
^{33}Na	R. Klapisch	[147]	CERN	Switzerland	1972	10.10
^{34}Na	M. Langevin	[148]	CERN	Switzerland	1983	10.10
^{35}Na	M. Langevin	[148]	CERN	Switzerland	1983	10.10
^{36}Na						
^{37}Na	M. Notani	[137]	RIKEN	Japan	2002	15.1
^{19}Mg	I. Mukha	[149]	Darmstadt	Germany	2007	16.2
^{20}Mg	R.G.H. Robertson	[78]	Juelich	Germany	1974	9.8
^{21}Mg	R. McPherson	[150]	McGill	Canada	1964	9.7
^{22}Mg	F. Ajzenberg-Selove	[151]	Los Alamos	USA	1961	9.8
^{23}Mg	M.G. White	[152]	Princeton	USA	1939	4.2
^{24}Mg	A.J. Dempster	[153]	Chicago	USA	1920	3.1
^{25}Mg	A.J. Dempster	[153]	Chicago	USA	1920	3.1
^{26}Mg	A.J. Dempster	[153]	Chicago	USA	1920	3.1
^{27}Mg	E. Fermi	[142]	Rome	Italy	1934	4.1
^{28}Mg	R.K. Sheline	[154]	Chicago	USA	1953	8.1
^{29}Mg	A.G. Artukh	[155]	Dubna	Russia	1971	14.2
^{30}Mg	A.G. Artukh	[155]	Dubna	Russia	1971	14.2
^{31}Mg	G.W. Butler	[133]	Los Alamos	USA	1977	10.6
^{32}Mg	G.W. Butler	[133]	Los Alamos	USA	1977	10.6
^{33}Mg	G.D. Westfall	[104]	Berkeley	USA	1979	15.1
^{34}Mg	G.D. Westfall	[104]	Berkeley	USA	1979	15.1
^{35}Mg	D. Guillemaud-Mueller	[123]	GANIL	France	1989	15.1
^{36}Mg	D. Guillemaud-Mueller	[123]	GANIL	France	1989	15.1

(continued)

Table 17.1 (continued)

Isotope	First author	Ref.	Laboratory	Country	Year	Table
^{37}Mg	H. Sakurai	[135]	RIKEN	Japan	1996	15.1
^{38}Mg	M. Notani	[137]	RIKEN	Japan	2002	15.1
^{39}Mg						
^{40}Mg	T. Baumann	[156]	Michigan State	USA	2007	15.1
^{22}Al	M.D. Cable	[157]	Berkeley	USA	1982	9.7
^{23}Al	J. Cerny	[139]	Berkeley	USA	1969	9.8
^{24}Al	N.W. Glass	[158]	UCLA	USA	1953	9.1
^{25}Al	J.L.W. Churchill	[159]	Aldermaston	UK	1953	9.1
^{26}Al	O.R. Frisch	[160]	London	UK	1934	4.1
^{27}Al	F.W. Aston	[161]	Cambridge	UK	1922	3.2
^{28}Al	I. Curie	[162]	Paris	France	1934	4.1
^{29}Al	H.A. Bethe	[163]	Purdue	USA	1939	4.2
^{30}Al	E.L. Robinson	[164]	Purdue	USA	1961	7.3
^{31}Al	A.G. Artukh	[155]	Dubna	Russia	1971	14.2
^{32}Al	A.G. Artukh	[155]	Dubna	Russia	1971	14.2
^{33}Al	A.G. Artukh	[155]	Dubna	Russia	1971	14.2
^{34}Al	G.W. Butler	[133]	Los Alamos	USA	1977	10.6
^{35}Al	T.J.M. Symons	[134]	Berkeley	USA	1979	15.1
^{36}Al	G.D. Westfall	[104]	Berkeley	USA	1979	15.1
^{37}Al	G.D. Westfall	[104]	Berkeley	USA	1979	15.1
^{38}Al	D. Guillemaud-Mueller	[123]	GANIL	France	1989	15.1
^{39}Al	D. Guillemaud-Mueller	[123]	GANIL	France	1989	15.1
^{40}Al	M. Notani	[137]	RIKEN	Japan	2002	15.1
^{41}Al	M. Notani	[137]	RIKEN	Japan	2002	15.1
^{42}Al	T. Baumann	[156]	Michigan State	USA	2007	15.1
^{43}Al	T. Baumann	[156]	Michigan State	USA	2007	15.1
^{22}Si	M.G.Saint-Laurent	[165]	GANIL	France	1987	15.4
^{23}Si	M. Langevin	[166]	GANIL	France	1986	15.4
^{24}Si	J. Äystö	[167]	Berkeley	USA	1979	9.7
^{25}Si	R. Barton	[168]	McGill	Canada	1963	9.7
^{26}Si	E.L. Robinson	[169]	Purdue	USA	1960	9.1
^{27}Si	G. Kuerti	[170]	Rochester	USA	1939	4.2
^{28}Si	F.W. Aston	[72]	Cambridge	UK	1920	3.2
^{29}Si	F.W. Aston	[72]	Cambridge	UK	1920	3.2
^{30}Si	R.S. Mulliken	[171]	Harvard	USA	1924	3.5
^{31}Si	E. Fermi	[142]	Rome	Italy	1934	4.1
^{32}Si	M. Lindner	[172]	Berkeley	USA	1953	10.2
^{33}Si	A.G. Artukh	[155]	Dubna	Russia	1971	14.2
^{34}Si	A.G. Artukh	[155]	Dubna	Russia	1971	14.2

(continued)

Table 17.1 (continued)

Isotope	First author	Ref.	Laboratory	Country	Year	Table
^{35}Si	A.G. Artukh	[155]	Dubna	Russia	1971	14.2
^{36}Si	A.G. Artukh	[155]	Dubna	Russia	1971	14.2
^{37}Si	P. Auger	[173]	Orsay	France	1979	14.2
^{38}Si	G.D. Westfall	[104]	Berkeley	USA	1979	15.1
^{39}Si	G.D. Westfall	[104]	Berkeley	USA	1979	15.1
^{40}Si	D. Guillemaud-Mueller	[123]	GANIL	France	1989	15.1
^{41}Si	D. Guillemaud-Mueller	[123]	GANIL	France	1989	15.1
^{42}Si	M. Lewitowicz	[174]	GANIL	France	1990	15.1
^{43}Si	M. Notani	[137]	RIKEN	Japan	2002	15.1
^{44}Si	O.B. Tarasov	[175]	Michigan State	USA	2007	15.1
^{26}P	M.D. Cable	[176]	Berkeley	USA	1983	9.7
^{27}P	W. Benenson	[177]	Michigan State	USA	1977	9.8
^{28}P	N.W. Glass	[158]	UCLA	USA	1953	9.1
^{29}P	M.G. White	[178]	Princeton	USA	1941	4.2
^{30}P	I. Curie	[96]	Paris	France	1934	4.1
^{31}P	F.W. Aston	[72]	Cambridge	UK	1920	3.2
^{32}P	E. Fermi	[142]	Rome	Italy	1934	4.1
^{33}P	R.K. Sheline	[179]	Chicago	USA	1951	8.1
^{34}P	W. Zünti	[180]	Zurich	Switzerland	1945	4.4
^{35}P	A.G. Artukh	[155]	Dubna	Russia	1971	14.2
^{36}P	A.G. Artukh	[155]	Dubna	Russia	1971	14.2
^{37}P	A.G. Artukh	[155]	Dubna	Russia	1971	14.2
^{38}P	A.G. Artukh	[155]	Dubna	Russia	1971	14.2
^{39}P	G.W. Butler	[133]	Los Alamos	USA	1977	10.6
^{40}P	P. Auger	[173]	Orsay	France	1979	14.2
^{41}P	G.D. Westfall	[104]	Berkeley	USA	1979	15.1
^{42}P	G.D. Westfall	[104]	Berkeley	USA	1979	15.1
^{43}P	D. Guillemaud-Mueller	[123]	GANIL	France	1989	15.1
^{44}P	D. Guillemaud-Mueller	[123]	GANIL	France	1989	15.1
^{45}P	M. Lewitowicz	[174]	GANIL	France	1990	15.1
^{46}P	M. Lewitowicz	[174]	GANIL	France	1990	15.1
^{27}S	M. Langevin	[166]	GANIL	France	1986	15.4
^{28}S	C.L. Morris	[181]	Los Alamos	USA	1982	8.2
^{29}S	J.C. Hardy	[182]	McGill	Canada	1964	9.7
^{30}S	E.L. Robinson	[183]	Purdue	USA	1961	9.1
^{31}S	L.D.P. King	[184]	Purdue	USA	1940	4.2
^{32}S	F.W. Aston	[72]	Cambridge	UK	1920	3.2
^{33}S	F.W. Aston	[185]	Cambridge	UK	1926	3.3
^{34}S	F.W. Aston	[185]	Cambridge	UK	1926	3.3

(continued)

Table 17.1 (continued)

Isotope	First author	Ref.	Laboratory	Country	Year	Table
^{35}S	E.B. Andersen	[186]	Aarhus	Denmark	1936	4.1
^{36}S	A.O. Nier	[187]	Harvard	USA	1938	3.4
^{37}S	W. Zünti	[180]	Zurich	Switzerland	1945	4.4
^{38}S	D.R. Nethaway	[188]	Berkeley	USA	1958	9.3
^{39}S	A.G. Artukh	[155]	Dubna	Russia	1971	14.2
^{40}S	A.G. Artukh	[155]	Dubna	Russia	1971	14.2
^{41}S	P. Auger	[173]	Orsay	France	1979	14.2
^{42}S	P. Auger	[173]	Orsay	France	1979	14.2
^{43}S	G.D. Westfall	[104]	Berkeley	USA	1979	15.1
^{44}S	G.D. Westfall	[104]	Berkeley	USA	1979	15.1
^{45}S	D. Guillemaud-Mueller	[123]	GANIL	France	1989	15.1
^{46}S	D. Guillemaud-Mueller	[123]	GANIL	France	1989	15.1
^{47}S	D. Guillemaud-Mueller	[123]	GANIL	France	1989	15.1
^{48}S	M. Lewitowicz	[174]	GANIL	France	1990	15.1
^{29}Cl	I. Mukha	[189]	Darmstadt	Germany	2015	16.2
^{30}Cl						
^{31}Cl	W. Benenson	[177]	Michigan State	USA	1977	9.8
^{32}Cl	N.W. Glass	[158]	UCLA	USA	1953	9.1
^{33}Cl	J.B. Hoag	[190]	Berkeley	USA	1940	4.3
^{34}Cl	O.R. Frisch	[160]	London	UK	1934	4.1
^{35}Cl	F.W. Aston	[82]	Cambridge	UK	1919	3.2
^{36}Cl	D.C. Grahame	[191]	Berkeley	USA	1941	4.4
^{37}Cl	F.W. Aston	[82]	Cambridge	UK	1919	3.2
^{38}Cl	J.W. Kennedy	[192]	Berkeley	USA	1940	4.4
^{39}Cl	R.N.H. Haslam	[193]	Saskatoon	Canada	1949	8.1
^{40}Cl	H. Morinaga	[194]	Purdue	USA	1956	7.3
^{41}Cl	A.G. Artukh	[155]	Dubna	Russia	1971	14.2
^{42}Cl	A.G. Artukh	[155]	Dubna	Russia	1971	14.2
^{43}Cl	E. Kashy	[195]	Michigan State	USA	1976	9.8
^{44}Cl	G.D. Westfall	[104]	Berkeley	USA	1979	15.1
^{45}Cl	G.D. Westfall	[104]	Berkeley	USA	1979	15.1
^{46}Cl	D. Guillemaud-Mueller	[123]	GANIL	France	1989	15.1
^{47}Cl	D. Guillemaud-Mueller	[123]	GANIL	France	1989	15.1
^{48}Cl	D. Guillemaud-Mueller	[123]	GANIL	France	1989	15.1
^{49}Cl	D. Guillemaud-Mueller	[123]	GANIL	France	1989	15.1
^{50}Cl	O.B. Tarasov	[196]	Michigan State	USA	2009	15.1
^{51}Cl	O.B. Tarasov	[196]	Michigan State	USA	2009	15.1
^{30}Ar	I. Mukha	[189]	Darmstadt	Germany	2015	16.2

(continued)

Table 17.1 (continued)

Isotope	First author	Ref.	Laboratory	Country	Year	Table
^{31}Ar	M. Langevin	[166]	GANIL	France	1986	15.4
^{32}Ar	E. Hagberg	[197]	CERN	Switzerland	1977	10.7
^{33}Ar	P.L. Reeder	[198]	Brookhaven	USA	1964	9.7
^{34}Ar	R.G. Miller	[199]	Caltech	USA	1966	9.1
^{35}Ar	L.D.P. King	[184]	Purdue	USA	1940	4.2
^{36}Ar	F.W. Aston	[97]	Cambridge	UK	1920	3.2
^{37}Ar	P.K. Weimer	[200]	Ohio State	USA	1941	4.2
^{38}Ar	P. Zeeman	[201]	Amsterdam	Netherlands	1934	3.4
^{39}Ar	A.R. Brosi	[202]	Oak Ridge	USA	1950	7.2
^{40}Ar	F.W. Aston	[97]	Cambridge	UK	1920	3.2
^{41}Ar	A.H. Snell	[203]	Berkeley	USA	1936	4.3
^{42}Ar	S. Katcoff	[204]	Brookhaven	USA	1952	7.2
^{43}Ar	P.G. Hansen	[205]	CERN	Switzerland	1969	10.7
^{44}Ar	R.E. Larson	[206]	Naval Research Lab.	USA	1969	8.1
^{45}Ar	N.A. Jelley	[207]	Berkeley	USA	1974	9.8
^{46}Ar	N.A. Jelley	[207]	Berkeley	USA	1974	9.8
^{47}Ar	D. Guillemaud-Mueller	[208]	GANIL	France	1985	15.1
^{48}Ar	S. Grévy	[209]	GANIL	France	2004	15.1
^{49}Ar	D. Guillemaud-Mueller	[123]	GANIL	France	1989	15.1
^{50}Ar	D. Guillemaud-Mueller	[123]	GANIL	France	1989	15.1
^{51}Ar	D. Guillemaud-Mueller	[123]	GANIL	France	1989	15.1
^{52}Ar	O.B. Tarasov	[196]	Michigan State	USA	2009	15.1
^{53}Ar	O.B. Tarasov	[196]	Michigan State	USA	2009	15.1
^{35}K	W. Benenson	[210]	Michigan State	USA	1976	9.8
^{36}K	R.E. Berg	[211]	Michigan State	USA	1967	9.1
^{37}K	C.R. Sun	[212]	UCLA	USA	1958	9.1
^{38}K	D.G. Hurst	[213]	Berkeley	USA	1937	4.3
^{39}K	A.J. Dempster	[35]	Chicago	USA	1918	3.1
^{40}K	A.O. Nier	[214]	Minnesota	USA	1935	3.4
^{41}K	F.W. Aston	[215]	Cambridge	UK	1921	3.2
^{42}K	G. Hevesy	[216]	Copenhagen	Denmark	1935	4.1
^{43}K	R. Overstreet	[217]	Berkeley	USA	1949	9.3
^{44}K	B.L. Cohen	[218]	Oak Ridge	USA	1954	7.3
^{45}K	H. Morinaga	[219]	Amsterdam	Netherlands	1964	9.5
^{46}K	A. Marinov	[220]	Argonne	USA	1965	9.8
^{47}K	T. Kuroyanagi	[221]	Tokai	Japan	1964	8.1
^{48}K	R. Klapisch	[147]	CERN	Switzerland	1972	10.10
^{49}K	R. Klapisch	[147]	CERN	Switzerland	1972	10.10
^{50}K	R. Klapisch	[147]	CERN	Switzerland	1972	10.10

(continued)

Table 17.1 (continued)

Isotope	First author	Ref.	Laboratory	Country	Year	Table
^{51}K	M. Langevin	[222]	CERN	Switzerland	1983	10.10
^{52}K	M. Langevin	[222]	CERN	Switzerland	1983	10.10
^{53}K	M. Langevin	[222]	CERN	Switzerland	1983	10.10
^{54}K	M. Langevin	[222]	CERN	Switzerland	1983	10.10
^{55}K	O.B. Tarasov	[196]	Michigan State	USA	2009	15.1
^{56}K	O.B. Tarasov	[196]	Michigan State	USA	2009	15.1
^{35}Ca	J. Äystö	[223]	Berkeley	USA	1985	9.7
^{36}Ca	R.E. Tribble	[224]	Texas A&M	USA	1977	9.8
^{37}Ca	J.C. Hardy	[225]	McGill	Canada	1964	9.7
	P. L. Reeder	[198]	Brookhaven	USA	1964	9.7
^{38}Ca	J.C. Hardy	[226]	Oxford	UK	1966	9.8
^{39}Ca	O. Huber	[227]	Zurich	Switzerland	1943	4.6
^{40}Ca	A.J. Dempster	[228]	Chicago	USA	1922	3.1
^{41}Ca	W.L. Davidson Jr.	[229]	Yale	USA	1939	4.2
^{42}Ca	F.W. Aston	[230]	Cambridge	UK	1934	3.3
^{43}Ca	F.W. Aston	[230]	Cambridge	UK	1934	3.3
^{44}Ca	A.J. Dempster	[228]	Chicago	USA	1922	3.1
^{45}Ca	H. Walke	[231]	Berkeley	USA	1940	4.3
^{46}Ca	A.O. Nier	[187]	Harvard	USA	1938	3.4
^{47}Ca	R.E. Batzel	[232]	Berkeley	USA	1951	10.2
^{48}Ca	A.O. Nier	[187]	Harvard	USA	1938	3.4
^{49}Ca	E. der Mateosian	[233]	Argonne	USA	1950	7.2
^{50}Ca	Y. Shida	[234]	Kawasaki	Japan	1964	9.1
^{51}Ca	A. Huck	[235]	CERN	Switzerland	1980	10.10
^{52}Ca	A. Huck	[236]	CERN	Switzerland	1985	10.10
^{53}Ca	M. Langevin	[222]	CERN	Switzerland	1983	10.10
^{54}Ca	M. Bernas	[237]	Darmstadt	Germany	1997	15.2
^{55}Ca	M. Bernas	[237]	Darmstadt	Germany	1997	15.2
^{56}Ca	M. Bernas	[237]	Darmstadt	Germany	1997	15.2
^{57}Ca	O.B. Tarasov	[196]	Michigan State	USA	2009	15.1
^{58}Ca	O.B. Tarasov	[196]	Michigan State	USA	2009	15.1
^{39}Sc	C.L. Woods	[238]	Canberra	Australia	1988	16.2
^{40}Sc	N.W. Glass	[239]	UCLA	USA	1955	9.1
^{41}Sc	D.R. Elliott	[240]	Purdue	USA	1941	4.2
^{42}Sc	H. Morinaga	[241]	Purdue	USA	1955	9.1
^{43}Sc	O.R. Frisch	[141]	Copenhagen	Denmark	1935	4.1
^{44}Sc	H. Walke	[242]	Berkeley	USA	1937	4.3

(continued)

Table 17.1 (continued)

Isotope	First author	Ref.	Laboratory	Country	Year	Table
^{45}Sc	F.W. Aston	[243]	Cambridge	UK	1923	3.2
^{46}Sc	G. Hevesy	[244]	Copenhagen	Denmark	1936	4.1
^{47}Sc	C.T. Hibdon	[245]	Ohio State	USA	1945	9.4
^{48}Sc	H. Walke	[246]	Berkeley	USA	1937	4.4
^{49}Sc	H. Walke	[247]	Berkeley	USA	1940	4.3
^{50}Sc	A. Poularikas	[248]	Arkansas	USA	1959	7.3
^{51}Sc	J.R. Erskine	[249]	Argonne	USA	1966	9.8
^{52}Sc	H. Breuer	[250]	Berkeley	USA	1980	14.2
^{53}Sc	H. Breuer	[250]	Berkeley	USA	1980	14.2
^{54}Sc	X.L. Tu	[251]	Los Alamos	USA	1990	10.6
^{55}Sc	X.L. Tu	[251]	Los Alamos	USA	1990	10.6
^{56}Sc	M. Bernas	[237]	Darmstadt	Germany	1997	15.2
^{57}Sc	M. Bernas	[237]	Darmstadt	Germany	1997	15.2
^{58}Sc	M. Bernas	[237]	Darmstadt	Germany	1997	15.2
^{59}Sc	O.B. Tarasov	[196]	Michigan State	USA	2009	15.1
^{60}Sc	O.B. Tarasov	[196]	Michigan State	USA	2009	15.1
^{61}Sc	O.B. Tarasov	[196]	Michigan State	USA	2009	15.1
^{39}Ti	C. Détraz	[252]	GANIL	France	1990	15.4
^{40}Ti	C.L. Morris	[181]	Los Alamos	USA	1982	8.2
^{41}Ti	P.L. Reeder	[198]	Brookhaven	USA	1964	9.7
^{42}Ti	H.C. Bryant	[116]	Los Alamos	USA	1964	9.8
^{43}Ti	A.D. Schelberg	[253]	Indiana	USA	1948	9.1
^{44}Ti	R.A. Sharp	[254]	Harvard	USA	1954	9.4
^{45}Ti	J.S.V. Allen	[255]	Ohio State	USA	1941	4.2
^{46}Ti	F.W. Aston	[230]	Cambridge	UK	1934	3.3
^{47}Ti	F.W. Aston	[230]	Cambridge	UK	1934	3.3
^{48}Ti	F.W. Aston	[243]	Cambridge	UK	1923	3.2
^{49}Ti	F.W. Aston	[230]	Cambridge	UK	1934	3.3
^{50}Ti	F.W. Aston	[230]	Cambridge	UK	1934	3.3
^{51}Ti	L. Seren	[256]	Argonne	USA	1947	7.1
^{52}Ti	D.C. Williams	[257]	Los Alamos	USA	1966	9.8
^{53}Ti	L.A. Parks	[258]	Argonne	USA	1977	11.2
^{54}Ti	D. Guerreau	[259]	Orsay	France	1980	14.2
^{55}Ti	H. Breuer	[250]	Berkeley	USA	1980	14.2
^{56}Ti	H. Breuer	[250]	Berkeley	USA	1980	14.2
^{57}Ti	D. Guillemaud-Mueller	[208]	GANIL	France	1985	15.1
^{58}Ti	M. Weber	[260]	Darmstadt	Germany	1992	15.1
^{59}Ti	M. Bernas	[237]	Darmstadt	Germany	1997	15.2

(continued)

Table 17.1 (continued)

Isotope	First author	Ref.	Laboratory	Country	Year	Table
^{60}Ti	M. Bernas	[237]	Darmstadt	Germany	1997	15.2
^{61}Ti	M. Bernas	[237]	Darmstadt	Germany	1997	15.2
^{62}Ti	O.B. Tarasov	[196]	Michigan State	USA	2009	15.1
^{63}Ti	O.B. Tarasov	[196]	Michigan State	USA	2009	15.1
^{64}Ti	O.B. Tarasov	[261]	Michigan State	USA	2013	15.1
^{43}V	F. Pougheon	[262]	GANIL	France	1987	15.4
^{44}V	J. Cerny	[263]	Brookhaven	USA	1971	11.5
^{45}V	K.P. Jackson	[264]	Chalk River	Canada	1974	11.11
^{46}V	W.M. Martin	[265]	McGill	Canada	1952	9.1
^{47}V	J.J. O'Connor	[266]	Ohio State	USA	1942	4.2
^{48}V	H. Walke	[267]	Berkeley	USA	1937	4.3
^{49}V	L.A. Turner	[268]	Princeton	USA	1940	4.2
^{50}V	D.C. Hess Jr.	[7]	Argonne	USA	1949	3.4
	W.T. Leland	[8]	Minnesota	USA	1949	3.4
^{51}V	F.W. Aston	[243]	Cambridge	UK	1923	3.2
^{52}V	E. Amaldi	[269]	Rome	Italy	1934	4.1
^{53}V	I. Kumabe	[270]	Arkansas	USA	1960	7.3
^{54}V	T.W. Ward	[271]	Arkansas	USA	1970	7.4
^{55}V	A.M. Nathan	[272]	Brookhaven	USA	1977	11.2
^{56}V	D. Guerreau	[259]	Orsay	France	1980	14.2
^{57}V	H. Breuer	[250]	Berkeley	USA	1980	14.2
^{58}V	H. Breuer	[250]	Berkeley	USA	1980	14.2
^{59}V	D. Guillemaud-Mueller	[208]	GANIL	France	1985	15.1
^{60}V	D. Guillemaud-Mueller	[208]	GANIL	France	1985	15.1
^{61}V	M. Weber	[260]	Darmstadt	Germany	1992	15.1
^{62}V	M. Bernas	[237]	Darmstadt	Germany	1997	15.2
^{63}V	M. Bernas	[237]	Darmstadt	Germany	1997	15.2
^{64}V	M. Bernas	[237]	Darmstadt	Germany	1997	15.2
^{65}V	O.B. Tarasov	[196]	Michigan State	USA	2009	15.1
^{66}V	O.B. Tarasov	[196]	Michigan State	USA	2009	15.1
^{67}V	O.B. Tarasov	[261]	Michigan State	USA	2013	15.1
^{42}Cr	B. Blank	[273]	Darmstadt	Germany	1996	15.4
^{43}Cr	V. Borrel	[274]	GANIL	France	1992	15.4
^{44}Cr	F. Pougheon	[262]	GANIL	France	1987	15.4
^{45}Cr	K.P. Jackson	[264]	Chalk River	Canada	1974	11.11
^{46}Cr	J. Zioni	[275]	Jerusalem	Israel	1972	11.2
^{47}Cr	I.D. Proctor	[276]	Michigan State	USA	1972	9.8
^{48}Cr	G. Rudstam	[277]	Berkeley	USA	1952	10.2
^{49}Cr	J.J. O'Connor	[266]	Ohio State	USA	1942	4.2

(continued)

Table 17.1 (continued)

Isotope	First author	Ref.	Laboratory	Country	Year	Table
^{50}Cr	F.W. Aston	[278]	Cambridge	UK	1930	3.3
^{51}Cr	H. Walke	[279]	Berkeley	USA	1940	4.3
^{52}Cr	F.W. Aston	[243]	Cambridge	UK	1923	3.2
^{53}Cr	F.W. Aston	[278]	Cambridge	UK	1930	3.3
^{54}Cr	F.W. Aston	[278]	Cambridge	UK	1930	3.3
^{55}Cr	A. Flammersfeld	[280]	Mainz	Germany	1952	7.3
^{56}Cr	B.J. Dropesky	[281]	Los Alamos	USA	1960	9.5
^{57}Cr	C.N. Davids	[282]	Argonne	USA	1978	11.2
^{58}Cr	D. Guerreau	[259]	Orsay	France	1980	14.2
^{59}Cr	D. Guerreau	[259]	Orsay	France	1980	14.2
^{60}Cr	H. Breuer	[250]	Berkeley	USA	1980	14.2
^{61}Cr	D. Guillemaud-Mueller	[208]	GANIL	France	1985	15.1
^{62}Cr	D. Guillemaud-Mueller	[208]	GANIL	France	1985	15.1
^{63}Cr	M. Weber	[260]	Darmstadt	Germany	1992	15.1
^{64}Cr	M. Weber	[260]	Darmstadt	Germany	1992	15.1
^{65}Cr	M. Bernas	[237]	Darmstadt	Germany	1997	15.2
^{66}Cr	M. Bernas	[237]	Darmstadt	Germany	1997	15.2
^{67}Cr	M. Bernas	[237]	Darmstadt	Germany	1997	15.2
^{68}Cr	O.B. Tarasov	[196]	Michigan State	USA	2009	15.1
^{69}Cr	O.B. Tarasov	[261]	Michigan State	USA	2013	15.1
^{70}Cr	O.B. Tarasov	[261]	Michigan State	USA	2013	15.1
^{46}Mn	F. Pougheon	[262]	GANIL	France	1987	15.4
^{47}Mn	F. Pougheon	[262]	GANIL	France	1987	15.4
^{48}Mn	T. Sekine	[283]	Darmstadt	Germany	1987	11.3
^{49}Mn	J. Cerny	[284]	Harwell	UK	1970	11.11
^{50}Mn	W.M. Martin	[265]	McGill	Canada	1952	9.1
^{51}Mn	J.J. Livingood	[285]	Berkeley	USA	1938	4.3
^{52}Mn	J.J. Livingood	[285]	Berkeley	USA	1938	4.3
^{53}Mn	J.R. Wilkinson	[286]	Berkeley	USA	1955	9.3
^{54}Mn	J.J. Livingood	[285]	Berkeley	USA	1938	4.3
^{55}Mn	F.W. Aston	[243]	Cambridge	UK	1923	3.2
^{56}Mn	E. Amaldi	[269]	Rome	Italy	1934	4.1
^{57}Mn	B.L. Cohen	[287]	Oak Ridge	USA	1954	7.3
^{58}Mn	D.M. Chittenden II	[288]	Arkansas	USA	1961	7.3
^{59}Mn	E. Kashy	[195]	Michigan State	USA	1976	9.8
^{60}Mn	E.B. Norman	[289]	Argonne	USA	1978	11.2
^{61}Mn	D. Guerreau	[259]	Orsay	France	1980	14.2
^{62}Mn	E. Runte	[290]	Darmstadt	Germany	1983	14.3
^{63}Mn	U. Bosch	[291]	Darmstadt	Germany	1985	14.3

(continued)

Table 17.1 (continued)

Isotope	First author	Ref.	Laboratory	Country	Year	Table
^{64}Mn	D. Guillemaud-Mueller	[208]	GANIL	France	1985	15.1
^{65}Mn	D. Guillemaud-Mueller	[208]	GANIL	France	1985	15.1
^{66}Mn	M. Weber	[260]	Darmstadt	Germany	1992	15.1
^{67}Mn	M. Bernas	[237]	Darmstadt	Germany	1997	15.2
^{68}Mn	M. Bernas	[237]	Darmstadt	Germany	1997	15.2
^{69}Mn	M. Bernas	[237]	Darmstadt	Germany	1997	15.2
^{70}Mn	O.B. Tarasov	[196]	Michigan State	USA	2009	15.1
^{71}Mn	T. Ohnishi	[292]	RIKEN	Japan	2010	15.2
^{72}Mn	O.B. Tarasov	[261]	Michigan State	USA	2013	15.1
^{45}Fe	B. Blank	[273]	Darmstadt	Germany	1996	16.3
^{46}Fe	V. Borrel	[274]	GANIL	France	1992	15.4
^{47}Fe	V. Borrel	[274]	GANIL	France	1992	15.4
^{48}Fe	F. Pougheon	[262]	GANIL	France	1987	15.4
^{49}Fe	J. Cerny	[284]	Harwell	UK	1970	11.11
^{50}Fe	R.E. Tribble	[293]	Texas A&M	USA	1977	9.8
^{51}Fe	I.D. Proctor	[276]	Michigan State	USA	1972	9.8
^{52}Fe	D.R. Miller	[294]	Berkeley	USA	1948	10.2
^{53}Fe	J.J. Livingood	[295]	Berkeley	USA	1938	4.3
^{54}Fe	F.W. Aston	[296]	Cambridge	UK	1924	3.2
^{55}Fe	J.J. Livingood	[297]	Berkeley	USA	1939	4.3
^{56}Fe	F.W. Aston	[298]	Cambridge	UK	1922	3.2
^{57}Fe	F.W. Aston	[299]	Cambridge	UK	1935	3.3
^{58}Fe	J. deGier	[300]	Amsterdam	Netherlands	1935	3.4
^{59}Fe	J.J. Livingood	[295]	Berkeley	USA	1938	4.3
^{60}Fe	J.-C. Roy	[301]	Pittsburgh	USA	1957	10.2
^{61}Fe	E. Ricci	[302]	Buenos Aires	Argentina	1957	9.5
^{62}Fe	E.M. Franz	[303]	Brookhaven	USA	1975	7.4
^{63}Fe	D. Guerreau	[259]	Orsay	France	1980	14.2
^{64}Fe	D. Guerreau	[259]	Orsay	France	1980	14.2
^{65}Fe	D. Guillemaud-Mueller	[208]	GANIL	France	1985	15.1
^{66}Fe	D. Guillemaud-Mueller	[208]	GANIL	France	1985	15.1
^{67}Fe	D. Guillemaud-Mueller	[208]	GANIL	France	1985	15.1
^{68}Fe	D. Guillemaud-Mueller	[208]	GANIL	France	1985	15.1
^{69}Fe	M. Weber	[260]	Darmstadt	Germany	1992	15.1
^{70}Fe	M. Bernas	[237]	Darmstadt	Germany	1997	15.2
^{71}Fe	M. Bernas	[237]	Darmstadt	Germany	1997	15.2
^{72}Fe	M. Bernas	[237]	Darmstadt	Germany	1997	15.2
^{73}Fe	T. Ohnishi	[292]	RIKEN	Japan	2010	15.2
^{74}Fe	T. Ohnishi	[292]	RIKEN	Japan	2010	15.2

(continued)

Table 17.1 (continued)

Isotope	First author	Ref.	Laboratory	Country	Year	Table
^{75}Fe	O.B. Tarasov	[261]	Michigan State	USA	2013	15.1
^{50}Co	F. Pougheon	[262]	GANIL	France	1987	15.4
^{51}Co	F. Pougheon	[262]	GANIL	France	1987	15.4
^{52}Co	F. Pougheon	[262]	GANIL	France	1987	15.4
^{53}Co	K.P. Jackson	[304]	Harwell	UK	1970	11.11
^{54}Co	W.M. Martin	[265]	McGill	Canada	1952	9.1
^{55}Co	J.J. Livingood	[305]	Berkeley	USA	1938	4.3
^{56}Co	J.J. Livingood	[306]	Berkeley	USA	1941	4.3
^{57}Co	J.J. Livingood	[306]	Berkeley	USA	1941	4.3
^{58}Co	J.J. Livingood	[306]	Berkeley	USA	1941	4.3
^{59}Co	F.W. Aston	[243]	Cambridge	UK	1923	3.2
^{60}Co	J.J. Livingood	[306]	Berkeley	USA	1941	4.3
^{61}Co	T.J. Parmley	[307]	Berkeley	USA	1947	7.3
^{62}Co	T.J. Parmley	[308]	Berkeley	USA	1949	7.3
^{63}Co	H. Morinaga	[309]	Tohoku	Japan	1960	8.1
^{64}Co	T.E. Ward	[310]	Arkansas	USA	1969	7.4
^{65}Co	R.T. Kouzes	[311]	Princeton	USA	1978	9.8
^{66}Co	U. Bosch	[291]	Darmstadt	Germany	1985	14.3
^{67}Co	U. Bosch	[291]	Darmstadt	Germany	1985	14.3
^{68}Co	D. Guillemaud-Mueller	[208]	GANIL	France	1985	15.1
^{69}Co	D. Guillemaud-Mueller	[208]	GANIL	France	1985	15.1
^{70}Co	D. Guillemaud-Mueller	[208]	GANIL	France	1985	15.1
^{71}Co	M. Weber	[260]	Darmstadt	Germany	1992	15.1
^{72}Co	M. Weber	[260]	Darmstadt	Germany	1992	15.1
^{73}Co	C. Engelmann	[312]	Darmstadt	Germany	1995	15.2
^{74}Co	C. Engelmann	[312]	Darmstadt	Germany	1995	15.2
^{75}Co	C. Engelmann	[312]	Darmstadt	Germany	1995	15.2
^{76}Co	T. Ohnishi	[292]	RIKEN	Japan	2010	15.2
^{77}Co	Z. Y. Xu	[313]	RIKEN	Japan	2014	15.2
^{48}Ni	B. Blank	[314]	GANIL	France	2000	16.3
^{49}Ni	B. Blank	[273]	Darmstadt	Germany	1996	15.4
^{50}Ni	B. Blank	[315]	Darmstadt	Germany	1994	15.4
^{51}Ni	F. Pougheon	[262]	GANIL	France	1987	15.4
^{52}Ni	F. Pougheon	[262]	GANIL	France	1987	15.4
^{53}Ni	D.J. Vieira	[316]	Berkeley	USA	1976	11.11
^{54}Ni	R.E. Tribble	[293]	Texas A&M	USA	1977	9.8
^{55}Ni	I.D. Proctor	[276]	Michigan State	USA	1972	9.8
^{56}Ni	W.J. Worthington Jr.	[317]	Berkeley	USA	1952	10.2
^{57}Ni	J.J. Livingood	[318]	Berkeley	USA	1938	4.3

(continued)

Table 17.1 (continued)

Isotope	First author	Ref.	Laboratory	Country	Year	Table
^{58}Ni	F.W. Aston	[319]	Cambridge	UK	1921	3.2
^{59}Ni	A.R. Brosi	[320]	Oak Ridge	USA	1951	7.2
^{60}Ni	F.W. Aston	[319]	Cambridge	UK	1921	3.2
^{61}Ni	F.W. Aston	[321]	Cambridge	UK	1934	3.3
^{62}Ni	F.W. Aston	[321]	Cambridge	UK	1934	3.3
^{63}Ni	A.R. Brosi	[320]	Oak Ridge	USA	1951	7.2
^{64}Ni	J. deGier	[322]	Amsterdam	Netherlands	1935	3.4
^{65}Ni	J.A. Swartout	[323]	Oak Ridge	USA	1946	7.1
^{66}Ni	R.H. Goeckermann	[324]	Berkeley	USA	1948	10.1
^{67}Ni	R.T. Kouzes	[325]	Princeton	USA	1978	9.8
^{68}Ni	T.S. Bhatia	[326]	Heidelberg	Germany	1977	14.1
^{69}Ni	Ph. Dessagne	[327]	Orsay	France	1984	14.1
^{70}Ni	P. Armbruster	[328]	Grenoble	France	1987	6.6
^{71}Ni	P. Armbruster	[328]	Grenoble	France	1987	6.6
^{72}Ni	P. Armbruster	[328]	Grenoble	France	1987	6.6
^{73}Ni	P. Armbruster	[328]	Grenoble	France	1987	6.6
^{74}Ni	P. Armbruster	[328]	Grenoble	France	1987	6.6
^{75}Ni	M. Weber	[260]	Darmstadt	Germany	1992	15.1
^{76}Ni	C. Engelmann	[312]	Darmstadt	Germany	1995	15.2
^{77}Ni	C. Engelmann	[312]	Darmstadt	Germany	1995	15.2
^{78}Ni	C. Engelmann	[312]	Darmstadt	Germany	1995	15.2
^{79}Ni	T. Ohnishi	[292]	RIKEN	Japan	2010	15.2
^{80}Ni	Z. Y. Xu	[313]	RIKEN	Japan	2014	15.2
^{55}Cu	F. Pougheon	[262]	GANIL	France	1987	15.4
^{56}Cu	F. Pougheon	[262]	GANIL	France	1987	15.4
^{57}Cu	D.J. Vieira	[316]	Berkeley	USA	1976	11.11
^{58}Cu	W.M. Martin	[265]	McGill	Canada	1952	9.1
^{59}Cu	C.E. Leith	[329]	Berkeley	USA	1947	9.3
^{60}Cu	C.E. Leith	[329]	Berkeley	USA	1947	9.3
^{61}Cu	L.N. Ridenour	[330]	Princeton	USA	1937	4.2
^{62}Cu	F.A. Heyn	[331]	Eindhoven	Netherlands	1936	4.4
^{63}Cu	F.W. Aston	[332]	Cambridge	UK	1923	3.2
^{64}Cu	S.N. Van Voorhis	[333]	Berkeley	USA	1936	4.3
^{65}Cu	F.W. Aston	[332]	Cambridge	UK	1923	3.2
^{66}Cu	W.Y. Chang	[334]	Cambridge	UK	1937	4.4
^{67}Cu	R.H. Goeckermann	[324]	Berkeley	USA	1948	10.1
^{68}Cu	A. Flammersfeld	[335]	Mainz	Germany	1953	7.3
^{69}Cu	J. van Klinken	[336]	Ames	USA	1966	8.1

(continued)

Table 17.1 (continued)

Isotope	First author	Ref.	Laboratory	Country	Year	Table
^{70}Cu	L.M. Taff	[337]	Groningen	Netherlands	1971	7.4
^{71}Cu	E. Runte	[290]	Darmstadt	Germany	1983	14.3
^{72}Cu	E. Runte	[290]	Darmstadt	Germany	1983	14.3
^{73}Cu	E. Runte	[290]	Darmstadt	Germany	1983	14.3
^{74}Cu	P. Armbruster	[328]	Grenoble	France	1987	6.6
^{75}Cu	P.L. Reeder	[338]	Brookhaven	USA	1985	6.5
^{76}Cu	P. Armbruster	[328]	Grenoble	France	1987	6.6
^{77}Cu	P. Armbruster	[328]	Grenoble	France	1987	6.6
^{78}Cu	K.-L. Kratz	[339]	CERN	Switzerland	1991	10.10
^{79}Cu	K.-L. Kratz	[339]	CERN	Switzerland	1991	10.10
^{80}Cu	C. Engelmann	[312]	Darmstadt	Germany	1995	15.2
^{81}Cu	T. Ohnishi	[292]	RIKEN	Japan	2010	15.2
^{82}Cu	T. Ohnishi	[292]	RIKEN	Japan	2010	15.2
^{54}Zn	B. Blank	[340]	GANIL	France	2005	16.3
^{55}Zn	J. Giovinazzo	[341]	GANIL	France	2001	15.4
^{56}Zn	J. Giovinazzo	[341]	GANIL	France	2001	15.4
^{57}Zn	D.J. Vieira	[316]	Berkeley	USA	1976	11.11
^{58}Zn	K.K. Seth	[342]	Los Alamos	USA	1986	8.2
^{59}Zn	J. Honkanen	[343]	Jyväskylä	Finland	1981	9.7
^{60}Zn	L. Lindner	[344]	Amsterdam	Netherlands	1955	9.5
^{61}Zn	L. Lindner	[344]	Amsterdam	Netherlands	1955	9.5
^{62}Zn	D.R. Miller	[294]	Berkeley	USA	1948	10.2
^{63}Zn	W. Bothe	[345]	Heidelberg	Germany	1937	4.6
^{64}Zn	A.J. Dempster	[228]	Chicago	USA	1922	3.1
^{65}Zn	J.J. Livingood	[346]	Berkeley	USA	1939	4.3
^{66}Zn	A.J. Dempster	[228]	Chicago	USA	1922	3.1
^{67}Zn	F.W. Aston	[347]	Cambridge	UK	1928	3.3
^{68}Zn	A.J. Dempster	[228]	Chicago	USA	1922	3.1
^{69}Zn	F.A. Heyn	[348]	Eindhoven	Netherlands	1937	4.4
^{70}Zn	A.J. Dempster	[228]	Chicago	USA	1922	3.1
^{71}Zn	J.M. LeBlanc	[349]	Argonne	USA	1955	7.2
^{72}Zn	J.M. Siegel	[350]	Oak Ridge	USA	1951	6.1
^{73}Zn	B.R. Erdal	[351]	CERN	Switzerland	1972	10.7
^{74}Zn	B.R. Erdal	[351]	CERN	Switzerland	1972	10.7
^{75}Zn	B. Grapengiesser	[352]	Studsvik	Sweden	1974	6.5
^{76}Zn	B. Grapengiesser	[352]	Studsvik	Sweden	1974	6.5
^{77}Zn	K. Aleklett	[353]	Studsvik	Sweden	1977	6.5
^{78}Zn	K. Aleklett	[353]	Studsvik	Sweden	1977	6.5
^{79}Zn	B. Ekström	[354]	Studsvik	Sweden	1986	6.5

(continued)

Table 17.1 (continued)

Isotope	First author	Ref.	Laboratory	Country	Year	Table
^{80}Zn	R.L. Gill	[355]	Brookhaven	USA	1986	6.5
^{81}Zn	K.-L. Kratz	[339]	CERN	Switzerland	1991	10.10
^{82}Zn	M. Bernas	[237]	Darmstadt	Germany	1997	15.2
^{83}Zn	M. Bernas	[237]	Darmstadt	Germany	1997	15.2
^{84}Zn	T. Ohnishi	[292]	RIKEN	Japan	2010	15.2
^{85}Zn	T. Ohnishi	[292]	RIKEN	Japan	2010	15.2
^{60}Ga	B. Blank	[356]	GANIL	France	1995	15.4
^{61}Ga	M.A.C. Hotchkis	[357]	Berkeley	USA	1987	11.11
^{62}Ga	R. Chiba	[358]	Tokyo	Japan	1978	9.1
^{63}Ga	M.J. Nurmia	[359]	Argonne	USA	1965	11.1
^{64}Ga	B.L. Cohen	[360]	Oak Ridge	USA	1953	9.4
^{65}Ga	L.W. Alvarez	[361]	Berkeley	USA	1938	4.3
^{66}Ga	W.B. Mann	[362]	Berkeley	USA	1937	4.3
^{67}Ga	L.W. Alvarez	[363]	Berkeley	USA	1938	4.3
^{68}Ga	W. Bothe	[345]	Heidelberg	Germany	1937	4.6
^{69}Ga	F.W. Aston	[243]	Cambridge	UK	1923	3.2
^{70}Ga	W. Bothe	[345]	Heidelberg	Germany	1937	4.6
^{71}Ga	F.W. Aston	[243]	Cambridge	UK	1923	3.2
^{72}Ga	R. Sagane	[364]	Berkeley	USA	1939	4.4
^{73}Ga	J.M. Siegel	[350]	Oak Ridge	USA	1951	6.1
^{74}Ga	H. Morinaga	[194]	Purdue	USA	1956	7.3
^{75}Ga	H. Morinaga	[309]	Tohoku	Japan	1960	8.1
^{76}Ga	K. Takahashi	[365]	Tokyo	Japan	1961	7.3
^{77}Ga	L. Wish	[366]	Naval Rad. Def. Lab.	USA	1968	6.4
^{78}Ga	P. del Marmol	[367]	Mol	Belgium	1972	6.4
^{79}Ga	B. Grapengiesser	[352]	Studsvik	Sweden	1974	6.5
^{80}Ga	B. Grapengiesser	[352]	Studsvik	Sweden	1974	6.5
^{81}Ga	G. Rudstam	[368]	Studsvik	Sweden	1976	6.5
^{82}Ga	G. Rudstam	[368]	Studsvik	Sweden	1976	6.5
^{83}Ga	G. Rudstam	[368]	Studsvik	Sweden	1976	6.5
^{84}Ga	K.-L. Kratz	[339]	CERN	Switzerland	1991	10.10
^{85}Ga	M. Bernas	[237]	Darmstadt	Germany	1997	15.2
^{86}Ga	M. Bernas	[237]	Darmstadt	Germany	1997	15.2
^{87}Ga	T. Ohnishi	[292]	RIKEN	Japan	2010	15.2
^{59}Ge	A. A. Ciemny	[369]	Michigan State	USA	2015	15.4
^{60}Ge	A. Stolz	[370]	Michigan State	USA	2005	15.4
^{61}Ge	M.A.C. Hotchkis	[357]	Berkeley	USA	1987	11.11
^{62}Ge	M.F. Mohar	[371]	Michigan State	USA	1991	15.4

(continued)

Table 17.1 (continued)

Isotope	First author	Ref.	Laboratory	Country	Year	Table
^{63}Ge	M.F. Mohar	[371]	Michigan State	USA	1991	15.4
^{64}Ge	R.G.H. Robertson	[372]	Michigan State	USA	1972	9.5
^{65}Ge	R.G.H. Robertson	[372]	Michigan State	USA	1972	9.5
^{66}Ge	H.H. Hopkins	[373]	Berkeley	USA	1950	10.2
^{67}Ge	H.H. Hopkins	[373]	Berkeley	USA	1950	10.2
^{68}Ge	H.H. Hopkins	[374]	Berkeley	USA	1948	10.2
^{69}Ge	W.B. Mann	[375]	Berkeley	USA	1938	4.3
^{70}Ge	F.W. Aston	[376]	Cambridge	UK	1923	3.2
^{71}Ge	G.T. Seaborg	[377]	Berkeley	USA	1941	4.3
^{72}Ge	F.W. Aston	[376]	Cambridge	UK	1923	3.2
^{73}Ge	F.W. Aston	[378]	Cambridge	UK	1931	3.3
^{74}Ge	F.W. Aston	[376]	Cambridge	UK	1923	3.2
^{75}Ge	R. Sagane	[364]	Berkeley	USA	1939	4.4
^{76}Ge	F.W. Aston	[378]	Cambridge	UK	1931	3.3
^{77}Ge	R. Sagane	[364]	Berkeley	USA	1939	4.4
^{78}Ge	E.P. Steinberg	[379]	Argonne	USA	1951	6.1
^{79}Ge	M. Karras	[380]	Arkansas	USA	1970	7.4
^{80}Ge	P. del Marmol	[367]	Mol	Belgium	1972	6.4
^{81}Ge	P. del Marmol	[367]	Mol	Belgium	1972	6.4
^{82}Ge	P. del Marmol	[367]	Mol	Belgium	1972	6.4
^{83}Ge	P. del Marmol	[367]	Mol	Belgium	1972	6.4
^{84}Ge	P. del Marmol	[367]	Mol	Belgium	1972	6.4
^{85}Ge	J.P. Omtvedt	[381]	Studsvik	Sweden	1991	6.5
^{86}Ge	M. Bernas	[382]	Darmstadt	Germany	1994	15.2
^{87}Ge	M. Bernas	[237]	Darmstadt	Germany	1997	15.2
^{88}Ge	M. Bernas	[237]	Darmstadt	Germany	1997	15.2
^{89}Ge	M. Bernas	[237]	Darmstadt	Germany	1997	15.2
^{90}Ge	T. Ohnishi	[292]	RIKEN	Japan	2010	15.2
^{64}As	B. Blank	[356]	GANIL	France	1995	15.4
^{65}As	M.F. Mohar	[371]	Michigan State	USA	1991	15.4
^{66}As	D.E. Alburger	[383]	Brookhaven	USA	1978	11.2
^{67}As	M.J. Murphy	[384]	Argonne	USA	1980	11.2
^{68}As	T. Paradellis	[385]	McGill	Canada	1971	9.1
^{69}As	F.D.S. Butement	[386]	Harwell	UK	1955	9.5
^{70}As	H.H. Hopkins	[373]	Berkeley	USA	1950	10.2
^{71}As	R. Sagane	[364]	Berkeley	USA	1939	4.3
^{72}As	A.C.G. Mitchell	[387]	Indiana	USA	1947	9.4
^{73}As	D.A. McCown	[388]	Ohio State	USA	1948	9.4

(continued)

Table 17.1 (continued)

Isotope	First author	Ref.	Laboratory	Country	Year	Table
^{74}As	R. Sagane	[389]	Berkeley	USA	1938	4.3
^{75}As	F.W. Aston	[72]	Cambridge	UK	1920	3.2
^{76}As	E. Amaldi	[269]	Rome	Italy	1934	4.1
^{77}As	E.P. Steinberg	[379]	Argonne	USA	1951	6.1
^{78}As	A.H. Snell	[390]	Berkeley	USA	1937	4.4
^{79}As	F.D.S. Butement	[391]	Harwell	UK	1950	8.1
^{80}As	R.E. Meads	[392]	Oxford	UK	1959	7.3
^{81}As	H. Morinaga	[309]	Tohoku	Japan	1960	8.1
^{82}As	P.J. Mathew	[393]	Wellington	New Zealand	1968	7.4
^{83}As	P. del Marmol	[394]	Mol	Belgium	1968	6.4
^{84}As	P. del Marmol	[394]	Mol	Belgium	1968	6.4
^{85}As	P. del Marmol	[395]	Mol	Belgium	1967	6.4
^{86}As	J.V. Kratz	[396]	Mainz	Germany	1973	6.4
^{87}As	J.V. Kratz	[397]	Mainz	Germany	1970	6.4
^{88}As	M. Bernas	[382]	Darmstadt	Germany	1994	15.2
^{89}As	M. Bernas	[382]	Darmstadt	Germany	1994	15.2
^{90}As	M. Bernas	[237]	Darmstadt	Germany	1997	15.2
^{91}As	M. Bernas	[237]	Darmstadt	Germany	1997	15.2
^{92}As	M. Bernas	[237]	Darmstadt	Germany	1997	15.2
^{64}Se	A. Stolz	[370]	Michigan State	USA	2005	15.4
^{65}Se	J.C. Batchelder	[398]	Berkeley	USA	1993	11.11
^{66}Se	J.A. Winger	[399]	Michigan State	USA	1993	15.4
^{67}Se	M.F. Mohar	[371]	Michigan State	USA	1991	15.4
^{68}Se	C.J. Lister	[400]	Daresbury	UK	1990	11.4
^{69}Se	E. Nolte	[401]	Munich	Germany	1974	11.2
^{70}Se	H.H. Hopkins	[373]	Berkeley	USA	1950	10.2
^{71}Se	J. Beydon	[402]	Saclay	France	1957	11.1
^{72}Se	H.H. Hopkins	[374]	Berkeley	USA	1948	10.2
^{73}Se	W.S. Cowart	[403]	Ohio State	USA	1948	9.4
^{74}Se	F.W. Aston	[161]	Cambridge	UK	1922	3.2
^{75}Se	H.N. Friedlander	[404]	Argonne	USA	1947	7.1
^{76}Se	F.W. Aston	[161]	Cambridge	UK	1922	3.2
^{77}Se	F.W. Aston	[161]	Cambridge	UK	1922	3.2
^{78}Se	F.W. Aston	[161]	Cambridge	UK	1922	3.2
^{79}Se	A. Flammersfeld	[405]	Mainz	Germany	1950	7.3
^{80}Se	F.W. Aston	[161]	Cambridge	UK	1922	3.2
^{81}Se	H. Wäffler	[406]	Zurich	Switzerland	1948	8.1
^{82}Se	F.W. Aston	[161]	Cambridge	UK	1922	3.2
^{83}Se	A.H. Snell	[390]	Berkeley	USA	1937	4.3

(continued)

Table 17.1 (continued)

Isotope	First author	Ref.	Laboratory	Country	Year	Table
^{84}Se	J.E. Sattizahn	[407]	Los Alamos	USA	1960	6.3
^{85}Se	J.E. Sattizahn	[407]	Los Alamos	USA	1960	6.3
^{86}Se	T. Tamai	[408]	Kyoto	Japan	1973	6.4
^{87}Se	L. Tomlinson	[409]	Harwell	UK	1968	6.4
^{88}Se	P. del Marmol	[410]	Mol	Belgium	1970	6.4
^{89}Se	L. Tomlinson	[411]	Harwell	UK	1971	6.4
^{90}Se	M. Bernas	[382]	Darmstadt	Germany	1994	15.2
^{91}Se	M. Asghar	[412]	Grenoble	France	1975	6.6
^{92}Se	M. Bernas	[237]	Darmstadt	Germany	1997	15.2
^{93}Se	M. Bernas	[237]	Darmstadt	Germany	1997	15.2
^{94}Se	M. Bernas	[237]	Darmstadt	Germany	1997	15.2
^{95}Se	T. Ohnishi	[292]	RIKEN	Japan	2010	15.2
^{69}Br	A. M. Rogers	[413]	Michigan State	USA	2011	16.2
^{70}Br	D.E. Alburger	[383]	Brookhaven	USA	1978	11.2
^{71}Br	E. Hagberg	[414]	Chalk River	Canada	1982	11.3
^{72}Br	E. Nolte	[415]	Heidelberg	Germany	1970	11.2
^{73}Br	G. Murray	[416]	Manchester	UK	1970	11.1
^{74}Br	J.M. Hollander	[417]	Berkeley	USA	1953	11.1
^{75}Br	L.L. Woodward	[418]	Ohio State	USA	1948	9.4
^{76}Br	S.C. Fultz	[419]	Ohio State	USA	1952	9.4
^{77}Br	L.L. Woodward	[418]	Ohio State	USA	1948	9.4
^{78}Br	W. Bothe	[420]	Heidelberg	Germany	1937	4.6
^{79}Br	F.W. Aston	[72]	Cambridge	UK	1920	3.2
^{80}Br	W. Bothe	[421]	Heidelberg	Germany	1937	4.6
^{81}Br	F.W. Aston	[72]	Cambridge	UK	1920	3.2
^{82}Br	A.H. Snell	[390]	Berkeley	USA	1937	4.3
^{83}Br	A.H. Snell	[390]	Berkeley	USA	1937	4.3
^{84}Br	H.J. Born	[422]	Berlin	Germany	1943	4.4
^{85}Br	H.J. Born	[422]	Berlin	Germany	1943	4.4
^{86}Br	A.F. Stehney	[423]	Argonne	USA	1962	7.3
^{87}Br	H.J. Born	[422]	Berlin	Germany	1943	4.4
^{88}Br	N. Sugarman	[424]	Argonne	USA	1949	6.1
^{89}Br	G.J. Perlow	[425]	Argonne	USA	1959	6.3
^{90}Br	G.J. Perlow	[425]	Argonne	USA	1959	6.3
^{91}Br	K.-L. Kratz	[426]	Mainz	Germany	1974	6.4
^{92}Br	K.-L. Kratz	[426]	Mainz	Germany	1974	6.4
^{93}Br	K.-L. Kratz	[427]	Grenoble/CERN	France/Switz.	1988	6.5
^{94}Br	K.-L. Kratz	[427]	Grenoble/CERN	France/Switz.	1988	6.5
^{95}Br	M. Bernas	[237]	Darmstadt	Germany	1997	15.2

(continued)

Table 17.1 (continued)

Isotope	First author	Ref.	Laboratory	Country	Year	Table
^{96}Br	M. Bernas	[237]	Darmstadt	Germany	1997	15.2
^{97}Br	M. Bernas	[237]	Darmstadt	Germany	1997	15.2
^{98}Br	T. Ohnishi	[292]	RIKEN	Japan	2010	15.2
^{69}Kr	B. Blank	[356]	GANIL	France	1995	15.4
^{70}Kr	B. Blank	[356]	GANIL	France	1995	15.4
^{71}Kr	G.T. Ewan	[428]	CERN	Switzerland	1981	10.7
^{72}Kr	H. Schmeing	[429]	Chalk River	Canada	1973	11.1
^{73}Kr	P. Hornshoj	[430]	CERN	Switzerland	1972	10.7
^{74}Kr	F.D.S. Butement	[431]	Liverpool	UK	1960	10.2
^{75}Kr	F.D.S. Butement	[431]	Liverpool	UK	1960	10.2
^{76}Kr	A.A. Caretto Jr.	[432]	Rochester	USA	1954	10.2
^{77}Kr	L.L. Woodward	[433]	Ohio State	USA	1948	9.4
^{78}Kr	F.W. Aston	[97]	Cambridge	UK	1920	3.2
^{79}Kr	L.L. Woodward	[433]	Ohio State	USA	1948	9.4
^{80}Kr	F.W. Aston	[97]	Cambridge	UK	1920	3.2
^{81}Kr	J.H. Reynolds	[434]	Argonne	USA	1950	7.2
^{82}Kr	F.W. Aston	[97]	Cambridge	UK	1920	3.2
^{83}Kr	F.W. Aston	[97]	Cambridge	UK	1920	3.2
^{84}Kr	F.W. Aston	[97]	Cambridge	UK	1920	3.2
^{85}Kr	H.J. Born	[422]	Berlin	Germany	1943	4.4
^{86}Kr	F.W. Aston	[97]	Cambridge	UK	1920	3.2
^{87}Kr	H.J. Born	[422]	Berlin	Germany	1943	4.4
^{88}Kr	A. Langsdorf Jr.	[435]	Berkeley	USA	1939	4.7
^{89}Kr	O. Hahn	[436]	Berlin	Germany	1943	4.7
^{90}Kr	C.R. Dillard	[437]	Argonne	USA	1951	6.1
^{91}Kr	C.R. Dillard	[437]	Argonne	USA	1951	6.1
^{92}Kr	C.R. Dillard	[437]	Argonne	USA	1951	6.1
^{93}Kr	C.R. Dillard	[437]	Argonne	USA	1951	6.1
^{94}Kr	S. Amiel	[438]	Soreq	Israel	1972	6.5
^{95}Kr	M. Bernas	[382]	Darmstadt	Germany	1994	15.2
^{96}Kr	M. Bernas	[382]	Darmstadt	Germany	1994	15.2
^{97}Kr	M. Bernas	[237]	Darmstadt	Germany	1997	15.2
^{98}Kr	M. Bernas	[237]	Darmstadt	Germany	1997	15.2
^{99}Kr	M. Bernas	[237]	Darmstadt	Germany	1997	15.2
^{100}Kr	M. Bernas	[237]	Darmstadt	Germany	1997	15.2
^{101}Kr	T. Ohnishi	[292]	RIKEN	Japan	2010	15.2
^{73}Rb	J.C. Batchelder	[439]	Berkeley	USA	1993	16.2
^{74}Rb	J.M. D'Auria	[440]	CERN	Switzerland	1977	10.7

(continued)

Table 17.1 (continued)

Isotope	First author	Ref.	Laboratory	Country	Year	Table
^{75}Rb	H.L. Ravn	[441]	CERN	Switzerland	1975	10.7
^{76}Rb	J. Chaumont	[442]	CERN	Switzerland	1969	10.7
^{77}Rb	R. Arlt	[443]	Dubna	Russia	1972	10.8
^{78}Rb	C. J. Toeset	[444]	Amsterdam	Netherlands	1968	9.5
^{79}Rb	J. Beydon	[445]	Saclay	France	1957	11.1
^{80}Rb	R.W. Hoff	[446]	Berkeley	USA	1961	11.1
^{81}Rb	F.L. Reynolds	[447]	Berkeley	USA	1949	9.6
^{82}Rb	F.L. Reynolds	[447]	Berkeley	USA	1949	9.6
^{83}Rb	D. G. Karraker	[448]	Berkeley	USA	1950	9.6
^{84}Rb	W. C. Barber	[449]	Berkeley	USA	1947	9.3
^{85}Rb	F.W. Aston	[215]	Cambridge	UK	1921	3.2
^{86}Rb	A.C. Helmholz	[450]	Berkeley	USA	1941	4.3
^{87}Rb	F.W. Aston	[215]	Cambridge	UK	1921	3.2
^{88}Rb	F.A. Heyn	[451]	Eindhoven	Netherlands	1939	4.7
^{89}Rb	G.N. Glasoe	[452]	Columbia	USA	1940	4.7
^{90}Rb	O. Kofoed-Hansen	[453]	Copenhagen	Denmark	1951	6.1
^{91}Rb	O. Kofoed-Hansen	[453]	Copenhagen	Denmark	1951	6.1
^{92}Rb	K. Fritze	[454]	McMaster	Canada	1960	6.3
^{93}Rb	K. Fritze	[454]	McMaster	Canada	1960	6.3
^{94}Rb	K. Fritze	[455]	McMaster	Canada	1961	6.3
^{95}Rb	I. Amarel	[456]	Orsay	France	1967	10.9
^{96}Rb	I. Amarel	[456]	Orsay	France	1967	10.9
^{97}Rb	I. Amarel	[457]	Orsay	France	1969	10.9
^{98}Rb	B.L. Tracy	[458]	Grenoble	France	1971	10.9
^{99}Rb	B.L. Tracy	[458]	Grenoble	France	1971	10.9
^{100}Rb	E. Koglin	[459]	Grenoble	France	1978	6.5
^{101}Rb	K. Balog	[460]	CERN	Switzerland	1992	10.10
^{102}Rb	G. Lhersonneau	[461]	CERN	Switzerland	1995	10.10
^{103}Rb	T. Ohnishi	[292]	RIKEN	Japan	2010	15.2
^{73}Sr	J.C. Batchelder	[439]	Berkeley	USA	1993	11.11
^{74}Sr	B. Blank	[356]	GANIL	France	1995	15.4
^{75}Sr	M.F. Mohar	[371]	Michigan State	USA	1991	15.4
^{76}Sr	C.J. Lister	[400]	Daresbury	UK	1990	11.4
^{77}Sr	J.C. Hardy	[462]	Chalk River	Canada	1976	11.11
^{78}Sr	C.J. Lister	[463]	Brookhaven	USA	1982	11.4
^{79}Sr	I.M. Ladenbauer-Bellis	[464]	Yale	USA	1972	11.2
^{80}Sr	R.W. Hoff	[446]	Berkeley	USA	1961	11.1
^{81}Sr	S.V. Castner	[465]	Berkeley	USA	1952	9.3
^{82}Sr	S.V. Castner	[465]	Berkeley	USA	1952	9.3

(continued)

Table 17.1 (continued)

Isotope	First author	Ref.	Laboratory	Country	Year	Table
^{83}Sr	S.V. Castner	[465]	Berkeley	USA	1952	9.3
^{84}Sr	J.P. Blewett	[466]	Princeton	USA	1936	3.4
^{85}Sr	L.A. DuBridge	[467]	Rochester	USA	1940	4.2
^{86}Sr	F.W. Aston	[296]	Cambridge	UK	1924	3.2
^{87}Sr	F.W. Aston	[468]	Cambridge	UK	1931	3.3
^{88}Sr	F.W. Aston	[243]	Cambridge	UK	1923	3.2
^{89}Sr	D.W. Stewart	[469]	Michigan	USA	1937	4.2
^{90}Sr	R.W. Nottorf	[470]	Chicago	USA	1951	6.1
^{91}Sr	W. Seelmann-Eggebert	[471]	Berlin	Germany	1943	4.4
^{92}Sr	H. Götte	[472]	Berlin	Germany	1941	4.7
^{93}Sr	J.D. Knight	[473]	Los Alamos	USA	1959	6.3
^{94}Sr	J.D. Knight	[473]	Los Alamos	USA	1959	6.3
^{95}Sr	K. Fritze	[455]	McMaster	Canada	1961	6.3
^{96}Sr	M.-I. Macias-Marques	[474]	Orsay	France	1971	10.9
^{97}Sr	F.K. Wohn	[475]	Grenoble	France	1978	6.5
^{98}Sr	B.L. Tracy	[458]	Grenoble	France	1971	10.9
^{99}Sr	M. Asghar	[412]	Grenoble	France	1975	6.6
^{100}Sr	E. Koglin	[459]	Grenoble	France	1978	6.5
^{101}Sr	F.K. Wohn	[476]	Brookhaven	USA	1983	6.5
^{102}Sr	J.C. Hill	[477]	Brookhaven	USA	1986	6.5
^{103}Sr	M. Bernas	[237]	Darmstadt	Germany	1997	15.2
^{104}Sr	M. Bernas	[237]	Darmstadt	Germany	1997	15.2
^{105}Sr	M. Bernas	[237]	Darmstadt	Germany	1997	15.2
^{106}Sr	T. Ohnishi	[292]	RIKEN	Japan	2010	15.2
^{107}Sr	T. Ohnishi	[292]	RIKEN	Japan	2010	15.2
^{76}Y	P. Kienle	[478]	Darmstadt	Germany	2001	15.4
^{77}Y	Z. Janas	[479]	GANIL	France	1999	15.4
^{78}Y	S.J. Yennello	[480]	Michigan State	USA	1992	15.4
^{79}Y	H. Grawe	[481]	CERN	Switzerland	1992	10.7
^{80}Y	C.J. Lister	[482]	Brookhaven	USA	1981	11.2
^{81}Y	C.J. Lister	[482]	Brookhaven	USA	1981	11.2
^{82}Y	C. Deprun	[483]	Orsay	France	1980	11.2
^{83}Y	V. Maxia	[484]	Berkeley	USA	1962	11.1
^{84}Y	V. Maxia	[484]	Berkeley	USA	1962	11.1
^{85}Y	A.A. Caretto Jr.	[485]	Rochester	USA	1952	10.2
^{86}Y	E.K. Hyde	[486]	Berkeley	USA	1951	10.2
^{87}Y	L.A. DuBridge	[467]	Rochester	USA	1940	4.2
^{88}Y	R.J. Hayden	[487]	Chicago	USA	1948	6.1
^{89}Y	F.W. Aston	[243]	Cambridge	UK	1923	3.2

(continued)

Table 17.1 (continued)

Isotope	First author	Ref.	Laboratory	Country	Year	Table
^{90}Y	M.L. Pool	[488]	Michigan	USA	1937	4.4
^{91}Y	W. Seelmann-Eggebert	[471]	Berlin	Germany	1943	4.4
^{92}Y	R. Sagane	[489]	RIKEN	Japan	1940	4.4
^{93}Y	S. Katcoff	[490]	Los Alamos	USA	1948	6.1
^{94}Y	S. Katcoff	[490]	Los Alamos	USA	1948	6.1
^{95}Y	J.D. Knight	[473]	Los Alamos	USA	1959	6.3
^{96}Y	H. Gunther	[491]	Munich	Germany	1975	6.5
^{97}Y	J. Eidens	[492]	Juelich	Germany	1970	6.6
^{98}Y	J.W. Grüter	[493]	Juelich	Germany	1970	6.6
^{99}Y	M. Asghar	[412]	Grenoble	France	1975	6.6
^{100}Y	B. Pfeiffer	[494]	Grenoble	France	1977	6.6
^{101}Y	F.K. Wohn	[476]	Brookhaven	USA	1983	6.5
^{102}Y	K. Shizuma	[495]	Juelich	Germany	1983	6.6
^{103}Y	M. Bernas	[382]	Darmstadt	Germany	1994	15.2
^{104}Y	M. Bernas	[382]	Darmstadt	Germany	1994	15.2
^{105}Y	M. Bernas	[382]	Darmstadt	Germany	1994	15.2
^{106}Y	M. Bernas	[237]	Darmstadt	Germany	1997	15.2
^{107}Y	M. Bernas	[237]	Darmstadt	Germany	1997	15.2
^{108}Y	T. Ohnishi	[292]	RIKEN	Japan	2010	15.2
^{109}Y	T. Ohnishi	[292]	RIKEN	Japan	2010	15.2
^{78}Zr	P. Kienle	[478]	Darmstadt	Germany	2001	15.4
^{79}Zr	Z. Janas	[479]	GANIL	France	1999	15.4
^{80}Zr	C.J. Lister	[496]	Daresbury	UK	1987	11.4
^{81}Zr	W.X. Huang	[497]	Lanzhou	China	1997	11.11
^{82}Zr	C.F. Liang	[498]	Orsay	France	1982	10.8
^{83}Zr	M. Kaba	[499]	Munich	Germany	1974	11.1
^{84}Zr	G. Korschinek	[500]	Munich	Germany	1977	11.2
^{85}Zr	F.D.S. Butement	[501]	Liverpool	UK	1963	10.5
^{86}Zr	E.K. Hyde	[486]	Berkeley	USA	1951	10.2
^{87}Zr	B.E. Robertson	[502]	Ohio State	USA	1949	9.4
^{88}Zr	E.K. Hyde	[486]	Berkeley	USA	1951	10.2
^{89}Zr	R. Sagane	[503]	RIKEN	Japan	1938	4.4
^{90}Zr	F.W. Aston	[504]	Cambridge	UK	1924	3.2
^{91}Zr	F.W. Aston	[230]	Cambridge	UK	1934	3.3
^{92}Zr	F.W. Aston	[504]	Cambridge	UK	1924	3.2
^{93}Zr	E.P. Steinberg	[505]	Argonne	USA	1950	6.1
^{94}Zr	F.W. Aston	[504]	Cambridge	UK	1924	3.2
^{95}Zr	W.E. Grummitt	[506]	Chalk River	Canada	1946	6.1
^{96}Zr	F.W. Aston	[230]	Cambridge	UK	1934	3.3

(continued)

Table 17.1 (continued)

Isotope	First author	Ref.	Laboratory	Country	Year	Table
^{97}Zr	S. Katcoff	[507]	Argonne	USA	1951	6.1
^{98}Zr	K. Hübenthal	[508]	Grenoble	France	1967	6.4
^{99}Zr	J. Eidens	[492]	Juelich	Germany	1970	6.6
^{100}Zr	E. Cheifetz	[509]	Berkeley	USA	1970	13.1
^{101}Zr	R.L. Watson	[510]	Berkeley	USA	1970	13.1
^{102}Zr	R.L. Watson	[510]	Berkeley	USA	1970	13.1
^{103}Zr	M. Graefenstedt	[511]	Grenoble	France	1987	6.6
^{104}Zr	M.A.C. Hotchkis	[512]	Argonne	USA	1990	13.1
^{105}Zr	J. Äystö	[513]	Jyväskylä	Finland	1992	10.9
^{106}Zr	M. Bernas	[382]	Darmstadt	Germany	1994	15.2
^{107}Zr	M. Bernas	[382]	Darmstadt	Germany	1994	15.2
^{108}Zr	M. Bernas	[237]	Darmstadt	Germany	1997	15.2
^{109}Zr	M. Bernas	[237]	Darmstadt	Germany	1997	15.2
^{110}Zr	M. Bernas	[237]	Darmstadt	Germany	1997	15.2
^{111}Zr	T. Ohnishi	[292]	RIKEN	Japan	2010	15.2
^{112}Zr	T. Ohnishi	[292]	RIKEN	Japan	2010	15.2
^{82}Nb	S.J. Yennello	[480]	Michigan State	USA	1992	15.4
^{83}Nb	T. Kuroyanagi	[514]	Kyushu	Japan	1988	11.4
^{84}Nb	G. Korschinek	[500]	Munich	Germany	1977	11.2
^{85}Nb	T. Kuroyanagi	[514]	Kyushu	Japan	1988	11.4
^{86}Nb	I. Votsilka	[515]	Dubna	Russia	1974	10.4
^{87}Nb	T.A. Doron	[516]	Rochester	USA	1971	11.2
^{88}Nb	R.G. Korteling	[517]	Berkeley	USA	1964	10.5
^{89}Nb	R.M. Diamond	[518]	Harvard	USA	1954	9.4
^{90}Nb	L. Jacobson	[519]	Berkeley	USA	1951	9.2
^{91}Nb	L. Jacobson	[519]	Berkeley	USA	1951	9.2
^{92}Nb	R. Sagane	[520]	RIKEN	Japan	1938	4.4
^{93}Nb	F.W. Aston	[521]	Cambridge	UK	1932	3.3
^{94}Nb	R. Sagane	[520]	RIKEN	Japan	1938	4.4
^{95}Nb	W.E. Grummitt	[506]	Chalk River	Canada	1946	6.1
^{96}Nb	D.N. Kundu	[522]	Ohio State	USA	1949	9.4
^{97}Nb	S. Katcoff	[507]	Argonne	USA	1951	6.1
^{98}Nb	C.J. Orth	[523]	Los Alamos	USA	1960	6.3
^{99}Nb	R.B. Duffield	[524]	Illinois	USA	1950	8.1
^{100}Nb	K. Hübenthal	[525]	Grenoble	France	1967	7.4
^{101}Nb	J. Eidens	[492]	Juelich	Germany	1970	6.6
^{102}Nb	N. Trautmann	[526]	Mainz	Germany	1972	6.4
^{103}Nb	F.F. Hopkins	[527]	Austin	USA	1971	13.1
^{104}Nb	F.F. Hopkins	[527]	Austin	USA	1971	13.1

(continued)

Table 17.1 (continued)

Isotope	First author	Ref.	Laboratory	Country	Year	Table
^{105}Nb	K. Shizuma	[528]	Juelich/Grenoble	Germ./France	1984	6.6
^{106}Nb	H. Ahrens	[529]	Mainz	Germany	1976	6.4
^{107}Nb	J. Äystö	[513]	Jyväskylä	Finland	1992	10.9
^{108}Nb	M. Bernas	[382]	Darmstadt	Germany	1994	15.2
^{109}Nb	M. Bernas	[382]	Darmstadt	Germany	1994	15.2
^{110}Nb	M. Bernas	[382]	Darmstadt	Germany	1994	15.2
^{111}Nb	M. Bernas	[237]	Darmstadt	Germany	1997	15.2
^{112}Nb	M. Bernas	[237]	Darmstadt	Germany	1997	15.2
^{113}Nb	M. Bernas	[237]	Darmstadt	Germany	1997	15.2
^{114}Nb	T. Ohnishi	[292]	RIKEN	Japan	2010	15.2
^{115}Nb	T. Ohnishi	[292]	RIKEN	Japan	2010	15.2
^{83}Mo	Z. Janas	[479]	GANIL	France	1999	15.4
^{84}Mo	W. Gelletly	[530]	Daresbury	UK	1991	11.4
^{85}Mo	S.J. Yennello	[480]	Michigan State	USA	1992	15.4
^{86}Mo	C.J. Gross	[531]	Daresbury	UK	1991	11.4
^{87}Mo	G. Korschinek	[500]	Munich	Germany	1977	11.2
^{88}Mo	T.A. Doron	[516]	Rochester	USA	1971	11.2
^{89}Mo	R.C. Pardo	[532]	Michigan State	USA	1980	9.8
^{90}Mo	R.M. Diamond	[533]	Harvard	USA	1953	9.4
^{91}Mo	R. Sagane	[503]	RIKEN	Japan	1938	4.4
^{92}Mo	F.W. Aston	[534]	Cambridge	UK	1930	3.3
^{93}Mo	M.L. Wiedenbeck	[535]	Michigan	USA	1946	9.1
^{94}Mo	F.W. Aston	[534]	Cambridge	UK	1930	3.3
^{95}Mo	F.W. Aston	[534]	Cambridge	UK	1930	3.3
^{96}Mo	F.W. Aston	[534]	Cambridge	UK	1930	3.3
^{97}Mo	F.W. Aston	[534]	Cambridge	UK	1930	3.3
^{98}Mo	F.W. Aston	[534]	Cambridge	UK	1930	3.3
^{99}Mo	R. Sagane	[503]	RIKEN	Japan	1938	4.4
^{100}Mo	F.W. Aston	[534]	Cambridge	UK	1930	3.3
^{101}Mo	W. Maurer	[536]	Berlin	Germany	1941	4.4
^{102}Mo	D.R. Wiles	[537]	MIT	USA	1954	6.2
^{103}Mo	P. Kienle	[538]	Munich	Germany	1963	6.3
^{104}Mo	P. Kienle	[539]	Munich	Germany	1962	6.3
^{105}Mo	P. Kienle	[540]	Munich	Germany	1962	6.3
^{106}Mo	J.D. Hastings	[541]	Oak Ridge	USA	1969	6.4
^{107}Mo	N. Trautmann	[526]	Mainz	Germany	1972	6.4
^{108}Mo	N. Trautmann	[526]	Mainz	Germany	1972	6.4
^{109}Mo	J. Äystö	[513]	Jyväskylä	Finland	1992	10.9
^{110}Mo	J. Äystö	[513]	Jyväskylä	Finland	1992	10.9

(continued)

Table 17.1 (continued)

Isotope	First author	Ref.	Laboratory	Country	Year	Table
^{111}Mo	M. Bernas	[382]	Darmstadt	Germany	1994	15.2
^{112}Mo	M. Bernas	[382]	Darmstadt	Germany	1994	15.2
^{113}Mo	M. Bernas	[382]	Darmstadt	Germany	1994	15.2
^{114}Mo	M. Bernas	[237]	Darmstadt	Germany	1997	15.2
^{115}Mo	T. Ohnishi	[292]	RIKEN	Japan	2010	15.2
^{116}Mo	T. Ohnishi	[292]	RIKEN	Japan	2010	15.2
^{117}Mo	T. Ohnishi	[292]	RIKEN	Japan	2010	15.2
^{118}Mo	G. Lorusso	[542]	RIKEN	Japan	2015	15.2
^{86}Tc	S.J. Yennello	[480]	Michigan State	USA	1992	15.4
^{87}Tc	D. Rudolph	[543]	Daresbury	UK	1991	11.4
^{88}Tc	D. Rudolph	[543]	Daresbury	UK	1991	11.4
^{89}Tc	K. Heiguchi	[544]	Kyushu	Japan	1991	11.2
^{90}Tc	R. Iafigliola	[545]	McGill	Canada	1974	9.1
^{91}Tc	R. Iafigliola	[545]	McGill	Canada	1974	9.1
^{92}Tc	R. van Lieshout	[546]	Amsterdam	Netherlands	1964	9.5
^{93}Tc	D.N. Kundu	[547]	Ohio State	USA	1948	9.4
^{94}Tc	E.E. Motta	[548]	Carnegie Institute	USA	1948	9.4
^{95}Tc	J.E. Edwards	[549]	Ohio State	USA	1947	9.4
^{96}Tc	J.E. Edwards	[549]	Ohio State	USA	1947	9.4
^{97}Tc	E.E. Motta	[550]	Oak Ridge	USA	1946	7.1
^{98}Tc	G.E. Boyd	[551]	Oak Ridge	USA	1955	9.6
^{99}Tc	E. Segrè	[552]	Berkeley	USA	1938	4.3
^{100}Tc	G.E. Boyd	[553]	Oak Ridge	USA	1952	7.2
^{101}Tc	W. Maurer	[536]	Berlin	Germany	1941	4.4
^{102}Tc	J. Flegenheimer	[554]	Buenos Aires	Argentina	1954	6.2
^{103}Tc	J. Flegenheimer	[555]	Buenos Aires	Argentina	1957	7.3
^{104}Tc	J. Flegenheimer	[556]	Buenos Aires	Argentina	1956	7.3
^{105}Tc	J. Flegenheimer	[557]	Buenos Aires	Argentina	1955	6.2
^{106}Tc	A. von Beckmann	[558]	Karlsruhe	Germany	1965	6.3
^{107}Tc	A. von Beckmann	[558]	Karlsruhe	Germany	1965	6.3
^{108}Tc	W. John	[559]	Livermore	USA	1970	13.1
^{109}Tc	N. Trautmann	[560]	Mainz	Germany	1976	6.4
^{110}Tc	N. Trautmann	[560]	Mainz	Germany	1976	6.4
^{111}Tc	H. Penttilä	[561]	Jyväskylä	Finland	1988	10.9
^{112}Tc	J. Äystö	[562]	Jyväskylä	Finland	1990	10.9
^{113}Tc	J. Äystö	[513]	Jyväskylä	Finland	1992	10.9
^{114}Tc	M. Bernas	[382]	Darmstadt	Germany	1994	15.2
^{115}Tc	M. Bernas	[382]	Darmstadt	Germany	1994	15.2
^{116}Tc	M. Bernas	[237]	Darmstadt	Germany	1997	15.2

(continued)

Table 17.1 (continued)

Isotope	First author	Ref.	Laboratory	Country	Year	Table
^{117}Tc	M. Bernas	[237]	Darmstadt	Germany	1997	15.2
^{118}Tc	T. Ohnishi	[292]	RIKEN	Japan	2010	15.2
^{119}Tc	T. Ohnishi	[292]	RIKEN	Japan	2010	15.2
^{120}Tc	T. Ohnishi	[292]	RIKEN	Japan	2010	15.2
^{121}Tc	G. Lorusso	[542]	RIKEN	Japan	2015	15.2
^{87}Ru	K. Rykaczewski	[563]	GANIL	France	1995	15.4
^{88}Ru	M. Hencheck	[564]	Michigan State	USA	1994	15.4
^{89}Ru	S.J. Yennello	[480]	Michigan State	USA	1992	15.4
^{90}Ru	S. Zhou	[565]	Beijing	China	1991	11.2
^{91}Ru	E. Hagberg	[566]	Chalk River	Canada	1983	11.11
^{92}Ru	R. Arlt	[567]	Dubna	Russia	1971	10.4
^{93}Ru	T.A. Doron	[568]	Rochester	USA	1972	9.1
^{94}Ru	A. van der Wiel	[569]	Amsterdam	Netherlands	1952	9.4
^{95}Ru	D.T. Eggen	[570]	Ohio State	USA	1948	9.4
^{96}Ru	F.W. Aston	[571]	Cambridge	UK	1931	3.3
^{97}Ru	W.H. Sullivan	[572]	Chicago	USA	1946	9.2
^{98}Ru	H. Ewald	[573]	Berlin	Germany	1944	3.4
^{99}Ru	F.W. Aston	[571]	Cambridge	UK	1931	3.3
^{100}Ru	F.W. Aston	[571]	Cambridge	UK	1931	3.3
^{101}Ru	F.W. Aston	[571]	Cambridge	UK	1931	3.3
^{102}Ru	F.W. Aston	[571]	Cambridge	UK	1931	3.3
^{103}Ru	E. Bohr	[574]	Stockholm	Sweden	1945	4.4
^{104}Ru	F.W. Aston	[571]	Cambridge	UK	1931	3.3
^{105}Ru	E. Bohr	[574]	Stockholm	Sweden	1945	4.4
^{106}Ru	W.E. Grummitt	[506]	Chalk River	Canada	1946	6.1
^{107}Ru	L.E. Glendenin	[575]	Oak Ridge	USA	1951	6.1
^{108}Ru	G.B. Baró	[576]	Buenos Aires	Argentina	1955	6.2
^{109}Ru	K. Griffiths	[577]	McMaster	Canada	1967	6.4
^{110}Ru	R.L. Watson	[510]	Berkeley	USA	1970	13.1
^{111}Ru	W. John	[559]	Livermore	USA	1970	13.1
^{112}Ru	E. Cheifetz	[509]	Berkeley	USA	1970	13.1
^{113}Ru	H. Penttilä	[561]	Jyväskylä	Finland	1988	10.9
^{114}Ru	M. Leino	[578]	Jyväskylä	Finland	1991	10.9
^{115}Ru	J. Äystö	[513]	Jyväskylä	Finland	1992	10.9
^{116}Ru	M. Bernas	[382]	Darmstadt	Germany	1994	15.2
^{117}Ru	M. Bernas	[382]	Darmstadt	Germany	1994	15.2
^{118}Ru	M. Bernas	[382]	Darmstadt	Germany	1994	15.2
^{119}Ru	M. Bernas	[237]	Darmstadt	Germany	1997	15.2
^{120}Ru	T. Ohnishi	[292]	RIKEN	Japan	2010	15.2

(continued)

Table 17.1 (continued)

Isotope	First author	Ref.	Laboratory	Country	Year	Table
^{121}Ru	T. Ohnishi	[292]	RIKEN	Japan	2010	15.2
^{122}Ru	T. Ohnishi	[292]	RIKEN	Japan	2010	15.2
^{123}Ru	T. Ohnishi	[292]	RIKEN	Japan	2010	15.2
^{124}Ru	T. Ohnishi	[292]	RIKEN	Japan	2010	15.2
^{89}Rh	K. Rykaczewski	[563]	GANIL	France	1995	15.4
^{90}Rh	M. Hencheck	[564]	Michigan State	USA	1994	15.4
^{91}Rh	M. Hencheck	[564]	Michigan State	USA	1994	15.4
^{92}Rh	M. Hencheck	[564]	Michigan State	USA	1994	15.4
^{93}Rh	M. Hencheck	[564]	Michigan State	USA	1994	15.4
^{94}Rh	R.A. Zarifov	[579]	Almaty	Kaszakhstan	1979	9.1
^{95}Rh	A.H.W. Aten Jr.	[580]	Amsterdam	Netherlands	1967	9.5
^{96}Rh	A.H.W. Aten Jr.	[580]	Amsterdam	Netherlands	1967	9.5
^{97}Rh	A.H.W. Aten Jr.	[581]	Amsterdam	Netherlands	1955	9.5
^{98}Rh	A.H.W. Aten Jr.	[581]	Amsterdam	Netherlands	1955	9.5
^{99}Rh	C.L. Scoville	[582]	Ohio State	USA	1952	9.4
^{100}Rh	M. Lindner	[583]	Berkeley	USA	1948	10.2
^{101}Rh	M. Lindner	[583]	Berkeley	USA	1948	10.2
^{102}Rh	O. Minikawa	[584]	RIKEN	Japan	1941	4.4
^{103}Rh	F.W. Aston	[230]	Cambridge	UK	1934	3.3
^{104}Rh	E.C. Crittenden Jr.	[585]	Cornell	USA	1939	4.4
^{105}Rh	E. Bohr	[574]	Stockholm	Sweden	1945	4.4
^{106}Rh	W.E. Grummitt	[506]	Chalk River	Canada	1946	6.1
^{107}Rh	L.E. Glendenin	[575]	Oak Ridge	USA	1951	6.1
^{108}Rh	G.B. Baró	[576]	Buenos Aires	Argentina	1955	6.2
^{109}Rh	F.F. Hopkins	[586]	Austin	USA	1972	13.1
^{110}Rh	M. Karras	[587]	Arkansas	USA	1963	7.3
^{111}Rh	G. Franz	[588]	Mainz	Germany	1975	6.4
^{112}Rh	F.F. Hopkins	[527]	Austin	USA	1971	13.1
^{113}Rh	F.F. Hopkins	[527]	Austin	USA	1971	13.1
^{114}Rh	J. Äystö	[589]	Jyväskylä	Finland	1988	10.9
^{115}Rh	J. Äystö	[590]	Jyväskylä	Finland	1988	10.9
^{116}Rh	R.L. Watson	[510]	Berkeley	USA	1970	13.1
^{117}Rh	H. Penttilä	[591]	Louvain-la-Neuve	Belgium	1991	10.9
^{118}Rh	M. Bernas	[382]	Darmstadt	Germany	1994	15.2
^{119}Rh	M. Bernas	[382]	Darmstadt	Germany	1994	15.2
^{120}Rh	M. Bernas	[382]	Darmstadt	Germany	1994	15.2
^{121}Rh	M. Bernas	[382]	Darmstadt	Germany	1994	15.2
^{122}Rh	M. Bernas	[237]	Darmstadt	Germany	1997	15.2
^{123}Rh	T. Ohnishi	[292]	RIKEN	Japan	2010	15.2

(continued)

Table 17.1 (continued)

Isotope	First author	Ref.	Laboratory	Country	Year	Table
^{124}Rh	T. Ohnishi	[292]	RIKEN	Japan	2010	15.2
^{125}Rh	T. Ohnishi	[292]	RIKEN	Japan	2010	15.2
^{126}Rh	T. Ohnishi	[292]	RIKEN	Japan	2010	15.2
^{127}Rh	G. Lorusso	[542]	RIKEN	Japan	2015	15.2
^{91}Pd	K. Rykaczewski	[563]	GANIL	France	1995	15.4
^{92}Pd	M. Hencheck	[564]	Michigan State	USA	1994	15.4
^{93}Pd	M. Hencheck	[564]	Michigan State	USA	1994	15.4
^{94}Pd	W. Kurcewicz	[592]	Darmstadt	Germany	1982	11.11
^{95}Pd	E. Nolte	[593]	Munich	Germany	1980	11.11
^{96}Pd	N.K. Aras	[594]	Maryland	USA	1980	9.1
^{97}Pd	A.H.W. Aten Jr.	[595]	Amsterdam	Netherlands	1969	9.5
^{98}Pd	A.H.W. Aten Jr.	[581]	Amsterdam	Netherlands	1955	9.5
^{99}Pd	A.H.W. Aten Jr.	[581]	Amsterdam	Netherlands	1955	9.5
^{100}Pd	M. Lindner	[583]	Berkeley	USA	1948	10.2
^{101}Pd	M. Lindner	[583]	Berkeley	USA	1948	10.2
^{102}Pd	A.J. Dempster	[596]	Chicago	USA	1935	3.1
^{103}Pd	J.Y. Mei	[597]	Indiana	USA	1950	9.4
^{104}Pd	A.J. Dempster	[596]	Chicago	USA	1935	3.1
^{105}Pd	A.J. Dempster	[596]	Chicago	USA	1935	3.1
^{106}Pd	A.J. Dempster	[596]	Chicago	USA	1935	3.1
^{107}Pd	U. Schindewolf	[598]	MIT	USA	1958	7.3
^{108}Pd	A.J. Dempster	[596]	Chicago	USA	1935	3.1
^{109}Pd	J.D. Kraus	[599]	Michigan	USA	1937	4.2
^{110}Pd	A.J. Dempster	[596]	Chicago	USA	1935	3.1
^{111}Pd	J.D. Kraus	[599]	Michigan	USA	1937	4.2
^{112}Pd	J.A. Seiler	[600]	Argonne	USA	1951	6.1
^{113}Pd	H. G. Hicks	[601]	Berkeley	USA	1954	10.1
^{114}Pd	J. M. Alexander	[602]	MIT	USA	1958	10.1
^{115}Pd	J. M. Alexander	[602]	MIT	USA	1958	10.1
^{116}Pd	E. Cheifetz	[509]	Berkeley	USA	1970	13.1
^{117}Pd	H.V. Weiss	[603]	Naval Rad. Def. Lab.	USA	1968	6.4
^{118}Pd	H.V. Weiss	[604]	Naval Rad. Def. Lab.	USA	1969	6.4
^{119}Pd	H. Penttilä	[605]	Jyväskylä	Finland	1991	10.9
^{120}Pd	Z. Janas	[606]	Jyväskylä	Finland	1993	10.9
^{121}Pd	M. Bernas	[382]	Darmstadt	Germany	1994	15.2
^{122}Pd	M. Bernas	[382]	Darmstadt	Germany	1994	15.2
^{123}Pd	M. Bernas	[382]	Darmstadt	Germany	1994	15.2
^{124}Pd	M. Bernas	[237]	Darmstadt	Germany	1997	15.2
^{125}Pd	T. Ohnishi	[26]	RIKEN	Japan	2008	15.2

(continued)

Table 17.1 (continued)

Isotope	First author	Ref.	Laboratory	Country	Year	Table
^{126}Pd	T. Ohnishi	[26]	RIKEN	Japan	2008	15.2
^{127}Pd	T. Ohnishi	[292]	RIKEN	Japan	2010	15.2
^{128}Pd	T. Ohnishi	[292]	RIKEN	Japan	2010	15.2
^{129}Pd	G. Lorusso	[542]	RIKEN	Japan	2015	15.2
^{93}Ag	K. Rykaczewski	[563]	GANIL	France	1995	15.4
^{94}Ag	M. Hencheck	[564]	Michigan State	USA	1994	15.4
^{95}Ag	M. Hencheck	[564]	Michigan State	USA	1994	15.4
^{96}Ag	W. Kurcewicz	[592]	Darmstadt	Germany	1982	11.11
^{97}Ag	M. Huyse	[607]	Louvain-la-Neuve	Belgium	1978	11.3
^{98}Ag	M. Huyse	[607]	Louvain-la-Neuve	Belgium	1978	11.3
^{99}Ag	H. Bakhru	[608]	Yale	USA	1967	11.1
^{100}Ag	D.J. Hnatowich	[609]	CERN	Switzerland	1970	10.7
^{101}Ag	F.D.S. Butement	[610]	Liverpool	UK	1966	10.5
^{102}Ag	O. Ames	[611]	Princeton	USA	1960	9.1
^{103}Ag	B.C. Haldar	[612]	Rochester	USA	1954	10.2
^{104}Ag	F.A. Johnson	[613]	McGill	Canada	1955	9.5
^{105}Ag	T. Enns	[614]	Rochester	USA	1939	4.2
^{106}Ag	W. Bothe	[420]	Heidelberg	Germany	1937	4.6
^{107}Ag	F.W. Aston	[243]	Cambridge	UK	1923	3.2
^{108}Ag	W. Bothe	[420]	Heidelberg	Germany	1937	4.6
^{109}Ag	F.W. Aston	[243]	Cambridge	UK	1923	3.2
^{110}Ag	W. Bothe	[420]	Heidelberg	Germany	1937	4.6
^{111}Ag	J.D. Kraus	[599]	Michigan	USA	1937	4.2
^{112}Ag	M.L. Pool	[615]	Michigan	USA	1938	4.4
^{113}Ag	R.B. Duffield	[616]	Illinois	USA	1949	8.1
^{114}Ag	J. M. Alexander	[602]	MIT	USA	1958	10.1
^{115}Ag	R.B. Duffield	[616]	Illinois	USA	1949	8.1
^{116}Ag	J. M. Alexander	[602]	MIT	USA	1958	10.1
^{117}Ag	J. M. Alexander	[602]	MIT	USA	1958	10.1
^{118}Ag	H.V. Weiss	[617]	Naval Rad. Def. Lab.	USA	1968	6.4
^{119}Ag	Y. Kawase	[618]	Studsvik	Sweden	1975	6.5
^{120}Ag	B. Fogelberg	[619]	Studsvik	Sweden	1971	6.5
^{121}Ag	B. Fogelberg	[620]	Studsvik	Sweden	1982	6.5
^{122}Ag	L.L. Shih	[621]	Ames	USA	1978	6.5
^{123}Ag	E. Lund	[622]	Studsvik	Sweden	1976	6.5
^{124}Ag	J.C. Hill	[623]	Brookhaven	USA	1984	6.5
^{125}Ag	M. Bernas	[382]	Darmstadt	Germany	1994	15.2
^{126}Ag	M. Bernas	[382]	Darmstadt	Germany	1994	15.2

(continued)

Table 17.1 (continued)

Isotope	First author	Ref.	Laboratory	Country	Year	Table
^{127}Ag	V.N. Fedoseyev	[624]	CERN	Switzerland	1995	10.10
^{128}Ag	T. Kautzsch	[625]	CERN	Switzerland	2000	10.10
^{129}Ag	K.-L. Kratz	[626]	CERN	Switzerland	2000	10.10
^{130}Ag	T. Kautzsch	[625]	CERN	Switzerland	2000	10.10
^{131}Ag	H. Wang	[627]	RIKEN	Japan	2013	15.2
^{132}Ag	G. Lorusso	[542]	RIKEN	Japan	2015	15.2
^{95}Cd	C.B. Hinke	[628]	Darmstadt	Germany	2012	15.4
^{96}Cd	D. Bazin	[629]	Michigan State	USA	2008	15.4
^{97}Cd	T. Elmroth	[630]	CERN	Switzerland	1978	10.7
^{98}Cd	T. Elmroth	[630]	CERN	Switzerland	1978	10.7
^{99}Cd	T. Elmroth	[630]	CERN	Switzerland	1978	10.7
^{100}Cd	D.J. Hnatowich	[609]	CERN	Switzerland	1970	10.7
^{101}Cd	P.G. Hansen	[205]	CERN	Switzerland	1969	10.7
^{102}Cd	P.G. Hansen	[205]	CERN	Switzerland	1969	10.7
^{103}Cd	I.L. Preiss	[631]	Yale	USA	1960	11.1
^{104}Cd	F.A. Johnson	[613]	McGill	Canada	1955	9.5
^{105}Cd	J.R. Gum	[632]	Ohio State	USA	1950	9.4
^{106}Cd	F.W. Aston	[321]	Cambridge	UK	1934	3.3
^{107}Cd	A.C. Helmholz	[633]	Berkeley	USA	1946	7.3
^{108}Cd	F.W. Aston	[321]	Cambridge	UK	1934	3.3
^{109}Cd	M. Lindner	[634]	Berkeley	USA	1950	10.2
^{110}Cd	F.W. Aston	[635]	Cambridge	UK	1924	3.2
^{111}Cd	F.W. Aston	[635]	Cambridge	UK	1924	3.2
^{112}Cd	F.W. Aston	[635]	Cambridge	UK	1924	3.2
^{113}Cd	F.W. Aston	[635]	Cambridge	UK	1924	3.2
^{114}Cd	F.W. Aston	[635]	Cambridge	UK	1924	3.2
^{115}Cd	M. Goldhaber	[636]	Cambridge	UK	1939	4.4
^{116}Cd	F.W. Aston	[635]	Cambridge	UK	1924	3.2
^{117}Cd	M. Goldhaber	[636]	Cambridge	UK	1939	4.4
^{118}Cd	C.E. Gleit	[637]	MIT	USA	1961	10.1
^{119}Cd	C.E. Gleit	[638]	MIT	USA	1961	10.1
^{120}Cd	E. Cheifetz	[639]	Berkeley	USA	1971	13.1
^{121}Cd	H.V. Weiss	[640]	Naval Rad. Def. Lab.	USA	1965	6.3
^{122}Cd	O. Scheidemann	[641]	CERN	Switzerland	1973	10.7
^{123}Cd	P.L. Reeder	[642]	Brookhaven	USA	1983	6.5
^{124}Cd	B. Fogelberg	[643]	Studsvik	Sweden	1974	6.5
^{125}Cd	H. Gokturk	[644]	Studsvik	Sweden	1986	6.5
^{126}Cd	M.L. Gartner	[645]	Ames	USA	1978	6.5

(continued)

Table 17.1 (continued)

Isotope	First author	Ref.	Laboratory	Country	Year	Table
^{127}Cd	H. Gokturk	[644]	Studsvik	Sweden	1986	6.5
^{128}Cd	H. Gokturk	[644]	Studsvik	Sweden	1986	6.5
^{129}Cd	H. Gokturk	[644]	Studsvik	Sweden	1986	6.5
^{130}Cd	K.-L. Kratz	[646]	CERN	Switzerland	1986	10.10
^{131}Cd	M. Hannawald	[647]	CERN	Switzerland	2000	10.10
^{132}Cd	M. Hannawald	[647]	CERN	Switzerland	2000	10.10
^{133}Cd	T. Ohnishi	[292]	RIKEN	Japan	2010	15.2
^{134}Cd	G. Lorusso	[542]	RIKEN	Japan	2015	15.2
^{97}In	C.B. Hinke	[628]	Darmstadt	Germany	2012	15.4
^{98}In	R. Schneider	[648]	Darmstadt	Germany	1994	15.4
^{99}In	R. Schneider	[648]	Darmstadt	Germany	1994	15.4
^{100}In	W. Kurcewicz	[592]	Darmstadt	Germany	1982	11.11
^{101}In	M. Huyse	[649]	Louvain-la-Neuve	Belgium	1988	11.3
^{102}In	B. Beraud	[650]	Grenoble	France	1981	11.3
^{103}In	G. Lhersonneau	[651]	Louvain-la-Neuve	Belgium	1978	11.3
^{104}In	B.J. Varley	[652]	Manchester	UK	1977	11.2
^{105}In	J. Rivier	[653]	Grenoble	France	1975	9.5
^{106}In	R.C. Catura	[654]	UCLA	USA	1962	9.5
^{107}In	E.C. Mallary	[655]	Ohio State	USA	1949	9.4
^{108}In	E.C. Mallary	[655]	Ohio State	USA	1949	9.4
^{109}In	S.N. Ghoshal	[656]	Berkeley	USA	1948	9.6
^{110}In	S.W. Barnes	[657]	Rochester	USA	1939	4.2
^{111}In	D.J. Tendam	[658]	Purdue	USA	1947	9.4
^{112}In	D.J. Tendam	[658]	Purdue	USA	1947	9.4
^{113}In	M. Wehrli	[659]	Basel	Switzerland	1934	3.5
^{114}In	J.L. Lawson	[660]	Michigan	USA	1937	4.1
^{115}In	F.W. Aston	[661]	Cambridge	UK	1924	3.2
^{116}In	J.L. Lawson	[660]	Michigan	USA	1937	4.1
^{117}In	J.M. Cork	[662]	Michigan	USA	1937	4.2
^{118}In	R.B. Duffield	[663]	Illinois	USA	1949	8.1
^{119}In	R.B. Duffield	[663]	Illinois	USA	1949	8.1
^{120}In	C.L. McGinnis	[664]	Carnegie Institute	USA	1958	7.3
^{121}In	H. Yuta	[665]	Tohoku	Japan	1960	8.1
^{122}In	J. Kantele	[666]	Arkansas	USA	1963	7.3
^{123}In	H. Yuta	[665]	Tohoku	Japan	1960	8.1
^{124}In	M. Karras	[667]	Arkansas	USA	1964	7.3
^{125}In	K. Fritze	[668]	McMaster	Canada	1967	6.4
^{126}In	B. Grapengiesser	[352]	Studsvik	Sweden	1974	6.5
^{127}In	K. Aleklett	[669]	Studsvik	Sweden	1975	6.5

(continued)

Table 17.1 (continued)

Isotope	First author	Ref.	Laboratory	Country	Year	Table
^{128}In	K. Aleklett	[669]	Studsvik	Sweden	1975	6.5
^{129}In	K. Aleklett	[669]	Studsvik	Sweden	1975	6.5
^{130}In	A. Kerek	[670]	Studsvik	Sweden	1973	6.5
^{131}In	E. Lund	[622]	Studsvik	Sweden	1976	6.5
^{132}In	A. Kerek	[671]	Studsvik	Sweden	1973	6.5
^{133}In	P. Hoff	[672]	CERN	Switzerland	1996	10.10
^{134}In	P. Hoff	[672]	CERN	Switzerland	1996	10.10
^{135}In	I. Dillmann	[673]	CERN	Switzerland	2002	10.10
^{136}In	G. Lorusso	[542]	RIKEN	Japan	2015	15.2
^{137}In	G. Lorusso	[542]	RIKEN	Japan	2015	15.2
^{99}Sn	C. B. Hinke	[628]	Darmstadt	Germany	2012	15.4
^{100}Sn	R. Schneider	[648]	Darmstadt	Germany	1994	15.4
^{101}Sn	M. Lewitowicz	[674]	GANIL	France	1994	15.4
^{102}Sn	M. Lewitowicz	[674]	GANIL	France	1994	15.4
^{103}Sn	P. Tidemand-Petersson	[675]	Darmstadt	Germany	1981	11.11
^{104}Sn	K. Deneffe	[676]	Louvain-la-Neuve	Belgium	1985	11.3
^{105}Sn	P. Tidemand-Petersson	[675]	Darmstadt	Germany	1981	11.11
^{106}Sn	V.N. Burminskii	[677]	Almaty	Kaszakhstan	1975	9.1
^{107}Sn	H.C. Hseuh	[678]	St. Louis	USA	1976	9.1
^{108}Sn	T. Yamazaki	[679]	Berkeley	USA	1968	9.1
^{109}Sn	D.E. Khulelidze	[680]	Dubna	Russia	1965	9.5
^{110}Sn	G. Bassani	[681]	Minnesota	USA	1965	9.8
^{111}Sn	R.A. Hinshaw	[682]	Ohio State	USA	1949	9.4
^{112}Sn	F.W. Aston	[683]	Cambridge	UK	1927	3.3
^{113}Sn	J.J. Livingood	[684]	Berkeley	USA	1939	4.3
^{114}Sn	F.W. Aston	[683]	Cambridge	UK	1927	3.3
^{115}Sn	F.W. Aston	[683]	Cambridge	UK	1927	3.3
^{116}Sn	F.W. Aston	[685]	Cambridge	UK	1922	3.2
^{117}Sn	F.W. Aston	[685]	Cambridge	UK	1922	3.2
^{118}Sn	F.W. Aston	[685]	Cambridge	UK	1922	3.2
^{119}Sn	F.W. Aston	[685]	Cambridge	UK	1922	3.2
^{120}Sn	F.W. Aston	[685]	Cambridge	UK	1922	3.2
^{121}Sn	M. Lindner	[686]	Berkeley	USA	1948	9.3
^{122}Sn	F.W. Aston	[685]	Cambridge	UK	1922	3.2
^{123}Sn	W.E. Grummitt	[687]	Chalk River	Canada	1948	6.1
^{124}Sn	F.W. Aston	[685]	Cambridge	UK	1922	3.2
^{125}Sn	J.J. Livingood	[684]	Berkeley	USA	1939	4.3
^{126}Sn	B.J. Dropesky	[688]	Los Alamos	USA	1962	6.3
^{127}Sn	J.W. Barnes	[689]	Los Alamos	USA	1951	6.1

(continued)

Table 17.1 (continued)

Isotope	First author	Ref.	Laboratory	Country	Year	Table
^{128}Sn	I. Fränz	[690]	Buenos Aires	Argentina	1956	6.2
^{129}Sn	E. Hagebo	[691]	Amsterdam	Netherlands	1962	10.1
^{130}Sn	T. Izak	[692]	Soreq	Israel	1972	6.4
^{131}Sn	A.E. Greendale	[693]	Naval Rad. Def. Lab.	USA	1963	6.3
^{132}Sn	A.E. Greendale	[693]	Naval Rad. Def. Lab.	USA	1963	6.3
^{133}Sn	S. Borg	[694]	Studsvik	Sweden	1973	6.5
^{134}Sn	S. Shalev	[695]	Studsvik	Sweden	1974	6.5
^{135}Sn	M. Bernas	[382]	Darmstadt	Germany	1994	15.2
^{136}Sn	M. Bernas	[382]	Darmstadt	Germany	1994	15.2
^{137}Sn	M. Bernas	[382]	Darmstadt	Germany	1994	15.2
^{138}Sn	T. Ohnishi	[292]	RIKEN	Japan	2010	15.2
^{139}Sn	G. Lorusso	[542]	RIKEN	Japan	2015	15.2
^{103}Sb	K. Rykaczewski	[563]	GANIL	France	1995	15.4
^{104}Sb	K. Rykaczewski	[563]	GANIL	France	1995	15.4
^{105}Sb	R.J. Tighe	[696]	Berkeley	USA	1994	16.3
^{106}Sb	A. Plochocki	[697]	Darmstadt	Germany	1981	11.8
^{107}Sb	D. Seweryniak	[698]	Roskilde	Denmark	1994	11.4
^{108}Sb	K. Oxorn	[699]	McGill	Canada	1976	9.1
^{109}Sb	K. Oxorn	[699]	McGill	Canada	1976	9.1
^{110}Sb	K. Miyano	[700]	Tokyo	Japan	1972	9.1
^{111}Sb	M. Singh	[701]	UCLA	USA	1972	9.1
^{112}Sb	I.P. Selinov	[702]	Moscow	Russia	1959	9.5
^{113}Sb	I.P. Selinov	[703]	Moscow	Russia	1958	9.5
^{114}Sb	I.P. Selinov	[702]	Moscow	Russia	1959	9.5
^{115}Sb	I.P. Selinov	[703]	Moscow	Russia	1958	9.5
^{116}Sb	G.M. Temmer	[704]	Berkeley	USA	1949	9.6
^{117}Sb	K.D. Coleman	[705]	Ohio State	USA	1947	9.4
^{118}Sb	K.D. Coleman	[705]	Ohio State	USA	1947	9.4
^{119}Sb	K.D. Coleman	[705]	Ohio State	USA	1947	9.4
^{120}Sb	W. Bothe	[420]	Heidelberg	Germany	1937	4.6
^{121}Sb	F.W. Aston	[706]	Cambridge	UK	1922	3.2
^{122}Sb	J.J. Livingood	[707]	Berkeley	USA	1939	4.3
^{123}Sb	F.W. Aston	[706]	Cambridge	UK	1922	3.2
^{124}Sb	J.J. Livingood	[707]	Berkeley	USA	1939	4.1
^{125}Sb	C.W. Stanley	[708]	Los Alamos	USA	1951	6.1
^{126}Sb	I. Fränz	[709]	Buenos Aires	Argentina	1956	9.5
^{127}Sb	P. Abelson	[710]	Berkeley	USA	1939	4.7
^{128}Sb	I. Fränz	[690]	Buenos Aires	Argentina	1956	6.2

(continued)

Table 17.1 (continued)

Isotope	First author	Ref.	Laboratory	Country	Year	Table
^{129}Sb	P. Abelson	[710]	Berkeley	USA	1939	4.7
^{130}Sb	E. Hagebo	[691]	Amsterdam	Netherlands	1962	7.3
^{131}Sb	A.C. Pappas	[711]	Oslo	Norway	1956	6.3
^{132}Sb	A.C. Pappas	[711]	Oslo	Norway	1956	6.3
^{133}Sb	P.O. Strom	[712]	Naval Rad. Def. Lab.	USA	1966	6.3
^{134}Sb	L. Tomlinson	[713]	Harwell	UK	1967	6.4
^{135}Sb	C.E. Bemis	[714]	MIT	USA	1964	6.3
^{136}Sb	E. Lund	[622]	Studsvik	Sweden	1976	6.5
^{137}Sb	M. Bernas	[382]	Darmstadt	Germany	1994	15.2
^{138}Sb	M. Bernas	[382]	Darmstadt	Germany	1994	15.2
^{139}Sb	M. Bernas	[382]	Darmstadt	Germany	1994	15.2
^{140}Sb	T. Ohnishi	[292]	RIKEN	Japan	2010	15.2
^{141}Sb	G. Lorusso	[542]	RIKEN	Japan	2015	15.2
^{105}Te	D. Seweryniak	[715]	Argonne	USA	2006	11.10
^{106}Te	D. Schardt	[716]	Darmstadt	Germany	1981	11.8
^{107}Te	D. Schardt	[717]	Darmstadt	Germany	1979	11.8
^{108}Te	D.D. Bogdanov	[718]	Dubna	Russia	1974	11.4.3
^{109}Te	V.A. Karnaukhov	[719]	Dubna	Russia	1967	11.11
^{110}Te	R. Kirchner	[720]	Darmstadt	Germany	1977	11.8
^{111}Te	V.A. Karnaukhov	[719]	Dubna	Russia	1967	11.11
^{112}Te	M.E.J. Wigmans	[721]	Amsterdam	Netherlands	1976	9.6
^{113}Te	A. Charvet	[722]	Lyon	France	1974	9.1
^{114}Te	O. Rahmouni	[723]	Orsay	France	1968	10.8
^{115}Te	I.P. Selinov	[724]	Moscow	Russia	1960	9.5
^{116}Te	M.Ya. Kuznetsova	[725]	Dubna	Russia	1958	10.4
^{117}Te	M.Ya. Kuznetsova	[725]	Dubna	Russia	1958	10.4
^{118}Te	M. Lindner	[686]	Berkeley	USA	1948	9.3
^{119}Te	M. Lindner	[686]	Berkeley	USA	1948	9.3
^{120}Te	A.J. Dempster	[726]	Chicago	USA	1936	3.1
^{121}Te	G.T. Seaborg	[727]	Berkeley	USA	1939	4.3
^{122}Te	K.T. Bainbridge	[728]	Franklin Institute	USA	1932	3.4
^{123}Te	K.T. Bainbridge	[728]	Franklin Institute	USA	1932	3.4
^{124}Te	K.T. Bainbridge	[728]	Franklin Institute	USA	1932	3.4
^{125}Te	F.W. Aston	[378]	Cambridge	UK	1931	3.3
^{126}Te	F.W. Aston	[635]	Cambridge	UK	1924	3.2
^{127}Te	J.J. Livingood	[729]	Berkeley	USA	1938	4.1
^{128}Te	F.W. Aston	[635]	Cambridge	UK	1924	3.2
^{129}Te	G.T. Seaborg	[727]	Berkeley	USA	1939	4.3
^{130}Te	F.W. Aston	[635]	Cambridge	UK	1924	3.2

(continued)

Table 17.1 (continued)

Isotope	First author	Ref.	Laboratory	Country	Year	Table
^{131}Te	G.T. Seaborg	[727]	Berkeley	USA	1939	4.3
^{132}Te	S. Katcoff	[490]	Los Alamos	USA	1948	6.1
^{133}Te	C.S. Wu	[730]	Berkeley	USA	1940	4.1
^{134}Te	S. Katcoff	[490]	Los Alamos	USA	1948	6.1
^{135}Te	H.O. Denschlag	[731]	Mainz	Germany	1969	6.4
^{136}Te	B. Grapengiesser	[352]	Studsvik	Sweden	1974	6.5
^{137}Te	M. Asghar	[412]	Grenoble	France	1975	6.6
^{138}Te	M. Asghar	[412]	Grenoble	France	1975	6.6
^{139}Te	M. Bernas	[382]	Darmstadt	Germany	1994	15.2
^{140}Te	M. Bernas	[382]	Darmstadt	Germany	1994	15.2
^{141}Te	M. Bernas	[382]	Darmstadt	Germany	1994	15.2
^{142}Te	M. Bernas	[382]	Darmstadt	Germany	1994	15.2
^{143}Te	T. Ohnishi	[292]	RIKEN	Japan	2010	15.2
^{144}Te	G. Lorusso	[542]	RIKEN	Japan	2015	15.2
^{108}I	R.D. Page	[732]	Daresbury	UK	1991	11.10
^{109}I	T. Faestermann	[733]	Munich	Germany	1984	16.3
^{110}I	R. Kirchner	[720]	Darmstadt	Germany	1977	11.8
^{111}I	R. Kirchner	[720]	Darmstadt	Germany	1977	11.8
^{112}I	R. Kirchner	[720]	Darmstadt	Germany	1977	11.8
^{113}I	R. Kirchner	[720]	Darmstadt	Germany	1977	11.8
^{114}I	R. Kirchner	[720]	Darmstadt	Germany	1977	11.8
^{115}I	P.G. Hansen	[205]	CERN	Switzerland	1969	10.7
^{116}I	G.M. Gowdy	[734]	Oak Ridge	USA	1976	11.3
^{117}I	P.G. Hansen	[205]	CERN	Switzerland	1969	10.7
^{118}I	P. Aagaard	[735]	Uppsala	Sweden	1957	10.8
^{119}I	G.B. Rossi	[736]	Berkeley	USA	1954	11.1
^{120}I	P. Aagaard	[735]	Uppsala	Sweden	1957	10.8
^{121}I	L. Marquez	[737]	Berkeley	USA	1950	9.3
^{122}I	L. Marquez	[737]	Berkeley	USA	1950	9.3
^{123}I	A.C.G. Mitchell	[738]	Indiana	USA	1949	9.4
^{124}I	J.J. Livingood	[739]	Berkeley	USA	1938	4.3
^{125}I	L.E. Glendenin	[740]	MIT	USA	1947	9.4
^{126}I	J.J. Livingood	[729]	Berkeley	USA	1938	4.1
^{127}I	F.W. Aston	[741]	Cambridge	UK	1920	3.2
^{128}I	E. Fermi	[142]	Rome	Italy	1934	4.1
^{129}I	S. Katcoff	[742]	Oak Ridge	USA	1951	6.1
^{130}I	J.J. Livingood	[739]	Berkeley	USA	1938	4.3
^{131}I	G.T. Seaborg	[727]	Berkeley	USA	1939	4.3
^{132}I	S. Katcoff	[490]	Los Alamos	USA	1948	6.1

(continued)

Table 17.1 (continued)

Isotope	First author	Ref.	Laboratory	Country	Year	Table
^{133}I	C.S. Wu	[730]	Berkeley	USA	1940	4.1
^{134}I	S. Katcoff	[490]	Los Alamos	USA	1948	6.1
^{135}I	C.S. Wu	[730]	Berkeley	USA	1940	4.1
^{136}I	C.W. Stanley	[743]	Los Alamos	USA	1949	6.1
^{137}I	W. Riezler	[744]	Grundlsee	Austria	1943	4.1
^{138}I	N. Sugarman	[424]	Argonne	USA	1949	6.1
^{139}I	N. Sugarman	[424]	Argonne	USA	1949	6.1
^{140}I	H.-D. Schussler	[745]	Mainz	Germany	1972	6.4
^{141}I	K.-L. Kratz	[426]	Mainz	Germany	1974	6.4
^{142}I	K.-L. Kratz	[746]	Mainz	Germany	1975	6.4
^{143}I	M. Bernas	[382]	Darmstadt	Germany	1994	15.2
^{144}I	M. Bernas	[382]	Darmstadt	Germany	1994	15.2
^{145}I	T. Ohnishi	[292]	RIKEN	Japan	2010	15.2
^{109}Xe	S.N. Liddick	[747]	Oak Ridge	USA	2006	11.10
^{110}Xe	D. Schardt	[716]	Darmstadt	Germany	1981	11.8
^{111}Xe	D. Schardt	[717]	Darmstadt	Germany	1979	11.8
^{112}Xe	E. Roeckl	[748]	Darmstadt	Germany	1978	11.8
^{113}Xe	E. Hagberg	[749]	CERN	Switzerland	1973	10.7
^{114}Xe	R. Kirchner	[720]	Darmstadt	Germany	1977	11.8
^{115}Xe	P.G. Hansen	[205]	CERN	Switzerland	1969	10.7
^{116}Xe	P.G. Hansen	[205]	CERN	Switzerland	1969	10.7
^{117}Xe	P.G. Hansen	[205]	CERN	Switzerland	1969	10.7
^{118}Xe	G. Andersson	[750]	CERN	Switzerland	1965	10.7
^{119}Xe	G. Andersson	[750]	CERN	Switzerland	1965	10.7
^{120}Xe	G. Andersson	[750]	CERN	Switzerland	1965	10.7
^{121}Xe	B. Dropesky	[751]	Rochester	USA	1952	10.2
^{122}Xe	B. Dropesky	[751]	Rochester	USA	1952	10.2
^{123}Xe	B. Dropesky	[751]	Rochester	USA	1952	10.2
^{124}Xe	F.W. Aston	[161]	Cambridge	UK	1922	3.2
^{125}Xe	D.L. Anderson	[752]	Ohio State	USA	1950	9.4
^{126}Xe	F.W. Aston	[161]	Cambridge	UK	1922	3.2
^{127}Xe	D.L. Anderson	[752]	Ohio State	USA	1950	9.4
^{128}Xe	F.W. Aston	[685]	Cambridge	UK	1922	3.2
^{129}Xe	F.W. Aston	[741]	Cambridge	UK	1920	3.2
^{130}Xe	F.W. Aston	[685]	Cambridge	UK	1922	3.2
^{131}Xe	F.W. Aston	[741]	Cambridge	UK	1920	3.2
^{132}Xe	F.W. Aston	[741]	Cambridge	UK	1920	3.2
^{133}Xe	C.S. Wu	[730]	Berkeley	USA	1940	4.1
^{134}Xe	F.W. Aston	[741]	Cambridge	UK	1920	3.2

(continued)

Table 17.1 (continued)

Isotope	First author	Ref.	Laboratory	Country	Year	Table
^{135}Xe	C.S. Wu	[730]	Berkeley	USA	1940	4.1
^{136}Xe	F.W. Aston	[741]	Cambridge	UK	1920	3.2
^{137}Xe	W. Riezler	[744]	Grundlsee	Austria	1943	4.1
^{138}Xe	W. Seelmann-Eggebert	[753]	Berlin	Germany	1943	4.1
^{139}Xe	C.R. Dillard	[437]	Argonne	USA	1951	6.1
^{140}Xe	C.R. Dillard	[437]	Argonne	USA	1951	6.1
^{141}Xe	C.R. Dillard	[437]	Argonne	USA	1951	6.1
^{142}Xe	D.R. Wolfsberg	[754]	St. Louis	USA	1960	6.2
^{143}Xe	C.R. Dillard	[437]	Argonne	USA	1951	6.1
^{144}Xe	W. Borchers	[755]	CERN	Switzerland	1989	10.10
^{145}Xe	U.C. Bergmann	[756]	CERN	Switzerland	2003	10.10
^{146}Xe	W. Borchers	[755]	CERN	Switzerland	1989	10.10
^{147}Xe	M. Bernas	[382]	Darmstadt	Germany	1994	15.2
^{148}Xe	T. Ohnishi	[292]	RIKEN	Japan	2010	15.2
^{112}Cs	R.D. Page	[757]	Daresbury	UK	1994	16.3
^{113}Cs	T. Faestermann	[733]	Munich	Germany	1984	16.3
^{114}Cs	J.M. D'Auria	[758]	CERN	Switzerland	1978	10.7
^{115}Cs	J.M. D'Auria	[758]	CERN	Switzerland	1978	10.7
^{116}Cs	D.D. Bogdanov	[759]	Dubna	Russia	1975	11.11
^{117}Cs	H. L. Ravn	[760]	CERN	Switzerland	1972	10.7
^{118}Cs	J. Chaumont	[442]	CERN	Switzerland	1969	10.7
^{119}Cs	J. Chaumont	[442]	CERN	Switzerland	1969	10.7
^{120}Cs	J. Chaumont	[442]	CERN	Switzerland	1969	10.7
^{121}Cs	J. Chaumont	[442]	CERN	Switzerland	1969	10.7
^{122}Cs	J. Chaumont	[442]	CERN	Switzerland	1969	10.7
^{123}Cs	H.B. Mathur	[761]	Berkeley	USA	1954	10.2
^{124}Cs	J. Chaumont	[442]	CERN	Switzerland	1969	10.7
^{125}Cs	M.C. Michel	[762]	Berkeley	USA	1954	10.2
^{126}Cs	M.I. Kalkstein	[763]	Berkeley	USA	1954	11.1
^{127}Cs	R.W. Fink	[764]	Berkeley	USA	1950	9.6
^{128}Cs	R.W. Fink	[765]	Rochester	USA	1951	10.2
^{129}Cs	R.W. Fink	[764]	Berkeley	USA	1950	9.6
^{130}Cs	A.B. Smith	[766]	Indiana	USA	1952	9.4
^{131}Cs	F.C. Yu	[767]	Oak Ridge	USA	1947	7.2
^{132}Cs	A.H. Wapstra	[768]	Amsterdam	Netherlands	1953	9.4
^{133}Cs	F.W. Aston	[215]	Cambridge	UK	1921	3.2
^{134}Cs	D.C. Kalbfell	[769]	Berkeley	USA	1940	4.1
^{135}Cs	N. Sugarman	[770]	Los Alamos	USA	1949	6.1
^{136}Cs	L.E. Glendenin	[771]	Oak Ridge	USA	1951	6.1

(continued)

Table 17.1 (continued)

Isotope	First author	Ref.	Laboratory	Country	Year	Table
^{137}Cs	A. Turkevich	[772]	Argonne	USA	1951	7.1
^{138}Cs	W. Seelmann-Eggebert	[753]	Berlin	Germany	1943	4.1
^{139}Cs	F.A. Heyn	[451]	Eindhoven	Netherlands	1939	4.7
^{140}Cs	N. Sugarman	[773]	Argonne	USA	1950	6.1
^{141}Cs	A.C. Wahl	[774]	St. Louis	USA	1962	6.2
^{142}Cs	K. Fritze	[775]	McMaster	Canada	1962	6.3
^{143}Cs	K. Fritze	[775]	McMaster	Canada	1962	6.3
^{144}Cs	I. Amarel	[456]	Orsay	France	1967	10.9
^{145}Cs	B.L. Tracy	[458]	Grenoble	France	1971	10.9
^{146}Cs	B.L. Tracy	[458]	Grenoble	France	1971	10.9
^{147}Cs	F.K. Wohn	[475]	Grenoble	France	1978	6.5
^{148}Cs	E. Koglin	[459]	Grenoble	France	1978	6.5
^{149}Cs	H.L. Ravn	[776]	CERN	Switzerland	1979	10.10
^{150}Cs	H.L. Ravn	[776]	CERN	Switzerland	1979	10.10
^{151}Cs	H.L. Ravn	[776]	CERN	Switzerland	1979	10.10
^{152}Cs	H.L. Ravn	[776]	CERN	Switzerland	1979	10.10
^{114}Ba	A. Guglielmetti	[777]	Darmstadt	Germany	1995	11.3
^{115}Ba	Z. Janas	[778]	Darmstadt	Germany	1997	11.3
^{116}Ba	Z. Janas	[778]	Darmstadt	Germany	1997	11.3
^{117}Ba	D.D. Bogdanov	[779]	Dubna	Russia	1977	11.11
^{118}Ba	Z. Janas	[778]	Darmstadt	Germany	1997	11.3
^{119}Ba	D.D. Bogdanov	[780]	Dubna	Russia	1974	11.11
^{120}Ba	J. Conrad	[781]	Heidelberg	Germany	1974	11.4
^{121}Ba	D.D. Bogdanov	[759]	Dubna	Russia	1975	11.11
^{122}Ba	J. Conrad	[781]	Heidelberg	Germany	1974	11.4
^{123}Ba	I.L. Preiss	[782]	Yale	USA	1962	11.1
^{124}Ba	J.E. Clarkson	[783]	Berkeley	USA	1967	11.4
^{125}Ba	I.L. Preiss	[782]	Yale	USA	1962	11.1
^{126}Ba	M.I. Kalkstein	[763]	Berkeley	USA	1954	11.1
^{127}Ba	M. Lindner	[784]	Livermore	USA	1952	10.2
^{128}Ba	R.W. Fink	[785]	Berkeley	USA	1950	10.2
^{129}Ba	C.C. Thomas	[786]	Rochester	USA	1950	10.2
^{130}Ba	A.J. Dempster	[787]	Chicago	USA	1936	3.1
^{131}Ba	S. Katcoff	[788]	Argonne	USA	1947	7.1
^{132}Ba	A.J. Dempster	[787]	Chicago	USA	1936	3.1
^{133}Ba	J.M. Cork	[789]	Michigan	USA	1941	4.2
^{134}Ba	J.P. Blewett	[466]	Princeton	USA	1936	3.4
^{135}Ba	F.W. Aston	[468]	Cambridge	UK	1931	3.3

(continued)

Table 17.1 (continued)

Isotope	First author	Ref.	Laboratory	Country	Year	Table
^{136}Ba	F.W. Aston	[468]	Cambridge	UK	1931	3.3
^{137}Ba	F.W. Aston	[468]	Cambridge	UK	1931	3.3
^{138}Ba	F.W. Aston	[296]	Cambridge	UK	1924	3.2
^{139}Ba	M.L. Pool	[488]	Michigan	USA	1937	4.1
^{140}Ba	O. Hahn	[790]	Berlin	Germany	1939	4.7
^{141}Ba	S. Katcoff	[791]	Chicago	USA	1951	6.1
^{142}Ba	R.P. Schuman	[792]	Idaho Falls	USA	1959	6.3
^{143}Ba	A.C. Wahl	[774]	St. Louis	USA	1962	6.2
^{144}Ba	I. Amarel	[456]	Orsay	France	1967	10.9
^{145}Ba	B. Grapengiesser	[352]	Studsvik	Sweden	1974	6.5
^{146}Ba	J.B. Wilhelmy	[793]	Berkeley	USA	1970	13.1
^{147}Ba	F.K. Wohn	[475]	Grenoble	France	1978	6.5
^{148}Ba	G. Engler	[794]	Soreq	Israel	1979	6.5
^{149}Ba	G. Rudstam	[795]	Studsvik	Sweden	1993	6.5
^{150}Ba	M. Bernas	[382]	Darmstadt	Germany	1994	15.2
^{151}Ba	M. Bernas	[382]	Darmstadt	Germany	1994	15.2
^{152}Ba	T. Ohnishi	[292]	RIKEN	Japan	2010	15.2
^{117}La	F. Soramel	[796]	Legnaro	Italy	2001	16.3
^{118}La						
^{119}La						
^{120}La	J.M. Nitschke	[797]	Berkeley	USA	1984	11.11
^{121}La	T. Sekine	[798]	Tokai	Japan	1988	11.3
^{122}La	J.M. Nitschke	[797]	Berkeley	USA	1984	11.11
^{123}La	D.D. Bogdanov	[799]	Dubna	Russia	1978	11.3
^{124}La	D.D. Bogdanov	[799]	Dubna	Russia	1978	11.3
^{125}La	K.Nakai	[800]	Berkeley	USA	1973	11.4
^{126}La	R.K. Sheline	[801]	Berkeley	USA	1961	11.1
^{127}La	I.L. Preiss	[802]	Yale	USA	1963	11.1
^{128}La	R.K. Sheline	[801]	Berkeley	USA	1961	11.1
^{129}La	I.L. Preiss	[802]	Yale	USA	1963	11.1
^{130}La	R.K. Sheline	[801]	Berkeley	USA	1961	11.1
^{131}La	M.M. Gransden	[803]	McGill	Canada	1951	9.6
^{132}La	M.M. Gransden	[803]	McGill	Canada	1951	9.6
^{133}La	R.A. Naumann	[804]	Berkeley	USA	1950	9.6
^{134}La	B.J. Stover	[805]	Berkeley	USA	1951	10.2
^{135}La	J.B. Chubbuck	[806]	Berkeley	USA	1948	9.3
^{136}La	R.A. Naumann	[804]	Berkeley	USA	1950	9.6
^{137}La	M.G. Inghram	[807]	Argonne	USA	1948	7.2
^{138}La	M.G. Inghram	[808]	Argonne	USA	1947	3.4

(continued)

Table 17.1 (continued)

Isotope	First author	Ref.	Laboratory	Country	Year	Table
^{139}La	F.W. Aston	[296]	Cambridge	UK	1924	3.2
^{140}La	J. K. Marsh	[809]	Oxford	UK	1935	4.1
^{141}La	S. Katcoff	[791]	Chicago	USA	1951	6.1
^{142}La	A. Vanden Bosch	[810]	Amsterdam	Netherlands	1953	6.2
^{143}La	H. Gest	[811]	Oak Ridge	USA	1951	6.1
^{144}La	I. Amarel	[456]	Orsay	France	1967	10.9
^{145}La	P.O. Aronsson	[812]	Chalmers	Sweden	1974	6.2
^{146}La	P.O. Aronsson	[812]	Chalmers	Sweden	1974	6.2
^{147}La	G. Engler	[794]	Soreq	Israel	1979	6.5
^{148}La	H. Gabelmann	[813]	Grenoble	France	1982	6.5
^{149}La	G. Engler	[794]	Soreq	Israel	1979	6.5
^{150}La	G. Rudstam	[795]	Studsvik	Sweden	1993	6.5
^{151}La	M. Bernas	[382]	Darmstadt	Germany	1994	15.2
^{152}La	M. Bernas	[382]	Darmstadt	Germany	1994	15.2
^{153}La	M. Bernas	[382]	Darmstadt	Germany	1994	15.2
^{121}Ce	Z. Li	[814]	Lanzhou	China	1997	11.11
^{122}Ce	J.F. Smith	[815]	Argonne	USA	2005	11.4
^{123}Ce	J.M. Nitschke	[797]	Berkeley	USA	1984	11.11
^{124}Ce	D.D. Bogdanov	[799]	Dubna	Russia	1978	11.3
^{125}Ce	D.D. Bogdanov	[799]	Dubna	Russia	1978	11.3
^{126}Ce	D.D. Bogdanov	[799]	Dubna	Russia	1978	11.3
^{127}Ce	D.D. Bogdanov	[799]	Dubna	Russia	1978	11.3
^{128}Ce	D. Ward	[816]	Berkeley	USA	1968	11.4
^{129}Ce	J. Gizon	[817]	Berkeley	USA	1977	11.4
^{130}Ce	G. Gerschel	[818]	Orsay	France	1965	10.8
^{131}Ce	A.E. Norris	[819]	Brookhaven	USA	1966	9.5
^{132}Ce	W.R. Ware	[820]	Rochester	USA	1960	10.2
^{133}Ce	B.J. Stover	[805]	Berkeley	USA	1951	10.2
^{134}Ce	B.J. Stover	[805]	Berkeley	USA	1951	10.2
^{135}Ce	J.B. Chubbuck	[806]	Berkeley	USA	1948	9.3
^{136}Ce	A.J. Dempster	[787]	Chicago	USA	1936	3.1
^{137}Ce	J.B. Chubbuck	[806]	Berkeley	USA	1948	9.3
^{138}Ce	A.J. Dempster	[787]	Chicago	USA	1936	3.1
^{139}Ce	M.L. Pool	[821]	Ohio State	USA	1948	9.4
^{140}Ce	F.W. Aston	[504]	Cambridge	UK	1924	3.2
^{141}Ce	N.E. Ballou	[822]	Chicago	USA	1951	7.3
^{142}Ce	F.W. Aston	[504]	Cambridge	UK	1924	3.2
^{143}Ce	N.E. Ballou	[822]	Chicago	USA	1951	7.3
^{144}Ce	W.E. Grummitt	[506]	Chalk River	Canada	1946	6.1

(continued)

Table 17.1 (continued)

Isotope	First author	Ref.	Laboratory	Country	Year	Table
^{145}Ce	S.S. Markowitz	[823]	Brookhaven	USA	1954	6.3
^{146}Ce	A.A. Caretto Jr.	[824]	Brookhaven	USA	1953	6.3
^{147}Ce	D.C. Hoffman	[825]	Los Alamos	USA	1964	6.3
^{148}Ce	D.C. Hoffman	[825]	Los Alamos	USA	1964	6.3
^{149}Ce	F.F. Hopkins	[527]	Austin	USA	1971	13.1
^{150}Ce	J.B. Wilhelmy	[793]	Berkeley	USA	1970	13.1
^{151}Ce	F. Hoellinger	[826]	Strasbourg	France	1997	13.1
^{152}Ce	I. Tago	[827]	Kyoto	Japan	1990	6.5
^{153}Ce	M. Bernas	[382]	Darmstadt	Germany	1994	15.2
^{154}Ce	M. Bernas	[382]	Darmstadt	Germany	1994	15.2
^{155}Ce	M. Bernas	[382]	Darmstadt	Germany	1994	15.2
^{121}Pr	A.P. Robinson	[828]	Argonne	USA	2005	16.3
^{122}Pr						
^{123}Pr						
^{124}Pr	P.A. Wilmarth	[829]	Berkeley	USA	1986	11.11
^{125}Pr	A.N. Wilson	[830]	Argonne	USA	2002	11.4
^{126}Pr	J.M. Nitschke	[831]	Berkeley	USA	1983	11.11
^{127}Pr	A. Gizon	[832]	Grenoble	France	1995	11.3
^{128}Pr	P.A. Wilmarth	[833]	Berkeley	USA	1985	11.11
^{129}Pr	D.D. Bogdanov	[779]	Dubna	Russia	1977	11.3
^{130}Pr	D.D. Bogdanov	[779]	Dubna	Russia	1977	11.3
^{131}Pr	D.D. Bogdanov	[779]	Dubna	Russia	1977	11.3
^{132}Pr	A. Latuszynski	[834]	Dubna	Russia	1974	10.8
^{133}Pr	A.A. Abdurazakov	[835]	Dubna	Russia	1970	10.4
^{134}Pr	J.E. Clarkson	[783]	Berkeley	USA	1967	11.4
^{135}Pr	T.H. Handley	[836]	Oak Ridge	USA	1954	9.4
^{136}Pr	Zh. Zhelev	[837]	Dubna	Russia	1968	10.4
^{137}Pr	C. Dahlstrom	[838]	McGill	Canada	1958	9.6
^{138}Pr	B.J. Stover	[805]	Berkeley	USA	1951	9.3
^{139}Pr	B.J. Stover	[805]	Berkeley	USA	1951	9.3
^{140}Pr	M.L. Pool	[839]	Michigan	USA	1938	4.1
^{141}Pr	F.W. Aston	[296]	Cambridge	UK	1924	3.2
^{142}Pr	J.K. Marsh	[809]	Oxford	UK	1935	4.1
^{143}Pr	N.E. Ballou	[822]	Chicago	USA	1951	7.3
^{144}Pr	W.E. Grummitt	[506]	Chalk River	Canada	1946	6.1
^{145}Pr	S.S. Markowitz	[823]	Brookhaven	USA	1954	6.3
^{146}Pr	A.A. Caretto Jr.	[824]	Brookhaven	USA	1953	6.3
^{147}Pr	D.C. Hoffman	[825]	Los Alamos	USA	1964	6.3
^{148}Pr	D.C. Hoffman	[825]	Los Alamos	USA	1964	6.3
^{149}Pr	D.C. Hoffman	[825]	Los Alamos	USA	1964	8.1

(continued)

Table 17.1 (continued)

Isotope	First author	Ref.	Laboratory	Country	Year	Table
^{150}Pr	T.E. Ward	[840]	Arkansas	USA	1970	7.4
^{151}Pr	M. Graefenstedt	[841]	Grenoble	France	1990	6.6
^{152}Pr	J.C. Hill	[842]	Brookhaven	USA	1983	6.5
^{153}Pr	R.C. Greenwood	[843]	Idaho Falls	USA	1987	13.1
^{154}Pr	Y. Kawase	[844]	Kyoto	Japan	1988	6.5
^{155}Pr	J. Van Schelt	[845]	Argonne	USA	2012	13.1
^{125}Nd	S.-W. Xu	[846]	Lanzhou	China	1999	11.11
^{126}Nd						
^{127}Nd	J.M. Nitschke	[831]	Berkeley	USA	1983	11.11
^{128}Nd	C.J. Lister	[847]	Daresbury	UK	1985	11.4
^{129}Nd	D.D. Bogdanov	[779]	Dubna	Russia	1977	11.11
^{130}Nd	D.D. Bogdanov	[779]	Dubna	Russia	1977	11.3
^{131}Nd	D.D. Bogdanov	[779]	Dubna	Russia	1977	11.3
^{132}Nd	D.D. Bogdanov	[779]	Dubna	Russia	1977	11.3
^{133}Nd	D.D. Bogdanov	[779]	Dubna	Russia	1977	11.3
^{134}Nd	A.A. Abdurazakov	[835]	Dubna	Russia	1970	10.4
^{135}Nd	A.A. Abdurazakov	[835]	Dubna	Russia	1970	10.4
^{136}Nd	Zh. Zhelev	[837]	Dubna	Russia	1968	10.4
^{137}Nd	Ch. Droste	[848]	Dubna	Russia	1970	11.2
^{138}Nd	K.Ya. Gromov	[849]	Dubna	Russia	1965	10.4
^{139}Nd	B.J. Stover	[805]	Berkeley	USA	1951	9.3
^{140}Nd	G. Wilkinson	[850]	Berkeley	USA	1949	9.3
^{141}Nd	G. Wilkinson	[850]	Berkeley	USA	1949	9.3
^{142}Nd	F.W. Aston	[504]	Cambridge	UK	1924	3.2
^{143}Nd	F.W. Aston	[851]	Cambridge	UK	1933	3.3
^{144}Nd	F.W. Aston	[504]	Cambridge	UK	1924	3.2
^{145}Nd	F.W. Aston	[851]	Cambridge	UK	1933	3.3
^{146}Nd	F.W. Aston	[504]	Cambridge	UK	1924	3.2
^{147}Nd	J.A. Marinsky	[852]	Oak Ridge	USA	1947	6.1
^{148}Nd	A.J. Dempster	[853]	Chicago	USA	1937	3.1
^{149}Nd	M.L. Pool	[839]	Michigan	USA	1938	4.1
^{150}Nd	A.J. Dempster	[853]	Chicago	USA	1937	3.1
^{151}Nd	M.L. Pool	[839]	Michigan	USA	1938	4.1
^{152}Nd	A. Wakat	[854]	Michigan	USA	1969	6.4
^{153}Nd	R.C. Greenwood	[843]	Idaho Falls	USA	1987	13.1
^{154}Nd	J.B. Wilhelmy	[793]	Berkeley	USA	1970	13.1
^{155}Nd	K. Okano	[855]	Kyoto	Japan	1986	6.5
^{156}Nd	R.C. Greenwood	[843]	Idaho Falls	USA	1987	13.1
^{157}Nd	J. Van Schelt	[845]	Argonne	USA	2012	13.1

(continued)

Table 17.1 (continued)

Isotope	First author	Ref.	Laboratory	Country	Year	Table
^{158}Nd	J. Kurcewicz	[856]	Darmstadt	Germany	2012	15.2
^{159}Nd	J. Kurcewicz	[856]	Darmstadt	Germany	2012	15.2
^{160}Nd	J. Kurcewicz	[856]	Darmstadt	Germany	2012	15.2
^{161}Nd	J. Kurcewicz	[856]	Darmstadt	Germany	2012	15.2
^{128}Pm	S.-W. Xu	[846]	Lanzhou	China	1999	11.11
^{129}Pm	S.-W. Xu	[857]	Lanzhou	China	2004	11.2
^{130}Pm	P.A. Wilmarth	[833]	Berkeley	USA	1985	11.11
^{131}Pm	C.M. Parry	[858]	Chalk River	Canada	1998	11.4
^{132}Pm	D.D. Bogdanov	[779]	Dubna	Russia	1977	11.3
^{133}Pm	D.D. Bogdanov	[779]	Dubna	Russia	1977	11.3
^{134}Pm	D.D. Bogdanov	[779]	Dubna	Russia	1977	11.3
^{135}Pm	J. van Klinken	[859]	Groningen	Netherlands	1975	10.5
^{136}Pm	G.D. Alkhazov	[860]	Leningrad	Russia	1982	10.8
^{137}Pm	G.P. Nowicki	[861]	Karlsruhe	Germany	1975	10.5
^{138}Pm	J. Deslauriers	[862]	McGill	Canada	1981	10.5
^{139}Pm	H.-J. Bleyl	[863]	Karlsruhe	Germany	1967	10.5
^{140}Pm	A.H.W. Aten Jr.	[864]	Amsterdam	Netherlands	1966	9.5
^{141}Pm	V. Kistiakowsky Fischer	[865]	Berkeley	USA	1952	9.3
^{142}Pm	I. Gratot	[866]	Saclay	France	1959	9.5
^{143}Pm	V. Kistiakowsky Fischer	[865]	Berkeley	USA	1952	9.3
^{144}Pm	V. Kistiakowsky Fischer	[865]	Berkeley	USA	1952	9.3
^{145}Pm	F.D.S. Butement	[867]	Harwell	UK	1951	7.2
^{146}Pm	E.G. Funk Jr.	[868]	Oak Ridge	USA	1960	9.5
^{147}Pm	J.A. Marinsky	[852]	Oak Ridge	USA	1947	6.1
^{148}Pm	G.W. Parker	[869]	Oak Ridge	USA	1947	7.2
^{149}Pm	J.A. Marinsky	[852]	Oak Ridge	USA	1947	6.1
^{150}Pm	J. K. Long	[870]	Ohio State	USA	1952	9.4
^{151}Pm	W.C. Rutledge	[871]	Argonne	USA	1952	7.2
^{152}Pm	R.G. Wille	[872]	Arkansas	USA	1958	7.3
^{153}Pm	K. Kotajima	[873]	Tokai	Japan	1962	8.1
^{154}Pm	R.G. Wille	[872]	Arkansas	USA	1958	7.3
^{155}Pm	R.C. Greenwood	[874]	Idaho Falls	USA	1982	13.1
^{156}Pm	H. Mach	[875]	Brookhaven	USA	1986	6.5
^{157}Pm	R.C. Greenwood	[843]	Idaho Falls	USA	1987	13.1
^{158}Pm	R.C. Greenwood	[843]	Idaho Falls	USA	1987	13.1
^{159}Pm	S. Ichikawa	[876]	Tokai	Japan	2005	10.9
^{160}Pm	J. Kurcewicz	[856]	Darmstadt	Germany	2012	15.2
^{161}Pm	J. Kurcewicz	[856]	Darmstadt	Germany	2012	15.2

(continued)

Table 17.1 (continued)

Isotope	First author	Ref.	Laboratory	Country	Year	Table
^{162}Pm	J. Kurcewicz	[856]	Darmstadt	Germany	2012	15.2
^{163}Pm	J. Kurcewicz	[856]	Darmstadt	Germany	2012	15.2
^{129}Sm	S.-W. Xu	[846]	Lanzhou	China	1999	11.11
^{130}Sm	A.A. Sonzogni	[877]	Argonne	USA	1999	11.11
^{131}Sm	P.A. Wilmarth	[829]	Berkeley	USA	1986	11.11
^{132}Sm	R. Wadsworth	[878]	Daresbury	UK	1989	11.4
^{133}Sm	D.D. Bogdanov	[779]	Dubna	Russia	1977	11.11
^{134}Sm	D.D. Bogdanov	[779]	Dubna	Russia	1977	11.3
^{135}Sm	D.D. Bogdanov	[779]	Dubna	Russia	1977	11.3
^{136}Sm	M. Nowicki	[879]	Dubna	Russia	1982	11.3
^{137}Sm	N. Redon	[880]	Grenoble	France	1986	11.3
^{140}Sm	E. Herrmann	[882]	Dubna	Russia	1967	10.4
^{141}Sm	E. Herrmann	[882]	Dubna	Russia	1967	10.4
^{142}Sm	I. Gratot	[866]	Saclay	France	1959	9.5
^{143}Sm	E. Silva	[883]	Sao Paulo	Brazil	1956	8.1
^{144}Sm	F.W. Aston	[230]	Cambridge	UK	1934	3.3
^{145}Sm	M.G. Inghram	[884]	Argonne	USA	1947	7.1
^{146}Sm	D.C. Dunlavey	[885]	Berkeley	USA	1953	9.3
^{147}Sm	F.W. Aston	[851]	Cambridge	UK	1933	3.3
^{148}Sm	F.W. Aston	[851]	Cambridge	UK	1933	3.3
^{149}Sm	F.W. Aston	[851]	Cambridge	UK	1933	3.3
^{150}Sm	F.W. Aston	[230]	Cambridge	UK	1934	3.3
^{151}Sm	M.G. Inghram	[884]	Argonne	USA	1947	7.1
^{138}Sm	M. Nowicki	[879]	Dubna	Russia	1982	11.3
^{139}Sm	J. van Klinken	[881]	Karlsruhe	Germany	1971	10.5
^{152}Sm	F.W. Aston	[851]	Cambridge	UK	1933	3.3
^{153}Sm	M.L. Pool	[839]	Michigan	USA	1938	4.1
^{154}Sm	F.W. Aston	[851]	Cambridge	UK	1933	3.3
^{155}Sm	L. Winsberg	[886]	Los Alamos	USA	1951	6.1
^{156}Sm	L. Winsberg	[887]	Argonne	USA	1951	6.1
^{157}Sm	J.M. D'Auria	[888]	Simon Fraser	Canada	1973	7.4
^{158}Sm	J.B. Wilhelmy	[793]	Berkeley	USA	1970	13.1
^{159}Sm	H. Mach	[875]	Brookhaven	USA	1986	6.5
^{160}Sm	H. Mach	[875]	Brookhaven	USA	1986	6.5
^{161}Sm	S. Ichikawa	[889]	Tokai	Japan	1998	10.9
^{162}Sm	S. Ichikawa	[876]	Tokai	Japan	2005	10.9
^{163}Sm	J. Kurcewicz	[856]	Darmstadt	Germany	2012	15.2
^{164}Sm	J. Kurcewicz	[856]	Darmstadt	Germany	2012	15.2
^{165}Sm	J. Kurcewicz	[856]	Darmstadt	Germany	2012	15.2

(continued)

Table 17.1 (continued)

Isotope	First author	Ref.	Laboratory	Country	Year	Table
^{130}Eu	C.N. Davids	[890]	Argonne	USA	2004	16.3
^{131}Eu	C.N. Davids	[891]	Argonne	USA	1998	16.3
^{132}Eu						
^{133}Eu						
^{134}Eu	K.S. Vierinen	[892]	Berkeley	USA	1989	11.11
^{135}Eu	K.S. Vierinen	[892]	Berkeley	USA	1989	11.11
^{136}Eu	B.D. Kern	[893]	Oak Ridge	USA	1987	11.3
^{137}Eu	M. Nowicki	[879]	Dubna	Russia	1982	11.3
^{138}Eu	M. Nowicki	[879]	Dubna	Russia	1982	11.3
^{139}Eu	J. van Klinken	[859]	Groningen	Netherlands	1975	10.5
^{140}Eu	M. Nowicki	[879]	Dubna	Russia	1982	11.3
^{141}Eu	J. Deslauriers	[894]	McGill	Canada	1977	9.1
^{142}Eu	H.P. Malan	[895]	Karlsruhe	Germany	1966	9.5
^{143}Eu	K. Kotajima	[896]	Amsterdam	Netherlands	1965	9.5
^{144}Eu	R. Messlinger	[897]	Heidelberg	Germany	1965	9.1
^{145}Eu	R.W. Hoff	[898]	Berkeley	USA	1951	9.3
^{146}Eu	G.M. Gorodinskii	[899]	Dubna	Russia	1957	10.4
^{147}Eu	R.W. Hoff	[898]	Berkeley	USA	1951	9.3
^{148}Eu	R.W. Hoff	[898]	Berkeley	USA	1951	9.3
^{149}Eu	N.A. Antoneva	[900]	Dubna	Russia	1959	10.4
^{150}Eu	F.D.S. Butement	[391]	Harwell	UK	1950	8.1
^{151}Eu	F.W. Aston	[851]	Cambridge	UK	1933	3.3
^{152}Eu	M.L. Pool	[839]	Michigan	USA	1938	4.1
^{153}Eu	F.W. Aston	[851]	Cambridge	UK	1933	3.3
^{154}Eu	M.G. Inghram	[901]	Argonne	USA	1947	7.1
^{155}Eu	L. Winsberg	[902]	Oak Ridge	USA	1951	6.1
^{156}Eu	L. Winsberg	[887]	Argonne	USA	1951	6.1
^{157}Eu	L. Winsberg	[903]	Argonne	USA	1951	6.1
^{158}Eu	L. Winsberg	[903]	Argonne	USA	1951	6.1
^{159}Eu	T. Kuroyanagi	[904]	Tohoku	Japan	1961	8.1
^{160}Eu	J.M. D'Auria	[888]	Simon Fraser	Canada	1973	7.4
^{161}Eu	H. Mach	[875]	Brookhaven	USA	1986	6.5
^{162}Eu	R.C. Greenwood	[843]	Idaho Falls	USA	1987	13.1
^{163}Eu	H. Hayashi	[905]	Tokai	Japan	2007	10.9
^{164}Eu	H. Hayashi	[905]	Tokai	Japan	2007	10.9
^{165}Eu	H. Hayashi	[905]	Tokai	Japan	2007	10.9
^{166}Eu	J. Kurcewicz	[856]	Darmstadt	Germany	2012	15.2
^{167}Eu	J. Kurcewicz	[856]	Darmstadt	Germany	2012	15.2

(continued)

Table 17.1 (continued)

Isotope	First author	Ref.	Laboratory	Country	Year	Table
^{168}Eu	J. Kurcewicz	[856]	Darmstadt	Germany	2012	15.2
^{135}Gd	S. Xu	[906]	Lanzhou	China	1996	11.11
^{136}Gd						
^{137}Gd	S.-W. Xu	[846]	Lanzhou	China	1999	11.11
^{138}Gd	C.J. Lister	[847]	Daresbury	UK	1985	11.4
^{139}Gd	J.M. Nitschke	[831]	Berkeley	USA	1983	11.11
^{140}Gd	C.J. Lister	[847]	Daresbury	UK	1985	11.4
^{141}Gd	N. Redon	[880]	Grenoble	France	1986	11.3
^{142}Gd	S. Lunardi	[907]	Juelich	Germany	1986	10.5
^{143}Gd	K.L. Kosanke	[908]	Michigan State	USA	1975	9.1
^{144}Gd	K.A. Keller	[909]	Karlsruhe	Germany	1968	9.5
^{145}Gd	J.R. Grover	[910]	Brookhaven	USA	1959	9.5
^{146}Gd	G.M. Gorodinskii	[899]	Dubna	Russia	1957	10.4
^{147}Gd	V.S. Shirley	[911]	Berkeley	USA	1957	9.3
^{148}Gd	J.O. Rasmussen	[912]	Berkeley	USA	1953	9.3
^{149}Gd	R.W. Hoff	[898]	Berkeley	USA	1951	9.3
^{150}Gd	J.O. Rasmussen	[912]	Berkeley	USA	1953	9.3
^{151}Gd	R.E. Hein	[913]	Ames	USA	1950	9.4
^{152}Gd	A.J. Dempster	[914]	Chicago	USA	1938	3.1
^{153}Gd	M.G. Inghram	[884]	Argonne	USA	1947	7.1
^{154}Gd	A.J. Dempster	[914]	Chicago	USA	1938	3.1
^{155}Gd	F.W. Aston	[851]	Cambridge	UK	1933	3.3
^{156}Gd	F.W. Aston	[851]	Cambridge	UK	1933	3.3
^{157}Gd	F.W. Aston	[851]	Cambridge	UK	1933	3.3
^{158}Gd	F.W. Aston	[851]	Cambridge	UK	1933	3.3
^{159}Gd	F.D.S. Butement	[915]	Harwell	UK	1949	7.2
^{160}Gd	F.W. Aston	[851]	Cambridge	UK	1933	3.3
^{161}Gd	F.D.S. Butement	[915]	Harwell	UK	1949	7.2
^{162}Gd	M.A. Wahlgren	[916]	Savannah River	USA	1967	7.2
^{163}Gd	R.J. Gehrke	[917]	Idaho Falls	USA	1982	13.1
^{164}Gd	R.C. Greenwood	[918]	Idaho Falls	USA	1988	13.1
^{165}Gd	S. Ichikawa	[889]	Tokai	Japan	1998	10.9
^{166}Gd	S. Ichikawa	[876]	Tokai	Japan	2005	10.9
^{167}Gd	J. Kurcewicz	[856]	Darmstadt	Germany	2012	15.2
^{168}Gd	J. Kurcewicz	[856]	Darmstadt	Germany	2012	15.2
^{169}Gd	J. Kurcewicz	[856]	Darmstadt	Germany	2012	15.2
^{170}Gd	J. Kurcewicz	[856]	Darmstadt	Germany	2012	15.2
^{135}Tb	P.J. Woods	[919]	Argonne	USA	2004	16.3

(continued)

Table 17.1 (continued)

Isotope	First author	Ref.	Laboratory	Country	Year	Table
^{136}Tb						
^{137}Tb						
^{138}Tb						
^{139}Tb	Y. Xie	[920]	Lanzhou	China	1999	11.2
^{140}Tb	P.A. Wilmarth	[829]	Berkeley	USA	1986	11.11
^{141}Tb	P.A. Wilmarth	[829]	Berkeley	USA	1986	11.3
^{142}Tb	R.B. Firestone	[921]	Berkeley	USA	1991	11.3
^{143}Tb	T. Ollivier	[922]	Grenoble	France	1985	11.2
^{144}Tb	D.C. Sousa	[923]	Texas A&M	USA	1982	11.2
^{145}Tb	G.D. Alkhazov	[924]	Leningrad	Russia	1981	10.8
^{146}Tb	E. Newman	[925]	Oak Ridge	USA	1974	11.2
^{147}Tb	Y.Y. Chu	[926]	Brookhaven	USA	1969	10.5
^{148}Tb	K.S. Toth	[927]	Berkeley	USA	1960	11.1
^{149}Tb	J.O. Rasmussen	[928]	Berkeley	USA	1950	10.2
^{150}Tb	K.S. Toth	[929]	Uppsala	Sweden	1959	10.2
^{151}Tb	J.O. Rasmussen	[912]	Berkeley	USA	1953	9.3
^{152}Tb	K.S. Toth	[930]	Berkeley	USA	1959	9.3
^{153}Tb	J.W. Mihelich	[931]	Oak Ridge	USA	1957	9.5
^{154}Tb	G. Wilkinson	[932]	Berkeley	USA	1950	9.3
^{155}Tb	J.W. Mihelich	[931]	Oak Ridge	USA	1957	9.5
^{156}Tb	G. Wilkinson	[932]	Berkeley	USA	1950	9.3
^{157}Tb	R.A. Naumann	[933]	Berkeley	USA	1960	7.2
^{158}Tb	C.L. Hammer	[934]	Ames	USA	1957	8.1
^{159}Tb	F.W. Aston	[851]	Cambridge	UK	1933	3.3
^{160}Tb	W. Bothe	[935]	Heidelberg	Germany	1943	4.4
^{161}Tb	F.D.S. Butement	[915]	Harwell	UK	1949	7.2
^{162}Tb	V.T. Schneider	[936]	Karlsruhe	Germany	1965	7.3
^{163}Tb	L. Funke	[937]	Jena	Germany	1966	8.1
^{164}Tb	E. Monnand	[938]	Grenoble	France	1968	7.4
^{165}Tb	R.C. Greenwood	[939]	Idaho Falls	USA	1983	13.1
^{166}Tb	S. Ichikawa	[940]	Tokai	Japan	1996	10.9
^{167}Tb	M. Asai	[941]	Tokai	Japan	1999	10.9
^{168}Tb	M. Asai	[941]	Tokai	Japan	1999	10.9
^{169}Tb	J. Kurcewicz	[856]	Darmstadt	Germany	2012	15.2
^{170}Tb	J. Kurcewicz	[856]	Darmstadt	Germany	2012	15.2
^{171}Tb	J. Kurcewicz	[856]	Darmstadt	Germany	2012	15.2
^{172}Tb	J. Kurcewicz	[856]	Darmstadt	Germany	2012	15.2
^{139}Dy	S.-W. Xu	[846]	Lanzhou	China	1999	11.11
^{140}Dy	W. Krolas	[942]	Oak Ridge	USA	2002	11.4

(continued)

Table 17.1 (continued)

Isotope	First author	Ref.	Laboratory	Country	Year	Table
^{141}Dy	J.M. Nitschke	[797]	Berkeley	USA	1984	11.11
^{142}Dy	P.A. Wilmarth	[829]	Berkeley	USA	1986	11.11
^{143}Dy	J.M. Nitschke	[831]	Berkeley	USA	1983	11.11
^{144}Dy	N. Redon	[880]	Grenoble	France	1986	11.3
^{145}Dy	E. Nolte	[943]	Munich	Germany	1982	11.2
	G.D. Alkhazov	[860]	Leningrad	Russia	1982	10.8
^{146}Dy	G.D. Alkhazov	[944]	Leningrad	Russia	1981	10.8
^{147}Dy	K.S. Toth	[945]	Oak Ridge	USA	1975	11.2
^{148}Dy	A. Latuszynski	[834]	Dubna	Russia	1974	10.8
^{149}Dy	K.S. Toth	[946]	Berkeley	USA	1958	11.5
^{150}Dy	K.S. Toth	[946]	Berkeley	USA	1958	11.5
^{151}Dy	K.S. Toth	[946]	Berkeley	USA	1958	11.5
^{152}Dy	K.S. Toth	[946]	Berkeley	USA	1958	9.3
^{153}Dy	K.S. Toth	[946]	Berkeley	USA	1958	9.3
^{154}Dy	R.D. Macfarlane	[947]	Berkeley	USA	1961	9.3
^{155}Dy	K.S. Toth	[946]	Berkeley	USA	1958	9.3
^{156}Dy	D.C. Hess Jr.	[948]	Argonne	USA	1948	3.4
^{157}Dy	T.H. Handley	[949]	Oak Ridge	USA	1953	9.4
^{158}Dy	A.J. Dempster	[914]	Chicago	USA	1938	3.1
^{159}Dy	F.D.S. Butement	[950]	Harwell	UK	1951	7.2
^{160}Dy	A.J. Dempster	[914]	Chicago	USA	1938	3.1
^{161}Dy	F.W. Aston	[951]	Cambridge	UK	1934	3.3
^{162}Dy	F.W. Aston	[951]	Cambridge	UK	1934	3.3
^{163}Dy	F.W. Aston	[951]	Cambridge	UK	1934	3.3
^{164}Dy	F.W. Aston	[951]	Cambridge	UK	1934	3.3
^{165}Dy	J. K. Marsh	[809]	Oxford	UK	1935	4.1
	G. Hevesy	[952]	Copenhagen	Denmark	1935	4.1
^{166}Dy	B.H. Ketelle	[953]	Oak Ridge	USA	1949	7.2
^{167}Dy	R.G. Wille	[954]	Arkansas	USA	1960	7.3
^{168}Dy	R.J. Gehrke	[955]	Idaho Falls	USA	1982	13.1
^{169}Dy	R.M. Chasteler	[956]	Berkeley	USA	1990	14.3
^{170}Dy	P.-A. Söderström	[957]	Legnaro	Italy	2010	14.2
^{171}Dy	J. Kurcewicz	[856]	Darmstadt	Germany	2012	15.2
^{172}Dy	J. Kurcewicz	[856]	Darmstadt	Germany	2012	15.2
^{173}Dy	J. Kurcewicz	[856]	Darmstadt	Germany	2012	15.2
^{174}Dy	J. Kurcewicz	[856]	Darmstadt	Germany	2012	15.2
^{140}Ho	K. Rykaczewski	[958]	Oak Ridge	USA	1999	16.3
^{141}Ho	C.N. Davids	[891]	Argonne	USA	1998	16.3
^{142}Ho	S.-W. Xu	[959]	Lanzhou	China	2001	11.11

(continued)

Table 17.1 (continued)

Isotope	First author	Ref.	Laboratory	Country	Year	Table
^{143}Ho						
^{144}Ho	P.A. Wilmarth	[829]	Berkeley	USA	1986	11.11
^{145}Ho	L. Goettig	[960]	Daresbury	UK	1987	11.4
^{146}Ho	S.Z. Gui	[961]	Munich	Germany	1982	11.2
^{147}Ho	E. Nolte	[943]	Munich	Germany	1982	11.2
^{148}Ho	K.S. Toth	[962]	Texas A&M	USA	1979	11.2
^{149}Ho	K.S. Toth	[962]	Texas A&M	USA	1979	11.2
^{150}Ho	R.D. Macfarlane	[963]	Berkeley	USA	1963	11.6
^{151}Ho	R.D. Macfarlane	[963]	Berkeley	USA	1963	11.6
^{152}Ho	R.D. Macfarlane	[963]	Berkeley	USA	1963	11.6
^{153}Ho	R.D. Macfarlane	[963]	Berkeley	USA	1963	11.6
^{154}Ho	P. Lagarde	[964]	Orsay	France	1966	10.8
^{155}Ho	A.V. Kalyamin	[965]	Dubna	Russia	1959	10.4
^{156}Ho	J.W. Mihelich	[931]	Oak Ridge	USA	1957	9.5
^{157}Ho	Z.T. Zhelev	[966]	Dubna	Russia	1966	10.4
^{158}Ho	I.S. Dneprovskii	[967]	Dubna	Russia	1961	10.4
^{159}Ho	K.S. Toth	[968]	Berkeley	USA	1958	9.3
^{160}Ho	G. Wilkinson	[932]	Berkeley	USA	1950	9.3
^{161}Ho	T.H. Handley	[969]	Oak Ridge	USA	1954	9.4
^{162}Ho	J.W. Mihelich	[931]	Oak Ridge	USA	1957	9.5
^{163}Ho	C.L. Hammer	[934]	Ames	USA	1957	8.1
^{164}Ho	M.L. Pool	[839]	Michigan	USA	1938	4.1
^{165}Ho	F.W. Aston	[951]	Cambridge	UK	1934	3.3
^{166}Ho	G. Hevesy	[244]	Copenhagen	Denmark	1936	4.1
^{167}Ho	T.H. Handley	[970]	Oak Ridge	USA	1955	9.4
^{168}Ho	R.G. Wille	[954]	Arkansas	USA	1960	7.3
^{169}Ho	K. Miyano	[971]	Tokai	Japan	1963	8.1
^{170}Ho	R.G. Wille	[954]	Arkansas	USA	1960	7.3
^{171}Ho	R.M. Chasteler	[972]	Berkeley	USA	1989	14.3
^{172}Ho	K. Becker	[973]	Darmstadt	Germany	1991	14.3
^{173}Ho	J. Kurcewicz	[856]	Darmstadt	Germany	2012	15.2
^{174}Ho	J. Kurcewicz	[856]	Darmstadt	Germany	2012	15.2
^{175}Ho	J. Kurcewicz	[856]	Darmstadt	Germany	2012	15.2
^{176}Ho	J. Kurcewicz	[856]	Darmstadt	Germany	2012	15.2
^{144}Er	M. Karny	[974]	Oak Ridge	USA	2003	11.11
^{145}Er	K.S. Vierinen	[975]	Berkeley	USA	1989	11.3
^{146}Er	K.S. Toth	[976]	Berkeley	USA	1993	11.3
^{147}Er	G. de Angelis	[977]	Legnaro	Italy	1992	11.4

(continued)

Table 17.1 (continued)

Isotope	First author	Ref.	Laboratory	Country	Year	Table
^{148}Er	E. Nolte	[943]	Munich	Germany	1982	11.2
^{149}Er	K.S. Toth	[978]	Berkeley	USA	1984	11.11
^{150}Er	E. Nolte	[943]	Munich	Germany	1982	11.2
^{151}Er	K.S. Toth	[979]	Oak Ridge	USA	1970	10.5
^{152}Er	R.D. Macfarlane	[980]	Berkeley	USA	1963	11.6
^{153}Er	R.D. Macfarlane	[980]	Berkeley	USA	1963	11.6
^{154}Er	R.D. Macfarlane	[980]	Berkeley	USA	1963	11.6
^{155}Er	K.S. Toth	[981]	Oak Ridge	USA	1969	10.5
^{156}Er	D. Ward	[982]	Berkeley	USA	1967	11.4
^{157}Er	Z.T. Zhelev	[966]	Dubna	Russia	1966	10.4
^{158}Er	I.S. Dneprovskii	[967]	Dubna	Russia	1961	10.4
^{159}Er	A.A. Abdurazakov	[983]	Dubna	Russia	1962	10.4
^{160}Er	M.C. Michel	[762]	Berkeley	USA	1954	10.2
^{161}Er	T.H. Handley	[969]	Oak Ridge	USA	1954	9.4
^{162}Er	A.J. Dempster	[914]	Chicago	USA	1938	3.1
^{163}Er	T.H. Handley	[984]	Oak Ridge	USA	1953	9.4
^{164}Er	A.J. Dempster	[914]	Chicago	USA	1938	3.1
^{165}Er	F.D.S. Butement	[985]	Harwell	UK	1950	9.4
^{166}Er	F.W. Aston	[951]	Cambridge	UK	1934	3.3
^{167}Er	F.W. Aston	[951]	Cambridge	UK	1934	3.3
^{168}Er	F.W. Aston	[951]	Cambridge	UK	1934	3.3
^{169}Er	A. Bisi	[986]	Harwell	UK	1956	7.2
^{170}Er	F.W. Aston	[951]	Cambridge	UK	1934	3.3
^{171}Er	M.L. Pool	[839]	Michigan	USA	1938	4.1
^{172}Er	D.R. Nethaway	[987]	Berkeley	USA	1956	7.2
^{173}Er	V. Pursiheimo	[988]	Helsinki	Finland	1972	7.4
^{174}Er	R.M. Chasteler	[972]	Berkeley	USA	1989	14.3
^{175}Er	X. Zhang	[989]	Lanzhou	China	1996	7.4
^{176}Er	J. Kurcewicz	[856]	Darmstadt	Germany	2012	15.2
^{177}Er	J. Kurcewicz	[856]	Darmstadt	Germany	2012	15.2
^{178}Er	J. Kurcewicz	[856]	Darmstadt	Germany	2012	15.2
^{145}Tm	J.C. Batchelder	[990]	Oak Ridge	USA	1998	16.3
^{146}Tm	K. Livingston	[991]	Daresbury	UK	1993	16.3
^{147}Tm	O. Klepper	[992]	Darmstadt	Germany	1982	16.3
^{148}Tm	E. Nolte	[943]	Munich	Germany	1982	11.2
^{149}Tm	K.S. Toth	[993]	Berkeley	USA	1987	11.3
^{150}Tm	E. Nolte	[943]	Munich	Germany	1982	11.2
^{151}Tm	H. Helppi	[994]	Argonne	USA	1982	11.4
^{152}Tm	C.F. Liang	[995]	Orsay	France	1980	10.8

(continued)

Table 17.1 (continued)

Isotope	First author	Ref.	Laboratory	Country	Year	Table
^{153}Tm	R.D. Macfarlane	[996]	Berkeley	USA	1964	11.6
^{154}Tm	R.D. Macfarlane	[996]	Berkeley	USA	1964	11.6
^{155}Tm	K.S. Toth	[997]	Oak Ridge	USA	1971	11.7
^{156}Tm	K.S. Toth	[997]	Oak Ridge	USA	1971	11.7
^{157}Tm	J.C. Putaux	[998]	Orsay	France	1974	10.8
^{158}Tm	F.W.N. de Boer	[999]	Amsterdam	Netherlands	1970	10.5
^{159}Tm	C. Ekström	[1000]	Uppsala	Sweden	1971	10.5
^{160}Tm	F.W.N. de Boer	[999]	Amsterdam	Netherlands	1970	10.5
^{161}Tm	B. Harmatz	[1001]	Oak Ridge	USA	1959	9.5
^{162}Tm	A. Abdumalikov	[1002]	Dubna	Russia	1963	10.4
^{163}Tm	B. Harmatz	[1001]	Oak Ridge	USA	1959	9.5
^{164}Tm	A.A. Abdurazakov	[1003]	Dubna	Russia	1960	10.4
^{165}Tm	T.H. Handley	[984]	Oak Ridge	USA	1953	9.4
^{166}Tm	G. Wilkinson	[1004]	Berkeley	USA	1948	9.3
^{167}Tm	G. Wilkinson	[1004]	Berkeley	USA	1948	9.3
^{168}Tm	G. Wilkinson	[1005]	Berkeley	USA	1949	9.3
^{169}Tm	F.W. Aston	[951]	Cambridge	UK	1934	3.3
^{170}Tm	E. Neuninger	[1006]	Wien	Austria	1936	4.1
^{171}Tm	S. DeBenedetti	[1007]	Oak Ridge	USA	1948	7.2
^{172}Tm	D.R. Nethaway	[987]	Berkeley	USA	1956	7.2
^{173}Tm	T. Kuroyanagi	[904]	Tohoku	Japan	1961	8.1
^{174}Tm	R.G. Wille	[954]	Arkansas	USA	1960	7.3
^{175}Tm	T. Kuroyanagi	[904]	Tohoku	Japan	1961	8.1
^{176}Tm	K. Takahashi	[365]	Tokyo	Japan	1961	7.3
^{177}Tm	K. Rykaczewski	[1008]	Darmstadt	Germany	1989	14.3
^{178}Tm	J. Kurcewicz	[856]	Darmstadt	Germany	2012	15.2
^{179}Tm	J. Kurcewicz	[856]	Darmstadt	Germany	2012	15.2
^{180}Tm	J. Kurcewicz	[856]	Darmstadt	Germany	2012	15.2
^{181}Tm	J. Kurcewicz	[856]	Darmstadt	Germany	2012	15.2
^{149}Yb	S.-W. Xu	[1009]	Lanzhou	China	2001	11.11
^{150}Yb						
^{151}Yb	P. Kleinheinz	[1010]	Darmstadt	Germany	1985	11.3
^{152}Yb	E. Nolte	[1011]	Munich	Germany	1982	11.4
^{153}Yb	E. Hagberg	[1012]	CERN	Switzerland	1977	10.7
^{154}Yb	R.D. Macfarlane	[996]	Berkeley	USA	1964	11.6
^{155}Yb	R.D. Macfarlane	[996]	Berkeley	USA	1964	11.6
^{156}Yb	K.S. Toth	[979]	Oak Ridge	USA	1970	10.5
^{157}Yb	K.S. Toth	[979]	Oak Ridge	USA	1970	10.5
^{158}Yb	D. Ward	[982]	Berkeley	USA	1967	11.4

(continued)

Table 17.1 (continued)

Isotope	First author	Ref.	Laboratory	Country	Year	Table
^{159}Yb	W. Trautmann	[1013]	Munich	Germany	1975	11.4
^{160}Yb	D. Ward	[982]	Berkeley	USA	1967	11.4
^{161}Yb	A. Latuszynski	[834]	Dubna	Russia	1974	10.8
^{162}Yb	A. Abdumalikov	[1002]	Dubna	Russia	1963	10.4
^{163}Yb	P. Paris	[1014]	Orsay	France	1967	10.8
^{164}Yb	F.D.S. Butement	[1015]	Liverpool	UK	1960	10.2
^{165}Yb	P. Paris	[1016]	Orsay	France	1964	10.8
^{166}Yb	M.C. Michel	[762]	Berkeley	USA	1954	10.2
^{167}Yb	T.H. Handley	[1017]	Oak Ridge	USA	1954	9.4
^{168}Yb	A.J. Dempster	[914]	Chicago	USA	1938	3.1
^{169}Yb	W. Bothe	[1018]	Berlin	Germany	1946	4.1
^{170}Yb	A.J. Dempster	[914]	Chicago	USA	1938	3.1
^{171}Yb	F.W. Aston	[951]	Cambridge	UK	1934	3.3
^{172}Yb	F.W. Aston	[951]	Cambridge	UK	1934	3.3
^{173}Yb	F.W. Aston	[951]	Cambridge	UK	1934	3.3
^{174}Yb	F.W. Aston	[951]	Cambridge	UK	1934	3.3
^{175}Yb	H. Atterling	[1019]	Stockholm	Sweden	1945	4.1
^{176}Yb	F.W. Aston	[951]	Cambridge	UK	1934	3.3
^{177}Yb	H. Atterling	[1019]	Stockholm	Sweden	1945	4.1
^{178}Yb	C.J. Orth	[1020]	Los Alamos	USA	1973	9.5
^{179}Yb	R. Kirchner	[1021]	Darmstadt	Germany	1982	14.3
^{180}Yb	E. Runte	[1022]	Darmstadt	Germany	1987	14.3
^{181}Yb	J. Kurcewicz	[856]	Darmstadt	Germany	2012	15.2
^{182}Yb	J. Kurcewicz	[856]	Darmstadt	Germany	2012	15.2
^{183}Yb	J. Kurcewicz	[856]	Darmstadt	Germany	2012	15.2
^{184}Yb	J. Kurcewicz	[856]	Darmstadt	Germany	2012	15.2
^{185}Yb	J. Kurcewicz	[856]	Darmstadt	Germany	2012	15.2
^{150}Lu	P.J. Sellin	[1023]	Daresbury	UK	1993	16.3
^{151}Lu	S. Hofmann	[1024]	Darmstadt	Germany	1982	16.3
^{152}Lu	K.S. Toth	[1025]	Berkeley	USA	1987	11.3
^{153}Lu	J.M. Nitschke	[1026]	Berkeley	USA	1989	11.3
^{154}Lu	S. Hofmann	[1027]	Darmstadt	Germany	1981	11.9
^{155}Lu	R.D. Macfarlane	[1028]	Berkeley	USA	1965	11.6
^{156}Lu	R.D. Macfarlane	[1028]	Berkeley	USA	1965	11.6
^{157}Lu	E. Hagberg	[1012]	CERN	Switzerland	1977	10.7
^{158}Lu	G.D. Alkhazov	[1029]	Leningrad	Russia	1979	10.8
^{159}Lu	G.D. Alkhazov	[1030]	Leningrad	Russia	1980	10.8
^{160}Lu	G.D. Alkhazov	[1029]	Leningrad	Russia	1979	10.8
^{161}Lu	R. Anholt	[1031]	Yale	USA	1973	11.4

(continued)

Table 17.1 (continued)

Isotope	First author	Ref.	Laboratory	Country	Year	Table
^{162}Lu	C. Burman	[1032]	Yale	USA	1978	11.2
^{163}Lu	G.D. Alkhazov	[1029]	Leningrad	Russia	1979	10.8
^{164}Lu	R.C. Hunter	[1033]	Oak Ridge	USA	1977	11.2
^{165}Lu	B.J. Meijer	[1034]	Amsterdam	Netherlands	1973	10.5
^{166}Lu	R. Arlt	[1035]	Dubna	Russia	1969	10.4
^{167}Lu	P.M. Aron	[1036]	Dubna	Russia	1958	10.4
^{168}Lu	R.G. Wilson	[1037]	Ohio State	USA	1960	9.1
^{169}Lu	W.E. Nervik	[1038]	Berkeley	USA	1955	10.2
^{170}Lu	G. Wilkinson	[1039]	Berkeley	USA	1951	9.3
^{171}Lu	G. Wilkinson	[1039]	Berkeley	USA	1951	9.3
^{172}Lu	G. Wilkinson	[1039]	Berkeley	USA	1951	9.3
^{173}Lu	G. Wilkinson	[1039]	Berkeley	USA	1951	9.3
^{174}Lu	G. Wilkinson	[1039]	Berkeley	USA	1951	9.3
^{175}Lu	F.W. Aston	[951]	Cambridge	UK	1934	3.3
^{176}Lu	J. K. Marsh	[809]	Oxford	UK	1935	3.5
^{177}Lu	H. Atterling	[1019]	Stockholm	Sweden	1945	4.1
^{178}Lu	T. Stribel	[1040]	Hechingen	Germany	1957	7.3
^{179}Lu	T. Kuroyanagi	[904]	Tohoku	Japan	1961	8.1
^{180}Lu	S.C. Gujrathi	[1041]	Simon Fraser	Canada	1971	7.4
^{181}Lu	R. Kirchner	[1021]	Darmstadt	Germany	1982	14.3
^{182}Lu	R. Kirchner	[1021]	Darmstadt	Germany	1982	14.3
^{183}Lu	K. Rykaczewski	[1042]	Darmstadt	Germany	1983	14.3
^{184}Lu	K. Rykaczewski	[1008]	Darmstadt	Germany	1989	14.3
^{185}Lu	J. Kurcewicz	[856]	Darmstadt	Germany	2012	15.2
^{186}Lu	J. Kurcewicz	[856]	Darmstadt	Germany	2012	15.2
^{187}Lu	J. Kurcewicz	[856]	Darmstadt	Germany	2012	15.2
^{188}Lu	J. Kurcewicz	[856]	Darmstadt	Germany	2012	15.2
^{154}Hf	S. Hofmann	[1027]	Darmstadt	Germany	1981	11.9
^{155}Hf	S. Hofmann	[1027]	Darmstadt	Germany	1981	11.9
^{156}Hf	S. Hofmann	[1043]	Darmstadt	Germany	1979	11.9
^{157}Hf	R.D. Macfarlane	[1028]	Berkeley	USA	1965	11.6
^{158}Hf	R.D. Macfarlane	[1028]	Berkeley	USA	1965	11.6
^{159}Hf	K.S. Toth	[1044]	Oak Ridge	USA	1973	11.7
^{160}Hf	K.S. Toth	[1044]	Oak Ridge	USA	1973	11.7
^{161}Hf	K.S. Toth	[1044]	Oak Ridge	USA	1973	11.7
^{162}Hf	U.J. Schrewe	[1045]	Chalk River	Canada	1982	11.2
^{163}Hf	U.J. Schrewe	[1045]	Chalk River	Canada	1982	11.2
^{164}Hf	H. Bruchertseifer	[1046]	Dubna	Russia	1981	11.1

(continued)

Table 17.1 (continued)

Isotope	First author	Ref.	Laboratory	Country	Year	Table
^{165}Hf	H. Bruchertseifer	[1046]	Dubna	Russia	1981	11.1
^{166}Hf	F.S. Stephens	[1047]	Berkeley	USA	1965	11.4
^{167}Hf	R. Arlt	[1035]	Dubna	Russia	1969	10.4
^{168}Hf	E.R. Merz	[1048]	Pittsburgh	USA	1961	10.2
^{169}Hf	R. Arlt	[1035]	Dubna	Russia	1969	10.4
^{170}Hf	E.R. Merz	[1048]	Pittsburgh	USA	1961	10.2
^{171}Hf	G. Wilkinson	[1039]	Berkeley	USA	1951	9.3
^{172}Hf	G. Wilkinson	[1039]	Berkeley	USA	1951	9.3
^{173}Hf	G. Wilkinson	[1039]	Berkeley	USA	1951	9.3
^{174}Hf	A.J. Dempster	[1049]	Chicago	USA	1939	3.1
^{175}Hf	G. Wilkinson	[1050]	Berkeley	USA	1949	9.3
^{176}Hf	F.W. Aston	[230]	Cambridge	UK	1934	3.3
^{177}Hf	F.W. Aston	[230]	Cambridge	UK	1934	3.3
^{178}Hf	F.W. Aston	[230]	Cambridge	UK	1934	3.3
^{179}Hf	F.W. Aston	[230]	Cambridge	UK	1934	3.3
^{180}Hf	F.W. Aston	[230]	Cambridge	UK	1934	3.3
^{181}Hf	G. Hevesy	[1051]	Copenhagen	Denmark	1935	4.1
^{182}Hf	W.H. Hutchin	[1052]	Livermore	USA	1961	7.2
^{183}Hf	O.O. Gatti	[1053]	Buenos Aires	Argentina	1956	7.3
^{184}Hf	T.E. Ward	[1054]	Brookhaven	USA	1973	9.6
^{185}Hf	S. Yuan	[1055]	Lanzhou	China	1993	7.4
^{186}Hf	S. Yuan	[1056]	Lanzhou	China	1998	14.1
^{187}Hf	J. Benlliure	[1057]	Darmstadt	Germany	1999	15.3
^{188}Hf	J. Benlliure	[1057]	Darmstadt	Germany	1999	15.3
^{189}Hf	N. Alkhomashi	[1058]	Darmstadt	Germany	2009	15.3
^{190}Hf	J. Kurcewicz	[856]	Darmstadt	Germany	2012	15.2
^{155}Ta	R.D. Page	[1059]	Jyväskylä	Finland	2007	16.3
^{156}Ta	R.D. Page	[1060]	Daresbury	UK	1992	16.3
^{157}Ta	S. Hofmann	[1043]	Darmstadt	Germany	1979	16.3
^{158}Ta	S. Hofmann	[1043]	Darmstadt	Germany	1979	11.9
^{159}Ta	S. Hofmann	[1043]	Darmstadt	Germany	1979	11.9
^{160}Ta	S. Hofmann	[1043]	Darmstadt	Germany	1979	11.9
^{161}Ta	S. Hofmann	[1043]	Darmstadt	Germany	1979	11.9
^{162}Ta	C.F. Liang	[1061]	Orsay	France	1985	10.8
^{163}Ta	C.F. Liang	[1061]	Orsay	France	1985	10.8
^{164}Ta	B. Eichler	[1062]	Dubna	Russia	1982	11.1
^{165}Ta	H. Bruchertseifer	[1063]	Dubna	Russia	1982	11.1
^{166}Ta	R.E. Leber	[1064]	Yale	USA	1977	11.2
^{167}Ta	C.F. Liang	[498]	Orsay	France	1982	10.8

(continued)

Table 17.1 (continued)

Isotope	First author	Ref.	Laboratory	Country	Year	Table
^{168}Ta	R. Arlt	[1065]	Dubna	Russia	1969	10.4
^{169}Ta	R. Arlt	[1065]	Dubna	Russia	1969	10.4
^{170}Ta	R. Arlt	[1065]	Dubna	Russia	1969	10.4
^{171}Ta	R. Arlt	[1065]	Dubna	Russia	1969	10.4
^{172}Ta	H. Abou-Leila	[1066]	Orsay	France	1964	10.8
^{173}Ta	K.T. Faler	[1067]	Berkeley	USA	1960	11.1
^{174}Ta	K.T. Faler	[1067]	Berkeley	USA	1960	11.1
^{175}Ta	K.T. Faler	[1067]	Berkeley	USA	1960	11.1
^{176}Ta	G. Wilkinson	[1004]	Berkeley	USA	1948	9.3
^{177}Ta	G. Wilkinson	[1004]	Berkeley	USA	1948	9.3
^{178}Ta	G. Wilkinson	[1068]	Berkeley	USA	1950	9.3
^{179}Ta	G. Wilkinson	[1068]	Berkeley	USA	1950	9.3
^{180}Ta	O. Oldenberg	[1069]	Berkeley	USA	1938	3.5
^{181}Ta	F.W. Aston	[521]	Cambridge	UK	1932	3.3
^{182}Ta	O. Oldenberg	[1069]	Berkeley	USA	1938	4.1
^{183}Ta	F.D.S. Butement	[391]	Harwell	UK	1950	8.1
^{184}Ta	F.D.S. Butement	[1070]	Harwell	UK	1955	7.3
^{185}Ta	F.D.S. Butement	[391]	Harwell	UK	1950	8.1
^{186}Ta	A.J. Poe	[1071]	Harwell	UK	1955	7.3
^{187}Ta	J. Benlliure	[1057]	Darmstadt	Germany	1999	15.3
^{188}Ta	J. Benlliure	[1057]	Darmstadt	Germany	1999	15.3
^{189}Ta	J. Benlliure	[1057]	Darmstadt	Germany	1999	15.3
^{190}Ta	N. Alkhomashi	[1058]	Darmstadt	Germany	2009	15.3
^{191}Ta	N. Alkhomashi	[1058]	Darmstadt	Germany	2009	15.3
^{192}Ta	N. Alkhomashi	[1058]	Darmstadt	Germany	2009	15.3
^{193}Ta	J. Kurcewicz	[856]	Darmstadt	Germany	2012	15.2
^{194}Ta	J. Kurcewicz	[856]	Darmstadt	Germany	2012	15.2
^{157}W	L. Bianco	[1072]	Jyväskylä	Finland	2010	11.10
^{158}W	S. Hofmann	[1027]	Darmstadt	Germany	1981	11.9
^{159}W	S. Hofmann	[1027]	Darmstadt	Germany	1981	11.9
^{160}W	S. Hofmann	[1043]	Darmstadt	Germany	1979	11.9
^{161}W	D.A. Eastham	[1073]	Manchester	UK	1973	11.7
^{162}W	D.A. Eastham	[1073]	Manchester	UK	1973	11.7
^{163}W	D.A. Eastham	[1073]	Manchester	UK	1973	11.7
^{164}W	D.A. Eastham	[1073]	Manchester	UK	1973	11.7
^{165}W	K.S. Toth	[1074]	Oak Ridge	USA	1975	11.7
^{166}W	K.S. Toth	[1074]	Oak Ridge	USA	1975	11.7
^{167}W	J. Gerl	[1075]	Canberra	Australia	1985	11.4
^{168}W	F.S. Stephens	[1076]	Berkeley	USA	1971	11.4

(continued)

Table 17.1 (continued)

Isotope	First author	Ref.	Laboratory	Country	Year	Table
^{169}W	J. Recht	[1077]	Berlin	Germany	1985	11.4
^{170}W	E. Nadjakov	[1078]	Dubna	Russia	1971	11.1
^{171}W	H.F.R. Arciszewski	[1079]	Louvain-la-Neuve	Belgium	1983	11.4
^{172}W	F.S. Stephens	[1047]	Berkeley	USA	1965	11.4
^{173}W	A. Santoni	[1080]	Orsay	France	1963	10.8
^{174}W	F.S. Stephens	[1047]	Berkeley	USA	1965	11.4
^{175}W	A. Santoni	[1080]	Orsay	France	1963	10.8
^{176}W	G. Wilkinson	[1068]	Berkeley	USA	1950	9.3
^{177}W	G. Wilkinson	[1068]	Berkeley	USA	1950	9.3
^{178}W	G. Wilkinson	[1068]	Berkeley	USA	1950	9.3
^{179}W	G. Wilkinson	[1068]	Berkeley	USA	1950	9.3
^{180}W	A.J. Dempster	[1081]	Chicago	USA	1937	3.1
^{181}W	G. Wilkinson	[1082]	Berkeley	USA	1947	9.3
^{182}W	F.W. Aston	[1083]	Cambridge	UK	1930	3.3
^{183}W	F.W. Aston	[1083]	Cambridge	UK	1930	3.3
^{184}W	F.W. Aston	[1083]	Cambridge	UK	1930	3.3
^{185}W	O. Minakawa	[1084]	RIKEN	Japan	1940	4.1
^{186}W	F.W. Aston	[1083]	Cambridge	UK	1930	3.3
^{187}W	O. Minakawa	[1084]	RIKEN	Japan	1940	4.1
^{188}W	M. Lindner	[1085]	Washington State	USA	1951	7.2
^{189}W	J. Flegenheimer	[1086]	Buenos Aires	Argentina	1963	7.3
^{190}W	P.E. Haustein	[1087]	Brookhaven	USA	1976	7.4
^{191}W	J. Benlliure	[1057]	Darmstadt	Germany	1999	15.3
^{192}W	J. Benlliure	[1057]	Darmstadt	Germany	1999	15.3
^{193}W	N. Alkhomashi	[1058]	Darmstadt	Germany	2009	15.3
^{194}W	N. Alkhomashi	[1058]	Darmstadt	Germany	2009	15.3
^{195}W	J. Kurcewicz	[856]	Darmstadt	Germany	2012	15.2
^{196}W	J. Kurcewicz	[856]	Darmstadt	Germany	2012	15.2
^{197}W	J. Kurcewicz	[856]	Darmstadt	Germany	2012	15.2
^{159}Re	D.T. Joss	[1088]	Jyväskylä	Finland	2006	16.3
^{160}Re	R.D. Page	[1060]	Daresbury	UK	1992	16.3
^{161}Re	S. Hofmann	[1043]	Darmstadt	Germany	1979	16.3
^{162}Re	S. Hofmann	[1043]	Darmstadt	Germany	1979	11.9
^{163}Re	S. Hofmann	[1043]	Darmstadt	Germany	1979	11.9
^{164}Re	S. Hofmann	[1043]	Darmstadt	Germany	1979	11.9
^{165}Re	S. Hofmann	[1027]	Darmstadt	Germany	1981	11.9
^{166}Re	U.J. Schrewe	[1089]	Darmstadt	Germany	1978	11.7
^{167}Re	F. Meissner	[1090]	Berlin	Germany	1992	11.7
^{168}Re	F. Meissner	[1090]	Berlin	Germany	1992	11.7

(continued)

Table 17.1 (continued)

Isotope	First author	Ref.	Laboratory	Country	Year	Table
^{169}Re	C. Cabot	[1091]	Orsay	France	1978	11.7
^{170}Re	E.E. Berlovich	[1092]	Leningrad	Russia	1974	10.5
^{171}Re	E. Runte	[1093]	Berlin	Germany	1987	11.2
^{172}Re	E.E. Berlovich	[1094]	Leningrad	Russia	1972	10.5
^{173}Re	A. Szymanski	[1095]	Manchester	UK	1986	11.2
^{174}Re	E.E. Berlovich	[1094]	Leningrad	Russia	1972	10.5
^{175}Re	E. Nadjakov	[1096]	Dubna	Russia	1967	11.1
^{176}Re	E. Nadjakov	[1096]	Dubna	Russia	1967	11.1
^{177}Re	B.C. Haldar	[1097]	Rochester	USA	1957	10.2
^{178}Re	B.C. Haldar	[1097]	Rochester	USA	1957	10.2
^{179}Re	B. Harmatz	[1098]	Oak Ridge	USA	1960	9.5
^{180}Re	V. Kistiakowsky Fischer	[1099]	Berkeley	USA	1955	9.1
^{181}Re	C.J. Gallagher Jr.	[1100]	Berkeley	USA	1957	9.3
^{182}Re	G. Wilkinson	[1101]	Berkeley	USA	1950	9.3
^{183}Re	B.J. Stover	[1102]	Berkeley	USA	1950	9.3
^{184}Re	K. Fajans	[1103]	Michigan	USA	1940	4.1
^{185}Re	F.W. Aston	[1104]	Cambridge	UK	1931	3.3
^{186}Re	K. Sinma	[1105]	RIKEN	Japan	1939	4.1
^{187}Re	F.W. Aston	[1104]	Cambridge	UK	1931	3.3
^{188}Re	K. Sinma	[1105]	RIKEN	Japan	1939	4.1
^{189}Re	B. Crasemann	[1106]	Brookhaven	USA	1963	7.3
^{190}Re	A.H.W. Aten Jr.	[1107]	Amsterdam	Netherlands	1955	9.5
^{191}Re	B. Crasemann	[1106]	Brookhaven	USA	1963	7.3
^{192}Re	J. Blachot	[1108]	Grenoble	France	1965	7.3
^{193}Re	J. Benlliure	[1057]	Darmstadt	Germany	1999	15.3
^{194}Re	J. Benlliure	[1057]	Darmstadt	Germany	1999	15.3
^{195}Re	T. Kurtukian-Nieto	[1109]	Darmstadt	Germany	2008	15.3
^{196}Re	S.J. Steer	[1110]	Darmstadt	Germany	2008	15.3
^{197}Re	S.J. Steer	[1111]	Darmstadt	Germany	2011	15.3
^{198}Re	J. Kurcewicz	[856]	Darmstadt	Germany	2012	15.2
^{199}Re	J. Kurcewicz	[856]	Darmstadt	Germany	2012	15.2
^{161}Os	L. Bianco	[1072]	Jyväskylä	Finland	2010	11.10
^{162}Os	S. Hofmann	[1112]	Darmstadt	Germany	1989	11.9
^{163}Os	S. Hofmann	[1027]	Darmstadt	Germany	1981	11.9
^{164}Os	S. Hofmann	[1027]	Darmstadt	Germany	1981	11.9
^{165}Os	C. Cabot	[1091]	Orsay	France	1978	11.7
^{166}Os	C. Cabot	[1113]	Orsay	France	1977	11.7
^{167}Os	C. Cabot	[1113]	Orsay	France	1977	11.7
^{168}Os	C. Cabot	[1113]	Orsay	France	1977	11.7

(continued)

Table 17.1 (continued)

Isotope	First author	Ref.	Laboratory	Country	Year	Table
^{169}Os	K.S. Toth	[1114]	Oak Ridge	USA	1972	11.7
^{170}Os	K.S. Toth	[1115]	Oak Ridge	USA	1972	11.7
^{171}Os	K.S. Toth	[1115]	Oak Ridge	USA	1972	11.7
^{172}Os	J. Borggreen	[1116]	Berkeley	USA	1971	11.6
^{173}Os	J. Borggreen	[1116]	Berkeley	USA	1971	11.6
^{174}Os	J. Borggreen	[1116]	Berkeley	USA	1971	11.6
^{175}Os	E.E. Berlovich	[1094]	Leningrad	Russia	1972	10.5
^{176}Os	R. Arlt	[1117]	Dubna	Russia	1970	10.4
^{177}Os	R. Arlt	[1117]	Dubna	Russia	1970	10.4
^{178}Os	J. Burde	[1118]	Berkeley	USA	1967	11.4
^{179}Os	B.N. Belyaev	[1119]	Dubna	Russia	1968	11.1
^{180}Os	B.N. Belyaev	[1120]	Dubna	Russia	1967	11.1
^{181}Os	K.J. Hofstetter	[1121]	Argonne/Oak Ridge	USA	1966	9.5
^{182}Os	B.J. Stover	[1102]	Berkeley	USA	1950	9.3
^{183}Os	B.J. Stover	[1102]	Berkeley	USA	1950	9.3
^{184}Os	A.O. Nier	[1122]	Harvard	USA	1937	3.4
^{185}Os	L.J. Goodman	[1123]	Ohio State	USA	1947	9.4
^{186}Os	F.W. Aston	[571]	Cambridge	UK	1931	3.3
^{187}Os	F.W. Aston	[571]	Cambridge	UK	1931	3.3
^{188}Os	F.W. Aston	[571]	Cambridge	UK	1931	3.3
^{189}Os	F.W. Aston	[571]	Cambridge	UK	1931	3.3
^{190}Os	F.W. Aston	[571]	Cambridge	UK	1931	3.3
^{191}Os	E. Zingg	[1124]	Zurich	Switzerland	1940	4.1
^{192}Os	F.W. Aston	[571]	Cambridge	UK	1931	3.3
^{193}Os	E. Zingg	[1124]	Zurich	Switzerland	1940	4.1
^{194}Os	M. Lindner	[1125]	Washington State	USA	1951	7.2
^{195}Os	G. B. Baró	[1126]	Buenos Aires	Argentina	1957	7.3
^{196}Os	P.E. Haustein	[1127]	Brookhaven	USA	1977	7.4
^{197}Os	Y. Xu	[11]	Lanzhou	China	2003	7.4
^{198}Os	S.J. Steer	[1110]	Darmstadt	Germany	2008	15.3
^{199}Os	S.J. Steer	[1110]	Darmstadt	Germany	2008	15.3
^{200}Os	A.I. Morales	[1128]	Darmstadt	Germany	2011	15.3
^{201}Os	A.I. Morales	[1128]	Darmstadt	Germany	2011	15.3
^{202}Os	J. Kurcewicz	[856]	Darmstadt	Germany	2012	15.2
^{203}Os	J. Kurcewicz	[856]	Darmstadt	Germany	2012	15.2
^{164}Ir	M.C. Drummond	[1129]	Jyväskylä	Finland	2014	16.3
^{165}Ir	C.N. Davids	[1130]	Argonne	USA	1997	16.3
^{166}Ir	S. Hofmann	[1027]	Darmstadt	Germany	1981	16.3
^{167}Ir	S. Hofmann	[1027]	Darmstadt	Germany	1981	16.3

(continued)

Table 17.1 (continued)

Isotope	First author	Ref.	Laboratory	Country	Year	Table
^{168}Ir	C. Cabot	[1091]	Orsay	France	1978	11.7
^{169}Ir	C. Cabot	[1091]	Orsay	France	1978	11.7
^{170}Ir	C. Cabot	[1113]	Orsay	France	1977	11.7
^{171}Ir	A. Siivola	[1131]	Berkeley	USA	1967	11.6
^{172}Ir	A. Siivola	[1131]	Berkeley	USA	1967	11.6
^{173}Ir	A. Siivola	[1131]	Berkeley	USA	1967	11.6
^{174}Ir	A. Siivola	[1131]	Berkeley	USA	1967	11.6
^{175}Ir	A. Siivola	[1131]	Berkeley	USA	1967	11.6
^{176}Ir	A. Siivola	[1131]	Berkeley	USA	1967	11.6
^{177}Ir	A. Siivola	[1131]	Berkeley	USA	1967	11.6
^{178}Ir	A.I. Akhmadzhanov	[1132]	Dubna	Russia	1972	11.2
^{179}Ir	U. Bosch-Wicke	[1133]	Berlin	Germany	1992	11.2
^{180}Ir	A.I. Akhmadzhanov	[1132]	Dubna	Russia	1972	11.2
^{181}Ir	A.I. Akhmadzhanov	[1132]	Dubna	Russia	1972	11.2
^{182}Ir	R.M. Diamond	[1134]	Berkeley	USA	1961	11.1
^{183}Ir	A.K. Lavrukhina	[1135]	Dubna	Russia	1961	10.4
^{184}Ir	V.I. Baranov	[1136]	Dubna	Russia	1960	10.4
^{185}Ir	R.M. Diamond	[1137]	Berkeley	USA	1958	9.3
^{186}Ir	R.M. Diamond	[1137]	Berkeley	USA	1958	9.3
^{187}Ir	R.M. Diamond	[1137]	Berkeley	USA	1958	9.3
^{188}Ir	T.C. Chu	[1138]	Berkeley	USA	1950	9.3
^{189}Ir	W.G. Smith	[1139]	Berkeley	USA	1955	10.2
^{190}Ir	L.J. Goodman	[1123]	Ohio State	USA	1947	9.4
^{191}Ir	B. Venkatesachar	[1140]	Bangalore	India	1935	3.5
^{192}Ir	E. McMillan	[1141]	Berkeley	USA	1937	4.1
^{193}Ir	B. Venkatesachar	[1140]	Bangalore	India	1935	3.5
^{194}Ir	E. McMillan	[1141]	Berkeley	USA	1937	4.1
^{195}Ir	D. Christian	[1142]	Ames	USA	1952	8.1
^{196}Ir	H. Vonach	[1143]	Wien	Austria	1966	7.3
^{197}Ir	D. Christian	[1142]	Ames	USA	1952	8.1
^{198}Ir	A. Szalay	[1144]	Debrecen	Hungary	1973	7.4
^{199}Ir	K. Zhao	[1145]	Daresbury	UK	1993	14.1
^{200}Ir	S.J. Steer	[1110]	Darmstadt	Germany	2008	15.3
^{201}Ir	S.J. Steer	[1110]	Darmstadt	Germany	2008	15.3
^{202}Ir	S.J. Steer	[1110]	Darmstadt	Germany	2008	15.3
^{203}Ir	A.I. Morales	[1128]	Darmstadt	Germany	2011	15.3
^{204}Ir	A. I. Morales	[1128]	Darmstadt	Germany	2011	15.3
^{205}Ir	J. Kurcewicz	[856]	Darmstadt	Germany	2012	15.2
^{166}Pt	C.R. Bingham	[1146]	Argonne	USA	1996	11.10

(continued)

Table 17.1 (continued)

Isotope	First author	Ref.	Laboratory	Country	Year	Table
^{167}Pt	C.R. Bingham	[1146]	Argonne	USA	1996	11.10
^{168}Pt	S. Hofmann	[1027]	Darmstadt	Germany	1981	11.9
^{169}Pt	S. Hofmann	[1027]	Darmstadt	Germany	1981	11.9
^{170}Pt	S. Hofmann	[1027]	Darmstadt	Germany	1981	11.9
^{171}Pt	S. Hofmann	[1027]	Darmstadt	Germany	1981	11.9
^{172}Pt	S. Della Negra	[1147]	Orsay	France	1981	11.7
^{173}Pt	A. Siivola	[1148]	Berkeley	USA	1966	11.6
^{174}Pt	A. Siivola	[1148]	Berkeley	USA	1966	11.6
^{175}Pt	A. Siivola	[1148]	Berkeley	USA	1966	11.6
^{176}Pt	A. Siivola	[1148]	Berkeley	USA	1966	11.6
^{177}Pt	A. Siivola	[1148]	Berkeley	USA	1966	11.6
^{178}Pt	A. Siivola	[1148]	Berkeley	USA	1966	11.6
^{179}Pt	A. Siivola	[1148]	Berkeley	USA	1966	11.6
^{180}Pt	A. Siivola	[1148]	Berkeley	USA	1966	11.6
^{181}Pt	A. Siivola	[1148]	Berkeley	USA	1966	11.6
^{182}Pt	G. Graeffe	[1149]	Uppsala	Sweden	1963	10.5
^{183}Pt	G. Graeffe	[1149]	Uppsala	Sweden	1963	10.5
^{184}Pt	G. Graeffe	[1149]	Uppsala	Sweden	1963	10.5
^{185}Pt	G. Albouy	[1150]	Orsay	France	1960	10.8
^{186}Pt	V.I. Baranov	[1136]	Dubna	Russia	1960	10.4
^{187}Pt	V.I. Baranov	[1136]	Dubna	Russia	1960	10.4
^{188}Pt	R.A. Naumann	[1151]	Columbia/Harvard	USA	1954	9.4
^{189}Pt	W.G. Smith	[1139]	Berkeley	USA	1955	10.2
^{190}Pt	H.E. Duckworth	[6]	Middletown	USA	1949	3.4
^{191}Pt	G. Wilkinson	[1152]	Berkeley	USA	1948	9.3
^{192}Pt	A.J. Dempster	[1153]	Chicago	USA	1935	3.1
^{193}Pt	G. Wilkinson	[1152]	Berkeley	USA	1948	9.3
^{194}Pt	B. Fuchs	[1154]	Berlin	Germany	1935	3.5
^{195}Pt	B. Fuchs	[1154]	Berlin	Germany	1935	3.5
^{196}Pt	B. Fuchs	[1154]	Berlin	Germany	1935	3.5
^{197}Pt	J.M. Cork	[1155]	Berkeley	USA	1936	4.3
^{198}Pt	A.J. Dempster	[1153]	Chicago	USA	1935	3.1
^{199}Pt	E. McMillan	[1141]	Berkeley	USA	1937	4.1
^{200}Pt	L.P. Roy	[1156]	Chalk River	Canada	1957	7.2
^{201}Pt	J. Facetti	[1157]	Mayaguez	Puerto Rico	1962	7.2
^{202}Pt	S. Shi	[1158]	Shanghai	China	1992	7.4
^{203}Pt	S.J. Steer	[1110]	Darmstadt	Germany	2008	15.3
^{204}Pt	S.J. Steer	[1110]	Darmstadt	Germany	2008	15.3
^{205}Pt	H. Alvarez-Pol	[1159]	Darmstadt	Germany	2010	15.3

(continued)

Table 17.1 (continued)

Isotope	First author	Ref.	Laboratory	Country	Year	Table
^{206}Pt	J. Kurcewicz	[856]	Darmstadt	Germany	2012	15.2
^{207}Pt	J. Kurcewicz	[856]	Darmstadt	Germany	2012	15.2
^{208}Pt	J. Kurcewicz	[856]	Darmstadt	Germany	2012	15.2
^{170}Au	H. Kettunen	[1160]	Jyväskylä	Finland	2004	16.3
^{171}Au	C.N. Davids	[1130]	Argonne	USA	1997	16.3
^{172}Au	P.J. Sellin	[1161]	Daresbury	UK	1993	11.10
^{173}Au	J.R.H. Schneider	[1162]	Darmstadt	Germany	1983	11.9
^{174}Au	J.R.H. Schneider	[1162]	Darmstadt	Germany	1983	11.9
^{175}Au	C. Cabot	[1163]	Orsay	France	1975	11.7
^{176}Au	C. Cabot	[1163]	Orsay	France	1975	11.7
^{177}Au	A. Siivola	[1164]	Berkeley	USA	1968	11.6
^{178}Au	A. Siivola	[1164]	Berkeley	USA	1968	11.6
^{179}Au	A. Siivola	[1164]	Berkeley	USA	1968	11.6
^{180}Au	J.P. Husson	[1165]	CERN	Switzerland	1977	10.7
^{181}Au	A. Siivola	[1164]	Berkeley	USA	1968	11.6
^{182}Au	P.G. Hansen	[1166]	CERN	Switzerland	1970	10.7
^{183}Au	A. Siivola	[1164]	Berkeley	USA	1968	11.6
^{184}Au	P.G. Hansen	[205]	CERN	Switzerland	1969	10.7
^{185}Au	G. Albouy	[1150]	Orsay	France	1960	10.8
^{186}Au	G. Albouy	[1150]	Orsay	France	1960	10.8
^{187}Au	W.G. Smith	[1139]	Berkeley	USA	1955	10.2
^{188}Au	W.G. Smith	[1139]	Berkeley	USA	1955	10.2
^{189}Au	W.G. Smith	[1139]	Berkeley	USA	1955	10.2
^{190}Au	G. Albouy	[1167]	Orsay	France	1959	10.8
^{191}Au	L.P. Gillon	[1168]	Harvard	USA	1954	10.2
^{192}Au	G. Wilkinson	[1152]	Berkeley	USA	1948	9.3
^{193}Au	G. Wilkinson	[1152]	Berkeley	USA	1948	9.3
^{194}Au	G. Wilkinson	[1152]	Berkeley	USA	1948	9.3
^{195}Au	G. Wilkinson	[1152]	Berkeley	USA	1948	9.3
^{196}Au	E. McMillan	[1141]	Berkeley	USA	1937	4.1
^{197}Au	A.J. Dempster	[596]	Chicago	USA	1935	3.1
^{198}Au	M.L. Pool	[488]	Michigan	USA	1937	4.1
^{199}Au	E. McMillan	[1141]	Berkeley	USA	1937	4.1
^{200}Au	F.D.S. Butement	[1169]	Harwell	UK	1951	8.1
^{201}Au	F.D.S. Butement	[1170]	Harwell	UK	1952	8.1
^{202}Au	T.E. Ward	[1171]	Arkansas	USA	1967	7.4
^{203}Au	F.D.S. Butement	[1170]	Harwell	UK	1952	8.1
^{204}Au	A. Pakkanen	[1172]	Jyväskylä	Finland	1972	7.4
^{205}Au	Ch. Wennemann	[1173]	Darmstadt	Germany	1994	14.3

(continued)

Table 17.1 (continued)

Isotope	First author	Ref.	Laboratory	Country	Year	Table
^{206}Au	A. I. Morales	[1128]	Darmstadt	Germany	2011	15.3
^{207}Au	H. Alvarez-Pol	[1159]	Darmstadt	Germany	2010	15.3
^{208}Au	H. Alvarez-Pol	[1159]	Darmstadt	Germany	2010	15.3
^{209}Au	H. Alvarez-Pol	[1159]	Darmstadt	Germany	2010	15.3
^{210}Au	H. Alvarez-Pol	[1159]	Darmstadt	Germany	2010	15.3
^{171}Hg	H. Kettunen	[1160]	Jyväskylä	Finland	2004	11.10
^{172}Hg	D. Seweryniak	[1174]	Argonne	USA	1999	11.10
^{173}Hg	D. Seweryniak	[1174]	Argonne	USA	1999	11.10
^{174}Hg	J. Uusitalo	[1175]	Jyväskylä	Finland	1997	11.10
^{175}Hg	J.R.H. Schneider	[1162]	Darmstadt	Germany	1983	11.9
^{176}Hg	J.R.H. Schneider	[1162]	Darmstadt	Germany	1983	11.9
^{177}Hg	C. Cabot	[1176]	Orsay	France	1975	11.7
^{178}Hg	P.G. Hansen	[1177]	CERN	Switzerland	1971	10.7
^{179}Hg	P.G. Hansen	[1166]	CERN	Switzerland	1970	10.7
^{180}Hg	P.G. Hansen	[1166]	CERN	Switzerland	1970	10.7
^{181}Hg	P.G. Hansen	[205]	CERN	Switzerland	1969	10.7
^{182}Hg	A.G. Demin	[1178]	Dubna	Russia	1968	11.6
^{183}Hg	P.G. Hansen	[205]	CERN	Switzerland	1969	10.7
^{184}Hg	P.G. Hansen	[205]	CERN	Switzerland	1969	10.7
^{185}Hg	G. Albouy	[1150]	Orsay	France	1960	10.8
^{186}Hg	G. Albouy	[1150]	Orsay	France	1960	10.8
^{187}Hg	G. Albouy	[1150]	Orsay	France	1960	10.8
^{188}Hg	N. Poffe	[1179]	Orsay	France	1960	10.8
^{189}Hg	N. Poffe	[1179]	Orsay	France	1960	10.8
^{190}Hg	G. Albouy	[1167]	Orsay	France	1959	10.8
^{191}Hg	L.P. Gillon	[1168]	Harvard	USA	1954	10.2
^{192}Hg	R.W. Fink	[1180]	Rochester	USA	1952	10.2
^{193}Hg	R.W. Fink	[1180]	Rochester	USA	1952	10.2
^{194}Hg	W.J. Tomlinson III	[1181]	Harvard	USA	1962	9.1
^{195}Hg	R.W. Fink	[1180]	Rochester	USA	1952	10.2
^{196}Hg	F.W. Aston	[1182]	Cambridge	UK	1927	3.3
^{197}Hg	R. Sherr	[1183]	Harvard	USA	1941	4.1
^{198}Hg	F.W. Aston	[1184]	Cambridge	UK	1925	3.3
^{199}Hg	F.W. Aston	[1184]	Cambridge	UK	1925	3.3
^{200}Hg	F.W. Aston	[1184]	Cambridge	UK	1925	3.3
^{201}Hg	F.W. Aston	[1184]	Cambridge	UK	1925	3.3
^{202}Hg	F.W. Aston	[97]	Cambridge	UK	1920	3.2
^{203}Hg	G. Friedlander	[1185]	Berkeley	USA	1943	4.1
^{204}Hg	F.W. Aston	[97]	Cambridge	UK	1920	3.2

(continued)

Table 17.1 (continued)

Isotope	First author	Ref.	Laboratory	Country	Year	Table
^{205}Hg	R.S. Krishnan	[1186]	Cambridge	UK	1940	4.2
^{206}Hg	M.J. Nurmia	[9]	Helsinki	Finland	1961	2.3
^{207}Hg	S. Mirzadeh	[1187]	Brookhaven	USA	1982	7.4
^{208}Hg	L. Zhang	[1188]	Lanzhou	China	1994	14.1
^{209}Hg	L. Zhang	[1189]	Lanzhou	China	1998	14.1
^{210}Hg	M. Pfützner	[1190]	Darmstadt	Germany	1998	15.3
^{211}Hg	H. Alvarez-Pol	[1159]	Darmstadt	Germany	2010	15.3
^{212}Hg	H. Alvarez-Pol	[1159]	Darmstadt	Germany	2010	15.3
^{213}Hg	H. Alvarez-Pol	[1159]	Darmstadt	Germany	2010	15.3
^{214}Hg	H. Alvarez-Pol	[1159]	Darmstadt	Germany	2010	15.3
^{215}Hg	H. Alvarez-Pol	[1159]	Darmstadt	Germany	2010	15.3
^{216}Hg	H. Alvarez-Pol	[1159]	Darmstadt	Germany	2010	15.3
^{176}Tl	H. Kettunen	[1160]	Jyväskylä	Finland	2004	16.3
^{177}Tl	G. Poli	[1191]	Argonne	USA	1999	16.3
^{178}Tl	M.P. Carpenter	[1192]	Argonne	USA	1997	11.10
^{179}Tl	J.R.H. Schneider	[1162]	Darmstadt	Germany	1983	11.9
^{180}Tl	Yu.A. Lazarev	[1193]	Dubna	Russia	1987	11.11
^{181}Tl	K.S. Toth	[1194]	Argonne	USA	1996	11.10
^{182}Tl	A. Bouldjedri	[1195]	CERN	Switzerland	1991	10.7
^{183}Tl	U.J. Schrewe	[1196]	Darmstadt	Germany	1980	11.8
^{184}Tl	K.S. Toth	[1197]	Oak Ridge	USA	1976	11.8
^{185}Tl	K.S. Toth	[1197]	Oak Ridge	USA	1976	11.8
^{186}Tl	J.H. Hamilton	[1198]	Oak Ridge	USA	1975	11.2
^{187}Tl	K.S. Toth	[1197]	Oak Ridge	USA	1976	11.8
^{188}Tl	J. Vandlik	[1199]	Dubna	Russia	1970	10.4
^{189}Tl	T.B. Vandlik	[1200]	Dubna	Russia	1972	10.8
^{190}Tl	J. Vandlik	[1199]	Dubna	Russia	1970	10.4
^{191}Tl	J. Vandlik	[1201]	Dubna	Russia	1974	10.4
^{192}Tl	G. Andersson	[1202]	Uppsala	Sweden	1961	10.2
^{193}Tl	K.F. Chackett	[1203]	Birmingham	UK	1960	11.1
^{194}Tl	B. Jung	[1204]	Uppsala	Sweden	1960	10.2
^{195}Tl	J.D. Knight	[1205]	Brookhaven	USA	1955	9.5
^{196}Tl	G. Andersson	[1206]	Uppsala	Sweden	1955	10.2
^{197}Tl	G. Andersson	[1206]	Uppsala	Sweden	1955	10.2
^{198}Tl	D.A. Orth	[1207]	Berkeley	USA	1949	9.3
^{199}Tl	D.A. Orth	[1207]	Berkeley	USA	1949	9.3
^{200}Tl	D.A. Orth	[1207]	Berkeley	USA	1949	9.3
^{201}Tl	H.M. Neumann	[1208]	Berkeley	USA	1950	10.2
^{202}Tl	R.S. Krishnan	[1186]	Cambridge	UK	1940	4.2
^{203}Tl	H. Schüler	[1209]	Berlin	Germany	1931	3.5

(continued)

Table 17.1 (continued)

Isotope	First author	Ref.	Laboratory	Country	Year	Table
^{204}Tl	G. Harbottle	[1210]	Brookhaven	USA	1953	7.2
^{205}Tl	H. Schüler	[1209]	Berlin	Germany	1931	3.5
^{206}Tl	P. Preiswerk	[1211]	Paris	France	1935	2.3
^{207}Tl	O. Hahn	[1212]	Berlin	Germany	1908	2.2
^{208}Tl	O. Hahn	[1213]	Berlin	Germany	1909	2.2
^{209}Tl	F. Hagemann	[1214]	Argonne	USA	1950	7.2
^{210}Tl	O. Hahn	[1215]	Berlin	Germany	1909	2.2
^{211}Tl	M. Pfützner	[1190]	Darmstadt	Germany	1998	15.3
^{212}Tl	M. Pfützner	[1190]	Darmstadt	Germany	1998	15.3
^{213}Tl	L. Chen	[1216]	Darmstadt	Germany	2010	15.3
^{214}Tl	H. Alvarez-Pol	[1159]	Darmstadt	Germany	2010	15.3
^{215}Tl	H. Alvarez-Pol	[1159]	Darmstadt	Germany	2010	15.3
^{216}Tl	H. Alvarez-Pol	[1159]	Darmstadt	Germany	2010	15.3
^{217}Tl	H. Alvarez-Pol	[1159]	Darmstadt	Germany	2010	15.3
^{179}Pb	A.N. Andreyev	[1217]	Darmstadt	Germany	2010	11.9
^{180}Pb	K.S. Toth	[1218]	Berkeley	USA	1996	11.5
^{181}Pb	K.S. Toth	[1219]	Berkeley	USA	1989	11.7
^{182}Pb	J.G. Keller	[1220]	Darmstadt	Germany	1986	11.9
^{183}Pb	U.J. Schrewe	[1196]	Darmstadt	Germany	1980	11.8
^{184}Pb	J.P. Dufour	[1221]	Orsay	France	1980	11.7
^{185}Pb	C. Cabot	[1163]	Orsay	France	1975	11.7
^{186}Pb	H. Gauvin	[1222]	Orsay	France	1972	11.7
^{187}Pb	H. Gauvin	[1222]	Orsay	France	1972	11.7
^{188}Pb	H. Gauvin	[1222]	Orsay	France	1972	11.7
^{189}Pb	H. Gauvin	[1222]	Orsay	France	1972	11.7
^{190}Pb	H. Gauvin	[1222]	Orsay	France	1972	11.7
^{191}Pb	Y. Le Beyec	[1223]	Berkeley	USA	1974	11.7
^{192}Pb	Y. Le Beyec	[1223]	Berkeley	USA	1974	11.7
^{193}Pb	J.O. Newton	[1224]	Manchester	UK	1974	11.2
^{194}Pb	B. Jung	[1204]	Uppsala	Sweden	1960	10.2
^{195}Pb	G. Andersson	[1225]	Uppsala	Sweden	1957	10.2
^{196}Pb	G. Andersson	[1225]	Uppsala	Sweden	1957	10.2
^{197}Pb	G. Andersson	[1206]	Uppsala	Sweden	1955	10.2
^{198}Pb	G. Andersson	[1206]	Uppsala	Sweden	1955	10.2
^{199}Pb	H.M. Neumann	[1208]	Berkeley	USA	1950	10.2
^{200}Pb	H.M. Neumann	[1208]	Berkeley	USA	1950	10.2
^{201}Pb	H.M. Neumann	[1208]	Berkeley	USA	1950	10.2
^{202}Pb	D. Maeder	[1226]	Amsterdam	Netherlands	1954	9.4
^{203}Pb	W. Maurer	[1227]	Berlin	Germany	1942	4.1

(continued)

Table 17.1 (continued)

Isotope	First author	Ref.	Laboratory	Country	Year	Table
^{204}Pb	H. Schüler	[1228]	Berlin	Germany	1932	3.5
^{205}Pb	J.R. Huizenga	[1229]	Argonne	USA	1954	9.6
^{206}Pb	F.W. Aston	[1182]	Cambridge	UK	1927	3.3
^{207}Pb	F.W. Aston	[1182]	Cambridge	UK	1927	3.3
^{208}Pb	F.W. Aston	[1182]	Cambridge	UK	1927	3.3
^{209}Pb	R.S. Krishnan	[1186]	Cambridge	UK	1940	4.2
^{210}Pb	K.A. Hofmann	[1230]	Munich	Germany	1900	2.1
^{211}Pb	A. Debierne	[1231]	Paris	France	1904	2.1
^{212}Pb	E. Rutherford	[3]	McGill	Canada	1900	2.1
^{213}Pb	F.D.S. Butement	[1232]	Liverpool	UK	1964	10.3
^{214}Pb	E. Rutherford	[4]	McGill	Canada	1904	2.1
^{215}Pb	M. Pfützner	[1190]	Darmstadt	Germany	1998	15.3
^{216}Pb	H. Alvarez-Pol	[1159]	Darmstadt	Germany	2010	15.3
^{217}Pb	H. Alvarez-Pol	[1159]	Darmstadt	Germany	2010	15.3
^{218}Pb	H. Alvarez-Pol	[1159]	Darmstadt	Germany	2010	15.3
^{219}Pb	H. Alvarez-Pol	[1159]	Darmstadt	Germany	2010	15.3
^{220}Pb	H. Alvarez-Pol	[1159]	Darmstadt	Germany	2010	15.3
^{184}Bi	A.N. Andreyev	[1233]	Darmstadt	Germany	2003	11.9
^{185}Bi	C.N. Davids	[1234]	Argonne	USA	1996	16.3
^{186}Bi	J.C. Batchelder	[1235]	Argonne	USA	1997	11.10
^{187}Bi	J.C. Batchelder	[1236]	Argonne	USA	1999	11.10
^{188}Bi	U.J. Schrewe	[1196]	Darmstadt	Germany	1980	11.8
^{189}Bi	H. Gauvin	[1237]	Orsay	France	1973	11.7
^{190}Bi	H. Gauvin	[1222]	Orsay	France	1972	11.7
^{191}Bi	H. Gauvin	[1222]	Orsay	France	1972	11.7
^{192}Bi	N.I. Tarantin	[1238]	Dubna	Russia	1971	11.8
^{193}Bi	N.I. Tarantin	[1238]	Dubna	Russia	1971	11.8
^{194}Bi	N.I. Tarantin	[1238]	Dubna	Russia	1971	11.8
^{195}Bi	N.I. Tarantin	[1238]	Dubna	Russia	1971	11.8
^{196}Bi	S. Chojnacki	[1239]	Dubna	Russia	1976	11.2
^{197}Bi	N.I. Tarantin	[1238]	Dubna	Russia	1971	11.8
^{198}Bi	H.M. Neumann	[1208]	Berkeley	USA	1950	10.2
^{199}Bi	H.M. Neumann	[1208]	Berkeley	USA	1950	10.2
^{200}Bi	H.M. Neumann	[1208]	Berkeley	USA	1950	10.2
^{201}Bi	H.M. Neumann	[1208]	Berkeley	USA	1950	10.2
^{202}Bi	D.G. Karraker	[1240]	Berkeley	USA	1951	10.2
^{203}Bi	H.M. Neumann	[1208]	Berkeley	USA	1950	10.2
^{204}Bi	J.J. Howland	[1241]	Berkeley	USA	1947	9.2
^{205}Bi	D.G. Karraker	[1240]	Berkeley	USA	1951	10.2

(continued)

Table 17.1 (continued)

Isotope	First author	Ref.	Laboratory	Country	Year	Table
^{206}Bi	J.J. Howland	[1241]	Berkeley	USA	1947	9.2
^{207}Bi	L.S. Germain	[1242]	Berkeley	USA	1950	9.3
^{208}Bi	J.A. Harvey	[1243]	MIT	USA	1953	9.8
^{209}Bi	F.W. Aston	[635]	Cambridge	UK	1924	3.2
^{210}Bi	E. Rutherford	[1244]	McGill	Canada	1905	2.2
^{211}Bi	E. Rutherford	[1245]	McGill	Canada	1905	2.2
^{212}Bi	E. Rutherford	[1245]	McGill	Canada	1905	2.2
^{213}Bi	F. Hagemann	[1246]	Argonne	USA	1947	7.1
^{214}Bi	E. Rutherford	[4]	McGill	Canada	1904	2.1
^{215}Bi	E.K. Hyde	[1247]	Berkeley	USA	1953	2.3
^{216}Bi	D.G. Burke	[1248]	CERN	Switzerland	1989	10.7
^{217}Bi	M. Pfützner	[1190]	Darmstadt	Germany	1998	15.3
^{218}Bi	M. Pfützner	[1190]	Darmstadt	Germany	1998	15.3
^{219}Bi	H. Alvarez-Pol	[1159]	Darmstadt	Germany	2010	15.3
^{220}Bi	H. Alvarez-Pol	[1159]	Darmstadt	Germany	2010	15.3
^{221}Bi	H. Alvarez-Pol	[1159]	Darmstadt	Germany	2010	15.3
^{222}Bi	H. Alvarez-Pol	[1159]	Darmstadt	Germany	2010	15.3
^{223}Bi	H. Alvarez-Pol	[1159]	Darmstadt	Germany	2010	15.3
^{224}Bi	H. Alvarez-Pol	[1159]	Darmstadt	Germany	2010	15.3
^{186}Po	A.N. Andreyev	[1249]	Darmstadt	Germany	2005	11.9
^{187}Po	A.N. Andreyev	[1249]	Darmstadt	Germany	2005	11.9
^{188}Po	A.N. Andreyev	[1250]	Darmstadt	Germany	1999	11.9
^{189}Po	A.N. Andreyev	[1250]	Darmstadt	Germany	1999	11.9
^{190}Po	J.C. Batchelder	[1251]	Berkeley	USA	1996	11.5
^{191}Po	A.B. Quint	[1252]	Darmstadt	Germany	1993	11.9
^{192}Po	S. Della Negra	[1253]	Orsay	France	1977	11.7
^{193}Po	A. Siivola	[1254]	Berkeley	USA	1967	11.6
^{194}Po	A. Siivola	[1254]	Berkeley	USA	1967	11.6
^{195}Po	A. Siivola	[1254]	Berkeley	USA	1967	11.6
^{196}Po	A. Siivola	[1254]	Berkeley	USA	1967	11.6
^{197}Po	C. Brun	[1255]	Orsay	France	1965	10.5
^{198}Po	C. Brun	[1255]	Orsay	France	1965	10.5
^{199}Po	C. Brun	[1255]	Orsay	France	1965	10.5
^{200}Po	D.G. Karraker	[1256]	Berkeley	USA	1951	10.2
^{201}Po	D.G. Karraker	[1256]	Berkeley	USA	1951	10.2
^{202}Po	D.G. Karraker	[1240]	Berkeley	USA	1951	10.2
^{203}Po	D.G. Karraker	[1240]	Berkeley	USA	1951	10.2
^{204}Po	D.G. Karraker	[1240]	Berkeley	USA	1951	10.2
^{205}Po	D.G. Karraker	[1240]	Berkeley	USA	1951	10.2

(continued)

Table 17.1 (continued)

Isotope	First author	Ref.	Laboratory	Country	Year	Table
^{206}Po	J.J. Howland	[1241]	Berkeley	USA	1947	9.2
^{207}Po	J.J. Howland	[1241]	Berkeley	USA	1947	9.2
^{208}Po	J.J. Howland	[1241]	Berkeley	USA	1947	9.2
^{209}Po	E.L. Kelly	[1257]	Berkeley	USA	1949	9.1
^{210}Po	P. Curie	[1258]	Paris	France	1898	2.1
^{211}Po	E. Marsden	[1259]	Manchester	UK	1913	2.3
^{212}Po	O. Hahn	[1260]	McGill	Canada	1906	2.2
^{213}Po	F. Hagemann	[1246]	Argonne	USA	1947	7.1
^{214}Po	K. Fajans	[1261]	Karlsruhe	Germany	1912	2.2
^{215}Po	H. Geiger	[1262]	Manchester	UK	1911	2.2
^{216}Po	H. Geiger	[1263]	Manchester	UK	1910	2.2
^{217}Po	F.F. Momyer	[1264]	Berkeley	USA	1956	10.3
^{218}Po	E. Rutherford	[4]	McGill	Canada	1904	2.1
^{219}Po	M. Pfützner	[1190]	Darmstadt	Germany	1998	15.3
^{220}Po	M. Pfützner	[1190]	Darmstadt	Germany	1998	15.3
^{221}Po	L. Chen	[1216]	Darmstadt	Germany	2010	15.3
^{222}Po	L. Chen	[1216]	Darmstadt	Germany	2010	15.3
^{223}Po	H. Alvarez-Pol	[1159]	Darmstadt	Germany	2010	15.3
^{224}Po	H. Alvarez-Pol	[1159]	Darmstadt	Germany	2010	15.3
^{225}Po	H. Alvarez-Pol	[1159]	Darmstadt	Germany	2010	15.3
^{226}Po	H. Alvarez-Pol	[1159]	Darmstadt	Germany	2010	15.3
^{227}Po	H. Alvarez-Pol	[1159]	Darmstadt	Germany	2010	15.3
^{191}At	H. Kettunen	[1265]	Jyväskylä	Finland	2003	11.10
^{192}At	A.N. Andreyev	[1266]	Darmstadt	Germany	2006	11.9
^{193}At	H. Kettunen	[1265]	Jyväskylä	Finland	2003	11.10
^{194}At	A.N. Andreyev	[1267]	Darmstadt	Germany	2009	11.9
^{195}At	Y. Tagaya	[1268]	RIKEN	Japan	1999	11.10
^{196}At	W. Treytl	[1269]	Berkeley	USA	1967	11.6
^{197}At	W. Treytl	[1269]	Berkeley	USA	1967	11.6
^{198}At	W. Treytl	[1269]	Berkeley	USA	1967	11.6
^{199}At	W. Treytl	[1269]	Berkeley	USA	1967	11.6
^{200}At	R.W. Hoff	[1270]	Berkeley	USA	1963	11.5
^{201}At	R.W. Hoff	[1270]	Berkeley	USA	1963	11.5
^{202}At	R.M. Latimer	[1271]	Berkeley	USA	1961	11.5
^{203}At	G.W. Barton	[1272]	Berkeley	USA	1951	10.2
^{204}At	R.M. Latimer	[1271]	Berkeley	USA	1961	11.5
^{205}At	G.W. Barton	[1272]	Berkeley	USA	1951	10.2
^{206}At	R.M. Latimer	[1271]	Berkeley	USA	1961	11.5
^{207}At	G.W. Barton	[1272]	Berkeley	USA	1951	10.2

(continued)

Table 17.1 (continued)

Isotope	First author	Ref.	Laboratory	Country	Year	Table
^{208}At	E.K. Hyde	[1273]	Berkeley	USA	1950	10.3
^{209}At	G.W. Barton	[1272]	Berkeley	USA	1951	10.2
^{210}At	E.L. Kelly	[1257]	Berkeley	USA	1949	9.1
^{211}At	D.R. Corson	[1274]	Berkeley	USA	1940	4.3
^{212}At	M.M. Winn	[1275]	Birmingham	UK	1954	9.1
^{213}At	R.L. Hahn	[1276]	Oak Ridge	USA	1968	10.3
^{214}At	W.W. Meinke	[1277]	Berkeley	USA	1949	10.3
^{215}At	B. Karlik	[1278]	Wien	Austria	1944	2.3
^{216}At	A. Ghiorso	[1279]	Berkeley	USA	1948	10.3
^{217}At	F. Hagemann	[1246]	Argonne	USA	1947	7.1
^{218}At	H. Hulubei	[1280]	Paris	France	1939	2.3
^{219}At	E.K. Hyde	[1247]	Berkeley	USA	1953	2.3
^{220}At	C.F. Liang	[1281]	Orsay	France	1989	10.8
^{221}At	D.G. Burke	[1248]	CERN	Switzerland	1989	10.7
^{222}At	D.G. Burke	[1248]	CERN	Switzerland	1989	10.7
^{223}At	D.G. Burke	[1248]	CERN	Switzerland	1989	10.7
^{224}At	L. Chen	[1216]	Darmstadt	Germany	2010	15.3
^{225}At	H. Alvarez-Pol	[1159]	Darmstadt	Germany	2010	15.3
^{226}At	H. Alvarez-Pol	[1159]	Darmstadt	Germany	2010	15.3
^{227}At	H. Alvarez-Pol	[1159]	Darmstadt	Germany	2010	15.3
^{228}At	H. Alvarez-Pol	[1159]	Darmstadt	Germany	2010	15.3
^{229}At	H. Alvarez-Pol	[1159]	Darmstadt	Germany	2010	15.3
^{193}Rn	A.N. Andreyev	[1282]	Darmstadt	Germany	2006	11.9
^{194}Rn	A.N. Andreyev	[1282]	Darmstadt	Germany	2006	11.9
^{195}Rn	H. Kettunen	[1283]	Jyväskylä	Finland	2001	11.10
^{196}Rn	K. Morita	[1284]	RIKEN	Japan	1995	11.10
^{197}Rn	K. Morita	[1284]	RIKEN	Japan	1995	11.10
^{198}Rn	F. Calaprice	[1285]	CERN	Switzerland	1984	10.7
^{199}Rn	A.C. DiRienzo	[1286]	Brookhaven	USA	1980	11.10
^{200}Rn	P. Hornshoj	[1287]	CERN	Switzerland	1971	10.7
^{201}Rn	K. Valli	[1288]	Berkeley	USA	1967	11.6
^{202}Rn	K. Valli	[1288]	Berkeley	USA	1967	11.6
^{203}Rn	K. Valli	[1288]	Berkeley	USA	1967	11.6
^{204}Rn	K. Valli	[1288]	Berkeley	USA	1967	11.6
^{205}Rn	K. Valli	[1288]	Berkeley	USA	1967	11.6
^{206}Rn	W.E. Burcham	[1289]	Birmingham	UK	1954	11.5
^{207}Rn	W.E. Burcham	[1289]	Birmingham	UK	1954	11.5

(continued)

Table 17.1 (continued)

Isotope	First author	Ref.	Laboratory	Country	Year	Table
^{208}Rn	F.F. Momyer	[1290]	Berkeley	USA	1955	10.3
^{209}Rn	F.F. Momyer	[1291]	Berkeley	USA	1952	10.3
^{210}Rn	F.F. Momyer	[1291]	Berkeley	USA	1952	10.3
^{211}Rn	F.F. Momyer	[1291]	Berkeley	USA	1952	10.3
^{212}Rn	E.K. Hyde	[1273]	Berkeley	USA	1950	10.3
^{213}Rn	H. Rotter	[1292]	Dubna	Russia	1967	11.5
^{214}Rn	D.F. Torgerson	[1293]	Yale	USA	1970	11.6
^{215}Rn	W.W. Meinke	[1294]	Berkeley	USA	1952	10.3
^{216}Rn	W.W. Meinke	[1277]	Berkeley	USA	1949	10.3
^{217}Rn	W.W. Meinke	[1277]	Berkeley	USA	1949	10.3
^{218}Rn	M.H. Studier	[1295]	Berkeley	USA	1948	9.2
^{219}Rn	A. Debierne	[1296]	Paris	France	1903	2.1
^{220}Rn	E. Rutherford	[1297]	McGill	Canada	1900	2.1
^{221}Rn	F.F. Momyer	[1264]	Berkeley	USA	1956	10.3
^{222}Rn	P. Curie	[1298]	Paris	France	1899	2.1
^{223}Rn	F.D.S. Butement	[5]	Liverpool	UK	1964	10.3
^{224}Rn	F.D.S. Butement	[5]	Liverpool	UK	1964	10.3
^{225}Rn	P.G. Hansen	[205]	CERN	Switzerland	1969	10.7
^{226}Rn	P.G. Hansen	[205]	CERN	Switzerland	1969	10.7
^{227}Rn	M.J.G. Borge	[1299]	CERN	Switzerland	1986	10.7
^{228}Rn	M.J.G. Borge	[1300]	CERN	Switzerland	1989	10.7
^{229}Rn	D. Neidherr	[1301]	CERN	Switzerland	2009	10.7
^{230}Rn	H. Alvarez-Pol	[1159]	Darmstadt	Germany	2010	15.3
^{231}Rn	H. Alvarez-Pol	[1159]	Darmstadt	Germany	2010	15.3
^{197}Fr	Z. Kalaninova	[1302]	Darmstadt	Germany	2013	11.9
^{198}Fr	J. Uusitalo	[1303]	Jyväskylä	Finland	2013	11.10
^{199}Fr	Y. Tagaya	[1268]	RIKEN	Japan	1999	11.10
^{200}Fr	K. Morita	[1284]	RIKEN	Japan	1995	11.10
^{201}Fr	G.T. Ewan	[1304]	CERN	Switzerland	1980	10.7
^{202}Fr	G.T. Ewan	[1304]	CERN	Switzerland	1980	10.7
^{203}Fr	K. Valli	[1305]	Berkeley	USA	1967	11.6
^{204}Fr	R.D. Griffioen	[1306]	Berkeley	USA	1964	11.6
^{205}Fr	R.D. Griffioen	[1306]	Berkeley	USA	1964	11.6
^{206}Fr	R.D. Griffioen	[1306]	Berkeley	USA	1964	11.6
^{207}Fr	R.D. Griffioen	[1306]	Berkeley	USA	1964	11.6
^{208}Fr	R.D. Griffioen	[1306]	Berkeley	USA	1964	11.6
^{209}Fr	R.D. Griffioen	[1306]	Berkeley	USA	1964	11.6
^{210}Fr	R.D. Griffioen	[1306]	Berkeley	USA	1964	11.6
^{211}Fr	R.D. Griffioen	[1306]	Berkeley	USA	1964	11.6

(continued)

Table 17.1 (continued)

Isotope	First author	Ref.	Laboratory	Country	Year	Table
^{212}Fr	E.K. Hyde	[1273]	Berkeley	USA	1950	10.3
^{213}Fr	R.D. Griffioen	[1306]	Berkeley	USA	1964	11.6
^{214}Fr	H. Rotter	[1292]	Dubna	Russia	1967	11.5
^{215}Fr	J. Borggreen	[1307]	Berkeley	USA	1970	11.6
^{216}Fr	J. Borggreen	[1307]	Berkeley	USA	1970	11.6
^{217}Fr	R.L. Hahn	[1276]	Oak Ridge	USA	1968	10.3
^{218}Fr	W.W. Meinke	[1277]	Berkeley	USA	1949	10.3
^{219}Fr	A. Ghiorso	[1279]	Berkeley	USA	1948	10.3
^{220}Fr	A. Ghiorso	[1279]	Berkeley	USA	1948	10.3
^{221}Fr	F. Hagemann	[1246]	Argonne	USA	1947	7.1
^{222}Fr	L. Westgaard	[1]	CERN	Switzerland	1975	10.7
^{223}Fr	M. Perey	[1308]	Paris	France	1939	2.3
^{224}Fr	P.G. Hansen	[205]	CERN	Switzerland	1969	10.7
^{225}Fr	P.G. Hansen	[205]	CERN	Switzerland	1969	10.7
^{226}Fr	P.G. Hansen	[205]	CERN	Switzerland	1969	10.7
^{227}Fr	R. Klapisch	[147]	CERN	Switzerland	1972	10.7
^{228}Fr	R. Klapisch	[147]	CERN	Switzerland	1972	10.7
^{229}Fr	H.L. Ravn	[441]	CERN	Switzerland	1975	10.7
^{230}Fr	W. Kurcewicz	[1309]	CERN	Switzerland	1987	10.7
^{231}Fr	P. Hill	[1310]	CERN	Switzerland	1985	10.7
^{232}Fr	K.A. Mezlev	[1311]	Leningrad	Russia	1990	10.8
^{233}Fr	H. Alvarez-Pol	[1159]	Darmstadt	Germany	2010	15.3
^{201}Ra	J. Uusitalo	[1312]	Jyväskylä	Finland	2005	11.10
^{202}Ra	J. Uusitalo	[1312]	Jyväskylä	Finland	2005	11.10
^{203}Ra	M. Leino	[1313]	Jyväskylä	Finland	1996	11.10
^{204}Ra	M.J. Leddy	[1314]	Argonne	USA	1995	11.10
^{205}Ra	F.P. Heßberger	[1315]	Darmstadt	Germany	1987	11.9
^{206}Ra	K. Valli	[1316]	Berkeley	USA	1967	11.6
^{207}Ra	K. Valli	[1316]	Berkeley	USA	1967	11.6
^{208}Ra	K. Valli	[1316]	Berkeley	USA	1967	11.6
^{209}Ra	K. Valli	[1316]	Berkeley	USA	1967	11.6
^{210}Ra	K. Valli	[1316]	Berkeley	USA	1967	11.6
^{211}Ra	K. Valli	[1316]	Berkeley	USA	1967	11.6
^{212}Ra	K. Valli	[1316]	Berkeley	USA	1967	11.6
^{213}Ra	F.F. Momyer	[1290]	Berkeley	USA	1955	10.3
^{214}Ra	H. Rotter	[1292]	Dubna	Russia	1967	11.5
^{215}Ra	H. Rotter	[1292]	Dubna	Russia	1967	11.5
^{216}Ra	T. Nomura	[1317]	RIKEN	Japan	1972	11.5
^{217}Ra	D.F. Torgerson	[1293]	Yale	USA	1970	11.6

(continued)

Table 17.1 (continued)

Isotope	First author	Ref.	Laboratory	Country	Year	Table
^{218}Ra	D.F. Torgerson	[1293]	Yale	USA	1970	11.6
^{219}Ra	W.W. Meinke	[1294]	Berkeley	USA	1952	10.3
^{220}Ra	W.W. Meinke	[1277]	Berkeley	USA	1949	10.3
^{221}Ra	W.W. Meinke	[1277]	Berkeley	USA	1949	10.3
^{222}Ra	M.H. Studier	[1295]	Berkeley	USA	1948	9.2
^{223}Ra	T. Godlewski	[1318]	McGill	Canada	1905	2.2
^{224}Ra	E. Rutherford	[1319]	McGill	Canada	1902	2.1
^{225}Ra	F. Hagemann	[1246]	Argonne	USA	1947	7.1
^{226}Ra	P. Curie	[1320]	Paris	France	1898	2.1
^{227}Ra	J.P. Butler	[1321]	Chalk River	Canada	1953	7.2
^{228}Ra	O. Hahn	[1322]	Berlin	Germany	1907	2.2
^{229}Ra	H.L. Ravn	[441]	CERN	Switzerland	1975	10.7
^{230}Ra	J. Gilat	[1323]	Brookhaven	USA	1978	7.4
^{231}Ra	P. Hill	[1310]	CERN	Switzerland	1985	10.7
^{232}Ra	S.A. Ahmad	[1324]	CERN	Switzerland	1983	10.7
^{233}Ra	K.A. Mezlev	[1311]	Leningrad	Russia	1990	10.8
^{234}Ra	K.A. Mezlev	[1311]	Leningrad	Russia	1990	10.8
^{205}Ac	Z. Y. Zhang	[1325]	Lanzhou	China	2014	11.10
^{206}Ac	K. Eskola	[1326]	Jyväskylä	Finland	1998	11.10
^{207}Ac	M. Leino	[1327]	Jyväskylä	Finland	1994	11.10
^{208}Ac	M. Leino	[1327]	Jyväskylä	Finland	1994	11.10
^{209}Ac	K. Valli	[1328]	Berkeley	USA	1968	11.6
^{210}Ac	K. Valli	[1328]	Berkeley	USA	1968	11.6
^{211}Ac	K. Valli	[1328]	Berkeley	USA	1968	11.6
^{212}Ac	K. Valli	[1328]	Berkeley	USA	1968	11.6
^{213}Ac	K. Valli	[1328]	Berkeley	USA	1968	11.6
^{214}Ac	K. Valli	[1328]	Berkeley	USA	1968	11.6
^{215}Ac	K. Valli	[1328]	Berkeley	USA	1968	11.6
^{216}Ac	H. Rotter	[1292]	Dubna	Russia	1967	11.5
^{217}Ac	T. Nomura	[1317]	RIKEN	Japan	1972	11.5
^{218}Ac	J. Borggreen	[1307]	Berkeley	USA	1970	11.6
^{219}Ac	J. Borggreen	[1307]	Berkeley	USA	1970	11.6
^{220}Ac	J. Borggreen	[1307]	Berkeley	USA	1970	11.6
^{221}Ac	R.L. Hahn	[1276]	Oak Ridge	USA	1968	10.3
^{222}Ac	W.W. Meinke	[1277]	Berkeley	USA	1949	10.3
^{223}Ac	A. Ghiorso	[1279]	Berkeley	USA	1948	10.3
^{224}Ac	A. Ghiorso	[1279]	Berkeley	USA	1948	10.3
^{225}Ac	F. Hagemann	[1246]	Argonne	USA	1947	7.1
^{226}Ac	W.W. Meinke	[1329]	Berkeley	USA	1950	9.3

(continued)

Table 17.1 (continued)

Isotope	First author	Ref.	Laboratory	Country	Year	Table
^{227}Ac	M. Curie	[1330]	Paris	France	1911	2.2
^{228}Ac	O. Hahn	[1331]	Berlin	Germany	1908	2.2
^{229}Ac	F. Depocas	[1332]	Chalk River	Canada	1952	7.2
^{230}Ac	K. Chayawattanangkur	[1333]	Mainz	Germany	1973	8.1
^{231}Ac	K. Chayawattanangkur	[1333]	Mainz	Germany	1973	8.1
^{232}Ac	K.-L. Gippert	[1334]	Darmstadt	Germany	1986	14.3
^{233}Ac	Y.Y. Chu	[1335]	Brookhaven	USA	1983	10.3
^{234}Ac	K.-L. Gippert	[1334]	Darmstadt	Germany	1986	14.3
^{235}Ac	J. Taieb	[1336]	Darmstadt	Germany	2003	15.3
^{236}Ac	L. Chen	[1216]	Darmstadt	Germany	2010	15.3
^{208}Th	J.A. Heredia	[1337]	Darmstadt	Germany	2010	11.9
^{209}Th	H. Ikezoe	[1338]	Tokai	Japan	1996	11.10
^{210}Th	J. Uusitalo	[1339]	Jyväskylä	Finland	1995	11.10
^{211}Th	J. Uusitalo	[1339]	Jyväskylä	Finland	1995	11.10
^{212}Th	D. Vermeulen	[1340]	Darmstadt	Germany	1980	11.9
^{213}Th	K. Valli	[1341]	Berkeley	USA	1968	11.6
^{214}Th	K. Valli	[1341]	Berkeley	USA	1968	11.6
^{215}Th	K. Valli	[1341]	Berkeley	USA	1968	11.6
^{216}Th	K. Valli	[1341]	Berkeley	USA	1968	11.6
^{217}Th	K. Valli	[1341]	Berkeley	USA	1968	11.6
^{218}Th	K. Hiruta	[1342]	RIKEN	Japan	1973	11.5
^{219}Th	O. Häusser	[1343]	Chalk River	Canada	1973	11.5
^{220}Th	O. Häusser	[1343]	Chalk River	Canada	1973	11.5
^{221}Th	D.F. Torgerson	[1293]	Yale	USA	1970	11.6
^{222}Th	D.F. Torgerson	[1293]	Yale	USA	1970	11.6
^{223}Th	W.W. Meinke	[1294]	Berkeley	USA	1952	10.3
^{224}Th	W.W. Meinke	[1277]	Berkeley	USA	1949	10.3
^{225}Th	W.W. Meinke	[1277]	Berkeley	USA	1949	10.3
^{226}Th	M.H. Studier	[1295]	Berkeley	USA	1948	9.2
^{227}Th	O. Hahn	[1344]	McGill	Canada	1906	2.2
^{228}Th	O. Hahn	[1345]	London	UK	1905	2.2
^{229}Th	F. Hagemann	[1246]	Argonne	USA	1947	7.1
^{230}Th	B.B. Boltwood	[1346]	Yale	USA	1907	2.2
^{231}Th	G.N. Antonoff	[1347]	Manchester	UK	1911	2.2
^{232}Th	G.C. Schmidt	[1348]	Erlangen	Germany	1898	2.1
^{233}Th	O. Hahn	[1349]	Berlin	Germany	1935	4.1
^{234}Th	W. Crookes	[1350]	London	UK	1900	2.1
^{235}Th	N. Trautmann	[1351]	Mainz	Germany	1969	7.4

(continued)

Table 17.1 (continued)

Isotope	First author	Ref.	Laboratory	Country	Year	Table
^{236}Th	N. Kaffrell	[1352]	Mainz	Germany	1973	8.1
^{237}Th	S. Yuan	[1353]	Lanzhou	China	1993	7.4
^{238}Th	J. He	[1354]	Lanzhou	China	1999	14.1
^{212}Pa	S. Mitsuoka	[1355]	Tokai	Japan	1997	11.10
^{213}Pa	V. Ninov	[1356]	Darmstadt	Germany	1995	11.9
^{214}Pa	V. Ninov	[1356]	Darmstadt	Germany	1995	11.9
^{215}Pa	K.-H. Schmidt	[1357]	Darmstadt	Germany	1979	11.9
^{216}Pa	G.Ya. Sung-Ching-Yang	[1358]	Dubna	Russia	1972	11.5
^{217}Pa	K. Valli	[1341]	Berkeley	USA	1968	11.6
^{218}Pa	K.-H. Schmidt	[1357]	Darmstadt	Germany	1979	11.9
^{219}Pa	Z. Liu	[1359]	Darmstadt	Germany	2005	15.4
^{220}Pa	Z. Liu	[1359]	Darmstadt	Germany	2005	15.4
^{221}Pa	R. Hingmann	[1360]	Darmstadt	Germany	1983	11.9
^{222}Pa	J. Borggreen	[1307]	Berkeley	USA	1970	11.6
^{223}Pa	J. Borggreen	[1307]	Berkeley	USA	1970	11.6
^{224}Pa	J. Borggreen	[1307]	Berkeley	USA	1970	11.6
^{225}Pa	R.L. Hahn	[1276]	Oak Ridge	USA	1968	10.3
^{226}Pa	W.W. Meinke	[1277]	Berkeley	USA	1949	10.3
^{227}Pa	A. Ghiorso	[1279]	Berkeley	USA	1948	10.3
^{228}Pa	A. Ghiorso	[1279]	Berkeley	USA	1948	10.3
^{229}Pa	E.K. Hyde	[1361]	Berkeley	USA	1949	9.2
^{230}Pa	M.H. Studier	[1295]	Berkeley	USA	1948	9.2
^{231}Pa	O. Hahn	[1362]	Berlin	Germany	1918	2.3
^{232}Pa	J.W. Gofman	[1363]	Berkeley	USA	1949	9.2
^{233}Pa	L. Meitner	[1364]	Berlin	Germany	1938	4.1
^{234}Pa	K. Fajans	[1365]	Karlsruhe	Germany	1913	2.3
^{235}Pa	W.W. Meinke	[1329]	Berkeley	USA	1950	9.3
^{236}Pa	G. Wolzak	[1366]	Amsterdam	Netherlands	1963	9.5
^{237}Pa	W.W.T. Crane	[1367]	Berkeley	USA	1954	9.3
^{238}Pa	N. Trautmann	[1368]	Mainz	Germany	1968	7.4
^{239}Pa	S. Yuan	[1369]	Lanzhou	China	1995	14.1
^{216}U	L. Ma	[1370]	Lanzhou	China	2015	11.10
^{217}U	O.N. Malyshev	[1371]	Dubna	Russia	2000	11.10
^{218}U	A.N. Andreyev	[1372]	Dubna	Russia	1992	11.10
^{219}U	A.N. Andreyev	[1373]	Dubna	Russia	1993	11.10
^{220}U						
^{221}U	J. Khuyagbaatar	[1374]	Darmstadt	Germany	2015	11.10
^{222}U	R. Hingmann	[1360]	Darmstadt	Germany	1983	11.9

(continued)

Table 17.1 (continued)

Isotope	First author	Ref.	Laboratory	Country	Year	Table
^{223}U	A.N. Andreyev	[1375]	Dubna	Russia	1991	11.10
^{224}U	A.N. Andreyev	[1375]	Dubna	Russia	1991	11.10
^{225}U	A.N. Andreyev	[1376]	Dubna	Russia	1989	11.10
^{226}U	V.E. Viola	[1377]	Maryland	USA	1973	10.3
^{227}U	W.W. Meinke	[1294]	Berkeley	USA	1952	10.3
^{228}U	W.W. Meinke	[1277]	Berkeley	USA	1949	10.3
^{229}U	W.W. Meinke	[1277]	Berkeley	USA	1949	10.3
^{230}U	M.H. Studier	[1295]	Berkeley	USA	1948	9.2
^{231}U	D.W. Osborne	[1378]	Berkeley	USA	1949	9.2
^{232}U	J.W. Gofman	[1363]	Berkeley	USA	1949	9.2
^{233}U	G.T. Seaborg	[1379]	Berkeley	USA	1947	7.1
^{234}U	H. Geiger	[1380]	Manchester	UK	1912	2.2
^{235}U	A.J. Dempster	[1381]	Chicago	USA	1935	2.3
^{236}U	A. Ghiorso	[1382]	Argonne	USA	1951	7.1
^{237}U	Y. Nishina	[1383]	RIKEN	Japan	1940	4.1
^{238}U	H. Becquerel	[1384]	Paris	France	1896	2.1
^{239}U	L. Meitner	[1385]	Berlin	Germany	1937	4.1
^{240}U	J.D. Knight	[1386]	Los Alamos	USA	1953	7.2
^{241}U						
^{242}U	P.E. Haustein	[1387]	Brookhaven	USA	1979	7.4
^{225}Np	A.V. Yeremin	[1388]	Dubna	Russia	1994	5.4
^{226}Np	V. Ninov	[1389]	Darmstadt	Germany	1990	5.4
^{227}Np	A.N. Andreyev	[1390]	Dubna	Russia	1990	5.4
^{228}Np	S.A. Kreek	[1391]	Berkeley	USA	1994	5.3
^{229}Np	R.L. Hahn	[1276]	Oak Ridge	USA	1968	5.3
^{230}Np	R.L. Hahn	[1276]	Oak Ridge	USA	1968	5.3
^{231}Np	L.B. Magnusson	[1392]	Berkeley	USA	1950	5.3
^{232}Np	L.B. Magnusson	[1392]	Berkeley	USA	1950	5.3
^{233}Np	L.B. Magnusson	[1392]	Berkeley	USA	1950	5.3
^{234}Np	E.K. Hyde	[1393]	Berkeley	USA	1949	5.1
^{235}Np	R.A. James	[1394]	Berkeley	USA	1949	5.1
^{236}Np	R.A. James	[1394]	Berkeley	USA	1949	5.1
^{237}Np	A.C. Wahl	[1395]	Berkeley	USA	1948	5.1
^{238}Np	J.W. Kennedy	[1396]	Berkeley	USA	1949	5.1
^{239}Np	E. McMillan	[1397]	Berkeley	USA	1940	5.1
^{240}Np	J.D. Knight	[1386]	Los Alamos	USA	1953	5.2
^{241}Np	R. Vandenbosch	[1398]	Argonne	USA	1959	5.3
^{242}Np	P.E. Haustein	[1387]	Brookhaven	USA	1979	5.3

(continued)

Table 17.1 (continued)

Isotope	First author	Ref.	Laboratory	Country	Year	Table
^{243}Np	E.R. Flynn	[1399]	Los Alamos	USA	1979	5.3
^{244}Np	K.J. Moody	[1400]	Darmstadt	Germany	1987	5.4
^{228}Pu	A.N. Andreyev	[1401]	Dubna	Russia	1994	5.4
^{229}Pu	A.N. Andreyev	[1401]	Dubna	Russia	1994	5.4
^{230}Pu	A.N. Andreyev	[1402]	Dubna	Russia	1990	5.4
^{231}Pu	C.A. Laue	[10]	Berkeley	USA	1999	5.3
^{232}Pu	U. Jäger	[1403]	Karlsruhe	Germany	1973	5.3
^{233}Pu	T.D. Thomas	[1404]	Berkeley	USA	1957	5.3
^{234}Pu	E.K. Hyde	[1393]	Berkeley	USA	1949	5.1
^{235}Pu	T.D. Thomas	[1404]	Berkeley	USA	1957	5.3
^{236}Pu	R.A. James	[1394]	Berkeley	USA	1949	5.1
^{237}Pu	R.A. James	[1394]	Berkeley	USA	1949	5.1
^{238}Pu	J.W. Kennedy	[1396]	Berkeley	USA	1949	5.1
^{239}Pu	J.W. Kennedy	[1405]	Berkeley	USA	1946	5.1
^{240}Pu	R.A. James	[1394]	Berkeley	USA	1949	5.1
^{241}Pu	G.T. Seaborg	[1406]	Berkeley	USA	1949	5.1
^{242}Pu	S.G. Thompson	[1407]	Berkeley	USA	1950	5.2
^{243}Pu	J.C. Sullivan	[1408]	Argonne	USA	1951	5.2
^{244}Pu	M.H. Studier	[1409]	Argonne	USA	1954	5.2
^{245}Pu	C.I. Browne	[1410]	Los Alamos	USA	1955	5.2
	P.R. Fields	[1411]	Argonne	USA	1955	5.2
^{246}Pu	D. Engelkemeir	[1412]	Argonne	USA	1955	5.1
^{247}Pu	Yu.S. Popov	[1413]	Dimitrovgrad	Russia	1983	5.2
^{223}Am	Devaraja	[1414]	Darmstadt	Germany	2015	5.4
^{224}Am						
^{225}Am						
^{226}Am						
^{227}Am						
^{228}Am						
^{229}Am	Devaraja	[1414]	Darmstadt	Germany	2015	5.4
^{230}Am						
^{231}Am						
^{232}Am	V.I. Kuznetsov	[1415]	Dubna	Russia	1967	5.4
^{233}Am	M. Sakama	[1416]	Tokai	Japan	2000	5.4
^{234}Am	V.I. Kuznetsov	[1417]	Dubna	Russia	1967	5.4
^{235}Am	J. Guo	[1418]	Beijing	China	1996	5.3
^{236}Am	K. Tsukada	[1419]	Tokai	Japan	1998	5.4
^{237}Am	S.M. Polikanov	[1420]	Copenhagen	Denmark	1970	5.3

(continued)

Table 17.1 (continued)

Isotope	First author	Ref.	Laboratory	Country	Year	Table
^{238}Am	K. Street	[1421]	Berkeley	USA	1950	5.3
^{239}Am	G.T. Seaborg	[1406]	Berkeley	USA	1949	5.1
^{240}Am	G.T. Seaborg	[1406]	Berkeley	USA	1949	5.1
^{241}Am	G.T. Seaborg	[1406]	Berkeley	USA	1949	5.1
^{242}Am	W. M. Manning	[1422]	Argonne	USA	1949	5.1
^{243}Am	K. Street	[1421]	Argonne	USA	1950	5.2
^{244}Am	K. Street	[1421]	Argonne	USA	1950	5.2
^{245}Am	C.I. Browne	[1410]	Los Alamos	USA	1955	5.2
	P.R. Fields	[1411]	Argonne	USA	1955	5.2
^{246}Am	D. Engelkemeir	[1412]	Argonne	USA	1955	5.1
^{247}Am	C.J. Orth	[1423]	Los Alamos	USA	1967	5.3
^{233}Cm	J. Khuyagbaatar	[1424]	Darmstadt	Germany	2010	5.4
^{234}Cm						
^{235}Cm						
^{236}Cm	J. Khuyagbaatar	[1424]	Darmstadt	Germany	2010	5.4
^{237}Cm	S. Ichikawa	[1425]	Tokai	Japan	2002	5.4
^{238}Cm	S.A. Kreek	[1426]	Berkeley	USA	1994	5.3
^{239}Cm	Z. Qin	[1427]	Darmstadt	Germany	2008	5.4
^{240}Cm	G.T. Seaborg	[1428]	Berkeley	USA	1949	5.1
^{241}Cm	G.H. Higgins	[1429]	Berkeley	USA	1952	5.3
^{242}Cm	G.T. Seaborg	[1428]	Berkeley	USA	1949	5.1
^{243}Cm	F.L. Reynolds	[1430]	Berkeley	USA	1950	5.2
^{244}Cm	F.L. Reynolds	[1430]	Berkeley	USA	1950	5.2
^{245}Cm	C.M. Stevens	[1431]	Argonne	USA	1954	5.2
^{246}Cm	C.M. Stevens	[1431]	Argonne	USA	1954	5.2
^{247}Cm	C.M. Stevens	[1431]	Argonne	USA	1954	5.2
^{248}Cm	P.R. Fields	[1432]	Argonne	USA	1956	5.1
^{249}Cm	P.R. Fields	[1432]	Argonne	USA	1956	5.1
^{250}Cm	Comb.Radiochem.Group	[1433]	Livermore	USA	1966	5.1
^{251}Cm	R.W. Lougheed	[1434]	Livermore	USA	1978	5.1
^{233}Bk	Devaraja	[1414]	Darmstadt	Germany	2015	5.4
^{234}Bk						
^{235}Bk						
^{236}Bk						
^{237}Bk						
^{238}Bk	S.A. Kreek	[1426]	Berkeley	USA	1994	5.3
^{239}Bk						
^{240}Bk	Yu.P. Gangrskii	[1435]	Dubna	Russia	1980	5.4

(continued)

Table 17.1 (continued)

Isotope	First author	Ref.	Laboratory	Country	Year	Table
^{241}Bk	M. Asai	[1436]	Tokai	Japan	2003	5.4
^{242}Bk	K.L. Wolf	[1437]	Argonne	USA	1972	5.3
^{243}Bk	S.G. Thompson	[1438]	Berkeley	USA	1950	5.3
^{244}Bk	K.L. Wolf	[1437]	Argonne	USA	1972	5.3
^{245}Bk	E.K. Hulet	[1439]	Berkeley	USA	1951	5.3
^{246}Bk	E.K. Hulet	[1440]	Berkeley	USA	1954	5.3
^{247}Bk	J. Milsted	[1441]	Argonne	USA	1965	5.3
^{248}Bk	E.K. Hulet	[1442]	Berkeley	USA	1956	5.3
^{249}Bk	H. Diamond	[1443]	Argonne	USA	1954	5.2
^{250}Bk	A. Ghiorso	[1444]	Berkeley	USA	1954	5.2
^{251}Bk	H. Diamond	[1445]	Argonne	USA	1967	5.2
^{237}Cf	Yu.A. Lazarev	[1446]	Dubna	Russia	1995	5.4
^{238}Cf	Yu.A. Lazarev	[1446]	Dubna	Russia	1995	5.4
^{239}Cf	G. Münzenberg	[1447]	Darmstadt	Germany	1981	5.4
^{240}Cf	R.J. Silva	[1448]	Oak Ridge	USA	1970	5.4
^{241}Cf	R.J. Silva	[1448]	Oak Ridge	USA	1970	5.4
^{242}Cf	T. Sikkeland	[1449]	Berkeley	USA	1967	5.4
	P.R. Fields	[1450]	Argonne	USA	1967	5.3
^{243}Cf	T. Sikkeland	[1451]	Berkeley	USA	1967	5.4
	P.R. Fields	[1450]	Argonne	USA	1967	5.3
^{244}Cf	A. Chetham-Strode	[1452]	Berkeley	USA	1956	5.3
^{245}Cf	A. Chetham-Strode	[1452]	Berkeley	USA	1956	5.3
^{246}Cf	A. Ghiorso	[1453]	Berkeley	USA	1951	5.4
^{247}Cf	A. Ghiorso	[1454]	Berkeley	USA	1954	5.4
^{248}Cf	A. Ghiorso	[1454]	Berkeley	USA	1954	5.4
^{249}Cf	A. Ghiorso	[1444]	Berkeley	USA	1954	5.2
^{250}Cf	A. Ghiorso	[1444]	Berkeley	USA	1954	5.2
^{251}Cf	H. Diamond	[1443]	Argonne	USA	1954	5.2
^{252}Cf	A. Ghiorso	[1444]	Berkeley	USA	1954	5.2
^{253}Cf	G.R. Choppin	[1455]	Berkeley	USA	1954	5.2
^{254}Cf	B.G. Harvey	[1456]	Berkeley	USA	1955	5.2
^{255}Cf	R.W. Lougheed	[1457]	Livermore	USA	1981	5.1
^{256}Cf	D.C. Hoffman	[1458]	Los Alamos	USA	1980	5.3
^{241}Es	V. Ninov	[1459]	Darmstadt	Germany	1996	5.4
^{242}Es	Yu.A. Lazarev	[1460]	Dubna	Russia	1994	5.4
^{243}Es	P. Eskola	[1461]	Berkeley	USA	1973	5.4
^{244}Es	P. Eskola	[1461]	Berkeley	USA	1973	5.4
^{245}Es	V.L. Mikheev	[1462]	Dubna	Russia	1967	5.4

(continued)

Table 17.1 (continued)

Isotope	First author	Ref.	Laboratory	Country	Year	Table
^{246}Es	A. Ghiorso	[1454]	Berkeley	USA	1954	5.4
^{247}Es	V.L. Mikheev	[1462]	Dubna	Russia	1967	5.4
^{248}Es	A. Chetham-Strode	[1463]	Berkeley	USA	1956	5.3
^{249}Es	B.G. Harvey	[1464]	Berkeley	USA	1956	5.3
^{250}Es	B.G. Harvey	[1464]	Berkeley	USA	1956	5.3
^{251}Es	B.G. Harvey	[1464]	Berkeley	USA	1956	5.3
^{252}Es	B.G. Harvey	[1464]	Berkeley	USA	1956	5.3
^{253}Es	S.G. Thompson	[1465]	Berkeley	USA	1954	5.2
^{254}Es	P.R. Fields	[1466]	Argonne	USA	1954	5.2
^{255}Es	G.R. Choppin	[1455]	Berkeley	USA	1954	5.2
^{256}Es	R.W. Lougheed	[1457]	Livermore	USA	1981	5.2
^{257}Es	Yu.S. Popov	[1467]	Dimitrovgrad	Russia	1987	5.2
^{241}Fm	J. Khuyagbaatar	[1468]	Darmstadt	Germany	2008	5.4
^{242}Fm	G.M. Ter-Akopian	[1469]	Dubna	Russia	1975	5.4
^{243}Fm	G. Münzenberg	[1447]	Darmstadt	Germany	1981	5.4
^{244}Fm	M.J. Nurmia	[1470]	Berkeley	USA	1967	5.4
^{245}Fm	M.J. Nurmia	[1470]	Berkeley	USA	1967	5.4
^{246}Fm	G.N. Akapev	[1471]	Dubna	Russia	1966	5.4
^{247}Fm	G.N. Flerov	[1472]	Dubna	Russia	1967	5.4
^{248}Fm	A. Ghiorso	[1473]	Berkeley	USA	1958	5.4
^{249}Fm	V.P. Perelygin	[1474]	Moscow	Russia	1960	5.4
^{250}Fm	H. Atterling	[1475]	Stockholm	Sweden	1954	5.4
^{251}Fm	S. Amiel	[1476]	Berkeley	USA	1957	5.3
^{252}Fm	A.M. Friedman	[1477]	Argonne	USA	1956	5.3
^{253}Fm	S. Amiel	[1478]	Berkeley	USA	1957	5.3
^{254}Fm	B.G. Harvey	[1479]	Berkeley	USA	1954	5.2
^{255}Fm	G.R. Choppin	[1455]	Berkeley	USA	1954	5.2
^{256}Fm	G.R. Choppin	[1480]	Berkeley	USA	1955	5.2
^{257}Fm	E.K. Hulet	[1481]	Berkeley	USA	1964	5.2
^{258}Fm	E.K. Hulet	[1482]	Berkeley	USA	1971	5.1
^{259}Fm	E.K. Hulet	[1483]	Los Alamos	USA	1980	5.3
^{245}Md	V. Ninov	[1459]	Darmstadt	Germany	1996	5.4
^{246}Md	V. Ninov	[1459]	Darmstadt	Germany	1996	5.4
^{247}Md	G. Münzenberg	[1447]	Darmstadt	Germany	1981	5.4
^{248}Md	P. Eskola	[1484]	Berkeley	USA	1973	5.4
^{249}Md	P. Eskola	[1484]	Berkeley	USA	1973	5.4
^{250}Md	P. Eskola	[1484]	Berkeley	USA	1973	5.4
^{251}Md	P. Eskola	[1484]	Berkeley	USA	1973	5.4

(continued)

Table 17.1 (continued)

Isotope	First author	Ref.	Laboratory	Country	Year	Table
^{252}Md	P. Eskola	[1484]	Berkeley	USA	1973	5.4
^{253}Md	B. Kadkhodayan	[1485]	Berkeley	USA	1992	5.4
^{254}Md	P.R. Fields	[1486]	Argonne	USA	1970	5.3
^{255}Md	L. Phillips	[1487]	Berkeley	USA	1958	5.3
^{256}Md	A. Ghiorso	[1488]	Berkeley	USA	1955	5.3
^{257}Md	T. Sikkeland	[1489]	Berkeley	USA	1965	5.4
^{258}Md	P.R. Fields	[1486]	Argonne	USA	1970	5.3
^{259}Md	J.F. Wild	[1490]	Berkeley	USA	1982	5.4
^{260}Md	E.K. Hulet	[1491]	Berkeley	USA	1989	5.4
^{250}No	Yu.Ts. Oganessian	[1492]	Dubna	Russia	2001	12.2
^{251}No	A. Ghiorso	[1493]	Berkeley	USA	1967	12.1
^{252}No	V.L. Mikheev	[1494]	Dubna	Russia	1967	12.1
^{253}No	V.L. Mikheev	[1494]	Dubna	Russia	1967	12.1
^{254}No	E.D. Donets	[1495]	Dubna	Russia	1966	12.1
	B.A. Zager	[1496]	Dubna	Russia	1966	12.1
^{255}No	V.A. Druin	[1497]	Dubna	Russia	1967	12.1
^{256}No	E.D. Donets	[1498]	Dubna	Russia	1963	12.1
^{257}No	A. Ghiorso	[1493]	Berkeley	USA	1967	12.1
^{258}No	E.K. Hulet	[1491]	Berkeley	USA	1989	12.1
^{259}No	R.J. Silva	[1499]	Oak Ridge	USA	1973	12.1
^{260}No	L.P. Somerville	[1500]	Berkeley	USA	1985	12.1
^{252}Lr	F.P. Heßberger	[1501]	Darmstadt	Germany	2001	12.2
^{253}Lr	F.P. Heßberger	[1502]	Darmstadt	Germany	1985	12.2
^{254}Lr	G. Münzenberg	[1503]	Darmstadt	Germany	1981	12.2
^{255}Lr	V. A. Druin	[1504]	Dubna	Russia	1971	12.1
^{256}Lr	E.D. Donets	[1505]	Dubna	Russia	1965	12.1
^{257}Lr	K. Eskola	[1506]	Berkeley	USA	1971	12.1
^{258}Lr	K. Eskola	[1506]	Berkeley	USA	1971	12.1
^{259}Lr	K. Eskola	[1506]	Berkeley	USA	1971	12.1
^{260}Lr	K. Eskola	[1506]	Berkeley	USA	1971	12.1
^{261}Lr						
^{262}Lr						
^{263}Lr						
^{264}Lr						
^{265}Lr						
^{266}Lr	J. Khuyagbaatar	[1507]	Darmstadt	Germany	2014	12.3
^{253}Rf	F.P. Heßberger	[1508]	Darmstadt	Germany	1997	12.2
^{254}Rf	F.P. Heßberger	[1508]	Darmstadt	Germany	1997	12.2

(continued)

Table 17.1 (continued)

Isotope	First author	Ref.	Laboratory	Country	Year	Table
^{255}Rf	Yu.Ts. Oganessian	[1509]	Dubna	Russia	1975	12.1
^{256}Rf	Yu.Ts. Oganessian	[1509]	Dubna	Russia	1975	12.1
^{257}Rf	A. Ghiorso	[1510]	Berkeley	USA	1969	12.1
^{258}Rf	A. Ghiorso	[1510]	Berkeley	USA	1969	12.1
^{259}Rf	A. Ghiorso	[1510]	Berkeley	USA	1969	12.1
^{260}Rf	L.P. Somerville	[1500]	Berkeley	USA	1985	12.1
^{261}Rf	A. Ghiorso	[1511]	Berkeley	USA	1970	12.1
^{262}Rf	L.P. Somerville	[1500]	Berkeley	USA	1985	12.1
^{263}Rf	J.V. Kratz	[1512]	Villigen	Switzerland	2003	12.1
^{264}Rf						
^{265}Rf	P.A. Ellison	[1513]	Berkeley	USA	2010	12.3
^{266}Rf						
^{267}Rf	Yu.Ts. Oganessian	[1514]	Dubna	Russia	2004	12.3
^{256}Db	F.P. Heßberger	[1501]	Darmstadt	Germany	2001	12.2
^{257}Db	F.P. Heßberger	[1502]	Darmstadt	Germany	1985	12.2
^{258}Db	G. Münzenberg	[1503]	Darmstadt	Germany	1981	12.2
^{259}Db	Z.G. Gan	[1515]	Lanzhou	China	2001	12.1
^{260}Db	A. Ghiorso	[1516]	Berkeley	USA	1970	12.1
^{261}Db	G.N. Flerov	[1517]	Dubna	Russia	1970	12.1
^{262}Db	A. Ghiorso	[1518]	Berkeley	USA	1971	12.1
^{263}Db	J.V. Kratz	[1519]	Berkeley	USA	1992	12.1
^{264}Db						
^{265}Db						
^{266}Db	Yu.Ts. Oganessian	[1520]	Dubna	Russia	2007	12.3
^{267}Db	Yu.Ts. Oganessian	[1521]	Dubna	Russia	2004	12.3
^{268}Db	Yu.Ts. Oganessian	[1521]	Dubna	Russia	2004	12.3
^{269}Db						
^{270}Db	Yu.Ts. Oganessian	[1522]	Dubna	Russia	2010	12.3
^{258}Sg	F.P. Heßberger	[1508]	Darmstadt	Germany	1997	12.2
^{259}Sg	G. Münzenberg	[1523]	Darmstadt	Germany	1985	12.2
^{260}Sg	A.G. Demin	[1524]	Dubna	Russia	1984	12.1
^{261}Sg	G. Münzenberg	[1525]	Darmstadt	Germany	1984	12.2
^{262}Sg	S. Hofmann	[1526]	Darmstadt	Germany	2001	12.2
^{263}Sg	A. Ghiorso	[1527]	Berkeley	USA	1974	12.1
^{264}Sg	K.E. Gregorich	[1528]	Berkeley	USA	2006	12.2
^{265}Sg	Yu.A. Lazarev	[1529]	Dubna	Russia	1994	12.2
^{266}Sg	J. Dvorak	[1530]	Darmstadt	Germany	2006	12.1
^{267}Sg	J. Dvorak	[1531]	Darmstadt	Germany	2008	12.1

(continued)

Table 17.1 (continued)

Isotope	First author	Ref.	Laboratory	Country	Year	Table
^{268}Sg						
^{269}Sg	P.A. Ellison	[1513]	Berkeley	USA	2010	12.3
^{270}Sg						
^{271}Sg	Yu.Ts. Oganessian	[1514]	Dubna	Russia	2004	12.3
^{260}Bh	S.L. Nelson	[1532]	Berkeley	USA	2008	12.2
^{261}Bh	G. Münzenberg	[1533]	Darmstadt	Germany	1989	12.2
^{262}Bh	G. Münzenberg	[1503]	Darmstadt	Germany	1981	12.2
^{263}Bh						
^{264}Bh	S. Hofmann	[1534]	Darmstadt	Germany	1995	12.2
^{265}Bh	Z.G. Gan	[1535]	Lanzhou	China	2004	12.1
^{266}Bh	P.A. Wilk	[1536]	Berkeley	USA	2000	12.1
^{267}Bh	P.A. Wilk	[1536]	Berkeley	USA	2000	12.1
^{268}Bh						
^{269}Bh						
^{270}Bh	Yu.Ts. Oganessian	[1520]	Dubna	Russia	2007	12.3
^{271}Bh	Yu.Ts. Oganessian	[16]	Dubna	Russia	2013	12.3
^{272}Bh	Yu.Ts. Oganessian	[1521]	Dubna	Russia	2004	12.3
^{273}Bh						
^{274}Bh	Yu.Ts. Oganessian	[1522]	Dubna	Russia	2010	12.3
^{263}Hs	I. Dragojevic	[1537]	Berkeley	USA	2009	12.2
^{264}Hs	G. Münzenberg	[1538]	Darmstadt	Germany	1986	12.2
^{265}Hs	G. Münzenberg	[1525]	Darmstadt	Germany	1984	12.2
^{266}Hs	S. Hofmann	[1526]	Darmstadt	Germany	2001	12.2
^{267}Hs	Yu. A. Lazarev	[1539]	Dubna	Russia	1995	12.2
^{268}Hs	K. Nishio	[1540]	Darmstadt	Germany	2010	12.2
^{269}Hs	S. Hofmann	[1541]	Darmstadt	Germany	1996	12.2
^{270}Hs	J. Dvorak	[1530]	Darmstadt	Germany	2006	12.1
^{271}Hs	J. Dvorak	[1531]	Darmstadt	Germany	2008	12.1
^{272}Hs						
^{273}Hs	P.A. Ellison	[1513]	Berkeley	USA	2010	12.3
^{274}Hs						
^{275}Hs	Yu.Ts. Oganessian	[1514]	Dubna	Russia	2004	12.3
^{276}Hs						
^{277}Hs	Ch. E. Düllmann	[1542]	Darmstadt	Germany	2010	12.3
^{266}Mt	G. Münzenberg	[1543]	Darmstadt	Germany	1982	12.2
^{267}Mt						
^{268}Mt	S. Hofmann	[1534]	Darmstadt	Germany	1995	12.2
^{269}Mt						

(continued)

Table 17.1 (continued)

Isotope	First author	Ref.	Laboratory	Country	Year	Table
^{270}Mt	K. Morita	[1544]	RIKEN	Japan	2004	12.2
^{271}Mt						
^{272}Mt						
^{273}Mt						
^{274}Mt	Yu.Ts. Oganessian	[1520]	Dubna	Russia	2007	12.3
^{275}Mt	Yu.Ts. Oganessian	[1521]	Dubna	Russia	2004	12.3
^{276}Mt	Yu.Ts. Oganessian	[1521]	Dubna	Russia	2004	12.3
^{277}Mt	Yu.Ts. Oganessian	[1545]	Dubna	Russia	2013	12.3
^{278}Mt	Yu.Ts. Oganessian	[1522]	Dubna	Russia	2010	12.3
^{267}Ds	A. Ghiorso	[1546]	Berkeley	USA	1995	12.2
^{268}Ds						
^{269}Ds	S. Hofmann	[1547]	Darmstadt	Germany	1995	12.2
^{270}Ds	S. Hofmann	[1526]	Darmstadt	Germany	2001	12.2
^{271}Ds	S. Hofmann	[1534]	Darmstadt	Germany	1995	12.2
^{272}Ds						
^{273}Ds	Yu. A. Lazarev	[1548]	Dubna	Russia	1996	12.2
^{274}Ds						
^{275}Ds						
^{276}Ds						
^{277}Ds	P.A. Ellison	[1513]	Berkeley	USA	2010	12.3
^{278}Ds						
^{279}Ds	Yu.Ts. Oganessian	[1549]	Dubna	Russia	2004	12.3
^{280}Ds						
^{281}Ds	Yu.Ts. Oganessian	[1549]	Dubna	Russia	2004	12.3
^{272}Rg	S. Hofmann	[1534]	Darmstadt	Germany	1995	12.2
^{273}Rg						
^{274}Rg	K. Morita	[1544]	RIKEN	Japan	2004	12.2
^{275}Rg						
^{276}Rg						
^{277}Rg						
^{278}Rg	Yu.Ts. Oganessian	[1520]	Dubna	Russia	2007	12.3
^{279}Rg	Yu.Ts. Oganessian	[1521]	Dubna	Russia	2004	12.3
^{280}Rg	Yu.Ts. Oganessian	[1521]	Dubna	Russia	2004	12.3
^{281}Rg	Yu.Ts. Oganessian	[1522]	Dubna	Russia	2010	12.3
^{282}Rg	Yu.Ts. Oganessian	[1522]	Dubna	Russia	2010	12.3
^{277}Cn	S. Hofmann	[1541]	Darmstadt	Germany	1996	12.2
^{278}Cn						
^{279}Cn						
^{280}Cn						

(continued)

Table 17.1 (continued)

Isotope	First author	Ref.	Laboratory	Country	Year	Table
^{281}Cn	P.A. Ellison	[1513]	Berkeley	USA	2010	12.3
^{282}Cn	Yu.Ts. Oganessian	[1549]	Dubna	Russia	2004	12.3
^{283}Cn	Yu.Ts. Oganessian	[1549]	Dubna	Russia	2004	12.3
^{284}Cn	Yu.Ts. Oganessian	[1549]	Dubna	Russia	2004	12.3
^{285}Cn	Yu.Ts. Oganessian	[1549]	Dubna	Russia	2004	12.3
^{278}113	K. Morita	[1544]	RIKEN	Japan	2004	12.2
^{279}113						
^{280}113						
^{281}113						
^{282}113	Yu.Ts. Oganessian	[1520]	Dubna	Russia	2007	12.3
^{283}113	Yu.Ts. Oganessian	[1521]	Dubna	Russia	2004	12.3
^{284}113	Yu.Ts. Oganessian	[1521]	Dubna	Russia	2004	12.3
^{285}113	Yu.Ts. Oganessian	[1522]	Dubna	Russia	2010	12.3
^{286}113	Yu.Ts. Oganessian	[1522]	Dubna	Russia	2010	12.3
^{284}Fl	V.K. Utyonkov	[1550]	Dubna	Russia	2015	12.3
^{285}Fl	P.A. Ellison	[1513]	Berkeley	USA	2010	12.3
^{286}Fl	Yu.Ts. Oganessian	[1549]	Dubna	Russia	2004	12.3
^{287}Fl	Yu.Ts. Oganessian	[1549]	Dubna	Russia	2004	12.3
^{288}Fl	Yu.Ts. Oganessian	[1549]	Dubna	Russia	2004	12.3
^{289}Fl	Yu.Ts. Oganessian	[1549]	Dubna	Russia	2004	12.3
^{287}115	Yu.Ts. Oganessian	[1521]	Dubna	Russia	2004	12.3
^{288}115	Yu.Ts. Oganessian	[1521]	Dubna	Russia	2004	12.3
^{289}115	Yu.Ts. Oganessian	[1522]	Dubna	Russia	2010	12.3
^{290}115	Yu.Ts. Oganessian	[1522]	Dubna	Russia	2010	12.3
^{290}Lv	Yu.Ts. Oganessian	[1549]	Dubna	Russia	2004	12.3
^{291}Lv	Yu.Ts. Oganessian	[1549]	Dubna	Russia	2004	12.3
^{292}Lv	Yu.Ts. Oganessian	[1514]	Dubna	Russia	2004	12.3
^{293}Lv	Yu.Ts. Oganessian	[1549]	Dubna	Russia	2004	12.3
^{293}117	Yu.Ts. Oganessian	[1522]	Dubna	Russia	2010	12.3
^{294}117	Yu.Ts. Oganessian	[1522]	Dubna	Russia	2010	12.3
^{294}118	Yu.Ts. Oganessian	[1551]	Dubna	Russia	2006	12.3

References

1. L. Westgaard et al., Z. Phys. A **275**, 127 (1975)
2. M. Thoennessen, Rep. Prog. Phys. **76**, 056301 (2013)
3. E. Rutherford, Phil. Mag. **49**, 161 (1900)
4. E. Rutherford, H.T. Barnes, Phil. Mag. **7**, 202 (1904)
5. F.D.S. Butement, V.J. Robinson, J. Inorg. Nucl. Chem. **26**, 1 (1964)
6. H.E. Duckworth, R.E. Black, R.F. Woodcock, Phys. Rev. **75**, 1438 (1949)
7. D.C. Hess Jr., M.G. Inghram, Phys. Rev. **76**, 1717 (1949)
8. W.T. Leland, Phys. Rev. **76**, 1722 (1949)
9. M.J. Nurmia et al., Nature **190**, 427 (1961)
10. C.A. Laue et al., Phys. Rev. C **59**, 3086 (1999)
11. Y. Xu et al., J. Radioanal. Nucl. Chem. **258**, 439 (2003)
12. J. Erler et al., Nature **486**, 509 (2012)
13. M. Thoennessen, Nucl. Data Sheets **118**, 85 (2014)
14. J.M. Gates et al., Phys. Rev. C **83**, 054618 (2011)
15. K. Morita et al., J. Phys. Soc. Jpn. **81**, 103201 (2012)
16. Y.T. Oganessian et al., Phys. Rev. C **87**, 014302 (2013)
17. V. Zagrebaev, W. Greiner, Phys. Rev. C **78**, 034610 (2008)
18. V. Zagrebaev, W. Greiner, Phys. Rev. C **83**, 044618 (2011)
19. W. Loveland, arXiv:1301.1759v1 (2013)
20. W. Loveland, Phys. Rev. C **76**, 014612 (2007)
21. H. Sakurai, A.I.P. Conf, Proc. **1269**, 84 (2010)
22. *Gesellschaft für Schwerionenforschung 2009 Green paper—the modularized version*, https://www.gsi.de/documents/DOC-2009-Nov-124-1.pdf
23. G. Bollen, A.I.P. Conf, Proc. **1224**, 432 (2010)
24. J. Wei et al., XXVI Linear Accelerator Conference, Tel Aviv, (2012) TU1A4
25. T. Kubo, Nucl. Instrum. Meth. B **204**, 97 (2003)
26. T. Ohnishi et al., J. Phys. Soc. Jpn. **77**, 083201 (2008)
27. H. Geissel et al., Nucl. Instrum. Meth. B **204**, 71 (2003)
28. L. Bandura et al., Nucl. Instrum. Meth. A **645**, 182 (2011)
29. J.C. Yang et al., Nucl. Instrum. Meth. B **317**, 263 (2013)
30. C.-B. Moon, AIP Adv. **4**, 041001 (2014)
31. L. Merminga et al., Proc. of IPAC 2011, San Sebastian, 4–9 Sep 2011, WEOBA01, http://accelconf.web.cern.ch/AccelConf/IPAC2011/papers/weoba01.pdf
32. http://www.eurisol.org
33. J. Chadwick, Nature **129**, 312 (1932)
34. R.P. Haddock et al., Phys. Rev. Lett. **14**, 318 (1965)
35. A.J. Dempster, Phys. Rev. **11**, 316 (1918)
36. H.C. Urey, F.G. Brickwedde, G.M. Murphy, Phys. Rev. **39**, 164 (1932)
37. M.L. Oliphant, P. Harteck, E. Rutherford, Nature **133**, 413 (1934)
38. U. Sennhauser et al., Phys. Lett. B **103**, 409 (1981)
39. A.A. Korsheninnikov et al., Phys. Rev. Lett. **87**, 092501 (2001)
40. D.V. Aleksandrov et al., Sov. J. Nucl. Phys. **39**, 323 (1984)
41. A.A. Korsheninnikov et al., Phys. Rev. Lett. **90**, 082501 (2003)
42. M.A. Tuve, N.P. Heydenburg, L.R. Hafstad, Phys. Rev. **50**, 806 (1936)
43. E. Rutherford, H. Geiger, Proc. Roy. Soc. A **81**, 162 (1908)
44. J.H. Williams, W.G. Shepherd, R.O. Haxby, Phys. Rev. **51**, 888 (1937)
45. T. Bjerge, K.J. Broström, Nature **138**, 400 (1936)
46. R.H. Stokes, P.G. Young, Phys. Rev. Lett. **18**, 611 (1967)
47. A.M. Poskanzer, R.A. Esterlund, R. McPherson, Phys. Rev. Lett. **15**, 1030 (1965)
48. K.K. Seth et al., Phys. Rev. Lett. **58**, 1930 (1987)
49. A.A. Korsheninnikov et al., Phys. Lett. B **326**, 31 (1994)
50. J. Cerny, C. Détraz, R.H. Pehl, Phys. Rev. Lett. **15**, 300 (1965)

51. N.P. Heydenburg, N.F. Ramsey, Phys. Rev. **60**, 42 (1941)
52. F.W. Aston, G.P. Thomson, Nature **106**, 827 (1921)
53. H.R. Crane et al., Phys. Rev. **47**, 971 (1935)
54. W.L. Gardner, N. Knable, B.J. Moyer, Phys. Rev. **83**, 1054 (1951)
55. K.H. Wilcox et al., Phys. Lett. B **59**, 142 (1975)
56. A.M. Poskanzer et al., Phys. Rev. Lett. **17**, 1271 (1966)
57. C.C. Hall et al., Phys. Rev. C **81**, 021302 (2010)
58. Yu. Aksyutina et al., Phys. Lett. B **666**, 430 (2008)
59. G.F. Bogdanov et al., J. Nucl. Energy II **8**, 148 (1958)
60. R.B. Roberts, N.P. Heydenburg, G.L. Locher, Phys. Rev. **53**, 1016 (1938)
61. J.D. Cockcroft, E.T.S. Walton, Nature **129**, 649 (1932)
62. G.P. Thomson, Nature **107**, 395 (1921)
63. M.L. Oliphant, A.E. Kempton, E. Rutherford, Proc. Roy. Soc. A **150**, 241 (1935)
64. M.J. Nurmia, R.W. Fink, Phys. Rev. Lett. **1**, 23 (1958)
65. D.V. Aleksandrov et al., Sov. J. Nucl. Phys. **37**, 474 (1983)
66. J.D. Bowman et al., Phys. Rev. Lett. **31**, 614 (1973)
67. J. Snyder et al., Phys. Rev. C **88**, 031303 (2013)
68. A. Spyrou et al., Phys. Rev. Lett. **108**, 102501 (2012)
69. R.L. McGrath, J. Cerny, E. Norbeck, Phys. Rev. Lett. **19**, 1442 (1967)
70. L.W. Alvarez, Phys. Rev. **80**, 519 (1950)
71. R.O. Haxby et al., Phys. Rev. **58**, 1035 (1940)
72. F.W. Aston, Nature **105**, 547 (1920)
73. H.R. Crane et al., Phys. Rev. **47**, 887 (1935)
74. S.K. Allison, P.G. Murphy, E. Norbeck, Phys. Rev. **102**, 1182 (1956)
75. R. Kalpakchieva et al., Eur. Phys. J. A **7**, 451 (2000)
76. A. Spyrou et al., Phys. Lett. B **683**, 129 (2010)
77. J.A. Musser, J.D. Stevenson, Phys. Rev. Lett. **53**, 2544 (1984)
78. R.G.H. Robertson et al., Phys. Rev. Lett. **32**, 1207 (1974)
79. J. Cerny et al., Phys. Rev. Lett. **13**, 726 (1964)
80. R. Sherr, H.R. Muether, M.G. White, Phys. Rev. **75**, 282 (1949)
81. H.R. Crane, C.C. Lauritsen, Phys. Rev. **45**, 430 (1934)
82. F.W. Aston, Nature **104**, 393 (1919)
83. A.S. King, R.T. Birge, Phys. Rev. **34**, 376 (1929)
84. A.S. King, R.T. Birge, Nature **124**, 127 (1929)
85. T.W. Bonner, W.M. Brubaker, Phys. Rev. **49**, 223 (1936)
86. E.L. Hudspeth, C.P. Swann, N.P. Heydenburg, Phys. Rev. **77**, 736 (1950)
87. S. Hinds et al., Phys. Rev. Lett. **6**, 113 (1961)
88. A.M. Poskanzer et al., Phys. Lett. B **27**, 414 (1968)
89. A.G. Artukh et al., Nucl. Phys. A **137**, 348 (1969)
90. J.D. Bowman et al., Phys. Rev. C **9**, 836 (1974)
91. J.D. Stevenson, P.B. Price, Phys. Rev. C **24**, 2102 (1981)
92. F. Pougheon et al., Europhys. Lett. **2**, 505 (1986)
93. A. Lépine-Szily et al., Phys. Rev. C **65**, 054318 (2002)
94. W. Benenson et al., Phys. Rev. C **9**, 2130 (1974)
95. L.W. Alvarez, Phys. Rev. **75**, 1815 (1949)
96. I. Curie, F. Joliot, Compt. Rend. Acad. Sci. **198**, 254 (1934)
97. F.W. Aston, Nature **105**, 8 (1920)
98. S.M. Naude, Phys. Rev. **34**, 1498 (1929)
99. W.D. Harkins, D.M. Gans, H.W. Newson, Phys. Rev. **44**, 945 (1933)
100. L.W. Alvarez, Phys. Rev. **75**, 1127 (1949)
101. L.F. Chase Jr. et al., Phys. Rev. Lett. **13**, 665 (1964)
102. T.D. Thomas et al., Phys. Lett. B **27**, 504 (1968)
103. A.G. Artukh et al., Phys. Lett. B **32**, 43 (1970)
104. G.D. Westfall et al., Phys. Rev. Lett. **43**, 1859 (1979)

105. M. Langevin et al., Phys. Lett. B **150**, 71 (1985)
106. G.J. KeKelis et al., Phys. Rev. C **17**, 1929 (1978)
107. R. McPherson et al., Phys. Rev. **140**, B1513 (1965)
108. M.S. Livingston, E. McMillan, Phys. Rev. **46**, 437 (1934)
109. P.M.S. Blackett, Proc. Roy. Soc. A **107**, 349 (1925)
110. W.F. Giauque, H.L. Johnston, J. Am. Chem. Soc. **51**, 1436 (1929)
111. M.E. Nahmias, R.J. Walen, Compt. Rend. Acad. Sci. **203**, 71 (1936)
112. N. Jarmie, M.G. Silbert, Phys. Rev. Lett. **3**, 50 (1959)
113. C.R. Hoffman et al., Phys. Rev. Lett. **100**, 152502 (2008)
114. E. Lunderberg et al., Phys. Rev. Lett. **108**, 142503 (2012)
115. V.Z. Goldberg et al., Phys. Lett. B **692**, 307 (2010)
116. H.C. Bryant et al., Nucl. Phys. **53**, 97 (1964)
117. L. Wertenstein, Nature **133**, 564 (1934)
118. L.A. DuBridge, S.W. Barnes, J.H. Buck, Phys. Rev. **51**, 995 (1937)
119. N. Jarmie, Phys. Rev. **99**, 1043 (1955)
120. F.J. Vaughn et al., Phys. Rev. Lett. **15**, 555 (1965)
121. A.G. Artukh et al., Phys. Lett. B **31**, 129 (1970)
122. G. Christian et al., Phys. Rev. Lett. **108**, 032501 (2012)
123. D. Guillemaud-Mueller et al., Z. Phys. A **332**, 189 (1989)
124. H. Sakurai et al., Phys. Lett. B **448**, 180 (1999)
125. F. Wamers et al., Phys. Rev. Lett. **112**, 132501 (2014)
126. R.J. Holt et al., Phys. Lett. B **69**, 55 (1977)
127. R. McPherson, J.C. Hardy, R.E. Bell, Phys. Lett. **11**, 65 (1964)
128. J.D. Gow, L.W. Alvarez, Phys. Rev. **94**, 365 (1954)
129. J.G. Fox et al., Phys. Rev. **55**, 1106 (1939)
130. F.W. Aston, Rep. B. Assoc. Adv. Sci. **83**, 403 (1913)
131. T.R. Hogness, H.M. Kvalnes, Phys. Rev. **32**, 942 (1928)
132. B.J. Dropesky, A.W. Schardt, Phys. Rev. **102**, 426 (1956)
133. G.W. Butler et al., Phys. Rev. Lett. **38**, 1380 (1977)
134. T.J.M. Symons et al., Phys. Rev. Lett. **42**, 40 (1979)
135. H. Sakurai et al., Phys. Rev. C **54**, R2802 (1996)
136. D. Guillemaud-Mueller et al., Phys. Rev. C **41**, 937 (1990)
137. M. Notani et al., Phys. Lett. B **542**, 49 (2002)
138. T. Zerguerras et al., Eur. Phys. J. A **20**, 389 (2004)
139. J. Cerny et al., Phys. Rev. Lett. **22**, 612 (1969)
140. E. Pollard, W.W. Watson, Phys. Rev. **58**, 12 (1940)
141. O.R. Frisch, Nature **136**, 220 (1935)
142. E. Fermi et al., Proc. R. Soc. A **146**, 483 (1934)
143. O. Huber et al., Helv. Phys. Acta **16**, 431 (1943)
144. M.J. Nurmia, R.W. Fink, Nucl. Phys. **8**, 139 (1958)
145. R. Klapisch et al., Phys. Rev. Lett. **20**, 740 (1968)
146. R. Klapisch et al., Phys. Rev. Lett. **23**, 652 (1969)
147. R. Klapisch et al., Phys. Rev. Lett. **29**, 1254 (1972)
148. M. Langevin et al., Phys. Lett. B **125**, 116 (1983)
149. I. Mukha et al., Phys. Rev. Lett. **99**, 182501 (2007)
150. R. McPherson, J.C. Hardy, Can. J. Phys. **43**, 1 (1964)
151. F. Ajzenberg-Selove, L. Cranberg, F.S. Dietrich, Phys. Rev. **124**, 1548 (1961)
152. M.G. White et al., Phys. Rev. **56**, 512 (1939)
153. A.J. Dempster, Science **52**, 559 (1920)
154. R.K. Sheline, N.R. Johnson, Phys. Rev. **89**, 520 (1953)
155. A.G. Artukh et al., Nucl. Phys. A **176**, 284 (1971)
156. T. Baumann et al., Nature **449**, 1022 (2007)
157. M.D. Cable et al., Phys. Rev. C **26**, 1778 (1982)
158. N.W. Glass, L.K. Jensen, Phys. Rev. **90**, 320 (1953)

159. J.L.W. Churchill, W.M. Jones, S.E. Hunt, Nature **172**, 460 (1953)
160. O.R. Frisch, Nature **133**, 721 (1934)
161. F.W. Aston, Nature **110**, 664 (1922)
162. I. Curie, F. Joliot, J. Phys. Radium **5**, 153 (1934)
163. H.A. Bethe, W.J. Henderson, Phys. Rev. **56**, 1060 (1939)
164. E.L. Robinson, O.E. Johnson, Phys. Rev. **123**, 1349 (1961)
165. M.G. Saint-Laurent et al., Phys. Rev. Lett. **59**, 33 (1987)
166. M. Langevin et al., Nucl. Phys. A **455**, 149 (1986)
167. J. Äystö et al., Phys. Lett. B **82**, 43 (1979)
168. R. Barton et al., Can. J. Phys. **41**, 2007 (1963)
169. E.L. Robinson, O.E. Johnson, Phys. Rev. **120**, 1321 (1960)
170. G. Kuerti, S.N. van Voorhis, Phys. Rev. **56**, 614 (1939)
171. R.S. Mulliken, Nature **113**, 423 (1924)
172. M. Lindner, Phys. Rev. **91**, 642 (1953)
173. P. Auger et al., Z. Phys. A **289**, 255 (1979)
174. M. Lewitowicz et al., Z. Phys. A **335**, 117 (1990)
175. O.B. Tarasov et al., Phys. Rev. C **75**, 064613 (2007)
176. M.D. Cable et al., Phys. Lett. B **123**, 25 (1983)
177. W. Benenson et al., Phys. Rev. C **15**, 1187 (1977)
178. M.G. White et al., Phys. Rev. **59**, 63 (1941)
179. R.K. Sheline, R.B. Holtzman, C.-Y. Fan, Phys. Rev. **83**, 919 (1951)
180. W. Zünti, E. Bleuler, Helv. Phys. Acta **18**, 263 (1945)
181. C.L. Morris et al., Phys. Rev. C **25**, 3218 (1982)
182. J.C. Hardy, R.I. Verrall, Phys. Lett. **13**, 148 (1964)
183. E.L. Robinson, J.I. Rhode, O.E. Johnson, Phys. Rev. **122**, 879 (1961)
184. L.D.P. King, D.R. Elliott, Phys. Rev. **58**, 846 (1940)
185. F.W. Aston, Nature **117**, 893 (1926)
186. E.B. Andersen, Z. Phys, Chemie **32B**, 237 (1936)
187. A.O. Nier, Phys. Rev. **53**, 282 (1938)
188. D.R. Nethaway, A.A. Caretto Jr., Phys. Rev. **109**, 504 (1958)
189. I. Mukha et al., Phys. Rev. Lett. **115**, 202501 (2015)
190. J.B. Hoag, Phys. Rev. **57**, 937 (1940)
191. D.C. Grahame, H.W. Walker, Phys. Rev. **60**, 909 (1941)
192. J.W. Kennedy, G.T. Seaborg, Phys. Rev. **57**, 843 (1940)
193. R.N.H. Haslam et al., Phys. Rev. **76**, 704 (1949)
194. H. Morinaga, Phys. Rev. **103**, 504 (1956)
195. E. Kashy et al., Phys. Rev. C **14**, 1773 (1976)
196. O.B. Tarasov et al., Phys. Rev. Lett. **102**, 142501 (2009)
197. E. Hagberg et al., Phys. Rev. Lett. **39**, 792 (1977)
198. P.L. Reeder, A.M. Poskanzer, R.A. Esterlund, Phys. Rev. Lett. **13**, 767 (1964)
199. R.G. Miller, R.W. Kavanagh, Phys. Lett. **22**, 461 (1966)
200. P.K. Weimer, J.D. Kurbatov, M.L. Pool, Phys. Rev. **60**, 469 (1941)
201. P. Zeeman, J. deGier, Proc. Akad. Soc. Amst. **37**, 127 (1934)
202. A.R. Brosi, H. Zeldes, B.H. Ketelle, Phys. Rev. **79**, 902 (1950)
203. A.H. Snell, Phys. Rev. **49**, 555 (1936)
204. S. Katcoff, Phys. Rev. **87**, 886 (1952)
205. P.G. Hansen et al., Phys. Lett. B **28**, 415 (1969)
206. R.E. Larson, C.M. Gordon, Nucl. Phys. A **133**, 237 (1969)
207. N.A. Jelley et al., Phys. Rev. C **9**, 2067 (1974)
208. D. Guillemaud-Mueller et al., Z. Phys. A **322**, 415 (1985)
209. S. Grévy et al., Phys. Lett. B **594**, 252 (2004)
210. W. Benenson et al., Phys. Rev. C **13**, 1479 (1976)
211. R.E. Berg, J.L. Snelgrove, E. Kashy, Phys. Rev. **153**, 1165 (1967)
212. C.R. Sun, B.T. Wright, Phys. Rev. **109**, 109 (1958)

213. D.G. Hurst, H. Walke, Phys. Rev. **51**, 1033 (1937)
214. A.O. Nier, Phys. Rev. **48**, 283 (1935)
215. F.W. Aston, Nature **107**, 72 (1921)
216. G. Hevesy, Nature **135**, 96 (1935)
217. R. Overstreet, L. Jacobson, P.R. Stout, Phys. Rev. **75**, 231 (1949)
218. B.L. Cohen, Phys. Rev. **94**, 117 (1954)
219. H. Morinaga, G. Wolzak, Phys. Lett. **11**, 148 (1964)
220. A. Marinov, J.R. Erskine, Phys. Lett. **14**, 46 (1965)
221. T. Kuroyanagi et al., Nucl. Phys. **50**, 417 (1964)
222. M. Langevin et al., Phys. Lett. B **130**, 251 (1983)
223. J. Äystö et al., Phys. Rev. Lett. **55**, 1384 (1985)
224. R.E. Tribble, J.D. Cossairt, R.A. Kenefick, Phys. Rev. C **15**, 2028 (1977)
225. J.C. Hardy, R.I. Verrall, Phys. Rev. Lett. **13**, 764 (1964)
226. J.C. Hardy, D.J. Skyrme, I.S. Towner, Phys. Lett. **23**, 487 (1966)
227. O. Huber et al., Helv. Phys. Acta **16**, 33 (1943)
228. A.J. Dempster, Phys. Rev. **20**, 631 (1922)
229. W.L. Davidson Jr., Phys. Rev. **56**, 1061 (1939)
230. F.W. Aston, Nature **133**, 684 (1934)
231. H. Walke, F.C. Thompson, J. Holt, Phys. Rev. **57**, 177 (1940)
232. R.E. Batzel, D.R. Miller, G.T. Seaborg, Phys. Rev. **84**, 671 (1951)
233. E. der Mateosian, M. Goldhaber, Phys. Rev. **79**, 192 (1950)
234. Y. Shida et al., Phys. Lett. **13**, 59 (1964)
235. A. Huck et al., Phys. Rev. C **22**, 2544 (1980)
236. A. Huck et al., Phys. Rev. C **31**, 2226 (1985)
237. M. Bernas et al., Phys. Lett. B **415**, 111 (1997)
238. C.L. Woods et al., Nucl. Phys. A **484**, 145 (1988)
239. N.W. Glass, J.R. Richardson, Phys. Rev. **98**, 1251 (1955)
240. D.R. Elliott, L.D.P. King, Phys. Rev. **59**, 403 (1941)
241. H. Morinaga, Phys. Rev. **100**, 431 (1955)
242. H. Walke, Phys. Rev. **51**, 439 (1937)
243. F.W. Aston, Nature **112**, 449 (1923)
244. G. Hevesy, H. Levi, Det Kgl. Danske Vid. Selskab., Math.-fys. Med. **XIV**, No. 5 (1936)
245. C.T. Hibdon, M.L. Pool, Phys. Rev. **67**, 313 (1945)
246. H. Walke, Phys. Rev. **52**, 777 (1937)
247. H. Walke, Phys. Rev. **57**, 163 (1940)
248. A. Poularikas, R.W. Fink, Phys. Rev. **115**, 989 (1959)
249. J.R. Erskine, A. Marinov, J.P. Schiffer, Phys. Rev. **142**, 633 (1966)
250. H. Breuer et al., Phys. Rev. C **22**, 2454 (1980)
251. X.L. Tu et al., Z. Phys. A **337**, 361 (1990)
252. C. Détraz et al., Nucl. Phys. A **519**, 529 (1990)
253. A.D. Schelberg, M.B. Sampson, A.C.G. Mitchell, Rev. Sci. Instrum. **19**, 458 (1948)
254. R.A. Sharp, R.M. Diamond, Phys. Rev. **93**, 358 (1954)
255. J.S.V. Allen et al., Phys. Rev. **60**, 425 (1941)
256. L. Seren, H.N. Friedlander, S.H. Turkel, Phys. Rev. **72**, 888 (1947)
257. D.C. Williams, J.D. Knight, W.T. Leland, Phys. Lett. **22**, 162 (1966)
258. L.A. Parks, C.N. Davids, R.C. Pardo, Phys. Rev. C **15**, 730 (1977)
259. D. Guerreau et al., Z. Phys. A **295**, 105 (1980)
260. M. Weber et al., Z. Phys. A **343**, 67 (1992)
261. O.B. Tarasov et al., Phys. Rev. C **87**, 054612 (2013)
262. F. Pougheon et al., Z. Phys. A **327**, 17 (1987)
263. J. Cerny, D.R. Goosman, D.E. Alburger, Phys. Lett. B **37**, 380 (1971)
264. K.P. Jackson et al., Phys. Lett. B **49**, 341 (1974)
265. W.M. Martin, S.W. Breckon, Can. J. Phys. **30**, 643 (1952)
266. J.J. O'Connor, M.L. Pool, J.D. Kurbatov, Phys. Rev. **62**, 413 (1942)

267. H. Walke, Phys. Rev. **52**, 669 (1937)
268. L.A. Turner, Phys. Rev. **58**, 679 (1940)
269. E. Amaldi et al., Ric. Scient. **5**, 21 (1934)
270. I. Kumabe et al., Phys. Rev. **117**, 1568 (1960)
271. T.W. Ward, P.H. Pile, P.K. Kuroda, Nucl. Phys. A **148**, 225 (1970)
272. A.M. Nathan et al., Phys. Rev. C **16**, 1566 (1977)
273. B. Blank et al., Phys. Rev. Lett. **77**, 2893 (1996)
274. V. Borrel et al., Z. Phys. A **344**, 135 (1992)
275. J. Zioni et al., Nucl. Phys. A **181**, 465 (1972)
276. I.D. Proctor et al., Phys. Rev. Lett. **29**, 434 (1972)
277. G. Rudstam, P.C. Stevenson, R.L. Folger, Phys. Rev. **87**, 358 (1952)
278. F.W. Aston, Nature **126**, 200 (1930)
279. H. Walke, F.C. Thompson, J. Holt, Phys. Rev. **57**, 171 (1940)
280. A. Flammersfeld, W. Herr, Z. Naturforsch. **7a**, 649 (1952)
281. B.J. Dropesky, A.W. Schardt, T.T. Shull, Nucl. Phys. **16**, 357 (1960)
282. C.N. Davids et al., Phys. Rev. C **17**, 1815 (1978)
283. T. Sekine et al., Nucl. Phys. A **467**, 93 (1987)
284. J. Cerny et al., Phys. Rev. Lett. **24**, 1128 (1970)
285. J.J. Livingood, G.T. Seaborg, Phys. Rev. **54**, 391 (1938)
286. J.R. Wilkinson, R.K. Sheline, Phys. Rev. **99**, 752 (1955)
287. B.L. Cohen et al., Phys. Rev. **94**, 953 (1954)
288. D.M. Chittenden II, D.G. Gardner, R.W. Fink, Phys. Rev. **122**, 860 (1961)
289. E.B. Norman et al., Phys. Rev. C **17**, 2176 (1978)
290. E. Runte et al., Nucl. Phys. A **399**, 163 (1983)
291. U. Bosch et al., Phys. Lett. B **164**, 22 (1985)
292. T. Ohnishi et al., J. Phys. Soc. Jpn **79**, 073201 (2010)
293. R.E. Tribble et al., Phys. Rev. C **16**, 917 (1977)
294. D.R. Miller, R.C. Thompson, B.B. Cunningham, Phys. Rev. **74**, 347 (1948)
295. J.J. Livingood, G.T. Seaborg, Phys. Rev. **54**, 51 (1938)
296. F.W. Aston, Nature **113**, 856 (1924)
297. J.J. Livingood, G.T. Seaborg, Phys. Rev. **55**, 1268 (1939)
298. F.W. Aston, Nature **110**, 312 (1922)
299. F.W. Aston, Proc. R. Soc. A **149**, 396 (1935)
300. J. deGier, P. Zeeman, Proc. Akad. Soc. Amst. **38**, 959 (1935)
301. J.-C. Roy, T.P. Kohman, Can. J. Phys. **35**, 649 (1957)
302. E. Ricci, J. Pahissa Campá, N. Nussis, Publs. Com. Nacl. Energia Atomica, Ser. quim **1**, 129 (1957)
303. E.M. Franz et al., Phys. Rev. C **12**, 616 (1975)
304. K.P. Jackson et al., Phys. Lett. B **33**, 281 (1970)
305. J.J. Livingood, G.T. Seaborg, Phys. Rev. **53**, 847 (1938)
306. J.J. Livingood, G.T. Seaborg, Phys. Rev. **60**, 913 (1941)
307. T.J. Parmley, B.J. Moyer, Phys. Rev. **72**, 82 (1947)
308. T.J. Parmley, B.J. Moyer, R.C. Lilly, Phys. Rev. **75**, 619 (1949)
309. H. Morinaga et al., J. Phys. Soc. Jpn **15**, 213 (1960)
310. T.E. Ward, P.H. Pile, P.K. Kuroda, J. Inorg. Nucl. Chem. **31**, 2679 (1969)
311. R.T. Kouzes, D. Mueller, Nucl. Phys. A **307**, 71 (1978)
312. C. Engelmann et al., Z. Phys. A **352**, 351 (1995)
313. Z.Y. Xu et al., Phys. Rev. Lett. **113**, 032505 (2014)
314. B. Blank et al., Phys. Rev. Lett. **84**, 1116 (2000)
315. B. Blank et al., Phys. Rev. C **50**, 2398 (1994)
316. D.J. Vieira et al., Phys. Lett. B **60**, 261 (1976)
317. W.J. Worthington Jr., Phys. Rev. **87**, 158 (1952)
318. J.J. Livingood, G.T. Seaborg, Phys. Rev. **53**, 765 (1938)
319. F.W. Aston, Nature **107**, 520 (1921)

320. A.R. Brosi et al., Phys. Rev. **81**, 391 (1951)
321. F.W. Aston, Nature **134**, 178 (1934)
322. J. deGier, P. Zeeman, Proc. Akad. Soc. Amst. **38**, 810 (1935)
323. J.A. Swartout, G.E. Boyd, Phys. Rev. **70**, 232 (1946)
324. R.H. Goeckermann, I. Perlman, Phys. Rev. **73**, 1127 (1948)
325. R.T. Kouzes, D. Mueller, C. Yu, Phys. Rev. C **18**, 1587 (1978)
326. T.S. Bhatia et al., Z. Phys. A **281**, 65 (1977)
327. Ph Dessagne et al., Nucl. Phys. A **426**, 399 (1984)
328. P. Armbruster et al., Europhys. Lett. **4**, 793 (1987)
329. C.E. Leith, A. Bratenahl, B.J. Moyer, Phys. Rev. **72**, 732 (1947)
330. L.N. Ridenour, W.J. Henderson, Phys. Rev. **52**, 889 (1937)
331. F.A. Heyn, Nature **138**, 723 (1936)
332. F.W. Aston, Nature **112**, 162 (1923)
333. S.N. Van Voorhis, Phys. Rev. **50**, 895 (1936)
334. W.Y. Chang, M. Goldhaber, R. Sagane, Nature **139**, 962 (1937)
335. A. Flammersfeld, Z. Naturforsch. **8a**, 274 (1953)
336. J. van Klinken et al., Phys. Rev. **150**, 886 (1966)
337. L.M. Taff, B.K.S. Koene, J. van Klinken, Nucl. Phys. A **164**, 565 (1971)
338. P.L. Reeder et al., Phys. Rev. C **31**, 1029 (1985)
339. K.-L. Kratz et al., Z. Phys. A **340**, 419 (1991)
340. B. Blank et al., Phys. Rev. Lett. **94**, 232501 (2005)
341. J. Giovinazzo et al., Eur. Phys. J. A **11**, 247 (2001)
342. K.K. Seth et al., Phys. Lett. B **173**, 397 (1986)
343. J. Honkanen et al., Nucl. Phys. A **366**, 109 (1981)
344. L. Lindner, G.A. Brinkman, Physica **21**, 747 (1955)
345. W. Bothe, W. Gentner, Naturwiss. **25**, 191 (1937)
346. J.J. Livingood, G.T. Seaborg, Phys. Rev. **55**, 457 (1939)
347. F.W. Aston, Nature **122**, 345 (1928)
348. F.A. Heyn, Physica **4**, 1224 (1937)
349. J.M. LeBlanc, J.M. Cork, S.B. Burson, Phys. Rev. **97**, 750 (1955)
350. J.M. Siegel, L.E. Glendenin, Radiochemical Studies: The Fission Products, Paper 53, p. 549. Nat. Nucl. Energy Ser. IV, 9 (McGraw-Hill, New York, 1951)
351. B.R. Erdal et al., Nucl. Phys. A **194**, 449 (1972)
352. B. Grapengiesser, E. Lund, G. Rudstam, J. Inorg. Nucl. Chem. **36**, 2409 (1974)
353. K. Aleklett et al., Nucl. Phys. A **285**, 1 (1977)
354. B. Ekström et al., Physica Scripta **34**, 614 (1986)
355. R.L. Gill et al., Phys. Rev. Lett. **56**, 1874 (1986)
356. B. Blank et al., Phys. Rev. Lett. **74**, 4611 (1995)
357. M.A.C. Hotchkis et al., Phys. Rev. C **35**, 315 (1987)
358. R. Chiba et al., Phys. Rev. C **17**, 2219 (1978)
359. M.J. Nurmia, R.W. Fink, Phys. Lett. **14**, 136 (1965)
360. B.L. Cohen, Phys. Rev. **91**, 74 (1953)
361. L.W. Alvarez, Phys. Rev. **54**, 486 (1938)
362. W.B. Mann, Phys. Rev. **52**, 405 (1937)
363. L.W. Alvarez, Phys. Rev. **53**, 606 (1938)
364. R. Sagane, Phys. Rev. **55**, 31 (1939)
365. K. Takahashi et al., J. Phys. Soc. Jpn **16**, 1664 (1961)
366. L. Wish, Phys. Rev. **172**, 1262 (1968)
367. P. del Marmol, P. Fettweis, Nucl. Phys. A **194**, 140 (1972)
368. G. Rudstam, E. Lund, Phys. Rev. C **13**, 321 (1976)
369. A.A. Ciemny et al., Phys. Rev. C **92**, 014622 (2015)
370. A. Stolz et al., Phys. Lett. B **627**, 32 (2005)
371. M.F. Mohar et al., Phys. Rev. Lett. **66**, 1571 (1991)
372. R.G.H. Robertson, S.M. Austin, Phys. Rev. Lett. **29**, 130 (1972)

373. H.H. Hopkins, Phys. Rev. **77**, 717 (1950)
374. H.H. Hopkins, B.B. Cunningham, Phys. Rev. **73**, 1406 (1948)
375. W.B. Mann, Phys. Rev. **54**, 649 (1938)
376. F.W. Aston, Nature **111**, 771 (1923)
377. G.T. Seaborg, J.J. Livingood, G. Friedlander, Phys. Rev. **59**, 320 (1941)
378. F.W. Aston, Proc. R. Soc. A **132**, 487 (1931)
379. E.P. Steinberg, D. Engelkemeir, *Radiochemical Studies: The Fission Products, Paper 54*. Nat. Nucl. Energy Ser. IV, 9 (McGraw-Hill, New York, 1951), p. 566
380. M. Karras, T.E. Ward, H. Ihochi, Nucl. Phys. A **147**, 120 (1970)
381. J.P. Omtvedt et al., Z. Phys. A **338**, 241 (1991)
382. M. Bernas et al., Phys. Lett. B **331**, 19 (1994)
383. D.E. Alburger, Phys. Rev. C **18**, 1875 (1978)
384. M.J. Murphy, C.N. Davids, E.B. Norman, Phys. Rev. C **22**, 2204 (1980)
385. T. Paradellis, A.J. Houdayer, S.K. Mark, Nucl. Phys. A **174**, 617 (1971)
386. F.D.S. Butement, E.G. Prout, Phil. Mag. **46**, 357 (1955)
387. A.C.G. Mitchell, E.T. Jurney, M. Ramsey, Phys. Rev. **71**, 825 (1947)
388. D.A. McCown, L.L. Woodward, M.L. Pool, Phys. Rev. **74**, 1315 (1948)
389. R. Sagane, S. Kojima, M. Ikawa, Phys. Rev. **54**, 149 (1938)
390. A.H. Snell, Phys. Rev. **52**, 1007 (1937)
391. F.D.S. Butement, Nature **165**, 149 (1950)
392. R.E. Meads, J.E.G. McIldowie, Proc. Phys. Soc. **74**, 693 (1959)
393. P.J. Mathew, G.J. McCallum, Phys. Lett. B **28**, 106 (1968)
394. P. del Marmol, J. Inorg. Nucl. Chem. **30**, 2873 (1968)
395. P. del Marmol, M. Nève de Mévergnies, J. Inorg. Nucl. Chem. **29**, 273 (1967)
396. J.V. Kratz, H. Franz, G. Herrmann, J. Inorg. Nucl. Chem. **35**, 1407 (1973)
397. J.V. Kratz, G. Herrmann, J. Inorg. Nucl. Chem. **32**, 3713 (1970)
398. J.C. Batchelder et al., Phys. Rev. C **47**, 2038 (1993)
399. J.A. Winger et al., Phys. Rev. C **48**, 3097 (1993)
400. C.J. Lister et al., Phys. Rev. C **42**, R1191 (1990)
401. E. Nolte et al., Z. Phys. **268**, 267 (1974)
402. J. Beydon et al., Compt. Rend. Acad. Sci. **244**, 586 (1957)
403. W.S. Cowart et al., Phys. Rev. **73**, 1454 (1948)
404. H.N. Friedlander, L. Seren, S.H. Turkel, Phys. Rev. **72**, 23 (1947)
405. A. Flammersfeld, W. Herr, Z. Naturforsch. **5a**, 569 (1950)
406. H. Wäffler, O. Hirzel, Helv. Phys. Acta **21**, 200 (1948)
407. J.E. Sattizahn, J.D. Knight, J. Inorg. Nucl. Chem. **12**, 206 (1960)
408. T. Tamai et al., Inorg. Nucl. Chem. Lett. **9**, 1145 (1973)
409. L. Tomlinson, M.H. Hurdus, J. Inorg. Nucl. Chem. **30**, 1995 (1968)
410. P. del Marmol, D.C. Perricos, J. Inorg. Nucl. Chem. **32**, 705 (1970)
411. L. Tomlinson, M.H. Hurdus, J. Inorg. Nucl. Chem. **33**, 3609 (1971)
412. M. Asghar et al., Nucl. Phys. A **247**, 359 (1975)
413. A.M. Rogers et al., Phys. Rev. Lett. **106**, 252503 (2011)
414. E. Hagberg et al., Nucl. Phys. A **383**, 109 (1982)
415. E. Nolte et al., Phys. Lett. B **33**, 294 (1970)
416. G. Murray et al., Nucl. Phys. A **142**, 21 (1970)
417. J.M. Hollander, Phys. Rev. **92**, 518 (1953)
418. L.L. Woodward, D.A. McCown, M.L. Pool, Phys. Rev. **74**, 870 (1948)
419. S.C. Fultz, M.L. Pool, Phys. Rev. **86**, 347 (1952)
420. W. Bothe, W. Gentner, Naturwiss. **25**, 126 (1937)
421. W. Bothe, W. Gentner, Naturwiss. **25**, 90 (1937)
422. H.J. Born, W. Seelmann-Eggebert, Naturwiss. **31**, 86 (1943)
423. A.F. Stehney, E.P. Steinberg, Phys. Rev. **127**, 563 (1962)
424. N. Sugarman, J. Chem. Phys. **17**, 11 (1949)
425. G.J. Perlow, A.F. Stehney, Phys. Rev. **113**, 1269 (1959)

426. K.-L. Kratz, G. Herrmann, Nucl. Phys. A **229**, 179 (1974)
427. K.-L. Kratz et al., Z. Phys. A **330**, 229 (1988)
428. G.T. Ewan et al., Nucl. Phys. A **352**, 13 (1981)
429. H. Schmeing et al., Phys. Lett. B **44**, 449 (1973)
430. P. Hornshoj et al., Nucl. Phys. A **187**, 637 (1972)
431. F.D.S. Butement, G.G.J. Boswell, J. Inorg. Nucl. Chem. **16**, 10 (1960)
432. A.A. Caretto Jr., E.O. Wiig, Phys. Rev. **93**, 175 (1954)
433. L.L. Woodward, D.A. McCown, M.L. Pool, Phys. Rev. **74**, 761 (1948)
434. J.H. Reynolds, Phys. Rev. **79**, 886 (1950)
435. A. Langsdorf Jr., Phys. Rev. **56**, 205 (1939)
436. O. Hahn, F. Strassmann, Naturwiss. **31**, 249 (1943)
437. C.R. Dillard et al., *Radiochemical Studies: The Fission Products, Paper 68*. Nat. Nucl. Energy Ser. IV, 9 (McGraw-Hill, New York, 1951), p. 624
438. S. Amiel et al., Phys. Rev. C **5**, 270 (1972)
439. J.C. Batchelder et al., Phys. Rev. C **48**, 2593 (1993)
440. J.M. D'Auria et al., Phys. Lett. B **66**, 233 (1977)
441. H.L. Ravn et al., J. Inorg. Nucl. Chem. **37**, 383 (1975)
442. J. Chaumont et al., Phys. Lett. B **29**, 652 (1969)
443. R. Arlt et al., Nucl. Instrum. Meth. **102**, 253 (1972)
444. C.J. Toeset, A.H.W. Aten Jr., Radiochim. Acta **9**, 55 (1968)
445. J. Beydon et al., Nucl. Phys. **2**, 593 (1957)
446. R.W. Hoff, J.M. Hollander, M.C. Michel, J. Inorg. Nucl. Chem. **18**, 1 (1961)
447. F.L. Reynolds, D.G. Karraker, D.H. Templeton, Phys. Rev. **75**, 313 (1949)
448. D.G. Karraker, D.H. Templeton, Phys. Rev. **80**, 646 (1950)
449. W.C. Barber, Phys. Rev. **72**, 1156 (1947)
450. A.C. Helmholz, C. Pecher, P.R. Stout, Phys. Rev. **59**, 902 (1941)
451. F.A. Heyn, A.H.W. Aten, C.J. Bakker, Nature **143**, 516 (1939)
452. G.N. Glasoe, J. Steigman, Phys. Rev. **58**, 1 (1940)
453. O. Kofoed-Hansen, K.O. Nielsen, Phys. Rev. **82**, 96 (1951)
454. K. Fritze, T.J. Kennett, Can. J. Phys. **38**, 1614 (1960)
455. K. Fritze, T.J. Kennett, W.V. Prestwich, Can. J. Phys. **39**, 67 (1961)
456. I. Amarel et al., Phys. Lett. B **24**, 402 (1967)
457. I. Amarel, H. Gauvin, A. Johnson, J. Inorg. Nucl. Chem. **31**, 577 (1969)
458. B.L. Tracy et al., Phys. Lett. B **34**, 277 (1971)
459. E. Koglin et al., Z. Phys. A **288**, 319 (1978)
460. K. Balog et al., Z. Phys. A **342**, 125 (1992)
461. G. Lhersonneau et al., Z. Phys. A **351**, 357 (1995)
462. J.C. Hardy et al., Phys. Lett. B **63**, 27 (1976)
463. C.J. Lister et al., Phys. Rev. Lett. **49**, 308 (1982)
464. I.M. Ladenbauer-Bellis, H. Bakhru, B. Jones, Can. J. Phys. **50**, 3071 (1972)
465. S.V. Castner, D.H. Templeton, Phys. Rev. **88**, 1126 (1952)
466. J.P. Blewett, M.B. Sampson, Phys. Rev. **49**, 778 (1936)
467. L.A. DuBridge, J. Marshall, Phys. Rev. **58**, 7 (1940)
468. F.W. Aston, Nature **128**, 221 (1931)
469. D.W. Stewart, J.L. Lawson, J.M. Cork, Phys. Rev. **52**, 901 (1937)
470. R.W. Nottorf, *Radiochemical Studies: The Fission Products, Paper 77*. Nat. Nucl. Energy Ser. IV, 9 (McGraw-Hill, New York, 1951), p. 682
471. W. Seelmann-Eggebert, Naturwiss. **31**, 510 (1943)
472. H. Götte, Naturwiss. **29**, 496 (1941)
473. J.D. Knight et al., J. Inorg. Nucl. Chem. **10**, 183 (1959)
474. M.-I. Macias-Marques et al., J. Phys. (Paris) **32**, 237 (1971)
475. F.K. Wohn et al., Phys. Rev. C **17**, 2185 (1978)
476. F.K. Wohn et al., Phys. Rev. Lett. **51**, 873 (1983)
477. J.C. Hill et al., Phys. Rev. C **33**, 1727 (1986)

478. P. Kienle et al., Prog. Part. Nucl. Phys. **46**, 73 (2001)
479. Z. Janas et al., Phys. Rev. Lett. **82**, 295 (1999)
480. S.J. Yennello et al., Phys. Rev. C **46**, 2620 (1992)
481. H. Grawe et al., Z. Phys. A **341**, 247 (1992)
482. C.J. Lister et al., Phys. Rev. C **24**, 260 (1981)
483. C. Deprun et al., Z. Phys. A **295**, 103 (1980)
484. V. Maxia, W.H. Kelly, D.J. Horen, J. Inorg. Nucl. Chem. **24**, 1175 (1962)
485. A.A. Caretto Jr., E.O. Wiig, J. Am. Chem. Soc. **74**, 5235 (1952)
486. E.K. Hyde, G.D. O'Kelley, Phys. Rev. **82**, 944 (1951)
487. R.J. Hayden, Phys. Rev. **74**, 650 (1948)
488. M.L. Pool, J.M. Cork, R.L. Thornton, Phys. Rev. **52**, 239 (1937)
489. R. Sagane et al., Phys. Rev. **57**, 1179 (1940)
490. S. Katcoff, J.A. Miskel, C.W. Stanley, Phys. Rev. **74**, 631 (1948)
491. H. Gunther et al., Nucl. Phys. A **242**, 56 (1975)
492. J. Eidens, E. Roeckl, P. Armbruster, Nucl. Phys. A **141**, 289 (1970)
493. J.W. Grüter et al., Phys. Lett. B **33**, 474 (1970)
494. B. Pfeiffer et al., J. Phys. (Paris) **38**, 9 (1977)
495. K. Shizuma et al., Phys. Rev. C **27**, 2869 (1983)
496. C.J. Lister et al., Phys. Rev. Lett. **59**, 1270 (1987)
497. W.X. Huang et al., Z. Phys. A **359**, 349 (1997)
498. C.F. Liang et al., Z. Phys. A **309**, 185 (1982)
499. M. Kaba, K. Miyanon, Radiochim. Acta **21**, 203 (1974)
500. G. Korschinek et al., Z. Phys. A **281**, 409 (1977)
501. F.D.S. Butement, G.B. Briscoe, J. Inorg. Nucl. Chem. **25**, 151 (1963)
502. B.E. Robertson, W.E. Scott, M.L. Pool, Phys. Rev. **76**, 1649 (1949)
503. R. Sagane et al., Phys. Rev. **54**, 542 (1938)
504. F.W. Aston, Nature **114**, 273 (1924)
505. E.P. Steinberg, L.E. Glendenin, Phys. Rev. **78**, 624 (1950)
506. W.E. Grummitt, G. Wilkinson, Nature **158**, 163 (1946)
507. S. Katcoff, B. Finkle, *Radiochemical Studies: The Fission Products, Paper 83*. Nat. Nucl. Energy Ser. IV, 9 (McGraw-Hill, New York, 1951), p. 705
508. K. Hübenthal, Compt. Rend. Acad. Sci. **264**, 1468 (1967)
509. E. Cheifetz et al., Phys. Rev. Lett. **25**, 38 (1970)
510. R.L. Watson et al., Nucl. Phys. A **141**, 449 (1970)
511. M. Graefenstedt et al., Z. Phys. A **327**, 383 (1987)
512. M.A.C. Hotchkis et al., Phys. Rev. Lett. **64**, 3123 (1990)
513. J. Äystö et al., Phys. Rev. Lett. **69**, 1167 (1992)
514. T. Kuroyanagi et al., Nucl. Phys. A **484**, 264 (1988)
515. I. Votsilka et al., Bull. Acad. Sci. USSR Phys. Ser. **38**(1), 49 (1974)
516. T.A. Doron, M. Blann, Nucl. Phys. A **161**, 12 (1971)
517. R.G. Korteling, E.K. Hyde, Phys. Rev. **136**, B425 (1964)
518. R.M. Diamond, Phys. Rev. **95**, 410 (1954)
519. L. Jacobson, R. Overstreet, *Radiochemical Studies: The Fission Products, Paper 91*. Nat. Nucl. Energy Ser. IV, 9 (McGraw-Hill, New York, 1951), p. 735
520. R. Sagane et al., Phys. Rev. **54**, 970 (1938)
521. F.W. Aston, Nature **130**, 130 (1932)
522. D.N. Kundu, M.L. Pool, Phys. Rev. **75**, 1690 (1949)
523. C.J. Orth, R.K. Smith, J. Inorg. Nucl. Chem. **15**, 4 (1960)
524. R.B. Duffield, L. Hsiao, E.N. Sloth, Phys. Rev. **79**, 1011 (1950)
525. K. Hübenthal, E. Monnand, A. Moussa, Compt. Rend. Acad. Sci. **265**, 579 (1967)
526. N. Trautmann et al., Radiochim. Acta **18**, 86 (1972)
527. F.F. Hopkins et al., Phys. Rev. C **4**, 1927 (1971)
528. K. Shizuma et al., Z. Phys. A **315**, 65 (1984)
529. H. Ahrens et al., Phys. Rev. C **14**, 211 (1976)

530. W. Gelletly et al., Phys. Lett. B **253**, 287 (1991)
531. C.J. Gross et al., Phys. Rev. C **44**, 2253 (1991)
532. R.C. Pardo et al., Phys. Rev. C **21**, 462 (1980)
533. R.M. Diamond, Phys. Rev. **89**, 1149 (1953)
534. F.W. Aston, Nature **126**, 348 (1930)
535. M.L. Wiedenbeck, Phys. Rev. **70**, 435 (1946)
536. W. Maurer, W. Ramm, Naturwiss. **29**, 368 (1941)
537. D.R. Wiles, C.D. Coryell, Phys. Rev. **96**, 696 (1954)
538. P. Kienle et al., Radiochim. Acta **1**, 84 (1963)
539. P. Kienle et al., Naturwiss. **49**, 294 (1962)
540. P. Kienle et al., Naturwiss. **49**, 295 (1962)
541. J.D. Hastings, D.E. Troutner, R.L. Ferguson, Radiochim. Acta **11**, 51 (1969)
542. G. Lorusso et al., Phys. Rev. Lett. **114**, 192501 (2015)
543. D. Rudolph et al., J. Phys. G **17**, L113 (1991)
544. K. Heiguchi et al., Z. Phys. A **338**, 7 (1991)
545. R. Iafigliola et al., Can. J. Phys. **52**, 96 (1974)
546. R. van Lieshout et al., Phys. Lett. **9**, 164 (1964)
547. D.N. Kundu, M.L. Pool, Phys. Rev. **74**, 1775 (1948)
548. E.E. Motta, G.E. Boyd, Phys. Rev. **74**, 220 (1948)
549. J.E. Edwards, M.L. Pool, Phys. Rev. **72**, 384 (1947)
550. E.E. Motta, G.E. Boyd, A.R. Brosi, Phys. Rev. **71**, 210 (1946)
551. G.E. Boyd et al., Phys. Rev. **99**, 1030 (1955)
552. E. Segrè, G.T. Seaborg, Phys. Rev. **54**, 772 (1938)
553. G.E. Boyd, Q.V. Larson, G.W. Parker, Phys. Rev. **86**, 1051 (1952)
554. J. Flegenheimer, W. Seelmann-Eggebert, Z. Naturforsch. **9a**, 806 (1954)
555. J. Flegenheimer, D. Geithoff, Z. Naturforsch. **12a**, 351 (1957)
556. J. Flegenheimer, W. Seelmann-Eggebert, Z. Naturforsch. **11a**, 678 (1956)
557. J. Flegenheimer, G. B. Baró, A. Medina, Z. Naturforsch. **10a**, 798 (1955)
558. A. von Baeckmann, H. Feuerstein, Radiochim. Acta **4**, 111 (1965)
559. W. John, F.W. Guy, J.J. Wesolowski, Phys. Rev. C **2**, 1451 (1970)
560. N. Trautmann et al., Phys. Rev. C **13**, 872 (1976)
561. H. Penttilä et al., Phys. Rev. C **38**, 931 (1988)
562. J. Äystö et al., Nucl. Phys. A **515**, 365 (1990)
563. K. Rykaczewski et al., Phys. Rev. C **52**, 2310 (1995)
564. M. Hencheck et al., Phys. Rev. C **50**, 2219 (1994)
565. S. Zhou et al., Chin. J. Nucl. Phys. **13**, 193 (1991)
566. E. Hagberg et al., Nucl. Phys. A **395**, 152 (1983)
567. R. Arlt et al., JETP Lett. **13**, 397 (1971)
568. T.A. Doron, W.A. Lanford, J. Inorg. Nucl. Chem. **34**, 443 (1972)
569. A. van der Wiel, A.H.W. Aten Jr., Physica **18**, 356 (1952)
570. D.T. Eggen, M.L. Pool, Phys. Rev. **74**, 57 (1948)
571. F.W. Aston, Nature **127**, 233 (1931)
572. W.H. Sullivan, N.R. Sleight, E.M. Gladrow, Phys. Rev. **70**, 778 (1946)
573. H. Ewald, Z. Phys. **122**, 487 (1944)
574. E. Bohr, N. Hole, Arkiv Mat. Astron. Fysik **32A**, No. 15 (1945)
575. L.E. Glendenin, *Radiochemical Studies: The Fission Products, Paper 115*. Nat. Nucl. Energy Ser. IV, 9 (McGraw-Hill, New York, 1951), p. 849
576. G.B. Baró, P. Rey, W. Seelmann-Eggebert, Z. Naturforsch. **10a**, 81 (1955)
577. K. Griffiths, K. Fritze, Z. Anal, Chem. **226**, 122 (1967)
578. M. Leino et al., Phys. Rev. C **44**, 336 (1991)
579. R.A. Zarifov, O.D. Kovrigin, V.P. Burminskii, Bull. Acad. Sci. USSR Phys. Ser. **43**(1), 76 (1979)
580. A.H.W. Aten Jr., J.C. Kapteijn, Physica **33**, 705 (1967)
581. A.H.W. Aten Jr., T. De Vries-Hamerling, Physica **21**, 597 (1955)

582. C.L. Scoville, S.C. Fultz, M.L. Pool, Phys. Rev. **85**, 1046 (1952)
583. M. Lindner, I. Perlman, Phys. Rev. **73**, 1202 (1948)
584. O. Minikawa, Phys. Rev. **60**, 689 (1941)
585. E.C. Crittenden Jr., Phys. Rev. **56**, 709 (1939)
586. F.F. Hopkins et al., Phys. Rev. C **5**, 1015 (1972)
587. M. Karras, J. Kantele, Phys. Lett. **6**, 98 (1963)
588. G. Franz, G. Herrmann, Inorg. Nucl. Chem. Lett. **11**, 857 (1975)
589. J. Äystö et al., Nucl. Phys. A **480**, 104 (1988)
590. J. Äystö et al., Phys. Lett. B **201**, 211 (1988)
591. H. Penttilä et al., Phys. Rev. C **44**, 935 (1991)
592. W. Kurcewicz et al., Z. Phys. A **308**, 21 (1982)
593. E. Nolte, H. Hick, Phys. Lett. B **97**, 55 (1980)
594. N.K. Aras, P.W. Gallagher, W.B. Walters, J. Phys. G **6**, L195 (1980)
595. A.H.W. Aten Jr., J.C. Kapteijn, Radiochim. Acta **12**, 218 (1969)
596. A.J. Dempster, Nature **136**, 65 (1935)
597. J.Y. Mei, C.M. Huddleston, A.C.G. Mitchell, Phys. Rev. **79**, 429 (1950)
598. U. Schindewolf, Phys. Rev. **109**, 1280 (1958)
599. J.D. Kraus, J.M. Cork, Phys. Rev. **52**, 763 (1937)
600. J.A. Seiler, *Radiochemical Studies: The Fission Products, Paper 119*. Nat. Nucl. Energy Ser. IV, 9 (McGraw-Hill, New York, 1951), p. 860
601. H.G. Hicks, R.S. Gilbert, Phys. Rev. **94**, 371 (1954)
602. J.M. Alexander, U. Schindewolf, C.D. Coryell, Phys. Rev. **111**, 228 (1958)
603. H.V. Weiss, J.L. Elzie, J.M. Fresco, Phys. Rev. **172**, 1269 (1968)
604. H.V. Weiss et al., Phys. Rev. **188**, 1893 (1969)
605. H. Penttilä et al., Z. Phys. A **338**, 291 (1991)
606. Z. Janas et al., Nucl. Phys. A **552**, 340 (1993)
607. M. Huyse et al., Z. Phys. A **288**, 107 (1978)
608. H. Bakhru, R.I. Morse, I.L. Preiss, Nucl. Phys. A **100**, 145 (1967)
609. D.J. Hnatowich et al., J. Inorg. Nucl. Chem. **32**, 3137 (1970)
610. F.D.S. Butement, M.Y. Mirza, J. Inorg. Nucl. Chem. **28**, 303 (1966)
611. O. Ames et al., Phys. Rev. **118**, 1599 (1960)
612. B.C. Haldar, E.O. Wiig, Phys. Rev. **94**, 1713 (1954)
613. F.A. Johnson, Can. J. Phys. **33**, 841 (1955)
614. T. Enns, Phys. Rev. **56**, 872 (1939)
615. M.L. Pool, Phys. Rev. **53**, 116 (1938)
616. R.B. Duffield, J.D. Knight, Phys. Rev. **75**, 1613 (1949)
617. H.V. Weiss, J.M. Fresco, W.L. Reichert, Phys. Rev. **172**, 1266 (1968)
618. Y. Kawase et al., Nucl. Phys. A **241**, 237 (1975)
619. B. Fogelberg, A. Backlin, T. Nagarajan, Phys. Lett. B **36**, 334 (1971)
620. B. Fogelberg, P. Hoff, Nucl. Phys. A **391**, 445 (1982)
621. L.L. Shih, J.C. Hill, S.A. Williams, Phys. Rev. C **17**, 1163 (1978)
622. E. Lund, G. Rudstam, Phys. Rev. C **13**, 1544 (1976)
623. J.C. Hill et al., Phys. Rev. C **29**, 1078 (1984)
624. V.N. Fedoseyev et al., Z. Phys. A **353**, 9 (1995)
625. T. Kautzsch et al., Eur. Phys. J. A **9**, 201 (2000)
626. K.-L. Kratz et al., Hyperfine Interact. **129**, 185 (2000)
627. H. Wang et al., Chin. Phys. Lett. **30**, 042501 (2013)
628. C.B. Hinke et al., Nature **486**, 341 (2012)
629. D. Bazin et al., Phys. Rev. Lett. **101**, 252501 (2008)
630. T. Elmroth et al., Nucl. Phys. A **304**, 493 (1978)
631. I.L. Preiss, P.J. Estrup, R. Wolfgang, Nucl. Phys. **18**, 624 (1960)
632. J.R. Gum, M.L. Pool, Phys. Rev. **80**, 315 (1950)
633. A.C. Helmholz, Phys. Rev. **70**, 982 (1946)
634. M. Lindner, I. Perlman, Phys. Rev. **78**, 499 (1950)

635. F.W. Aston, Nature **114**, 717 (1924)
636. M. Goldhaber, R.D. Hill, L. Szilard, Phys. Rev. **55**, 47 (1939)
637. C.E. Gleit, C.D. Coryell, Phys. Rev. **122**, 229 (1961)
638. C.E. Gleit, C.D. Coryell, Phys. Rev. **124**, 1914 (1961)
639. E. Cheifetz et al., Phys. Rev. C **4**, 1913 (1971)
640. H.V. Weiss, Phys. Rev. **139**, B304 (1965)
641. O. Scheidemann, E. Hagebo, J. Inorg. Nucl. Chem. **35**, 3055 (1973)
642. P.L. Reeder, R.A. Warner, R.L. Gill, Phys. Rev. C **27**, 3002 (1983)
643. B. Fogelberg, T. Nagarajan, B. Grapengiesser, Nucl. Phys. A **230**, 214 (1974)
644. H. Gokturk et al., Z. Phys. A **324**, 117 (1986)
645. M.L. Gartner, J.C. Hill, Phys. Rev. C **18**, 1463 (1978)
646. K.-L. Kratz et al., Z. Phys. A **325**, 489 (1986)
647. M. Hannawald et al., Phys. Rev. C **62**, 054301 (2000)
648. R. Schneider et al., Z. Phys. A **348**, 241 (1994)
649. M. Huyse et al., Z. Phys. A **330**, 121 (1988)
650. B. Beraud et al., Z. Phys. A **299**, 279 (1981)
651. G. Lhersonneau et al., Phys. Rev. C **18**, 2688 (1978)
652. B.J. Varley, J.C. Cunnane, W. Gelletly, J. Phys. G **3**, 55 (1977)
653. J. Rivier, R. Moret, Radiochim. Acta **22**, 27 (1975)
654. R.C. Catura, J.R. Richardson, Phys. Rev. **126**, 646 (1962)
655. E.C. Mallary, M.L. Pool, Phys. Rev. **76**, 1454 (1949)
656. S.N. Ghoshal, Phys. Rev. **73**, 417 (1948)
657. S.W. Barnes, Phys. Rev. **56**, 414 (1939)
658. D.J. Tendam, H.L. Bradt, Phys. Rev. **72**, 1118 (1947)
659. M. Wehrli, Naturwiss. **22**, 289 (1934)
660. J.L. Lawson, J.M. Cork, Phys. Rev. **52**, 531 (1937)
661. F.W. Aston, Nature **113**, 192 (1924)
662. J.M. Cork, R.L. Thornton, Phys. Rev. **51**, 608 (1937)
663. R.B. Duffield, J.D. Knight, Phys. Rev. **75**, 1967 (1949)
664. C.L. McGinnis, Phys. Rev. **109**, 888 (1958)
665. H. Yuta, H. Morinaga, Nucl. Phys. **16**, 119 (1960)
666. J. Kantele, M. Karras, Phys. Rev. **129**, 270 (1963)
667. M. Karras, Phys. Rev. **135**, B1301 (1964)
668. K. Fritze, K. Griffiths, Radiochim. Acta **7**, 59 (1967)
669. K. Aleklett, G. Nyman, G. Rudstam, Nucl. Phys. A **246**, 425 (1975)
670. A. Kerek et al., Nucl. Phys. A **209**, 520 (1973)
671. A. Kerek et al., Phys. Lett. B **44**, 252 (1973)
672. P. Hoff et al., Phys. Rev. Lett. **77**, 1020 (1996)
673. I. Dillmann et al., Eur. Phys. J. A **13**, 281 (2002)
674. M. Lewitowicz et al., Phys. Lett. B **332**, 20 (1994)
675. P. Tidemand-Petersson et al., Z. Phys. A **302**, 343 (1981)
676. K. Deneffe et al., J. Phys. G **11**, L59 (1985)
677. V.N. Burminskii et al., JETP Lett. **22**, 54 (1975)
678. H.C. Hseuh, E.S. Macias, Phys. Rev. C **14**, 345 (1976)
679. T. Yamazaki, G.T. Ewan, S.G. Prussin, Phys. Rev. Lett. **20**, 1376 (1968)
680. D.E. Khulelidze, V.L. Chikhladze, V.G. Onufriev, Bull. Acad. Sci. USSR Phys. Ser. **29**, 732 (1965)
681. G. Bassani et al., Phys. Rev. **139**, B830 (1965)
682. R.A. Hinshaw, M.L. Pool, Phys. Rev. **76**, 358 (1949)
683. F.W. Aston, Nature **120**, 956 (1927)
684. J.J. Livingood, G.T. Seaborg, Phys. Rev. **55**, 667 (1939)
685. F.W. Aston, Nature **109**, 813 (1922)
686. M. Lindner, I. Perlman, Phys. Rev. **73**, 1124 (1948)
687. W.E. Grummitt, G. Wilkinson, Nature **161**, 520 (1948)

688. B.J. Dropesky, C.J. Orth, J. Inorg. Nucl. Chem. **24**, 1301 (1962)
689. J.W. Barnes, A.J. Freedman, Phys. Rev. **84**, 365 (1951)
690. I. Fränz, J. Rodriguez, R. Radicella, Z. Naturforsch. **11a**, 1037 (1956)
691. E. Hagebo, A. Kjelberg, A.C. Pappas, J. Inorg. Nucl. Chem. **24**, 117 (1962)
692. T. Izak, S. Amiel, J. Inorg. Nucl. Chem. **34**, 1469 (1972)
693. A.E. Greendale, D.L. Love, Anal. Chem. **35**, 1712 (1963)
694. S. Borg, G.B. Holm, B. Rydberg, Nucl. Phys. A **212**, 197 (1973)
695. S. Shalev, G. Rudstam, Nucl. Phys. A **230**, 153 (1974)
696. R.J. Tighe et al., Phys. Rev. C **49**, R2871 (1994)
697. A. Plochocki et al., Phys. Lett. B **106**, 285 (1981)
698. D. Seweryniak et al., Phys. Lett. B **321**, 323 (1994)
699. K. Oxorn, A.J. Houdayer, S.K. Mark, Z. Phys, A **279**, 289 (1976)
700. K. Miyano, H. Nakahara, C. Gil, J. Phys. Soc. Jpn. **33**, 1505 (1972)
701. M. Singh et al., Nucl. Phys. A **193**, 449 (1972)
702. I.P. Selinov et al., Sov. At. Energy **7**, 1011 (1959)
703. I.P. Selinov et al., Sov. At. Energy **5**, 1605 (1958)
704. G.M. Temmer, Phys. Rev. **76**, 424 (1949)
705. K.D. Coleman, M.L. Pool, Phys. Rev. **72**, 1070 (1947)
706. F.W. Aston, Nature **110**, 732 (1922)
707. J.J. Livingood, G.T. Seaborg, Phys. Rev. **55**, 414 (1939)
708. C.W. Stanley, L.E. Glendenin, *Radiochemical Studies: The Fission Products, Paper 134*. Nat. Nucl. Energy Ser. IV, 9 (McGraw-Hill, New York, 1951), p. 947
709. I. Fränz, R. Radicella, J. Rodriguez, Z. Naturforsch. **11a**, 1038 (1956)
710. P. Abelson, Phys. Rev. **55**, 876 (1939)
711. A.C. Pappas, D.R. Wiles, J. Inorg. Nucl. Chem. **2**, 69 (1956)
712. P.O. Strom et al., Phys. Rev. **144**, 984 (1966)
713. L. Tomlinson, M.H. Hurdus, Phys. Lett. B **25**, 545 (1967)
714. C.E. Bemis, G.E. Gordon, C.D. Coryell, J. Inorg. Nucl. Chem. **26**, 213 (1964)
715. D. Seweryniak et al., Phys. Rev. C **73**, 061301 (2006)
716. D. Schardt et al., Nucl. Phys. A **368**, 153 (1981)
717. D. Schardt et al., Nucl. Phys. A **326**, 65 (1979)
718. D.D. Bogdanov, V.A. Karnaukhov, L.A. Petrov, Sov. J. Nucl. Phys. **18**, 1 (1974)
719. V.A. Karnaukhov et al., Sov. J. Nucl. Phys. **4**, 327 (1967)
720. R. Kirchner et al., Phys. Lett. B **70**, 150 (1977)
721. M.E.J. Wigmans et al., Phys. Rev. C **14**, 243 (1976)
722. A. Charvet, R. Chery, R. Duffait, J. Phys. (Paris) **35**, L (1974)
723. O. Rahmouni, J. Phys. (Paris) **29**, 550 (1968)
724. I.P. Selinov et al., Sov. Phys. JETP **11**, 1191 (1960)
725. M. Ya, Kuznetsova et al., Sov. At. Energy **4**, 766 (1958)
726. A.J. Dempster, Phys. Rev. **50**, 186 (1936)
727. G.T. Seaborg, J.J. Livingood, J.W. Kennedy, Phys. Rev. **55**, 794 (1939)
728. K.T. Bainbridge, Phys. Rev. **39**, 1021 (1932)
729. J.J. Livingood, G.T. Seaborg, Phys. Rev. **53**, 1015 (1938)
730. C.S. Wu, Phys. Rev. **58**, 926 (1940)
731. H.O. Denschlag, J. Inorg. Nucl. Chem. **31**, 1873 (1969)
732. R.D. Page et al., Z. Phys. A **338**, 295 (1991)
733. T. Faestermann et al., Phys. Lett. B **137**, 23 (1984)
734. G.M. Gowdy et al., Phys. Rev. C **13**, 1601 (1976)
735. P. Aagaard et al., J. Inorg. Nucl. Chem. **5**, 105 (1957)
736. G.B. Rossi et al., Phys. Rev. **93**, 256 (1954)
737. L. Marquez, I. Perlman, Phys. Rev. **78**, 189 (1950)
738. A.C.G. Mitchell et al., Phys. Rev. **76**, 1450 (1949)
739. J.J. Livingood, G.T. Seaborg, Phys. Rev. **54**, 775 (1938)
740. L.E. Glendenin, R.R. Edwards, Phys. Rev. **71**, 742 (1947)

741. F.W. Aston, Nature **106**, 468 (1920)
742. S. Katcoff, O.A. Schaeffer, J.M. Hastings, Phys. Rev. **82**, 688 (1951)
743. C.W. Stanley, S. Katcoff, J. Chem. Phys. **17**, 653 (1949)
744. W. Riezler, Naturwiss. **31**, 326 (1943)
745. H.-D. Schussler, G. Herrmann, Radiochim. Acta **18**, 123 (1972)
746. K.-L. Kratz, W. Lauppe, G. Herrmann, Inorg. Nucl. Chem. Lett. **11**, 331 (1975)
747. S.N. Liddick et al., Phys. Rev. Lett. **97**, 802501 (2006)
748. E. Roeckl et al., Phys. Lett. B **78**, 393 (1978)
749. E. Hagberg et al., Nucl. Phys. A **208**, 309 (1973)
750. G. Andersson, G. Rudstam, G. Sorensen, Ark. Fysik **28**, 37 (1965)
751. B. Dropesky, E.O. Wiig, Phys. Rev. **88**, 683 (1952)
752. D.L. Anderson, M.L. Pool, Phys. Rev. **77**, 142 (1950)
753. W. Seelmann-Eggebert, Naturwiss. **31**, 491 (1943)
754. D.R. Wolfsberg et al., J. Inorg. Nucl. Chem. **12**, 201 (1960)
755. W. Borchers et al., Phys. Lett. B **216**, 7 (1989)
756. U.C. Bergmann et al., Nucl. Phys. A **714**, 21 (2003)
757. R.D. Page et al., Phys. Rev. Lett. **72**, 1798 (1994)
758. J.M. D'Auria et al., Nucl. Phys. A **301**, 397 (1978)
759. D.D. Bogdanov et al., Sov. J. Nucl. Phys. **21**, 123 (1975)
760. H.L. Ravn, S. Sundell, L. Westgaard, Phys. Lett. B **39**, 337 (1972)
761. H.B. Mathur, E.K. Hyde, Phys. Rev. **95**, 708 (1954)
762. M.C. Michel, D.H. Templeton, Phys. Rev. **93**, 1422 (1954)
763. M.I. Kalkstein, J.M. Hollander, Phys. Rev. **96**, 730 (1954)
764. R.W. Fink, F.L. Reynolds, D.H. Templeton, Phys. Rev. **77**, 614 (1950)
765. R.W. Fink, E.O. Wiig, J. Am. Chem. Soc. **73**, 2365 (1951)
766. A.B. Smith, A.C.G. Mitchell, R.S. Caird, Phys. Rev. **87**, 454 (1952)
767. F.C. Yu, D. Gideon, J.D. Kurbatov, Phys. Rev. **71**, 382 (1947)
768. A.H. Wapstra, N.F. Verster, M. Boelhouwer, Physica **19**, 138 (1953)
769. D.C. Kalbfell, R.A. Cooley, Phys. Rev. **58**, 91 (1940)
770. N. Sugarman, Phys. Rev. **75**, 1473 (1949)
771. L.E. Glendenin, *Radiochemical Studies: The Fission Products, Paper 159*. Nat. Nucl. Energy Ser. IV, 9 (McGraw-Hill, New York, 1951), p. 1092
772. A. Turkevich et al., *Radiochemical Studies: The Fission Products, Paper 153*. Nat. Nucl. Energy Ser. IV, 9 (McGraw-Hill, New York, 1951), p. 1070
773. N. Sugarman, H. Richter, J. Chem. Phys. **18**, 174 (1950)
774. A.C. Wahl et al., Phys. Rev. **126**, 1112 (1962)
775. K. Fritze, Can. J. Chem. **40**, 1344 (1962)
776. H.L. Ravn, Phys. Rep. **54**, 201 (1979)
777. A. Guglielmetti et al., Phys. Rev. C **52**, 740 (1995)
778. Z. Janas et al., Nucl. Phys. A **627**, 119 (1997)
779. D.D. Bogdanov et al., Nucl. Phys. A **275**, 229 (1977)
780. D.D. Bogdanov, V.A. Karnaukhov, L.A. Petrov, Sov. J. Nucl. Phys. **19**, 481 (1974)
781. J. Conrad et al., Nucl. Phys. A **234**, 157 (1974)
782. I.L. Preiss, P.M. Strudler, J. Inorg. Nucl. Chem. **24**, 589 (1962)
783. J.E. Clarkson et al., Nucl. Phys. A **93**, 272 (1967)
784. M. Lindner, R.N. Osborne, Phys. Rev. **88**, 1422 (1952)
785. R.W. Fink, D.H. Templeton, J. Am. Chem. Soc. **72**, 2818 (1950)
786. C.C. Thomas, E.O. Wiig, J. Am. Chem. Soc. **72**, 2818 (1950)
787. A.J. Dempster, Phys. Rev. **49**, 947 (1936)
788. S. Katcoff, Phys. Rev. **72**, 1160 (1947)
789. J.M. Cork, G.P. Smith, Phys. Rev. **60**, 480 (1941)
790. O. Hahn, F. Strassmann, Naturwiss. **27**, 89 (1939)
791. S. Katcoff, *Radiochemical Studies: The Fission Products, Paper 172*. Nat. Nucl. Energy Ser. IV, 9 (McGraw-Hill, New York, 1951), p. 1147

792. R.P. Schuman, E.H. Turk, R.L. Heath, Phys. Rev. **115**, 185 (1959)
793. J.B. Wilhelmy et al., Phys. Rev. Lett. **25**, 1122 (1970)
794. G. Engler et al., Phys. Rev. C **19**, 1948 (1979)
795. G. Rudstam, K. Aleklett, L. Sihver, At. Data Nucl. Data Tables **53**, 1 (1993)
796. F. Soramel et al., Phys. Rev. C **63**, 031304 (2001)
797. J.M. Nitschke et al., Z. Phys. A **316**, 249 (1984)
798. T. Sekine et al., Z. Phys. A **331**, 105 (1988)
799. D.D. Bogdanov et al., Nucl. Phys. A **307**, 421 (1978)
800. K. Nakai et al., Phys. Lett. B **44**, 443 (1973)
801. R.K. Sheline, T. Sikkeland, R.N. Chanda, Phys. Rev. Lett. **7**, 446 (1961)
802. I.L. Preiss, P.M. Strudler, R. Wolfgang, Phys. Rev. **129**, 1284 (1963)
803. M.M. Gransden, W.S. Boyle, Phys. Rev. **82**, 447 (1951)
804. R.A. Naumann, F.L. Reynolds, I. Perlman, Phys. Rev. **77**, 398 (1950)
805. B.J. Stover, Phys. Rev. **81**, 8 (1951)
806. J.B. Chubbuck, I. Perlman, Phys. Rev. **74**, 982 (1948)
807. M.G. Inghram, D.C. Hess Jr., Phys. Rev. **74**, 627 (1948)
808. M.G. Inghram, R.J. Hayden, D.C. Hess Jr., Phys. Rev. **72**, 967 (1947)
809. J.K. Marsh, S. Sugden, Nature **136**, 102 (1935)
810. A. Vanden, Bosch. Physica **19**, 374 (1953)
811. H. Gest and R. R. Edwards, *Radiochemical Studies: The Fission Products, Paper 171*. Nat. Nucl. Energy Ser. IV, 9 (McGraw-Hill, New York, 1951), p. 1144
812. P.O. Aronsson, G. Skarnemark, M. Skarestad, J. Inorg. Nucl. Chem. **36**, 1689 (1974)
813. H. Gabelmann et al., Z. Phys. A **308**, 359 (1982)
814. Z. Li et al., Phys. Rev. C **56**, 1157 (1997)
815. J.F. Smith et al., Phys. Lett. B **625**, 203 (2005)
816. D. Ward, R.M. Diamond, F.S. Stephens, Nucl. Phys. A **117**, 309 (1968)
817. J. Gizon et al., Nucl. Phys. A **290**, 272 (1977)
818. G. Gerschel et al., Nuovo Cimento **37**, 1756 (1965)
819. A.E. Norris, G. Friedlander, E.M. Franze, Nucl. Phys. **86**, 102 (1966)
820. W.R. Ware, E.O. Wiig, Phys. Rev. **117**, 191 (1960)
821. M.L. Pool, N.L. Krisberg, Phys. Rev. **73**, 1035 (1948)
822. N.E. Ballou, W.H. Burgus, *Radiochemical Studies: The Fission Products, Paper 177*. Nat. Nucl. Energy Ser. IV, 9 (McGraw-Hill, New York, 1951), p. 1168
823. S.S. Markowitz, W. Bernstein, S. Katcoff, Phys. Rev. **93**, 178 (1954)
824. A.A. Caretto Jr., S. Katcoff, Phys. Rev. **89**, 1267 (1953)
825. D.C. Hoffman, W.R. Daniels, J. Inorg. Nucl. Chem. **26**, 1769 (1964)
826. F. Hoellinger et al., Phys. Rev. C **56**, 1296 (1997)
827. I. Tago, Y. Kawase, K. Okano, Z. Phys, A **335**, 477 (1990)
828. A.P. Robinson et al., Phys. Rev. Lett. **95**, 032502 (2005)
829. P.A. Wilmarth et al., Z. Phys. A **325**, 485 (1986)
830. A.N. Wilson et al., Phys. Rev. C **66**, 021305 (2002)
831. J.M. Nitschke, M.D. Cable, W.D. Zeitz, Z. Phys, A **312**, 265 (1983)
832. A. Gizon et al., Z. Phys. A **351**, 361 (1995)
833. P.A. Wilmarth et al., Z. Phys. A **321**, 179 (1985)
834. A. Latuszynski et al., Nukleonika **20**, 1043 (1974)
835. A.A. Abdurazakov et al., Bull. Acad. Sci. USSR Phys. Ser. **34**, 706 (1970)
836. T.H. Handley, E.L. Olson, Phys. Rev. **96**, 1003 (1954)
837. Zh Zhelev et al., Bull. Acad. Sci. USSR Phys. Ser. **32**, 1497 (1968)
838. C. Dahlstrom, J.S. Foster, A.L. Thompson, Can. J. Phys. **36**, 1483 (1958)
839. M.L. Pool, L.L. Quill, Phys. Rev. **53**, 437 (1938)
840. T.E. Ward, N.A. Morcos, P.K. Kuroda, Phys. Rev. C **2**, 2410 (1970)
841. M. Graefenstedt et al., Z. Phys. A **336**, 247 (1990)
842. J.C. Hill, H. Yamamoto, A. Wolf, Phys. Rev. C **27**, 2857 (1983)
843. R.C. Greenwood et al., Phys. Rev. C **35**, 1965 (1987)

844. Y. Kawase, K. Okano, Z. Phys, A **330**, 231 (1988)
845. J. Van Schelt et al., Phys. Rev. C **85**, 045805 (2012)
846. S.-W. Xu et al., Phys. Rev. C **60**, 061302 (1999)
847. C.J. Lister et al., Phys. Rev. Lett. **55**, 810 (1985)
848. Ch. Droste et al., Nucl. Phys. A **152**, 579 (1970)
849. K. Ya, Gromov et al. Sov. Phys. JETP **20**, 1104 (1965)
850. G. Wilkinson, H.G. Hicks, Phys. Rev. **75**, 1687 (1949)
851. F.W. Aston, Nature **132**, 930 (1933)
852. J.A. Marinsky, L.E. Glendenin, C.D. Coryell, J. Am. Chem. Soc. **69**, 2781 (1947)
853. A.J. Dempster, Phys. Rev. **51**, 289 (1937)
854. A. Wakat, C. Griffin, Radiochem. Radioanal. Lett. **2**, 351 (1969)
855. K. Okano, Y. Kawase, K. Aoki, Radiochim. Acta **40**, 57 (1986)
856. J. Kurcewicz et al., Phys. Lett. B **717**, 371 (2012)
857. S.-W. Xu et al., Eur. Phys. J. A **21**, 75 (2004)
858. C.M. Parry et al., Phys. Rev. C **57**, 2215 (1998)
859. J. van Klinken, S.J. Feenstra, Phys. Rev. C **12**, 2111 (1975)
860. G.D. Alkhazov et al., Z. Phys. A **305**, 185 (1982)
861. G.P. Nowicki et al., Nucl. Phys. A **249**, 76 (1975)
862. J. Deslauriers, S.C. Gujrathi, S.K. Mark, Z. Phys, A **303**, 151 (1981)
863. H.-J. Bleyl, H. Münzel, G. Pfennig, Radiochim. Acta **8**, 200 (1967)
864. A.H.W. Aten Jr., J.C. Kapteijn, Physica **32**, 1159 (1966)
865. V. Kistiakowsky Fischer, Phys. Rev. **87**, 859 (1952)
866. I. Gratot et al., Nucl. Phys. **13**, 302 (1959)
867. F.D.S. Butement, Nature **167**, 400 (1951)
868. E.G. Funk Jr., J.W. Mihelich, C.F. Schwerdtfeger, Phys. Rev. **120**, 1781 (1960)
869. G.W. Parker et al., Phys. Rev. **72**, 85 (1947)
870. J.K. Long, M.L. Pool, Phys. Rev. **85**, 137 (1952)
871. W.C. Rutledge, J.M. Cork, S.B. Burson, Phys. Rev. **86**, 775 (1952)
872. R.G. Wille, R.W. Fink, Phys. Rev. **112**, 1950 (1958)
873. K. Kotajima, Nucl. Phys. **39**, 89 (1962)
874. R.C. Greenwood et al., Radiochim. Acta **30**, 57 (1982)
875. H. Mach et al., Phys. Rev. Lett. **56**, 1547 (1986)
876. S. Ichikawa et al., Phys. Rev. C **71**, 067302 (2005)
877. A.A. Sonzogni et al., Phys. Rev. Lett. **83**, 1116 (1999)
878. R. Wadsworth et al., Z. Phys. A **333**, 411 (1989)
879. M. Nowicki et al., Acta Phys. Pol. B **13**, 879 (1982)
880. N. Redon et al., Z. Phys. A **325**, 127 (1986)
881. J. van Klinken et al., Z. Phys. **246**, 369 (1971)
882. E. Herrmann, G. Pfrepper, D. Christov, Radiochim. Acta **7**, 10 (1967)
883. E. Silva, J. Goldemberg, Nuovo Cimento **3**, 12 (1956)
884. M.G. Inghram, R.J. Hayden, D.C. Hess Jr., Phys. Rev. **71**, 643 (1947)
885. D.C. Dunlavey, G.T. Seaborg, Phys. Rev. **92**, 206 (1953)
886. L. Winsberg, *Radiochemical Studies: The Fission Products, Paper 196*. Nat. Nucl. Energy Ser. IV, 9 (McGraw-Hill, New York, 1951), p. 1284
887. L. Winsberg, *Radiochemical Studies: The Fission Products, Paper 198*. Nat. Nucl. Energy Ser. IV, 9 (McGraw-Hill, New York, 1951), p. 1302
888. J.M. D'Auria, R.D. Guy, S.C. Gujrathi, Can. J. Phys. **51**, 686 (1973)
889. S. Ichikawa et al., Phys. Rev. C **58**, 1329 (1998)
890. C.N. Davids et al., Phys. Rev. C **69**, 011302 (2004)
891. C.N. Davids et al., Phys. Rev. Lett. **80**, 1849 (1998)
892. K.S. Vierinen et al., Nucl. Phys. A **499**, 1 (1989)
893. B.D. Kern et al., Phys. Rev. C **36**, 1514 (1987)
894. J. Deslauriers, S.C. Gujrathi, S.K. Mark, Z. Phys, A **283**, 33 (1977)
895. H.P. Malan, H. Münzel, G. Pfennig, Radiochim. Acta **5**, 24 (1966)

896. K. Kotajima, K.W. Brockman Jr., G. Wolzak, Nucl. Phys. **65**, 109 (1965)
897. R. Messlinger, H. Morinaga, C. Signorini, Phys. Lett. **19**, 133 (1965)
898. R.W. Hoff, J.O. Rasmussen, S.G. Thompson, Phys. Rev. **83**, 1068 (1951)
899. G.M. Gorodinskii et al., Bull. Acad. Sci. USSR Phys. Ser. **21**, 1611 (1957)
900. N.A. Antoneva et al., Bull. Acad. Sci. USSR Phys. Ser. **23**, 196 (1959)
901. M.G. Inghram, R.J. Hayden, Phys. Rev. **71**, 130 (1947)
902. L. Winsberg, *Radiochemical Studies: The Fission Products, Paper 198*. Nat. Nucl. Energy Ser. IV, 9 (McGraw-Hill, New York, 1951), p. 1311
903. L. Winsberg, *Radiochemical Studies: The Fission Products, Paper 197*. Nat. Nucl. Energy Ser. IV, 9 (McGraw-Hill, New York, 1951), p. 1292
904. T. Kuroyanagi et al., J. Phys. Soc. Jpn. **16**, 2393 (1961)
905. H. Hayashi et al., Eur. Phys. J. A **34**, 363 (2007)
906. S. Xu et al., Z. Phys. A **356**, 227 (1996)
907. S. Lunardi et al., Z. Phys. A **324**, 433 (1986)
908. K.L. Kosanke et al., Nucl. Instrum. Meth. **124**, 365 (1975)
909. K.A. Keller, H. Münzel, Radiochim. Acta **9**, 176 (1968)
910. J.R. Grover, Phys. Rev. **116**, 406 (1959)
911. V.S. Shirley, W.G. Smith, J.O. Rasmussen, Nucl. Phys. **4**, 395 (1957)
912. J.O. Rasmussen, S.G. Thompson, A. Ghiorso, Phys. Rev. **89**, 33 (1953)
913. R.E. Hein, A.F. Voigt, Phys. Rev. **79**, 783 (1950)
914. A.J. Dempster, Phys. Rev. **53**, 727 (1938)
915. F.D.S. Butement, Phys. Rev. **75**, 1276 (1949)
916. M.A. Wahlgren et al., Phys. Rev. **153**, 1310 (1967)
917. R.J. Gehrke et al., Radiochim. Acta **31**, 1 (1982)
918. R.C. Greenwood, M.A. Lee, R.A. Anderl, Radiochim. Acta **43**, 129 (1988)
919. P.J. Woods et al., Phys. Rev. C **69**, 051302 (2004)
920. Y. Xie et al., Eur. Phys. J. A **6**, 239 (1999)
921. R.B. Firestone et al., Phys. Rev. C **43**, 1066 (1991)
922. T. Ollivier et al., Z. Phys. A **320**, 695 (1985)
923. D.C. Sousa et al., Phys. Rev. C **25**, 1012 (1982)
924. G.D. Alkhazov et al., Acta Phys. Pol. B **12**, 175 (1981)
925. E. Newman et al., Phys. Rev. C **9**, 674 (1974)
926. Y.Y. Chu, E.M. Franz, G. Friedlander, Phys. Rev. **187**, 1529 (1969)
927. K.S. Toth, J.O. Rasmussen, J. Inorg. Nucl. Chem. **12**, 236 (1960)
928. J.O. Rasmussen et al., Phys. Rev. **80**, 475 (1950)
929. K.S. Toth et al., Phys. Rev. **116**, 118 (1959)
930. K.S. Toth, K.T. Faler, J.O. Rasmussen, Phys. Rev. **115**, 158 (1959)
931. J.W. Mihelich, B. Harmatz, T.H. Handley, Phys. Rev. **108**, 989 (1957)
932. G. Wilkinson, H.G. Hicks, Phys. Rev. **79**, 815 (1950)
933. R.A. Naumann, M.C. Michel, J. Inorg. Nucl. Chem. **16**, 163 (1960)
934. C.L. Hammer, M.G. Stewart, Phys. Rev. **106**, 1001 (1957)
935. W. Bothe, Naturwiss. **31**, 551 (1943)
936. V.T. Schneider, H. Münzel, Radiochim. Acta **4**, 171 (1965)
937. L. Funke et al., Nucl. Phys. **84**, 424 (1966)
938. E. Monnand, A. Moussa, J. Phys. (Paris) **29**, 545 (1968)
939. R.C. Greenwood et al., Phys. Rev. C **27**, 1266 (1983)
940. S. Ichikawa et al., Nucl. Instrum. Meth. A **374**, 330 (1996)
941. M. Asai et al., Phys. Rev. C **59**, 3060 (1999)
942. W. Krolas et al., Phys. Rev. C **65**, 031303 (2002)
943. E. Nolte et al., Z. Phys. A **306**, 223 (1982)
944. G.D. Alkhazov et al., Acta Phys. Pol. B **12**, 825 (1981)
945. K.S. Toth et al., Phys. Lett. B **56**, 29 (1975)
946. K.S. Toth, J.O. Rasmussen, Phys. Rev. **109**, 121 (1958)
947. R.D. MacFarlane, J. Inorg. Nucl. Chem. **19**, 9 (1961)

948. D.C. Hess Jr., M.G. Inghram, Phys. Rev. **74**, 1724 (1948)
949. T.H. Handley, E.L. Olson, Phys. Rev. **90**, 500 (1953)
950. F.D.S. Butement, Proc. Phys. Soc. A **64**, 428 (1951)
951. F.W. Aston, Nature **133**, 327 (1934)
952. G. Hevesy, H. Levi, Nature **136**, 103 (1935)
953. B.H. Ketelle, Phys. Rev. **76**, 1256 (1949)
954. R.G. Wille, R.W. Fink, Phys. Rev. **118**, 242 (1960)
955. R.J. Gehrke et al., Z. Phys. A **306**, 363 (1982)
956. R.M. Chasteler et al., Phys. Rev. C **42**, R1171 (1990)
957. P.-A. Söderström et al., Phys. Rev. C **81**, 034310 (2010)
958. K. Rykaczewski et al., Phys. Rev. C **60**, 011301 (1999)
959. S.-W. Xu et al., Phys. Rev. C **64**, 017301 (2001)
960. L. Goettig et al., Nucl. Phys. A **475**, 569 (1987)
961. S.Z. Gui, G. Colombo, E. Nolte, Z. Phys, A **305**, 297 (1982)
962. K.S. Toth et al., Phys. Rev. C **19**, 482 (1979)
963. R.D. Macfarlane, R.D. Griffioen, Phys. Rev. **130**, 1491 (1963)
964. P. Lagarde et al., J. Phys. (Paris) **27**, 116 (1966)
965. A.V. Kalyamin, I.Y. Levenberg, B.A. Yakovlev, Sov. At. Energy **6**, 435 (1959)
966. Z.T. Zhelev et al., Sov. J. Nucl. Phys. **2**, 682 (1966)
967. I.S. Dneprovskii, Sov. At. Energy **8**, 38 (1961)
968. K.S. Toth, J. Inorg. Nucl. Chem. **7**, 1 (1958)
969. T.H. Handley, E.M. Olson, Phys. Rev. **93**, 524 (1954)
970. T.H. Handley, W.S. Lyon, E.M. Olson, Phys. Rev. **98**, 688 (1955)
971. K. Miyano, T. Kuroyanagi, Nucl. Phys. **49**, 315 (1963)
972. R.M. Chasteler et al., Z. Phys. A **332**, 239 (1989)
973. K. Becker et al., Nucl. Phys. A **522**, 557 (1991)
974. M. Karny et al., Phys. Rev. Lett. **90**, 012502 (2003)
975. K.S. Vierinen et al., Phys. Rev. C **39**, 1972 (1989)
976. K.S. Toth et al., Phys. Rev. C **48**, 445 (1993)
977. G. de Angelis et al., Z. Phys. A **343**, 121 (1992)
978. K.S. Toth et al., Phys. Rev. C **30**, 712 (1984)
979. K.S. Toth et al., Phys. Rev. C **2**, 1480 (1970)
980. R.D. Macfarlane, R.D. Griffioen, Phys. Rev. **131**, 2176 (1963)
981. K.S. Toth et al., Phys. Rev. **183**, 1004 (1969)
982. D. Ward, F.S. Stephens, J.O. Newton, Phys. Rev. Lett. **19**, 1247 (1967)
983. A.A. Abdurazakov et al., Sov. Phys. JETP **14**, 1229 (1962)
984. T.H. Handley, E.M. Olson, Phys. Rev. **92**, 1260 (1953)
985. F.D.S. Butement, Proc. Phys. Soc. A **63**, 775 (1950)
986. A. Bisi, S. Terrani, L. Zappa, Nuovo Cimento **4**, 758 (1956)
987. D.R. Nethaway, M.C. Michel, W.E. Nervik, Phys. Rev. **103**, 147 (1956)
988. V. Pursiheimo, T. Tuurnala, T. Raunemaa, Z. Phys. **252**, 283 (1972)
989. X. Zhang et al., Z. Phys. A **353**, 353 (1996)
990. J.C. Batchelder et al., Phys. Rev. C **57**, R1042 (1998)
991. K. Livingston et al., Phys. Lett. B **312**, 46 (1993)
992. O. Klepper et al., Z. Phys. A **305**, 125 (1982)
993. K.S. Toth et al., Phys. Rev. C **36**, 826 (1987)
994. H. Helppi et al., Phys. Lett. B **115**, 11 (1982)
995. C.F. Liang et al., Z. Phys. A **297**, 303 (1980)
996. R.D. Macfarlane, Phys. Rev. **136**, B941 (1964)
997. K.S. Toth, R.L. Hahn, M.A. Ijaz, Phys. Rev. C **4**, 2223 (1971)
998. J.C. Putaux, J. Obert, P. Aguer, Nucl. Instrum. Meth. **121**, 615 (1974)
999. F.W.N. de Boer et al., Radiochim. Acta **13**, 118 (1970)
1000. C. Ekström, M. Olsmats, B. Wannberg, Nucl. Phys. A **170**, 649 (1971)
1001. B. Harmatz, T.H. Handley, J.W. Mihelich, Phys. Rev. **114**, 1082 (1959)

1002. A. Abdumalikov et al., Phys. Lett. **5**, 359 (1963)
1003. A.A. Abdurazakov et al., Nucl. Phys. **21**, 164 (1960)
1004. G. Wilkinson, H.G. Hicks, Phys. Rev. **74**, 1733 (1948)
1005. G. Wilkinson, H.G. Hicks, Phys. Rev. **75**, 1370 (1949)
1006. E. Neuninger, E. Rona, Wiener Akad. Anzeiger **73**, 159 (1936)
1007. S. DeBenedetti, F.K. McGowan, Phys. Rev. **74**, 728 (1948)
1008. K. Rykaczewski et al., Nucl. Phys. A **499**, 529 (1989)
1009. S.-W. Xu et al., Eur. Phys. J. A **12**, 1 (2001)
1010. P. Kleinheinz et al., Z. Phys. A **322**, 705 (1985)
1011. E. Nolte, G. Korschinek, Ch. Setzensack, Z. Phys, A **309**, 33 (1982)
1012. E. Hagberg et al., Nucl. Phys. A **293**, 1 (1977)
1013. W. Trautmann et al., Phys. Rev. Lett. **35**, 1694 (1975)
1014. P. Paris, F. Perrin, Compt. Rend. Acad. Sci. **265**, 510 (1967)
1015. F.D.S. Butement, P. Glentworth, J. Inorg. Nucl. Chem. **15**, 205 (1960)
1016. P. Paris, F. Perrin, Compt. Rend. Acad. Sci. **258**, 5855 (1964)
1017. T.H. Handley, E.M. Olson, Phys. Rev. **94**, 968 (1954)
1018. W. Bothe, Z. Naturforsch. **1**, 173 (1946)
1019. H. Atterling, E. Bohr, and Th. Sigurgeirsson, Arkiv Mat. Astron. Fysik **32A**, No. 2 (1945)
1020. C.J. Orth et al., Phys. Rev. C **8**, 718 (1973)
1021. R. Kirchner et al., Nucl. Phys. A **378**, 549 (1982)
1022. E. Runte et al., Z. Phys. A **328**, 119 (1987)
1023. P.J. Sellin et al., Phys. Rev. C **47**, 1933 (1993)
1024. S. Hofmann et al., Z. Phys. A **305**, 111 (1982)
1025. K.S. Toth et al., Phys. Rev. C **35**, 310 (1987)
1026. J.M. Nitschke et al., Z. Phys. A **334**, 111 (1989)
1027. S. Hofmann et al., Z. Phys. A **299**, 281 (1981)
1028. R.D. Macfarlane, Phys. Rev. **137**, B1448 (1965)
1029. G.D. Alkhazov et al., Z. Phys. A **291**, 397 (1979)
1030. G.D. Alkhazov et al., Z. Phys. A **295**, 305 (1980)
1031. R. Anholt, J.O. Rasmussen, I. Rezanka, Nucl. Phys. A **209**, 72 (1973)
1032. C. Burman, P. Sen, H. Bakhru, Can. J. Phys. **56**, 786 (1978)
1033. R.C. Hunter et al., Phys. Rev. C **16**, 384 (1977)
1034. B.J. Meijer, F.W.N. de Boer, P.F.A. Goudsmit, Radiochim. Acta **19**, 150 (1973)
1035. R. Arlt et al., Bull. Acad. Sci. USSR Phys. Ser. **33**, 1133 (1969)
1036. P.M. Aron et al., Bull. Acad. Sci. USSR Phys. Ser. **22**, 811 (1958)
1037. R.G. Wilson, M.L. Pool, Phys. Rev. **118**, 227 (1960)
1038. W.E. Nervik, G.T. Seaborg, Phys. Rev. **97**, 1092 (1955)
1039. G. Wilkinson, H.G. Hicks, Phys. Rev. **81**, 540 (1951)
1040. T. Stribel, Z. Naturforsch. **12a**, 519 (1957)
1041. S.C. Gujrathi, J.M. D'Auria, Nucl. Phys. A **161**, 410 (1971)
1042. K. Rykaczewski et al., Z. Phys. A **309**, 273 (1983)
1043. S. Hofmann et al., Z. Phys. A **291**, 53 (1979)
1044. K.S. Toth et al., Phys. Rev. C **7**, 2010 (1973)
1045. U.J. Schrewe et al., Phys. Rev. C **25**, 3091 (1982)
1046. H. Bruchertseifer, B. Eichler, Radiochem. Radioanal. Lett. **48**, 391 (1981)
1047. F.S. Stephens, N.L. Lark, R.M. Diamond, Nucl. Phys. **63**, 82 (1965)
1048. E.R. Merz, A.A. Caretto Jr., Phys. Rev. **122**, 1558 (1961)
1049. A.J. Dempster, Phys. Rev. **55**, 794 (1939)
1050. G. Wilkinson, H.G. Hicks, Phys. Rev. **75**, 696 (1949)
1051. G. Hevesy, Nature **135**, 580 (1935)
1052. W.H. Hutchin, M. Lindner, J. Inorg. Nucl. Chem. **16**, 369 (1961)
1053. O.O. Gatti, J. Flegenheimer, Z. Naturforsch, **11a**, 679 (1956)
1054. T.E. Ward, Y.Y. Chu, J.B. Cumming, Phys. Rev. C **8**, 340 (1973)
1055. S. Yuan et al., Z. Phys. A **344**, 355 (1993)

1056. S. Yuan et al., Phys. Rev. C **57**, 1506 (1998)
1057. J. Benlliure et al., Nucl. Phys. A **660**, 87 (1999)
1058. N. Alkhomashi et al., Phys. Rev. C **80**, 064308 (2009)
1059. R.D. Page et al., Phys. Rev. C **75**, 061302 (2007)
1060. R.D. Page et al., Phys. Rev. Lett. **68**, 1287 (1992)
1061. C.F. Liang et al., Z. Phys. A **321**, 695 (1985)
1062. B. Eichler et al., Radiochem. Radioanal. Lett. **53**, 161 (1982)
1063. H. Bruchertseifer et al., Radiochem. Radioanal. Lett. **51**, 319 (1982)
1064. R.E. Leber, P.E. Haustein, I.M. Ladenbauer-Bellis, J. Inorg. Nucl. Chem. **39**, 927 (1977)
1065. R. Arlt et al., Bull. Acad. Sci. USSR Phys. Ser. **33**, 1144 (1969)
1066. H. Abou-Leila, Compt. Rend. Acad. Sci. **259**, 3003 (1964)
1067. K.T. Faler, J.O. Rasmussen, Phys. Rev. **118**, 265 (1960)
1068. G. Wilkinson, Phys. Rev. **80**, 495 (1950)
1069. O. Oldenberg, Phys. Rev. **53**, 35 (1938)
1070. F.D.S. Butement, A.J. Poe, Phil. Mag. **46**, 482 (1955)
1071. A.J. Poe, Phil. Mag. **46**, 1165 (1955)
1072. L. Bianco et al., Phys. Lett. B **690**, 15 (2010)
1073. D.A. Eastham, I.S. Grant, Nucl. Phys. A **208**, 119 (1973)
1074. K.S. Toth et al., Phys. Rev. C **12**, 533 (1975)
1075. J. Gerl et al., Nucl. Phys. A **443**, 348 (1985)
1076. F.S. Stephens, J.R. Leigh, R.M. Diamond, Nucl. Phys. A **170**, 321 (1971)
1077. J. Recht et al., Nucl. Phys. A **440**, 366 (1985)
1078. E. Nadjakov et al., Bull. Acad. Sci. USSR Phys. Ser. **35**, 2004 (1971)
1079. H.F.R. Arciszewski et al., Nucl. Phys. A **401**, 531 (1983)
1080. A. Santoni, A. Caruette, J. Valentin, J. Phys. (Paris) **24**, 407 (1963)
1081. A.J. Dempster, Phys. Rev. **52**, 1074 (1937)
1082. G. Wilkinson, Nature **160**, 864 (1947)
1083. F.W. Aston, Nature **126**, 913 (1930)
1084. O. Minakawa, Phys. Rev. **57**, 1189 (1940)
1085. M. Lindner, J.S. Coleman, J. Am. Chem. Soc. **73**, 1610 (1951)
1086. J. Flegenheimer, G.B. Baró, M. Viirsoo, Radiochim. Acta **2**, 7 (1963)
1087. P.E. Haustein et al., Phys. Rev. C **14**, 645 (1976)
1088. D.T. Joss et al., Phys. Lett. B **641**, 34 (2006)
1089. U.J. Schrewe et al., Z. Phys. A **288**, 189 (1978)
1090. F. Meissner et al., Z. Phys. A **343**, 283 (1992)
1091. C. Cabot et al., Z. Phys. A **287**, 71 (1978)
1092. E.E. Berlovich et al., Bull. Acad. Sci. USSR Phys. Ser. **38**(8), 14 (1974)
1093. E. Runte et al., Z. Phys. A **328**, 373 (1987)
1094. E.E. Berlovich et al., Bull. Acad. Sci. USSR Phys. Ser. **36**, 2165 (1972)
1095. A. Szymanski et al., Radiochim. Acta **40**, 61 (1986)
1096. E. Nadjakov et al., Compt. Rend. Acad. Bulgare Sci. **20**, 533 (1967)
1097. B.C. Haldar, E.O. Wiig, Phys. Rev. **105**, 1285 (1957)
1098. B. Harmatz, T.H. Handley, J.W. Mihelich, Phys. Rev. **119**, 1345 (1960)
1099. V. Kistiakowsky Fischer, Phys. Rev. **99**, 764 (1955)
1100. C.J. Gallagher Jr., M. Sweeney, J.O. Rasmussen, Phys. Rev. **108**, 108 (1957)
1101. G. Wilkinson, H.G. Hicks, Phys. Rev. **77**, 314 (1950)
1102. B.J. Stover, Phys. Rev. **80**, 99 (1950)
1103. K. Fajans, W.H. Sullivan, Phys. Rev. **58**, 276 (1940)
1104. F.W. Aston, Nature **127**, 591 (1931)
1105. K. Sinma, F. Yamasaki, Phys. Rev. **55**, 320 (1939)
1106. B. Crasemann et al., Phys. Rev. **132**, 1681 (1963)
1107. A.H.W. Aten Jr., G.D. de Feyfer, Physica **21**, 543 (1955)
1108. J. Blachot, E. Monnand, A. Moussa, Compt. Rend. Acad. Sci. **261**, 1835 (1965)
1109. T. Kurtukian-Nieto, J. Benlliure, K.-H. Schmidt, Nucl. Instrum. Meth. A **589**, 472 (2008)

1110. S.J. Steer et al., Phys. Rev. C **78**, 061302 (2008)
1111. S.J. Steer et al., Phys. Rev. C **84**, 044313 (2011)
1112. S. Hofmann et al., Z. Phys. A **333**, 107 (1989)
1113. C. Cabot et al., Z. Phys. A **283**, 221 (1977)
1114. K.S. Toth et al., Phys. Rev. C **6**, 2297 (1972)
1115. K.S. Toth et al., Phys. Rev. C **5**, 2060 (1972)
1116. J. Borggreen, E.K. Hyde, Nucl. Phys. A **162**, 407 (1971)
1117. R. Arlt et al., Bull. Acad. Sci. USSR Phys. Ser. **34**, 619 (1970)
1118. J. Burde, R.M. Diamond, F.S. Stephens, Nucl. Phys. A **92**, 306 (1967)
1119. B.N. Belyaev et al., Bull. Acad. Sci. USSR Phys. Ser. **32**, 59 (1968)
1120. B.N. Belyaev et al., Sov. J. Nucl. Phys. **4**, 163 (1967)
1121. K.J. Hofstetter, P.J. Daly, Phys. Rev. **152**, 1050 (1966)
1122. A.O. Nier, Phys. Rev. **52**, 885 (1937)
1123. L.J. Goodman, M.L. Pool, Phys. Rev. **71**, 288 (1947)
1124. E. Zingg, Helv. Phys. Acta **13**, 219 (1940)
1125. M. Lindner, Phys. Rev. **84**, 240 (1951)
1126. G.B. Baró, P. Rey, Z. Naturforsch. **129**, 520 (1957)
1127. P.E. Haustein et al., Phys. Rev. C **16**, 1559 (1977)
1128. A.I. Morales et al., Phys. Rev. C **84**, 011601 (2011)
1129. M.C. Drummond et al., Phys. Rev. C **89**, 064309 (2014)
1130. C.N. Davids et al., Phys. Rev. C **55**, 2255 (1997)
1131. A. Siivola, Nucl. Phys. A **92**, 475 (1967)
1132. A.I. Akhmadzhanov et al., Bull. Acad. Sci. USSR Phys. Ser. **36**, 1820 (1972)
1133. U. Bosch-Wicke et al., Z. Phys. A **341**, 245 (1992)
1134. R.M. Diamond et al., Nucl. Phys. **25**, 248 (1961)
1135. A.K. Lavrukhina, T.V. Malysheva, B.A. Khotin, Sov. Phys. Dokl. **6**, 236 (1961)
1136. V.I. Baranov et al., Bull. Acad. Sci. USSR Phys. Ser. **24**, 1086 (1960)
1137. R.M. Diamond, J.M. Hollander, Nucl. Phys. **8**, 143 (1958)
1138. T.C. Chu, Phys. Rev. **79**, 582 (1950)
1139. W.G. Smith, J.M. Hollander, Phys. Rev. **98**, 1258 (1955)
1140. B. Venkatesachar, L. Sibaiya, Nature **136**, 437 (1935)
1141. E. McMillan, M. Kamen, S. Ruben, Phys. Rev. **52**, 375 (1937)
1142. D. Christian, R.F. Mitchell, D.S. Martin Jr., Phys. Rev. **86**, 946 (1952)
1143. H. Vonach, H. Munzer, P. Hille, Z. Phys. **195**, 343 (1966)
1144. A. Szalay, S. Uray, Radiochem. Radioanal. Lett. **14**, 135 (1973)
1145. K. Zhao et al., Chin. Phys. Lett. **10**, 265 (1993)
1146. C.R. Bingham et al., Phys. Rev. C **54**, R20 (1996)
1147. S. Della, Negra et al. Z. Phys. A **300**, 251 (1981)
1148. A. Siivola, Nucl. Phys. **84**, 385 (1966)
1149. G. Graeffe, Ann. Acad. Sci. Fenn., Ser. A **VI**, No. 128 (1963)
1150. G. Albouy, M.M. Gusakow, N. Poffe, J. Phys. Radium **21**, 751 (1960)
1151. R.A. Naumann, Phys. Rev. **96**, 90 (1954)
1152. G. Wilkinson, Phys. Rev. **73**, 252 (1948)
1153. A.J. Dempster, Nature **135**, 993 (1935)
1154. B. Fuchs, H. Kopfermann, Naturwiss. **23**, 372 (1935)
1155. J.M. Cork, E.O. Lawrence, Phys. Rev. **49**, 788 (1936)
1156. L.P. Roy, J.-C. Roy, J.S. Merritt, Phys. Rev. **105**, 1337 (1957)
1157. J. Facetti et al., Phys. Rev. **127**, 1690 (1962)
1158. S. Shi et al., Z. Phys. A **342**, 369 (1992)
1159. H. Alvarez-Pol et al., Phys. Rev. C **82**, 041602 (2010)
1160. H. Kettunen et al., Phys. Rev. C **69**, 054323 (2004)
1161. P.J. Sellin et al., Z. Phys. A **346**, 323 (1993)
1162. J.R.H. Schneider et al., Z. Phys. A **312**, 21 (1983)
1163. C. Cabot et al., Nucl. Phys. A **241**, 341 (1975)

1164. A. Siivola, Nucl. Phys. A **109**, 231 (1968)
1165. J. P. Husson, C. F. Liang, and C. Richard-Serre, J. Phys. (Paris) Lett. **38**, L (1977)
1166. P.G. Hansen et al., Nucl. Phys. A **148**, 249 (1970)
1167. G. Albouy et al., Compt. Rend. Acad. Sci. **249**, 407 (1959)
1168. L.P. Gillon, K. Gopalakrishnan, A. de Shalit, Phys. Rev. **93**, 124 (1954)
1169. F.D.S. Butement, Proc. Phys. Soc. A **64**, 395 (1951)
1170. F.D.S. Butement, R. Shillito, Proc. Phys. Soc. A **65**, 945 (1952)
1171. T.E. Ward et al., Phys. Rev. **164**, 1545 (1967)
1172. A. Pakkanen, T. Komppa, H. Helppi, Nucl. Phys. A **184**, 157 (1972)
1173. Ch. Wennemann et al., Z. Phys. A **347**, 185 (1994)
1174. D. Seweryniak et al., Phys. Rev. C **60**, 031304 (1999)
1175. J. Uusitalo et al., Z. Phys. A **358**, 375 (1997)
1176. C. Cabot et al., Compt. Rend. Acad. Sci. **281**, 453 (1975)
1177. P.G. Hansen et al., Nucl. Phys. A **160**, 445 (1971)
1178. A.G. Demin et al., Nucl. Phys. A **106**, 337 (1968)
1179. N. Poffe et al., J. Phys. Radium **21**, 343 (1960)
1180. R.W. Fink, E.O. Wiig, J. Am. Chem. Soc. **74**, 2457 (1952)
1181. W.J. Tomlinson III, H.H. Stroke, Phys. Rev. Lett. **8**, 436 (1962)
1182. F.W. Aston, Nature **120**, 224 (1927)
1183. R. Sherr, K.T. Bainbridge, H.H. Anderson, Phys. Rev. **60**, 473 (1941)
1184. F.W. Aston, Nature **116**, 208 (1925)
1185. G. Friedlander, C. Wu, Phys. Rev. **63**, 227 (1943)
1186. R.S. Krishnan, E.A. Nahum, Proc. Camb. Phil. Soc. **36**, 490 (1940)
1187. S. Mirzadeh, Y.Y. Chu, Radiochem. Radioanal. Lett. **51**, 279 (1982)
1188. L. Zhang et al., Phys. Rev. C **49**, R592 (1994)
1189. L. Zhang et al., Phys. Rev. C **58**, 156 (1998)
1190. M. Pfützner et al., Phys. Lett. B **444**, 32 (1998)
1191. G. Poli et al., Phys. Rev. C **59**, R2979 (1999)
1192. M.P. Carpenter et al., Phys. Rev. Lett. **78**, 3650 (1997)
1193. YuA Lazarev et al., Europhys. Lett. **4**, 893 (1987)
1194. K.S. Toth et al., Phys. Rev. C **53**, 2513 (1996)
1195. A. Bouldjedri et al., Z. Phys. A **339**, 311 (1991)
1196. U.J. Schrewe et al., Phys. Lett. B **91**, 46 (1980)
1197. K.S. Toth et al., Phys. Lett. B **63**, 150 (1976)
1198. J.H. Hamilton et al., Phys. Rev. Lett. **35**, 562 (1975)
1199. J. Vandlik et al., Bull. Acad. Sci. USSR Phys. Ser. **34**, 1472 (1970)
1200. T.B. Vandlik et al., JETP Lett. **15**, 271 (1972)
1201. J. Vandlik et al., Bull. Acad. Sci. USSR Phys. Ser. **38**(4), 26 (1974)
1202. G. Andersson, I.B. Häller, R. Ringh, J. Inorg. Nucl. Chem. **17**, 15 (1961)
1203. K.F. Chackett, G.A. Chackett, J. Inorg. Nucl. Chem. **13**, 1 (1960)
1204. B. Jung, G. Andersson, Nucl. Phys. **15**, 108 (1960)
1205. J.D. Knight, E.W. Baker, Phys. Rev. **100**, 1334 (1955)
1206. G. Andersson et al., Phil. Mag. **46**, 70 (1955)
1207. D.A. Orth et al., Phys. Rev. **75**, 1100 (1949)
1208. H.M. Neumann, I. Perlman, Phys. Rev. **78**, 191 (1950)
1209. H. Schüler, J.E. Keyston, Naturwiss. **19**, 320 (1931)
1210. G. Harbottle, Phys. Rev. **91**, 1234 (1953)
1211. P. Preiswerk, H. von Halban, Compt. Rend. Acad. Sci. **201**, 722 (1935)
1212. O. Hahn, L. Meitner, Phys. Z. **9**, 649 (1908)
1213. O. Hahn, L. Meitner, Verh. Deutsch. Phys. Gesell. **11**, 55 (1909)
1214. F. Hagemann, Phys. Rev. **79**, 534 (1950)
1215. O. Hahn, L. Meitner, Phys. Z. **10**, 697 (1909)
1216. L. Chen et al., Phys. Lett. B **691**, 234 (2010)
1217. A.N. Andreyev et al., J. Phys. G **37**, 035102 (2010)

1218. K.S. Toth et al., Z. Phys. A **355**, 225 (1996)
1219. K.S. Toth, D.M. Moltz, J.D. Robertson, Phys. Rev. C **39**, 1150 (1989)
1220. J.G. Keller et al., Nucl. Phys. A **452**, 173 (1986)
1221. J.P. Dufour et al., Z. Phys. A **294**, 107 (1980)
1222. H. Gauvin et al., Phys. Rev. Lett. **29**, 958 (1972)
1223. Y. Le Beyec et al., Phys. Rev. C **9**, 1091 (1974)
1224. J.O. Newton, F.S. Stephens, R.M. Diamond, Nucl. Phys. A **236**, 225 (1974)
1225. G. Andersson, E. Arbman, B. Jung, Ark. Fysik **11**, 297 (1957)
1226. D. Maeder, A.H. Wapstra, Phys. Rev. **93**, 1433 (1954)
1227. W. Maurer, W. Ramm, Z. Phys. **119**, 602 (1942)
1228. H. Schüler, E.G. Jones, Naturwiss. **20**, 171 (1932)
1229. J.R. Huizenga, C.M. Stevens, Phys. Rev. **96**, 548 (1954)
1230. K.A. Hofmann, E. Strauss, Ber. Deuts. Chem. Ges. **33**, 3126 (1900)
1231. A. Debierne, Compt. Rend. Acad. Sci. **138**, 411 (1904)
1232. F.D.S. Butement, V.J. Robinson, S.M. Qaim, J. Inorg. Nucl. Chem. **26**, 491 (1964)
1233. A.N. Andreyev et al., Eur. Phys. J. A **18**, 55 (2003)
1234. C.N. Davids et al., Phys. Rev. Lett. **76**, 592 (1996)
1235. J.C. Batchelder et al., Z. Phys. A **357**, 121 (1997)
1236. J.C. Batchelder et al., Eur. Phys. J. A **5**, 49 (1999)
1237. H. Gauvin et al., Nucl. Phys. A **208**, 360 (1973)
1238. N.I. Tarantin, A.P. Kabachenko, A.V. Demyanov, Sov. J. Nucl. Phys. **12**, 248 (1971)
1239. S. Chojnacki et al., Acta Phys. Pol. B **7**, 823 (1976)
1240. D.G. Karraker, D.H. Templeton, Phys. Rev. **81**, 510 (1951)
1241. J.J. Howland, D.H. Templeton, I. Perlman, Phys. Rev. **71**, 552 (1947)
1242. L.S. Germain, Phys. Rev. **80**, 937 (1950)
1243. J.A. Harvey, Can. J. Phys. **31**, 278 (1953)
1244. E. Rutherford, Nature **71**, 341 (1905)
1245. E. Rutherford, Phil. Trans. Roy. Soc. Lond. A **204**, 169 (1905)
1246. F. Hagemann et al., Phys. Rev. **72**, 252 (1947)
1247. E.K. Hyde, A. Ghiorso, Phys. Rev. **90**, 267 (1953)
1248. D.G. Burke et al., Z. Phys. A **333**, 131 (1989)
1249. A.N. Andreyev et al., Phys. Rev. C **72**, 014612 (2005)
1250. A.N. Andreyev et al., Eur. Phys. J. A **6**, 381 (1999)
1251. J.C. Batchelder et al., Phys. Rev. C **54**, 949 (1996)
1252. A.B. Quint et al., Z. Phys. A **346**, 119 (1993)
1253. S. Della Negra, B. Lagarde, and Y. Le Beyec, J. Phys. (Paris) Lett. **38**, L393 (1977)
1254. A. Siivola, Nucl. Phys. A **101**, 129 (1967)
1255. C. Brun, Y. Le Beyec, M. Lefort, Phys. Lett. **16**, 286 (1965)
1256. D.G. Karraker, A. Ghiorso, D.H. Templeton, Phys. Rev. **83**, 390 (1951)
1257. E.L. Kelly, E. Segrè, Phys. Rev. **75**, 999 (1949)
1258. P. Curie, M. Curie, Compt. Rend. Acad. Sci. **127**, 175 (1898)
1259. E. Marsden, R.H. Wilson, Nature **92**, 29 (1913)
1260. O. Hahn, Phys. Z. **7**, 412 (1906)
1261. K. Fajans, Phys. Z. **13**, 699 (1912)
1262. H. Geiger, Phil. Mag. **22**, 201 (1911)
1263. H. Geiger, E. Marsden, Phys. Z. **11**, 7 (1910)
1264. F.F. Momyer, E.K. Hyde, Phys. Rev. **101**, 136 (1956)
1265. H. Kettunen et al., Eur. Phys. J. A **17**, 537 (2003)
1266. A.N. Andreyev et al., Phys. Rev. C **73**, 024317 (2006)
1267. A.N. Andreyev et al., Phys. Rev. C **79**, 064320 (2009)
1268. Y. Tagaya et al., Eur. Phys. J. A **5**, 123 (1999)
1269. W. Treytl, K. Valli, Nucl. Phys. A **97**, 405 (1967)
1270. R.W. Hoff, F. Asaro, I. Perlman, J. Inorg. Nucl. Chem. **25**, 1303 (1963)
1271. R.M. Latimer, G.E. Gordon, T.D. Thomas, J. Inorg. Nucl. Chem. **17**, 1 (1961)

1272. G.W. Barton, A. Ghiorso, I. Perlman, Phys. Rev. **82**, 13 (1951)
1273. E.K. Hyde, A. Ghiorso, G.T. Seaborg, Phys. Rev. **77**, 765 (1950)
1274. D.R. Corson, K.R. MacKenzie, E. Segrè, Phys. Rev. **58**, 672 (1940)
1275. M.M. Winn, Proc. Phys. Soc. A **67**, 949 (1954)
1276. R.L. Hahn, M.F. Roche, K.S. Toth, Nucl. Phys. A **113**, 206 (1968)
1277. W.W. Meinke, A. Ghiorso, G.T. Seaborg, Phys. Rev. **75**, 314 (1949)
1278. B. Karlik, T. Bernert, Naturwiss. **32**, 44 (1944)
1279. A. Ghiorso, W.W. Meinke, G.T. Seaborg, Phys. Rev. **74**, 695 (1948)
1280. H. Hulubei, Y. Cauchois, Compt. Rend. Acad. Sci. **209**, 39 (1939)
1281. C.F. Liang et al., J. Phys. G **15**, L31 (1989)
1282. A.N. Andreyev et al., Phys. Rev. C **74**, 064303 (2006)
1283. H. Kettunen et al., Phys. Rev. C **63**, 044315 (2001)
1284. K. Morita et al., Z. Phys. A **352**, 7 (1995)
1285. F. Calaprice et al., Phys. Rev. C **30**, 1671 (1984)
1286. A.C. DiRienzo et al., Phys. Rev. C **21**, 2101 (1980)
1287. P. Hornshoj et al., Nucl. Phys. A **163**, 277 (1971)
1288. K. Valli, M.J. Nurmia, E.K. Hyde, Phys. Rev. **159**, 1013 (1967)
1289. W.E. Burcham, Proc. Phys. Soc. A **67**, 555 (1954)
1290. F.F. Momyer, E.K. Hyde, J. Inorg. Nucl. Chem. **1**, 274 (1955)
1291. F.F. Momyer et al., Phys. Rev. **86**, 805 (1952)
1292. H. Rotter et al., Sov. J. Nucl. Phys. **4**, 178 (1967)
1293. D.F. Torgerson, R.D. MacFarlane, Nucl. Phys. A **149**, 641 (1970)
1294. W.W. Meinke, A. Ghiorso, G.T. Seaborg, Phys. Rev. **85**, 429 (1952)
1295. M.H. Studier, E.K. Hyde, Phys. Rev. **74**, 591 (1948)
1296. A. Debierne, Compt. Rend. Acad. Sci. **136**, 446 (1903)
1297. E. Rutherford, Phil. Mag. **49**, 1 (1900)
1298. P. Curie, M. Curie, Compt. Rend. Acad. Sci. **129**, 714 (1899)
1299. M.J.G. Borge et al., Z. Phys. A **325**, 429 (1985)
1300. M.J.G. Borge et al., Z. Phys. A **333**, 109 (1989)
1301. D. Neidherr et al., Phys. Rev. Lett. **102**, 112501 (2009)
1302. Z. Kalaninova et al., Phys. Rev. C **87**, 044355 (2013)
1303. J. Uusitalo et al., Phys. Rev. C **87**, 064304 (2013)
1304. G.T. Ewan et al., Z. Phys. A **296**, 223 (1980)
1305. K. Valli, E.K. Hyde, W. Treytl, J. Inorg. Nucl. Chem. **29**, 2503 (1967)
1306. R.D. Griffioen, R.D. MacFarlane, Phys. Rev. **133**, B1373 (1964)
1307. J. Borggreen, K. Valli, E.K. Hyde, Phys. Rev. C **2**, 1841 (1970)
1308. M. Perey, Compt. Rend. Acad. Sci. **208**, 97 (1939)
1309. W. Kurcewicz et al., Nucl. Phys. A **464**, 1 (1987)
1310. P. Hill et al., Z. Phys. A **320**, 531 (1985)
1311. K.A. Mezlev et al., Z. Phys. A **337**, 109 (1990)
1312. J. Uusitalo et al., Phys. Rev. C **71**, 024306 (2005)
1313. M. Leino et al., Z. Phys. A **355**, 157 (1996)
1314. M.J. Leddy et al., Phys. Rev. C **51**, R1047 (1995)
1315. F.P. Heßberger et al., Europhys. Lett. **3**, 895 (1987)
1316. K. Valli, W. Treytl, E.K. Hyde, Phys. Rev. **161**, 1284 (1967)
1317. T. Nomura et al., Phys. Lett. B **40**, 543 (1972)
1318. T. Godlewski, Nature **71**, 294 (1905)
1319. E. Rutherford, F. Soddy, Phil. Mag. **4**, 370 (1902)
1320. P. Curie, M. Curie, G. Bemont, Compt. Rend. Acad. Sci. **127**, 1215 (1898)
1321. J.P. Butler, J.S. Adam, Phys. Rev. **91**, 1219 (1953)
1322. O. Hahn, Ber. Deuts. Chem. Ges. **40**, 1462 (1907)
1323. J. Gilat, S. Katcoff, J. Inorg. Nucl. Chem. **40**, 369 (1978)
1324. S.A. Ahmad et al., Phys. Lett. B **133**, 47 (1983)
1325. Z.Y. Zhang et al., Phys. Rev. C **89**, 014308 (2014)

1326. K. Eskola et al., Phys. Rev. C **57**, 417 (1998)
1327. M. Leino et al., Z. Phys. A **348**, 151 (1994)
1328. K. Valli, W. Treytl, E.K. Hyde, Phys. Rev. **167**, 1094 (1968)
1329. W.W. Meinke, G.T. Seaborg, Phys. Rev. **78**, 475 (1950)
1330. M. Curie, Radium (Paris) **8**, 353 (1911)
1331. O. Hahn, Phys. Z. **9**, 246 (1908)
1332. F. Depocas, B.G. Harvey, Phys. Rev. **85**, 499 (1952)
1333. K. Chayawattanangkur, G. Herrmann, N. Trautmann, J. Inorg. Nucl. Chem. **35**, 3061 (1973)
1334. K.-L. Gippert et al., Nucl. Phys. A **453**, 1 (1986)
1335. Y.Y. Chu, M.L. Zhou, Phys. Rev. C **28**, 1379 (1983)
1336. J. Taieb et al., Nucl. Phys. A **724**, 413 (2003)
1337. J.A. Heredia et al., Eur. Phys. J. A **46**, 337 (2010)
1338. H. Ikezoe et al., Phys. Rev. C **54**, 2043 (1996)
1339. J. Uusitalo et al., Phys. Rev. C **52**, 113 (1995)
1340. D. Vermeulen et al., Z. Phys. A **294**, 149 (1980)
1341. K. Valli, E.K. Hyde, Phys. Rev. **176**, 1377 (1968)
1342. K. Hiruta et al., Phys. Lett. B **45**, 244 (1973)
1343. O. Häusser et al., Phys. Rev. Lett. **31**, 323 (1973)
1344. O. Hahn, Nature **73**, 559 (1906)
1345. O. Hahn, Proc. Roy. Soc. A **76**, 115 (1905)
1346. B.B. Boltwood, Nature **76**, 544 (1907)
1347. G.N. Antonoff, Phil. Mag. **22**, 419 (1911)
1348. G.C. Schmidt, Ann. Physik **65**, 141 (1898)
1349. O. Hahn, L. Meitner, Naturwiss. **23**, 320 (1935)
1350. W. Crookes, Proc. Roy. Soc. A **66**, 409 (1900)
1351. N. Trautmann, R. Denig, G. Herrmann, Radiochim. Acta **11**, 168 (1969)
1352. N. Kaffrell and N. Trautmann, Z. Naturforsch. **28a**, 541 (1973)
1353. S. Yuan et al., Z. Phys. A **346**, 187 (1993)
1354. J. He et al., Phys. Rev. C **59**, 520 (1999)
1355. S. Mitsuoka et al., Phys. Rev. C **55**, 1555 (1997)
1356. V. Ninov et al., Z. Phys. A **351**, 125 (1995)
1357. K.-H. Schmidt et al., Nucl. Phys. A **318**, 253 (1979)
1358. G. Ya, Sung-Ching-Yang, V. A. Druin, and A. S. Trofimov. Sov. J. Nucl. Phys. **14**, 725 (1972)
1359. Z. Liu et al., Nucl. Instrum. Meth. A **543**, 591 (2005)
1360. R. Hingmann et al., Z. Phys. A **313**, 141 (1983)
1361. E.K. Hyde et al., *The Transuranium Elements: Research Papers, Paper 19.17*. Nat. Nucl. Energy Ser. IV, 14b (McGraw-Hill, New York, 1949), p. 1439
1362. O. Hahn, L. Meitner, Phys. Z. **19**, 208 (1918)
1363. J.W. Gofman, G.T. Seaborg, *The Transuranium Elements: Research Papers, Paper 19.14*. Nat. Nucl. Energy Ser. IV, 14b (McGraw-Hill, New York, 1949), p. 1427
1364. L. Meitner, F. Strassmann, O. Hahn, Z. Phys. **109**, 538 (1938)
1365. K. Fajans, O. Göhring, Naturwiss. **1**, 339 (1913)
1366. G. Wolzak, H. Morinaga, Radiochim. Acta **1**, 225 (1963)
1367. W.W.T. Crane, G.M. Iddings, Phys. Rev. **95**, 1702 (1954)
1368. N. Trautmann et al., Z. Naturforsch. **23a**, 2127 (1968)
1369. S. Yuan et al., Z. Phys. A **352**, 235 (1995)
1370. L. Ma et al., Phys. Rev. C **91**, 051302 (2015)
1371. O.N. Malyshev et al., Eur. Phys. J. A **8**, 295 (2000)
1372. A.N. Andreyev et al., Z. Phys. A **342**, 123 (1992)
1373. A.N. Andreyev et al., Z. Phys. A **345**, 247 (1993)
1374. J. Khuyagbaatar et al., Phys. Rev. Lett. **115**, 242502 (2015)
1375. A.N. Andreyev et al., Sov. J. Nucl. Phys. **53**, 554 (1991)
1376. A.N. Andreyev et al., Sov. J. Nucl. Phys. **50**, 381 (1989)
1377. V.E. Viola, M.M. Minor, C.T. Roche, Nucl. Phys. A **217**, 372 (1973)

1378. D.W. Osborne, R.C. Thompson, Q. Van Winkle, *The Transuranium Elements: Research Papers, Paper 19.11*. Nat. Nucl. Energy Ser. IV, 14b (McGraw-Hill, New York, 1949), p. 1397
1379. G.T. Seaborg, J.W. Gofman, R.W. Stoughton, Phys. Rev. **71**, 378 (1947)
1380. H. Geiger, J.M. Nuttall, Phil. Mag. **23**, 439 (1912)
1381. A.J. Dempster, Nature **136**, 180 (1935)
1382. A. Ghiorso et al., Phys. Rev. **82**, 558 (1951)
1383. Y. Nishina et al., Phys. Rev. **57**, 1182 (1940)
1384. H. Becquerel, Compt. Rend. Acad. Sci. **122**, 420 (1896)
1385. L. Meitner, O. Hahn, F. Strassmann, Z. Phys. **106**, 249 (1937)
1386. J.D. Knight et al., Phys. Rev. **91**, 889 (1953)
1387. P.E. Haustein et al., Phys. Rev. C **19**, 2332 (1979)
1388. A.V. Yeremin et al., Nucl. Instrum. Meth. A **350**, 608 (1994)
1389. V. Ninov et al., Z. Phys. A **336**, 473 (1990)
1390. A.N. Andreyev et al., Z. Phys. A **337**, 229 (1990)
1391. S.A. Kreek et al., Phys. Rev. C **50**, 2288 (1994)
1392. L.B. Magnusson, S.G. Thompson, G.T. Seaborg, Phys. Rev. **78**, 363 (1950)
1393. E.K. Hyde, M.H. Studier, A. Ghiorso, *The Transuranium Elements: Research Papers, Paper 22.15*. Nat. Nucl. Energy Ser. IV, 14b (McGraw-Hill, New York, 1949), p. 1622
1394. R.A. James et al., *The Transuranium Elements: Research Papers, Paper 22.8*. Nat. Nucl. Energy Ser. IV, 14b (McGraw-Hill, New York, 1949), p. 1604
1395. A.C. Wahl, G.T. Seaborg, Phys. Rev. **73**, 940 (1948)
1396. J.W. Kennedy et al., *The Transuranium Elements: Research Papers, Paper 1.9*. Nat. Nucl. Energy Ser. IV, 14b (McGraw-Hill, New York, 1949), p. 79
1397. E. McMillan, P. Abelson, Phys. Rev. **57**, 1185 (1940)
1398. R. Vandenbosch, Phys. Rev. **113**, 259 (1959)
1399. E.R. Flynn, D.L. Hanson, R.A. Hardekopf, Phys. Rev. C **19**, 355 (1979)
1400. K.J. Moody et al., Z. Phys. A **328**, 417 (1987)
1401. A.N. Andreyev et al., Z. Phys. A **347**, 225 (1994)
1402. A.N. Andreyev et al., Z. Phys. A **337**, 231 (1990)
1403. U. Jäger, H. Münzel, G. Pfennig, Z. Phys. **258**, 337 (1973)
1404. T.D. Thomas et al., Phys. Rev. **106**, 1228 (1957)
1405. J.W. Kennedy et al., Phys. Rev. **70**, 555 (1946)
1406. G.T. Seaborg, R.A. James, L.O. Morgan, *The Transuranium Elements: Research Papers, Paper 22.1*. Nat. Nucl. Energy Ser. IV, 14b (McGraw-Hill, New York, 1949), p. 1525
1407. S.G. Thompson et al., Phys. Rev. **80**, 1108 (1950)
1408. J.C. Sullivan et al., Phys. Rev. **83**, 1267 (1951)
1409. M.H. Studier et al., Phys. Rev. **93**, 1433 (1954)
1410. C.I. Browne et al., J. Inorg. Nucl. Chem. **1**, 254 (1955)
1411. P.R. Fields et al., J. Inorg. Nucl. Chem. **1**, 262 (1955)
1412. D. Engelkemeir et al., J. Inorg. Nucl. Chem. **1**, 345 (1955)
1413. YuS Popov et al., Sov. Radiochem. **25**, 458 (1983)
1414. H.M. Devaraja et al., Phys. Lett. B **748**, 199 (2015)
1415. V.I. Kuznetsov, N.K. Skobelev, Sov. J. Nucl. Phys. **5**, 810 (1967)
1416. M. Sakama et al., Eur. Phys. J. A **9**, 303 (2000)
1417. V.I. Kuznetsov, N.K. Skobelev, G.N. Flerov, Sov. J. Nucl. Phys. **5**, 191 (1967)
1418. J. Guo et al., Z. Phys. A **355**, 111 (1996)
1419. K. Tsukada et al., Phys. Rev. C **57**, 2057 (1998)
1420. S.M. Polikanov, G. Sletten, Nucl. Phys. A **151**, 656 (1970)
1421. K. Street, A. Ghiorso, G.T. Seaborg, Phys. Rev. **79**, 530 (1950)
1422. W.M. Manning, L.B. Asprey, *The Transuranium Elements: Research Papers, Paper 22.7*. Nat. Nucl. Energy Ser. IV, 14b (McGraw-Hill, New York, 1949), p. 1595
1423. C.J. Orth et al., Phys. Rev. Lett. **19**, 128 (1967)
1424. J. Khuyagbaatar et al., Eur. Phys. J. A **46**, 59 (2010)

1425. S. Ichikawa et al., Nucl. Instrum. Meth. B **187**, 548 (2002)
1426. S.A. Kreek et al., Phys. Rev. C **49**, 1859 (1994)
1427. Z. Qin et al., Radiochim. Acta **96**, 455 (2008)
1428. G.T. Seaborg, R.A. James, A. Ghiorso, *The Transuranium Elements: Research Papers, Paper 22.2.* Nat. Nucl. Energy Ser. IV, 14b (McGraw-Hill, New York, 1949), p. 1554
1429. G.H. Higgins, K. Street, Phys. Rev. **86**, 252 (1952)
1430. F.L. Reynolds, E.K. Hulet, K. Street, Phys. Rev. **80**, 467 (1950)
1431. C.M. Stevens et al., Phys. Rev. **94**, 974 (1954)
1432. P.R. Fields et al., Phys. Rev. **102**, 180 (1956)
1433. Combined Radiochemistry Group et al., Phys. Rev. **148**, 1192 (1966)
1434. R.W. Lougheed et al., J. Inorg. Nucl. Chem. **40**, 1865 (1978)
1435. YuP Gangrskii et al., Sov. J. Nucl. Phys. **31**, 162 (1980)
1436. M. Asai et al., Eur. Phys. J. A **16**, 17 (2003)
1437. K.L. Wolf, K.P. Unik, Phys. Lett. B **38**, 405 (1972)
1438. S.G. Thompson, A. Ghiorso, G.T. Seaborg, Phys. Rev. **77**, 838 (1950)
1439. E.K. Hulet et al., Phys. Rev. **84**, 366 (1951)
1440. E.K. Hulet, S.G. Thompson, A. Ghiorso, Phys. Rev. **95**, 1703 (1954)
1441. J. Milsted, A.M. Friedman, C.M. Stevens, Nucl. Phys. **71**, 299 (1965)
1442. E.K. Hulet, Phys. Rev. **102**, 182 (1956)
1443. H. Diamond et al., Phys. Rev. **94**, 1083 (1954)
1444. A. Ghiorso et al., Phys. Rev. **94**, 1081 (1954)
1445. H. Diamond et al., J. Inorg. Nucl. Chem. **29**, 601 (1967)
1446. YuA Lazarev et al., Nucl. Phys. A **588**, 501 (1995)
1447. G. Münzenberg et al., Z. Phys. A **302**, 7 (1981)
1448. R.J. Silva et al., Phys. Rev. C **2**, 1948 (1970)
1449. T. Sikkeland, A. Ghiorso, Phys. Lett. B **24**, 331 (1967)
1450. P.R. Fields et al., Phys. Lett. B **24**, 340 (1967)
1451. T. Sikkeland et al., Phys. Lett. B **24**, 333 (1967)
1452. A. Chetham-Strode, G.R. Choppin, B.G. Harvey, Phys. Rev. **102**, 747 (1956)
1453. A. Ghiorso et al., Phys. Rev. **81**, 154 (1951)
1454. A. Ghiorso et al., Phys. Rev. **93**, 257 (1954)
1455. G.R. Choppin et al., Phys. Rev. **94**, 1080 (1954)
1456. B.G. Harvey et al., Phys. Rev. **99**, 337 (1955)
1457. R.W. Lougheed et al., J. Inorg. Nucl. Chem. **43**, 2239 (1981)
1458. D.C. Hoffman et al., Phys. Rev. C **21**, 972 (1980)
1459. V. Ninov et al., Z. Phys. A **356**, 11 (1996)
1460. YuA Lazarev et al., Nucl. Phys. A **580**, 113 (1994)
1461. P. Eskola et al., Physica Fennica **8**, 357 (1973)
1462. V.L. Mikheev, V.I. Ilyushchenko, M.B. Miller, Sov. J. Nucl. Phys. **5**, 35 (1967)
1463. A. Chetham-Strode, L.W. Holm, Phys. Rev. **104**, 1314 (1956)
1464. B.G. Harvey et al., Phys. Rev. **104**, 1315 (1956)
1465. S.G. Thompson et al., Phys. Rev. **93**, 908 (1954)
1466. P.R. Fields et al., Phys. Rev. **94**, 209 (1954)
1467. YuS Popov et al., Sov. Radiochem. **29**, 431 (1987)
1468. J. Khuyagbaatar et al., Eur. Phys. J. A **37**, 177 (2008)
1469. G.M. Ter-Akopian et al., Nucl. Phys. A **255**, 509 (1975)
1470. M.J. Nurmia et al., Phys. Lett. B **26**, 78 (1967)
1471. G.N. Akapev et al., Sov. At. Energy **21**, 908 (1966)
1472. G.N. Flerov et al., Sov. At. Energy **22**, 434 (1967)
1473. A. Ghiorso et al., Phys. Rev. Lett. **1**, 18 (1958)
1474. V.P. Perelygin, E.D. Donets, G.N. Flerov, Sov. Phys. JETP **10**, 1106 (1960)
1475. H. Atterling et al., Phys. Rev. **95**, 585 (1954)
1476. S. Amiel et al., Phys. Rev. **106**, 553 (1957)
1477. A.M. Friedman et al., Phys. Rev. **102**, 585 (1956)

1478. S. Amiel, Phys. Rev. **105**, 1412 (1957)
1479. B.G. Harvey et al., Phys. Rev. **93**, 1129 (1954)
1480. G.R. Choppin et al., Phys. Rev. **98**, 1519 (1955)
1481. E.K. Hulet et al., Phys. Rev. Lett. **13**, 343 (1964)
1482. E.K. Hulet et al., Phys. Rev. Lett. **26**, 523 (1971)
1483. E.K. Hulet et al., Phys. Rev. C **21**, 966 (1980)
1484. P. Eskola, Phys. Rev. C **7**, 280 (1973)
1485. B. Kadkhodayan et al., Radiochim. Acta **56**, 1 (1992)
1486. P.R. Fields et al., Nucl. Phys. A **154**, 407 (1970)
1487. L. Phillips et al., Phys. Rev. Lett. **1**, 215 (1958)
1488. A. Ghiorso et al., Phys. Rev. **98**, 1518 (1955)
1489. T. Sikkeland et al., Phys. Rev. **140**, B277 (1965)
1490. J.F. Wild et al., Phys. Rev. C **26**, 1531 (1982)
1491. E.K. Hulet et al., Phys. Rev. C **40**, 770 (1989)
1492. Yu. Ts, Oganessian et al. Phys. Rev. C **64**, 054606 (2001)
1493. A. Ghiorso, T. Sikkeland, M.J. Nurmia, Phys. Rev. Lett. **18**, 401 (1967)
1494. V.L. Mikheev et al., Sov. At. Energy **22**, 93 (1967)
1495. E.D. Donets, V.A. Shchegolev, V.A. Ermakov, Sov. At. Energy **20**, 257 (1966)
1496. B.A. Zager et al., Sov. At. Energy **20**, 264 (1966)
1497. V.A. Druin et al., Sov. At. Energy **22**, 135 (1967)
1498. E.D. Donets, V.A. Shchegolev, V.A. Ermakov, Sov. At. Energy **14**, 540 (1963)
1499. R.J. Silva et al., Nucl. Phys. A **216**, 97 (1973)
1500. L.P. Somerville et al., Phys. Rev. C **31**, 1801 (1985)
1501. F.P. Heßberger et al., Eur. Phys. J. A **12**, 57 (2001)
1502. F.P. Heßberger et al., Z. Phys. A **322**, 557 (1985)
1503. G. Münzenberg et al., Z. Phys. A **300**, 107 (1981)
1504. V.A. Druin, Sov. J. Nucl. Phys. **12**, 146 (1971)
1505. E.D. Donets, V.A. Shchegolev, V.A. Ermakov, Sov. At. Energy **19**, 995 (1965)
1506. K. Eskola et al., Phys. Rev. C **4**, 632 (1971)
1507. J. Khuyagbaatar et al., Phys. Rev. Lett. **112**, 172501 (2014)
1508. F.P. Heßberger et al., Z. Phys. A **359**, 415 (1997)
1509. Yu. Ts, Oganessian et al., Sov. At. Energy **38**, 492 (1975)
1510. A. Ghiorso et al., Phys. Rev. Lett. **22**, 1317 (1969)
1511. A. Ghiorso et al., Phys. Lett. B **32**, 95 (1970)
1512. J.V. Kratz et al., Radiochim. Acta **91**, 59 (2003)
1513. P.A. Ellison et al., Phys. Rev. Lett. **105**, 182701 (2010)
1514. Yu. Ts, Oganessian et al. Phys. Rev. C **70**, 064609 (2004)
1515. Z.G. Gan et al., Eur. Phys. J. A **10**, 21 (2001)
1516. A. Ghiorso et al., Phys. Rev. Lett. **24**, 1498 (1970)
1517. G.N. Flerov et al., Sov. At. Energy **29**, 967 (1970)
1518. A. Ghiorso et al., Phys. Rev. C **4**, 1850 (1971)
1519. J.V. Kratz et al., Phys. Rev. C **45**, 1064 (1992)
1520. Yu. Ts, Oganessian et al. Phys. Rev. C **76**, 011601 (2007)
1521. Yu. Ts, Oganessian et al. Phys. Rev. C **69**, 021601 (2004)
1522. Yu. Ts, Oganessian et al. Phys. Rev. Lett. **104**, 142502 (2010)
1523. G. Münzenberg et al., Z. Phys. A **322**, 227 (1985)
1524. A.G. Demin et al., Z. Phys. A **315**, 197 (1984)
1525. G. Münzenberg et al., Z. Phys. A **317**, 235 (1984)
1526. S. Hofmann et al., Eur. Phys. J. A **10**, 5 (2001)
1527. A. Ghiorso et al., Phys. Rev. Lett. **33**, 1490 (1974)
1528. K.E. Gregorich et al., Phys. Rev. C **74**, 044611 (2006)
1529. YuA Lazarev et al., Phys. Rev. Lett. **73**, 624 (1994)
1530. J. Dvorak et al., Phys. Rev. Lett. **97**, 242501 (2006)
1531. J. Dvorak et al., Phys. Rev. Lett. **100**, 132503 (2008)

1532. S.L. Nelson et al., Phys. Rev. Lett. **100**, 022501 (2008)
1533. G. Münzenberg et al., Z. Phys. A **333**, 163 (1989)
1534. S. Hofmann et al., Z. Phys. A **350**, 281 (1995)
1535. Z.G. Gan et al., Eur. Phys. J. A **20**, 385 (2004)
1536. P.A. Wilk et al., Phys. Rev. Lett. **85**, 2697 (2000)
1537. I. Dragojevic et al., Phys. Rev. C **79**, 011602 (2009)
1538. G. Münzenberg et al., Z. Phys. A **324**, 489 (1986)
1539. YuA Lazarev et al., Phys. Rev. Lett. **75**, 1903 (1995)
1540. K. Nishio et al., Phys. Rev. C **82**, 024611 (2010)
1541. S. Hofmann et al., Z. Phys. A **354**, 229 (1996)
1542. ChE Düllmann et al., Phys. Rev. Lett. **104**, 252701 (2010)
1543. G. Münzenberg et al., Z. Phys. A **309**, 89 (1982)
1544. K. Morita et al., J. Phys. Soc. Jpn. **73**, 2593 (2004)
1545. Yu. Ts, Oganessian et al. Phys. Rev. C **87**, 054621 (2013)
1546. A. Ghiorso et al., Phys. Rev. C **51**, R2293 (1995)
1547. S. Hofmann et al., Z. Phys. A **350**, 277 (1995)
1548. YuA Lazarev et al., Phys. Rev. C **54**, 620 (1996)
1549. Yu. Ts, Oganessian et al. Phys. Rev. C **69**, 054607 (2004)
1550. V.K. Utyonkov et al., Phys. Rev. C **92**, 034609 (2015)
1551. Yu. Ts, Oganessian et al. Phys. Rev. C **74**, 044602 (2006)

MIX
Papier aus verantwortungsvollen Quellen
Paper from responsible sources
FSC® C105338

If you have any concerns about our products,
you can contact us on
ProductSafety@springernature.com

In case Publisher is established outside the EU,
the EU authorized representative is:
Springer Nature Customer Service Center GmbH
Europaplatz 3, 69115 Heidelberg, Germany

Printed by Libri Plureos GmbH
in Hamburg, Germany